Modern Sensors, Transducers and Sensor Networks

Sergey Y. Yurish
Editor

Modern Sensors, Transducers and Sensor Networks

Book Series: Advances in Sensors: Reviews, Vol. 1

International Frequency Sensor Association Publishing

Editor
Sergey Y. Yurish,
Modern Sensors, Transducers and Sensor Networks
Advances in Sensors: Reviews, Vol. 1

ISBN–10: 84-615-9613-7
ISBN–13: 978-84-615-9613-3
BN-20120515-XX
BIC: TJFC

Content

Chapter 8

Chapter 9

Chapter 10

Contributors

Palash Kumar Basu Microelectronics Laboratory, Department of Electrical Communication Engineering, Indian Institute of Science, Bangalore 560012, India

Sukumar Basu IC Design and Fabrication Centre, Department of Electronics and Telecommunication Engineering, Jadavpur University, Kolkata 700032, India

Jeremy S. Bradbury University of Ontario, Institute of Technology, Oshawa, Ontario, Canada

James Brusey Cogent Computing Applied Research Centre, Coventry University, CV1 5FB, UK

Cheng-Hsin Chuang Department of Mechanical Engineering, Southern Taiwan University, Tainan, Taiwan

Tessa Daniel Cogent Computing Applied Research Centre, Coventry University, CV1 5FB, UK

Winncy Y. Du San Jose State University, San Jose, USA

Elena Gaura Cogent Computing Applied Research Centre, Coventry University, CV1 5FB, UK

Sarmishtha Ghoshal School of Materials Science and Engineering, Bengal Engineering and Science University, Howrah, 711 103, India

Michael J. Haji-Sheikh College of Engineering and Engineering Technology, Northern Illinois University, DeKalb, Illinois, USA

John Halloran Cogent Computing Applied Research Centre, Coventry University, CV1 5FB, UK

Ebtisam H. Hasan National Institute for Standards (NIS), Giza, Egypt

John Khalil Jacoub University of Ontario, Institute of Technology, Oshawa, Ontario, Canada

Tom J. Kazmierski School of Electronics and Computer Science University of Southampton, UK

Ivica Kostanic Civil Engineering and Electrical & Computer Engineering Florida Institute of Technology, 150 W. University Blvd., Melbourne, FL, 32901, USA

Frederic Kreit Mechanical & Aerospace Engineering, Florida Institute of Technology, 150 W. University Blvd., Melbourne, FL, 32901, USA

Gabriel Lapilli Mechanical & Aerospace Engineering, Florida Institute of Technology, 150 W. University Blvd., Melbourne, FL, 32901, USA

Ramiro Liscano University of Ontario, Institute of Technology, Oshawa, Ontario, Canada

Dwijesh Dutta Majumder Indian Statistical Institute, Kolkata, India

Debasis Mitra Department of Information Technology, National Institute of Technology, Durgapur, 713 209, West Bengal, India

Subhas C. Mukhopadhyay School of Engineering and Advanced Technology, Massey University Palmerston North, New Zealand

K. R. Nemade Department of Physics, Sant Gadge Baba Amravati University, Amravati, 44 602, India

Jean-Paul Pinelli Civil Engineering and Electrical & Computer Engineering Florida Institute of Technology, 150 W. University Blvd., Melbourne, FL, 32901, USA

Mohd Syaifudin Bin Abdul Rahman Malaysian Agricultural Research and Development Institute, Serdang 43300, Selangor, Malaysia

Sudip Roy Department of Computer Science and Engineering, Indian Institute of Technology, Kharagpur, 721 302, India

Chelakara Subramanian Mechanical & Aerospace Engineering, Florida Institute of Technology, 150 W. University Blvd., Melbourne, FL, 32901, USA

Scott W. Yelich San Jose State University, San Jose, USA

Pak Lam Yu School of Engineering and Advanced Technology, Massey University Palmerston North, New Zealand

Sergey Y. Yurish Research & Development Department, Technology Assistance BCNA 2010, S. L., Parc UPC-PMT, Edificio RDIT-K2M C/ Esteve Terradas, 1, 08860, Castelldefels, Barcelona, Spain

Chenxu Zhao School of Electronics and Computer Science University of Southampton, UK

Qing Zhu FUJIFILM Dimatix, Inc. 2250 Martin Avenue, Santa Clara, CA 95050, USA

Preface

Every research and development should be started from a state-of-the-art review. Such review is one of the most labor- and time-consuming parts of research, especially in such high technological areas as sensors, MEMS, NEMS and measurements. It is strongly necessary to take into account and reflect in the review the current stage of development, including existing sensing principles, methods of measurements, technologies and existing devices. Many PhD students and researchers working in the same area must make (and do it) the same type of work. A researcher must find appropriate references, to read it and make a critical analysis to determine what was done well before and what was not solved till now, and determine and formulate his future scientific aim and objectives.

The first book entitled 'Modern Sensors, Transducers and Sensor Networks' from the Advances in Sensors: Reviews book Series started by the IFSA Publishing in 2012 contains dozen collected sensor related state-of-the-art reviews written by 31 experts from academia and industry from 9 countries: Canada, Egypt, India, Malaysia, New Zealand, Spain, Taiwan, UK and USA.

This book ensures that PhD students and researchers will stay at the cutting edge of the field and get the right and effective start point and road map for the further researches and developments. By this way, they will be able to save more time for productive research activity and eliminate routine work.

Built upon the series Advances in Sensors: Reviews - a premier sensor review source, it presents an overview of highlights in the field. Coverage includes current developments in sensing nanomaterials, technologies, design, synthesis, modeling and applications of sensors, transducers and wireless sensor networks, signal detection and advanced signal processing, as well as new sensing principles and methods of measurements.

This volume is divided into three main sections: physical sensors, chemical sensors and biosensors, and sensor networks including sensor technology, sensor market reviews and applications. With this unique combination of information in each volume, the Advances in Sensors book Series will be of value for scientists and engineers in industry and at universities, to sensors developers, distributors, and users.

Sergey Y. Yurish

Barcelona, Spain

Editor
IFSA Publishing

Chapter 1
Introduction

Sergey Y. Yurish

1.1. Some Data about Sensor Market

We are living in an era of sensorization. Sensors are everywhere. It can be found in any devices, which we are using on the everyday basis: in mobile phone, car, computer, LCD full HD TV, various office equipment, climatic control, etc. Now sensors can be built-in into special clothing such as gloves, sport shoes, helmets, glasses and others. In the nearest future sensors and even wireless sensor networks will wearable and can be embedded even in T-shirts [1-6].

Modern sensor market is one of the hugest and fast growing markets, which is comparable with PC and communication devices markets. According to the new, a little bit conservative report (October 2011) from the *Global Industry Analysts, Inc.* the global sensors market will reach US $ 76.7 billion by 2017 [7]. The market will grow at a compound annual growth rate of 4 % in average (Table 1.1).

Table 1.1. Global Sensor and Smart Sensors Markets Size Prediction (annual growth rate of 4 %).

Year	2009	2010	2011	2012	2013	2014	2015	2016	2017
Global Sensor Market, billion $ US	–	56.3	62.8	65.7	68.5	71.4	74.2	75.45	76.7
European Sensor Market, billion $ US	12.5	–	–	–	–	–	–	19.0	–
Global Smart Sensors Market, billion $ US	–	–	–	–	–	–	6.0	–	6.7
European Smart Sensors Market, billion $ US	–	2.1	–	–	–	–	–	–	–

"-" information is not available.

S. Y. Yurish

Research & Development Department, Technology Assistance BCNA 2010, S. L.

Parc UPC-PMT, Edificio RDIT-K2M, C/ Esteve Terradas, 1, 08860 Castelldefels, Barcelona, Spain

Sensors are being used widely in most of the industries, including automotive, medical, industrial, entertainment, security, and defense due to increased usage of process controls and sensing elements in different sectors [7].

Smart and intelligent sensors and sensor systems are great of interest due to technological driven challenges [8]. The Global Industry Analysts, Inc. also reported that the global smart sensors market to reach US$ 6.7 billion by 2017 [9]. Growing acceptance of advanced technologies such as MEMS and NEMS also augurs well for the future of the market. While micromachining enhanced the scope of sensor fabrication applications, nanotechnology and micro-electromechanical technology led to improvements in sensor development, design, and the production of in-expensive compact sensors [9]. Advanced models of smart sensors and/or intelligent sensor systems will possess the capability to normalize, digitize, learn, fuse, sense, understand, adapt, and initiate appropriate action [10, 11]. Traditionally, at first, smart sensors were designed for various physical quantities [12-14]. Today a tendency for creation of various chemical and bio-smart-sensors is observed. Europe represents the largest market worldwide, followed by North America. Asia-Pacific is forecast to emerge into the fastest growing regional market trailing a projected CAGR of 5.5 %, over the analysis period. Growth in this region will primarily stem from increased acceptance of the technology as a result of lower costs, which in turn is a factor of faster technology diffusion in the region.

According to the 'Sensor Survey 2010 What is your topic of interest ?' from the Sensors Web Portal, Inc. (http://www.sensorsportal.com), among 520 respondents the topic 'Smart sensors and systems' is the most popular (see Fig. 1.1).

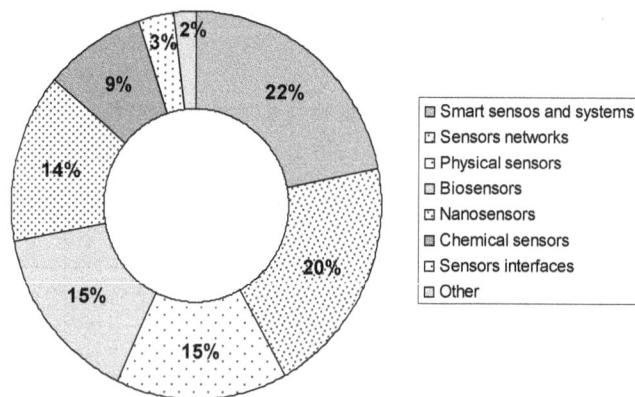

Fig. 1.1. Sensor Survey 2010: What is your topic of interest?
(http://www.sensorsportal.com).

The 'Other' includes sensor interfacing, sensor signal conditioners, image processing for sensor systems SAW sensors, underwater sensors and sensors for satellite remote sensing.

Traditional, popular sensors such as temperature, pressure, magnetic, flow, etc. also demonstrate steady growth. So, according to a new market research report 'Temperature Sensor Market, A Study of major Sensor types (ICs, Thermostat, Thermistor, Resistive Temperature Detectors (RTDs), Thermocouple) & Applications, Global Forecast & Analysis 2011–2016 published by *MarketsandMarkets*, the market size of temperature sensors in the year 2010 was US $3.27 billion and is expected to reach US $4.51 billion units by 2016, at an estimated CAGR of 5.6 %. In terms of volume, the unit shipment for temperature sensors was 2.02 billion units in the year 2010 and is expected to reach 3.54 billion units by 2016, at an estimated CAGR of 10 % from 2011 to 2016 [15]. Temperature sensors mainly utilize digital technology [8], which means better efficiency and sensing performance. Temperature sensors have a significant place in different industry verticals. The major applications of temperature sensors are in petrochemical industry, automotive industry, consumer electronics industry, metal industries, food and beverages industry, and healthcare. The demand for reliable, high performance and low cost sensors is increasing leading to the development of microtechnology and nanotechnology, offering opportunities like miniaturization, low power consumption, mass production, etc.

The demand for magnetic field sensors in the year 2010 was 3.67 billion units; which are expected to reach US $ 7.14 billion units in 2016 at a CAGR of 10.3 %. In value terms, the magnetic field sensors market stood at US $1.1 billion in year 2010 and is expected to reach US $ 2.0 billion by year 2016, at 8.7 % CAGR during the projected period [16].

According to new analysis from *Frost & Sullivan*, Global Flow Sensors and Transmitters Market, finds that the market earned revenues of US $ 4,847.6 million in 2010 and estimates this to reach US $ 6,423.8 million in 2017 [17]. Accuracy and reliability are the major selection/purchase criteria in this market.

International Frequency Sensor Association (IFSA) members' interests in particular sensors are shown in Fig. 1.2 [18]. There were 673 members from 67 countries in March 2012: 58 % are from industrial companies, 35 % are from universities and 7 % from research centers. As it is visible from the Fig. 1.2 IFSA members are most interested in Sensors Instrumentation, Temperature, Pressure, Optical, Biosensors, Gas and Chemical sensors.

1.2. About this Book

This book contains collected state-of-the-art reviews from various physical, chemical, biosensors and wireless sensors networks areas. The book has been organized by topics of high interest. In order to offer a fast and easy reading of the state of the art of each topic, every chapter in this book is independent and self-contained. The eleven chapters, which follow the introduction of this book, have the similar structure: first an introduction to specific topic under study; second particular field description including sensing applications.

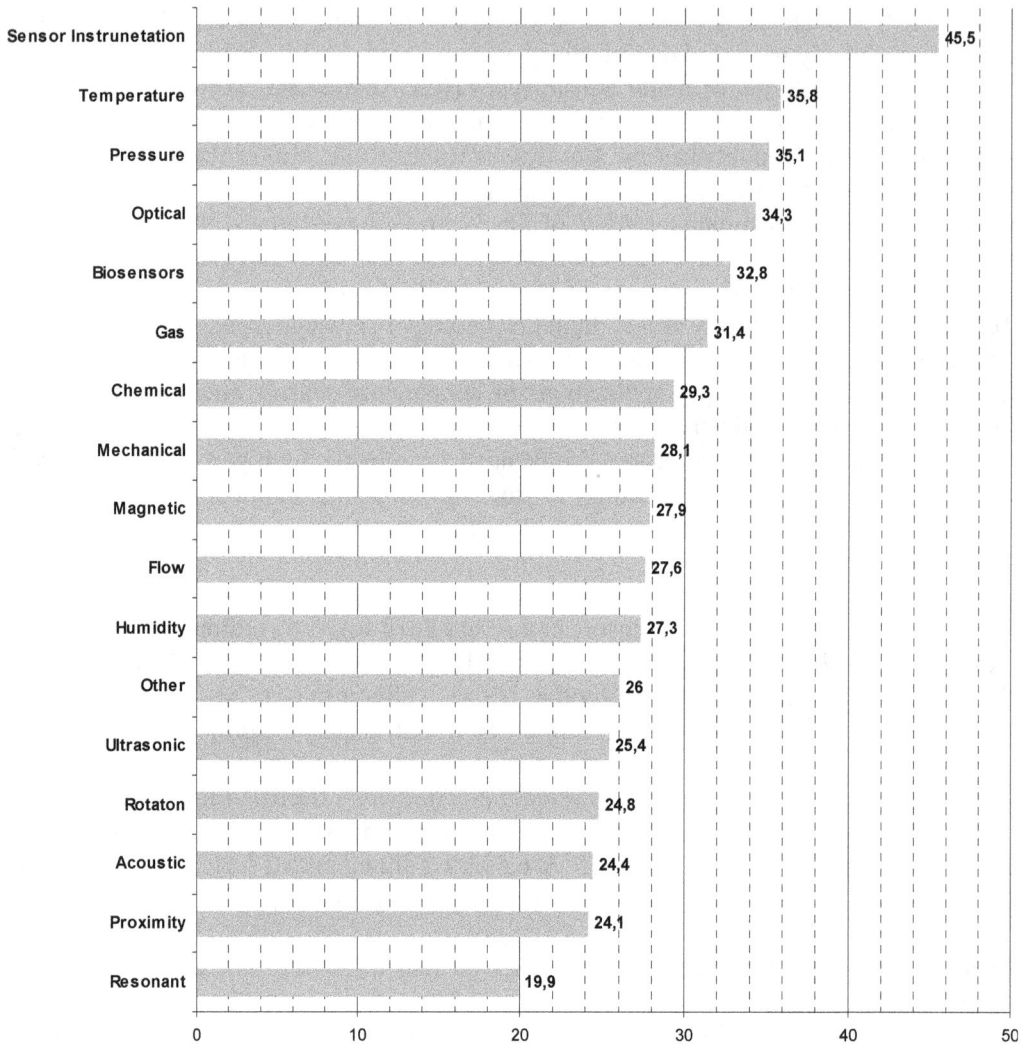

Fig. 1.2. IFSA Members' interest in particular type of sensors
(IFSA, March 2012).

In Chapter 2 a review of resistive and capacitive based sensing technologies is presented. Resistive and capacitive (RC) sensors are the most commonly used sensors. More than 30 % of modern sensors are direct or indirect applications of the RC sensing principles. Their reliability, simple construction, adjustable resolution, matured design and manufacturing processes, and maintenance-free technology have made RC sensors the preferred choice in sensor designs, and they are widely used in homeland security, industry, environment, space, traffic control, home automation, aviation, and medicine. The operating principles of resistive sensors are governed by several important laws and phenomena such as Wiedemann-Franz Law; Photoresistive-, Thermoresistive- Piezoresistive-, Magnetoresistive-, Chemoresistive-, and Bioimpedance-Effects. The applications of these principles are presented through a variety of examples including accelerometers, flame detectors, pressure/flow rate sensors, resistive temperature devices (RTDs), hygristors, gas sensors, and bio-impedance instrument. The capacitive sensors are classified and described through their three primary configurations: parallel (flat), cylindrical (coaxial), and spherical (concentric). Each configuration is discussed with respect to its geometric structure, function, and application in various sensor designs. Capacitance sensor arrays are also presented in this chapter.

Chapter 3 deals with an automated synthesis MEMS sensors and consists of two parts: 'Layout Synthesis of MEMS Component with Distributed Mechanical Dynamics' and 'Synthesis of a MEMS System with Associated Control Loop'. The first part presents a novel, holistic methodology for automated optimal layout synthesis of MEMS systems embedded in electronic control circuitry from user-defined high-level performance specifications and design constraints. The described approach is based on simulation-based optimization where the genetic-based synthesis of both mechanical layouts and associated electronic control loops is coupled with calculations of optimal design parameters. The underlying MEMS models include distributed mechanical dynamics described by partial differential equations to enable accurate performance prediction of critical mechanical components. The proposed genetic-based synthesis technique has been implemented in SystemC-A and named SystemC-AGNES. A practical case study of an automated design of a capacitive MEMS accelerometer with Sigma-Delta control demonstrates the operation of the SystemC-AGNES platform. This part of the Chapter focuses on the layout synthesis of mechanical components, while the full synthesis methodology including automated and optimal electronic control loop synthesis is outlined in the second part.

The second part of this Chapter presents a holistic synthesis approach to designing both the mechanical layout and associated electronic control in a mixed-domain MEMS sensor. It develops further the concepts presented in the first part which focuses on layout synthesis of the mechanical part only. A case study is discussed where the proposed genetic-based synthesis approach implemented in SystemC-AGNES is applied to a high order $\Sigma\Delta$ control system in an electromechanical MEMS accelerometer. The method efficiently and in an automated manner generates suitable configurations of the $\Sigma\Delta$ control loop by combining primitive components stored in a library and optimizes them according to user specifications. The synthesis results show that the proposed technique explores the configuration space effectively and it develops new circuit structures which have not been investigated before. The noise floors in the MEMS accelerometers synthesized by SystemC-AGNES are further reduced leading to an improvement of the SNR compared with a manually designed standard electromechanical $\Sigma\Delta$ MEMS accelerometer.

Sensors for food inspection are reported in the Chapter 4. Three types of novel interdigital sensors have been designed and fabricated. These sensors were used to assess different chemicals related to food poisoning. The sensors were designed to have different configurations and were constructed on different substrates. The performances of the sensors were evaluated for different configurations of the electrode structures as well as dielectric materials. Analyses of sensors have been carried out using analytical method and modeling using COMSOL Multiphysic. Initial experiments have been conducted to analyze the sensor's performance with two peptide derivatives namely Sarcosine and Proline. These peptides are closely related to the target molecule of domoic acid, a natural toxin in seafood. Experiments with endotoxin have been presented and the possibility of extending the sensors for detection of chemicals responsible for food poisoning has been discussed. This chapter will highlight the design and fabrication process and initial investigations on the sensors' performance based on Impedance Spectroscopy method.

Chapter 5 is focused on techniques for force measurements. This review of research literatures has documented the advantages and disadvantages of different systems.

Chapter 6 deals with typical 3S parameters of gas sensors namely sensitivity, selectivity and stability of metal oxides. Metal oxide gas sensors are considered as one of the basic technology for identification and measuring concentration of gases in atmosphere. Research on sensing materials has been focused on design higher performance and elevated efficiency gas sensing materials. Various experimental methodologies associated with gas sensors are discussed for their screening in this article. In the present review gas sensors based on inorganic materials are described in details. Factors deals with sensitivity and selectivity are discussed including inorganic materials such as Transition-metal oxides and Non-transition-metal oxides.

Chapter 7 specially covering microcantilever-based sensors in biological and chemical sensing applications. In comparison to conventional sensing techniques, the major advantages of microcantilevers include high sensitivity and quick response, direct detection (label free), low cost, versatility, array capability (small size and microfabricationable). The review covers the basic working principles and sensing mechanisms, the major types of microcantilevers, and the reported applications in various biological and chemical detections.

Chapter 8 devoted to nanomaterials and chemical sensors. Nanomaterials and nanosensors are two most important iconic words of the modern science and technology. Though nano technology is relatively a new area of research and development it will soon be included in the most modern electronic circuitry used for advanced computing systems. Since it will provide the potential link between the nanotechnology and the macroscopic world the development is primarily directed towards exploitation of nanotechnology to computer chip miniaturization and vast storage capacity. However, for implementation in the consumer products the present high cost of production must be overcome. There are different ways to make nanosensors e.g. top-down lithography, bottom-up assembly, and self molecular assembly. Consequently, nanomaterials and nanosensors have to be made compatible with the consumer technologies. The progress in detecting and sensing different chemical species with increased accuracy may transform the human society from uncertainty and inaccuracy to more precise and definite world of information. For example, extremely low concentrations of air pollutants or toxic materials in air & water around us can be accurately and economically detected in no time to save the human beings from the serious illnesses. Also, the medical sensors will help in diagnoses of the diseases, their treatment and in predicting the future profile of the individual so that the health insurance companies may exploit the opportunity to grant or to deny the health coverage. Other social issues like privacy invasion and security may be best monitored by the widespread use of the surveillance devices using nanosensors.

Advances in biosensors and biochips are presented in Chapter 9. This review is focused on a very important and emerging area of nano-bio sensors and biochips, which have prospects of numerous applications to nanomedicine. Various topics of biosensors and transducers based on quantum dots (QD), porous silicon (PS), and Si-nanoparticles are discussed. A detailed discussion on microarrays and microfluidic biochips along with their classification and various applications is also presented. Some powerful optical techniques like Fluorescence Resonance Energy Transfer (FRET), and Surface Enhanced Raman Spectroscopy (SERS) that are often deployed in conjunction with biosensors and biochips, as an interface mechanism, are also reviewed.

Distributed Information Extraction from Large-scale Wireless Sensor Networks is studied in Chapter 10. Regardless of the application domain and deployment scope, the ability to retrieve information is critical to the successful functioning of any wireless sensor network (WSN) system. In general, information extraction procedures can be categorized into three main approaches: agent-based, query-based and macroprogramming. While query-based systems are the most popular, macroprogramming techniques provide a more general-purpose approach to distributed computation. Finally, the agent-based approaches tailor the information extraction mechanism to the type of information needed and the configuration of the network it needs to be extracted from. This chapter offers an extensive survey of the literature in the area of WSN information extraction, covering in Sections 1 and 2 the three main approaches above. Section 3 highlights the open research questions and issues faced by deployable WSN system designers and discuss the potential benefits of both in-network processing and complex querying for large scale wireless informational systems.

Chapter 11 is a detailed review about software modeling techniques for wireless sensor networks. WSNs monitor environment phenomena and in some cases react in response to the observed phenomena. The distributed nature of WSNs and the interaction between software and hardware components makes it difficult to correctly design and develop WSN systems. One solution to the WSN design challenges is system modeling. In this chapter a survey of 9 WSN modeling techniques and show how each technique models different parts of the system such as sensor behavior, sensor data and hardware are presented. Furthermore, the chapter considers how each modeling technique represents the network behavior and network topology. It is also described the available supporting tools for each of the modeling techniques. Based on the survey, the modeling techniques and derive examples of the surveyed modeling techniques by using SensIV are classified.

The last chapter of the book, Chapter 12, deals with application of the multi-sensor wireless network system for hurricane monitoring. This wireless sensor network system was developed at Florida Institute of Technology to monitor wind induced pressure on low-rise residential building roofs during hurricane events. The system was tested to evaluate the performance of the sensors and their reliability to measure accurate pressure variations. The reliability of the pressure sensors is established by comparing measurements with secondary references and basic Bernoulli theory. The effects of sensor case, wind gusts, wind direction and structural vibration on the measured pressure are also presented. The system was tested in a wind tunnel, on top of a van on a highway road test, and at the University of Florida hurricane simulator. These tests revealed that the pressure readings were sensitive to mechanical vibrations and the sensor case shape, only when facing the windward direction. Some computational fluid dynamics analysis was also employed to verify the sensors performance and to develop reliable computational tools to simulate hurricane effects.

Each chapter has been written by different contributors. Many of contributors are members of the editorials board of different journals related to the field and some of them are IFSA members.

We hope that readers enjoy this book and that can be a valuable tool for those who involved in research and development of various sensors and sensor systems.

References

[1]. Vladimir Leonov, Yvonne van Andel, Ziyang Wang, Ruud J. M. Vullers and Chris Van Hoof, Micromachined Polycrystalline Si Thermopiles in a T-shirt, *Sensors & Transducers*, Vol. 127, Issue 4, April 2011, pp. 15-26.

[2]. Vladimir Leonov, Human Heat Generator for Energy Scavenging with Wearable Thermopiles, *Sensors & Transducers*, Vol. 126, Issue 3, March 2011, pp. 1-10.

[3]. Vladimir Leonov, Tom Torfs, Chris Van Hoof and Ruud J. M. Vullers, Smart Wireless Sensors Integrated in Clothing: an Electrocardiography System in a Shirt Powered Using Human Body Heat, *Sensors & Transducers*, Vol. 107, Issue 8, August 2009, pp. 165-176.

[4]. Mieke Van Bavel, Vladimir Leonov, Refet Firat Yazicioglu, Tom Torfs, Chris Van Hoof, Niels E. Posthuma and Ruud J. M. Vullers, Wearable Battery-free Wireless 2-channel EEG Systems Powered by Energy Scavengers, *Sensors & Transducers*, Vol. 94, Issue 7, July 2008, pp. 103-115.

[5]. Vladimir Leonov, Ziyang Wang, Paolo Fiorini and Chris Van Hoof, Modeling of Micromachined Thermopiles Powered from the Human Body for Energy Harvesting in Wearable Devices, *Sensors & Transducers*, Vol. 103, Issue 4, April 2009, pp. 29-43.

[6]. Ziyang Wang, Vladimir Leonov, Paolo Fiorini, Chris Van Hoof, Realization of a Wearable Miniaturized Thermoelectric Generator for Human Body Applications, *Sensors and Actuators A: Physical*, Vol. 156, Issue 1, EUROSENSORS XXII, 2008, November 2009, pp. 95-102.

[7]. Global Sensors Market to Reach US$76.7 Billion by 2017, According to New Report by Global Industry Analysts, Inc., Press Release, San Jose, CA USA, *PRWeb*, October 12, 2011.

[8]. Sergey Y. Yurish, Digital Sensors and Sensor Systems: Practical Design, *IFSA Publishing*, Barcelona, Spain, 2011.

[9]. Global Smart Sensors Market to Reach US$6.7 Billion by 2017, According to New Report by Global Industry Analysts, Inc., Press Release, San Jose, CA USA, *PRWeb*, 6 March 2012.

[10]. Vincenzo Di Lecce, Marco Calabrese, From Smart to Intelligent Sensors: A Case Study, *Sensors & Transducers*, Vol. 14-1, Special Issue, March 2012, pp. 1-17.

[11]. Sergey Y. Yurish, Sensors: Smart vs. Intelligent, Vol. 114, Issue 3, *Sensors & Transducers*, March 2010, pp. I-VI.

[12]. Nikolay V. Kirianaki, SergeyY. Yurish, Nestor O. Shpak, Vadim P. Deynega, Data Acquisition and Signal Processing for Smart Sensors, *John Wiley & Sons*, Chichester, UK, 2002.

[13.] Smart Sensors and MEMS, Ed. by Sergey Y. Yurish and Maria Teresa S. R. Gomes, *Springer*, 2005.

[14]. Smart Sensor Systems, Ed. by Gerard Meijer, *John Wiley & Sons*, Chichester, UK, 2008.

[15]. Global Temperature Sensors Market worth $4.51 Billion by 2016, *MarketsandMarkets,* Dallas, TX, USA, 2012.

[16]. Global Magnetic Field Sensors Market by Product (Angular, Revolution, Current, Position) by Technology (Hall Effect, Magneto Resistive & Inductive, Fluxgate, Squid), & Applications (Consumer, Automotive, Industrial, Aerospace & Defense) 2011 – 2016, Press Release, *Reportlinker*, 29 December 2011.

[17]. Global Flow Sensors and Transmitters Market, *Frost & Sullivan*, 23 September 2011, USA.

[18]. International Frequency Sensors Association (IFSA), http://www.sensorsportal.com

Chapter 2
Resistive and Capacitive Based Sensing Technologies

Winncy Y. Du and Scott W. Yelich

2.1. Introduction

Resistive and capacitive (RC) sensors have assisted mankind in analyzing, controlling, and monitoring thousands of functions for over a century. Some milestones in the development of resistive sensors include the discovery of the *Piezoresistive Effect* by Lord Kelvin in 1856, the use of platinum as the sensing element of a resistive thermometer by Charles William Siemens in 1871, the invention of a *Carbon Track Potentiometer* by Thomas Edison in 1872, and the patenting of a non-linear *rheostat* by Mary Hallock-Greenewalt in 1920. In automobiles, resistive sensors are used to measure air flow rate, throttle position, coolant/air temperature, oxygen volume, wheel speed, and so forth. In airplanes, highly reliable resistive sensors monitor engine functions, hydraulic systems, electronic devices, temperature and pressure readings, and have greatly increased aircraft safety. Measuring and analyzing electrical impedance characteristics of the human body (e.g., resistance and capacitance) have allowed doctors to diagnose certain diseases, monitor health conditions, and analyze the treatment results. The advantages of resistive sensors are their reliability, simple construction, adjustable resolution, matured design and manufacturing processes, and maintenance-free features. Electrical resistance is also the easiest electrical property to measure precisely over a wide range at moderate cost. These important features have often made resistive sensors the preferred choice in sensor designs.

The earliest capacitive sensing can be traced to 1600 when William Gilbert experimented with frictional electrical charges on objects. He found that electrical charges cause objects to attract or repel each other. This attracting or repelling force was greatly affected by the distance between the objects. In 1745, the first capacitor, the Leyden jar, was invented independently by Ewald Georg von Kleist and Pieter van Musschenbroek. Capacitive sensors are traditionally divided into two basic classifications: *passive* or *active*, based on whether or not there are any active electronic

Winncy Y. Du,
Department of Mechanical & Aerospace Engineering, San Jose State University, San Jose, USA

Sergey Y. Yurish (ed.), *Modern Sensors, Transducers and Sensor Networks*
© International Frequency Sensor Association Publishing, 2012

components (e.g., amplifiers) in the sensors. Passive sensors do not have any active electronics in the sensor, thus minimizing their sizes. Passive sensors have some advantages: they have greater flexibility in probe configuration, are more stable, and cost less than active sensors. Their disadvantages include lower driving frequency and lower bandwidth, which limits their applications. Active sensors usually have electronics packaged inside the sensors. They operate at much higher frequencies and bandwidths, and are particularly suitable in applications that involve stray electrical noise on the target. Their disadvantages include higher cost and less configuration flexibility [1]. Capacitive sensors are the most precise of all electrical sensors and are known for their extremely high sensitivity, high resolution, high stability, broad bandwidth, and drift-free measurement capability. Capacitive sensors have a wide range of applications including: precision movement detection, coating thickness gauging, liquid level and flow rate monitoring, diamond turning, chemical element selection, biocell recognition, and rotational alignment. They can also be used in severe environments (e.g., high temperature, strong magnetic fields and radiation) and in non-contact and non-intrusive applications.

RC sensors employ a broad range of theories and phenomena from the fields of physics, material science, electrochemistry, biology, and electronics [2, 3]. The progress in micro- and nano-machining technologies has significantly advanced traditional RC sensors to a new level – high sensitivity, rapid response, low power consumption, and miniaturization. RC sensing principles can also be combined with other sensing technologies, such as ultrasound, radio frequency (RF), complementary metal-oxide semiconductor (CMOS), or fiber-optics, to create more sophisticated and powerful hybrid sensors.

This chapter is organized as follows: Section 2.1 is an introduction; Section 2.2 overviews the principles, design, and application of resistive sensors; Section 2.3 presents the classification, principles, design, and application of capacitive sensors; and Section 2.4 gives the summary.

2.2. Resistive Sensing Technologies

2.2.1. Principles

A resistive sensor responses to a change in a physical or chemical parameter and induces a change in electrical resistance. The magnitude of the physical or chemical parameter (e.g., temperature, light intensity, strain, voltage, magnetic field, electrical current, or gas/liquid concentration) can be inferred from the measured resistance value. The basic principles behind resistive sensors are summarized in Table 2.1.

Table 2.1. Basic Principles of Resistive Sensors.

Ohm's Law The resistance, R, of a material passing an electric current, I, under an applied voltage, V, is: $$R = \frac{V}{I} \qquad (2.1)$$	*Electrical Resistance* The resistance, R, of a material is a function of both the specific resistivity, ρ, of the material and physical geometry (length, l, and cross sectional area, a): $$R = \rho \frac{l}{a} \qquad (2.2)$$
Photoresistive Effect When light (e.g., infrared, visible, or ultraviolet light) strikes photoconductive semiconductor materials, the resistance of the materials decreases. The conductivity of a semiconductor, σ, is described by: $$\sigma = q\,(\mu_e n + \mu_h p) \qquad (2.3)$$ q – charge of an electron (1.602×10^{-19} C); μ_e, μ_h – mobility of the free electrons and holes, respectively; n, p – density of electrons and holes, respectively.	*Piezoresistive Effect* In a thin strip of silicon, change in resistance vs. in-plane stresses in the longitudinal (parallel to the current) direction and transverse (perpendicular to the current) direction can be expressed as: $$\Delta R / R = \pi_L \sigma_L + \pi_I \sigma_I \qquad (2.4)$$ ΔR, R – change in resistance and the original resistance of the strip, respectively; σ_L, σ_T – longitudinal and transverse stress, respectively; π_L, π_T – piezoresistive coefficient along the longitudinal and transverse direction, respectively.
Wiedemann-Franz Law The ratio of the thermal conductivity to the electrical conductivity of a material is proportional to its absolute temperature. $$K / \sigma = LT \qquad (2.5)$$ K – thermal conductivity; σ – electrical conductivity; L – Lorenz Number; T – absolute temperature (in Kelvin).	*Thermoresistive Effect of Metals* The electrical resistance of a metal conductor increases as the temperature increases. The relationship between resistance, R, and temperature, t, is: $$R(t) = R_0 \left[1 + \alpha(t - t_0) + \beta(t - t_0)^2 + \gamma(t - t_0)^3 + \ldots \right] \qquad (2.6)$$ R_0 – resistance at the reference temperature t_0 (0°C or 25°C); α, β, γ, …, – temperature coefficients of the metal.

Thermoresistive Effect of Semiconductors
The electrical resistance of semiconductor materials decreases as temperature increases. The relationship between electrical resistance, R, and temperature, t, is:
 	
$$R(t) = R_0 e^{\left[\beta \left(\frac{1}{t} - \frac{1}{t_0} \right) \right]} \qquad (2.7)$$
 	
R_0 – resistance at the reference temperature t_0 (0°C or 25°C); β – temperature coefficient of the semiconductor material.
	Magnetoresistive (MR) Effect
	A conductor changes its electric resistance in the presence of an external magnetic field. For a strip of ferro-magnetic material, the resistance, R, is:
	
	$$R(\Theta) = \underbrace{\rho_\perp \frac{l}{bd}}_{R_0} + \underbrace{(\rho_{//} - \rho_\perp) \frac{l}{bd} \cos^2 \Theta}_{\Delta R} \qquad (2.8)$$
	
	Θ – angle between the applied (external) magnetic field and the sensor's easy axis (internal magnetic field); $\rho_\perp$, $\rho_{//}$ – specific resistivity of the strip perpendicular and parallel to the easy axis, respectively; l, b, d – length, width, and thickness of the strip, respectively.

Table 2.1. Basic Principles of Resistive Sensors (Cont.).

Chemoresistive Effect	*Bioresistive* and *Biopotential Principles*
When a target (e.g., gas) is in contact with a chemoresistive material, it interacts with the material both physically and chemically. Adsorption and/or reaction between the target gas and the material results in a variation of conductivity and/or electron concentration, causing a change in electrical resistance. The two main reactions are: Oxidation Process: $O_2 + 2e\text{-} \rightarrow 2O\text{-}_{ads}$ (2.9) Reduction Process: $X + O\text{-}_{ads} \rightarrow XO_{des} + e\text{-}$ (2.10) In (2.9), resistance increases since two electrons are taken away during reaction; while in (2.10), resistance decreases since one electron is returned during reaction.	Living organisms are composed of positive and negative ions in various quantities and concentrations. The migration of ions through a region causes a change in electrical resistance or difference in voltage potential between two points. The electrical model of a cell is often characterized by a three-element electrical model: R_e – extracellular liquid resistance; R_i – intracellular liquid resistance; C – cell membrane capacitance. Thus, the equivalent impedance, Z, of the cell is: $Z = R_e /\!/(Z_c + R_i) = R_e /\!/(\frac{1}{j\omega C} + R_i)$ (2.11) $j = \sqrt{-1}$; ω – the angular frequency.

2.2.2. Design and Applications

2.2.2.1. Potentiometric Sensors

A potentiometer (or *pot* for short) relates a change in length or position to a resistance change. When the position (either linear or angular position) of a movable contact on a fixed resistor changes, the resistance of the pot changes. Potentiometers are commonly used as linear or angular position sensors, air flow meters, wind direction detectors, or volume and tone controllers in stereo equipment. Fig. 2.1 shows a potentiometric pressure sensor designed by *SFIM SAGEM,* France. It has three terminals: a power input, a ground, and an output. When the pressure of the input liquid or gas expands the diaphragm, the wiper connected to the diaphragm will slide along the fixed resistor. The location of the wiper is an indicator of the magnitude of the pressure.

Fig. 2.2 shows a potentiometric air flow meter used in Toyota vehicles. It converts air flow volume to a vane opening angle that is measured by a potentiometer. The sensor's output voltage is then sent to the vehicle's electronic control unit (ECU) for determining the volume of air that is getting to the engine.

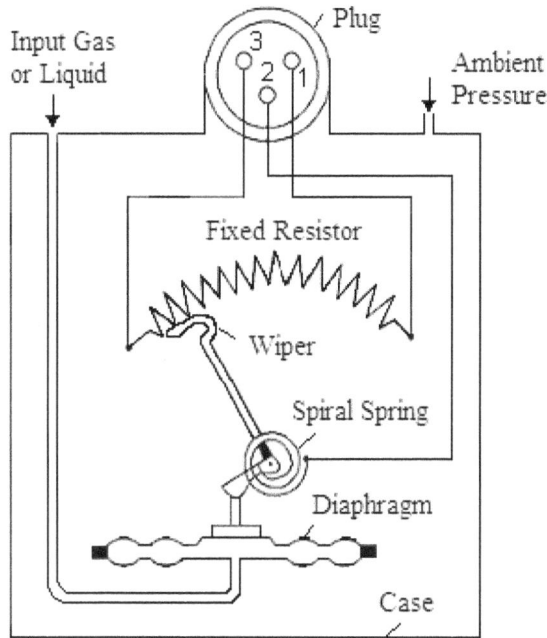

Fig. 2.1. A Potentiometric Pressure Sensor.

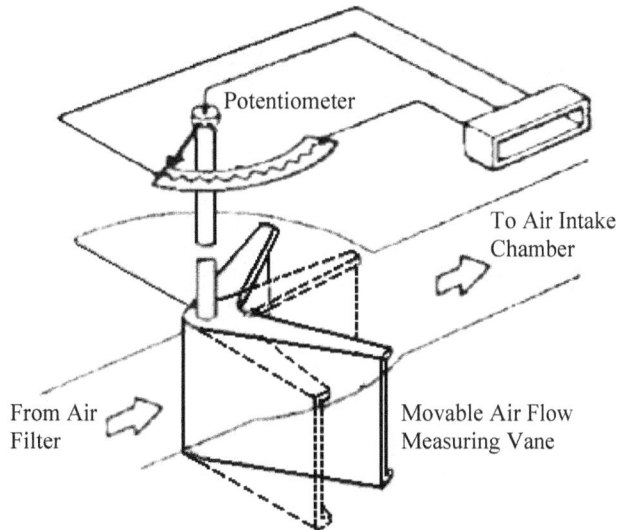

Fig. 2.2. A Potentiometric Air Flow Sensor Used in Toyota Vehicles
(Courtesy of *Toyota Corporation*, Japan).

The advantages of potentiometers are their high output signal level (no need for an amplifier), low cost, and adaptability to many applications. Disadvantages include their high hysteresis and sensitivity to vibration.

2.2.2.2. Photoresistive Sensors

A photoresistive sensor (or photoresistor) utilizes the *Photoresistive Effect* – the resistance decreases with increasing incident light intensity. Thus, it is also referred to as a *light-dependent resistor* (LDR), *photoconductor*, or *photocell*. A photoresistive sensor is made of a high-resistance semiconductor material. When light reaching the device is of high enough frequency, photons absorbed by the semiconductor give bound electrons enough energy to jump into the conduction band. The resulting free electrons (and their hole partners) conduct electricity, thereby lowering resistance. The semiconductor cadmium sulfide (CdS) is most sensitive to visible light, while lead selenide (PbSe) is most efficient in near-infrared light.

Fig. 2.3 shows a practical application of a photoresistor – a flame detector. It comprised of an ultraviolet (UV) sensitive photoresistor (cathode) and an anode [5]. When UV light from a flame is present in front of the UV sensitive photocathode, the voltage across the photocathode and the anode will force photoelectrons emitted from the photocathode to move towards the anode. A readout circuit detects the resistance change due to the presence of the flame.

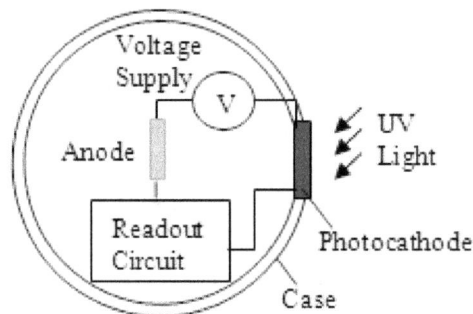

Fig. 2.3. Schematic of an UV Flame Detector.

Photoresistors are generally inexpensive. Their small size and ease of use make them popular in many applications, e.g., detecting fires, turning street lights on and off automatically based on the level of daylight, reading inventory bar codes, sensing motion, and measuring light intensity.

2.2.2.3. Piezoresistive Sensors

Piezoresistive sensors are designed based on a materials' piezoresistivity – defined as a change in electrical resistance of the material due to its mechanical stress or deformation. Piezoresistive sensors are commonly used to measure force, pressure, acceleration, vibration, and impact. Piezoresistive accelerometers have an advantage over piezoelectric accelerometers in that they can measure both static and dynamic accelerations (while piezoelectric type can be only used for dynamic acceleration measurement). Piezoresistive sensors have good high frequency response and high performance. However, they have relatively low voltage outputs that are easily affected by temperature variations and noise, thus compensation circuits are often required to correct sensor errors. Fig. 2.4 illustrates a catheter based medical device for intravascular blood pressure measurement [6]. Its diaphragm consists of a force transducing beam and a piezoresistor. The sensor chip is ultra-miniaturized (0.1 mm × 0.14 mm × 1.3 mm).

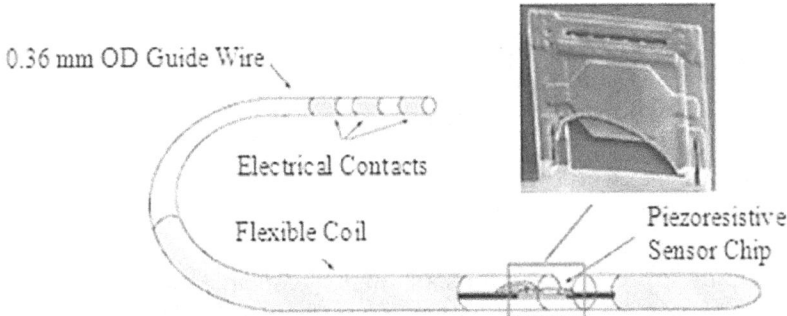

Fig. 2.4. A Piezoresistive Blood Pressure Sensor.

2.2.2.4. Thermoresistive Sensors

Resistance-based temperature sensors include *Resistance Temperature Detectors* (*RTDs*) and *Thermistors*. RTDs are *positive* temperature coefficient (PTC) sensors whose resistance increases with increasing temperature. RTDs can be constructed in either *wire-wound* (Fig. 2.5a) or *thin-film* types (Fig. 2.5b). A wire-wound RTD is the most common type. It is made by winding a very fine metal wire around an inert substrate (glass or ceramic). A thin-film RTD is produced through *Thin Film Technology* or *Thin Film Lithography* that deposits a thin film of metal (e.g., 1 μm Platinum) onto a ceramic substrate through cathodic atomization or "sputtering" [7].

The primary metals in RTD use are platinum, copper, and nickel, because they: (1) are available in near pure form, ensuring consistency in the manufacturing process; (2) have a very predictable, near linear temperature versus resistance relationship; and (3) can be processed into extremely fine wire. This is particularly important in wire-wound RTDs. RTDs are typically used for temperatures not exceeding 850 °C. Although slower in response than thermocouples, RTD sensors offer several advantages in industrial applications. They are especially recognized for excellent linearity throughout their temperature range (typically from -200 to +850 °C) with a high degree of accuracy, robustness, stability and repeatability. For a typical Platinum RTD, stability is rated at ±0.5 °C per year.

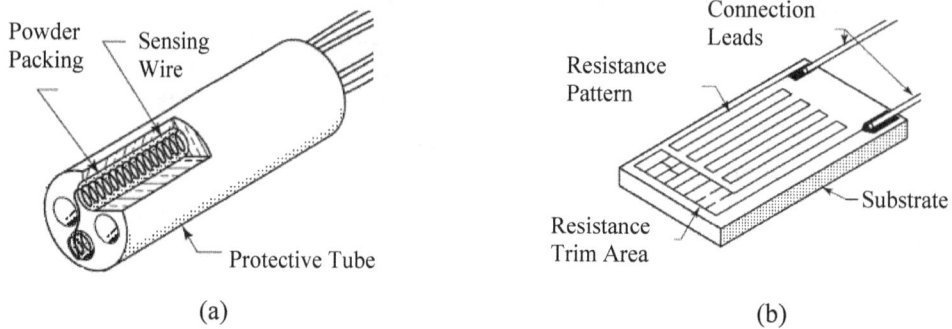

Fig. 2.5. Resistance Temperature Detectors (RTDs): (a) Wire-wound RTD;
(b) Thin-film RTD (Source: *RdF Corporation*, USA).

Thermistors (from the words ***therm**al* and res***istor***) are made from semiconductor materials whose resistance decreases with increasing temperature. Thus they are *negative* temperature coefficient (NTC) sensors. Due to their nonlinearity and exponential nature (Eq. 2.7), thermistors are limited to temperature measurements of less than 200 °C. Although Eq. 2.7 can be linearized, they generally cannot meet accuracy and linearity requirements over larger measurement spans. Their drift under alternating temperatures is also larger than RTD's. Thermistors are quite fragile and great care must be taken to mount them so that they are not exposed to shock or vibration. Thermistors have not gained the popularity of RTDs or thermocouples in industry due to their limited temperature range. Compared to wire-wound RTDs and thermocouples, thermistors are less expensive and much smaller in size. They also have a faster response, lower thermal mass, simpler electronic circuitry, and better sensitivity.

Fig. 2.6 shows a thermistor-based thermoanemometer for flow rate measurement. Two thermistors (R_0 and R_s) are immersed into a moving medium. R_0 measures the initial temperature of the flowing medium. A heater, located between R_0 and R_s, warms the medium and its temperature is then measured by R_s. The flow rate ΔQ can be derived for the medium based on the temperature difference (T_s-T_0) attained (*King's Law*) [8]:

$$\Delta Q = kl \left(1 + \sqrt{\frac{2\pi\rho c d v}{K}} \right)(T_s - T_0),\tag{2.9}$$

where K, c – thermal conductivity and specific heat of fluid at a given pressure, respectively; ρ – fluid density; l, d – length and diameter of the sensor, respectively; T_s, T_0 – surface temperature of sensor R_s and R_0, respectively; v – velocity of the medium.

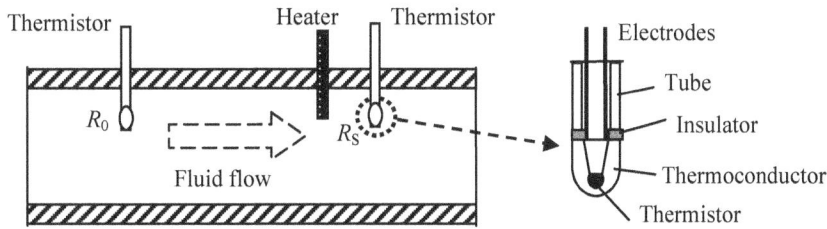

Fig. 2.6. A Thermoanemometer [5].

2.2.2.5. Magnetoresistive Sensors

Magnetoresistive sensors include: (1) *Ordinary Magnetoresistance* (OMR) in non-magnetic metals; (2) *Anisotropic Magnetoresistance* (AMR) in ferrromagnetic (FM) alloys; (3) *Giant Magnetoresistance* (GMR) in multiple alternating ferrromagnetic alloy and metallic layers; (4) *Tunneling Magnetoresistance* (TMR); (5) *Ballistic Magnetoresistance* (BMR); and (6) *Colossal Magnetoresistance* (CMR) in $La_{l-x}M_xMn$ O_{3+8} (M = Ca, Sr) perovskite structures. Table 2.2 compares the MR effects.

Table 2.2. Comparison of MR Effects.

MR Effect	$\Delta R /R$ (%)	B (Tesla)	Mechanism	Comments
OMR	< 1	1 T (metal)	Lorentz force on the electron trajectories.	No saturation at large magnetic field. Limited application.
AMR	1~3	0.5 mT ~ 1 T (depend on bulk or wire permalloy)	Electron spin-orbit interaction (leads to scattering of conducting electrons).	R directly related to the orientation of magnetization M and current I.
GMR	10~100	1-10 T	Spin-dependent electron transport.	In a FM-metal-FM alternating multi-layer structure. Extensive application in sensors and read-heads of magnetic hard disks.
TMR	≈ 100	≈ 4 T	Spin-polarized tunneling.	In FM-insulator-FM alternating layer structure. Temperature independent. Applied in magnetic random access memory (MRAM).
BMR	> 3000	< 0.1 T	Spin scattering across very narrow magnetic domain walls trapped at nano-sized constrictions.	Still in experimental phase. Nano- or atomic-scale point contacts between FM electrodes.
CMR	≈ 100,000	≈ 3 T	Insulation-to-metal phase transition at Curie temperature.	Still in study phase. Extremely temperature and doping dependent. Challenging to get useful, reproducible behavior at room temperature.

MR sensors are widely used in disk drives and controls, automotive engine and transmission systems, vehicle traction and stability controls, electronic compasses for navigation, precision positioning systems, angle and rotational speed measurement, current measurement, and earth-, magnetic-, and biomagnetic- field measurement.

Fig. 2.7 shows a magnetoresistive sensor array (made by *NVE Corporation*, USA) in a typical currency detection application. This sensor array is positioned approximately 1 mm from the currency path. The residual magnetization in the magnetic ink or magnetic strip in the currency is detected by the sensor array. This information is then analyzed to determine if the currency is genuine.

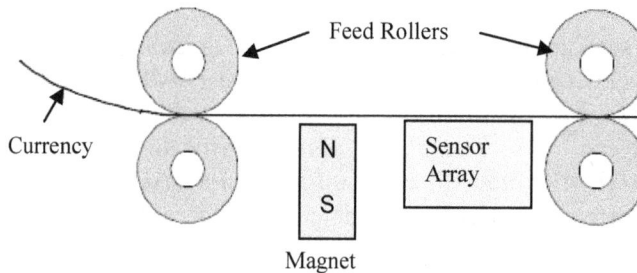

Fig. 2.7. A Magnetoresistive Sensor for Currency Detection.

2.2.2.6. Resistance-Based Chemical Sensors (Chemoresistors)

Resistive chemical sensors measure the change in electrical resistance or conductance of a sensing layer resulting from the interaction between the sensing layer and a chemical analyte. Materials used in these chemoresistive sensors are mixed metal oxide semiconductors (MMOS), organic macromolecule-metal complexes, conducting polymers and carbon black-polymer blends. A classical example is the tin oxide based semiconductor gas sensor. The tin oxide sensing layer is first activated by heating to >2500 °C to form a depletion layer where oxygen is chemisorbed on the surface. The conductivity of the activated sensor may be increased or decreased depending on the nature of the incoming gases. Reducing gases increase the conductivity and oxidizing gases decrease the conductivity of the sensor. The advantages of these semiconductor-based sensors are: easy fabrication (by sputtering), simple operation and low cost. The main disadvantages are their high-energy requirements and low selectivity. Recent research and development efforts have focused on increasing their energy efficiency and improving their selectivity. Hence, materials that operate at ambient temperature, such as conducting polymers and carbon black-polymer blends, have been extensively investigated. The array sensing approach combined with statistical algorithms, such as cluster analysis and principal component analysis, greatly improves the selectivity of this sensing technique.

Fig. 2.8 shows a catalytic gas sensor (pellistor). It consists of a very fine coil of wire suspended between two posts. The coil is embedded in a pellet of a ceramic material, and on the surface of the pellet (or "bead") there is a special catalyst layer. In operation, current is passed through the coil, which heats up the bead to a high temperature.

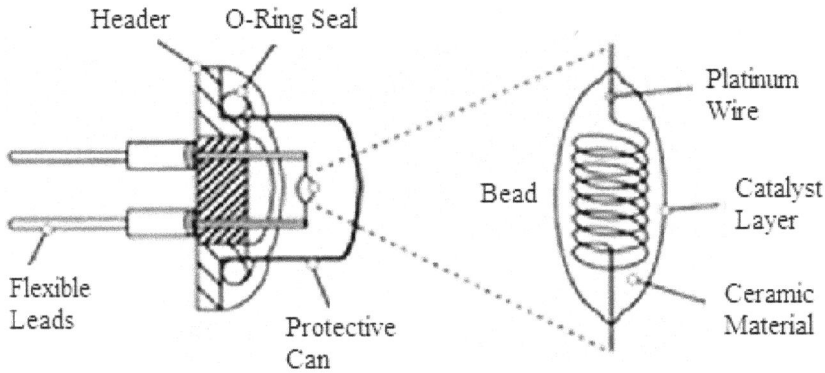

Fig. 2.8. A Catalytic Gas Sensor (Pellistor).

When a gas molecule comes into contact with the catalyst layer, the gas "burns" and heat is released which increases the temperature of the bead. This temperature rise causes the electrical resistance of the coil to increase.

A more versatile sensing system is based on the carbon black-polymer blend where the carbon particles give the electrical conductivity and the polymer provides the sensor function. The sensor response is a result of the polymer swelling, which causes the conductivity of the sensor to change. The main advantages of using carbon black-polymer blends as a sensing layer are that the sensor is reusable and the selectivity can be tailored by choosing polymers with desired functionalities. Fig. 2.9 shows a chemoresistor formed by depositing a thin film of non-conductive polymer, infused with carbon black particles, onto the metallic inter-digitated electrodes of a glass chip. The absorption of certain chemical vapor results in a measurable decrease in the electrical conductance of the sensing element. Such sensors are inexpensive and easily mass-produced. A potential application of this sensing technique could be illegal drug detection [9].

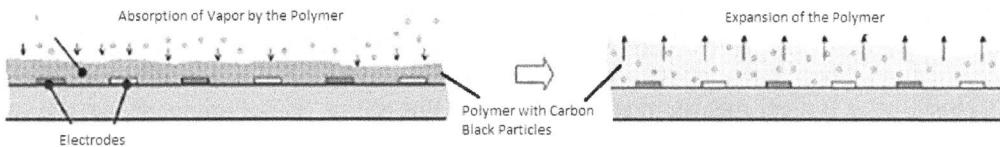

Fig. 2.9. Operation of a Chemoresistor.

2.2.2.7. Resistive Humidity Sensors (Hygristors)

Resistive humidity sensors measure the change in electrical resistance of a hygroscopic medium such as a conductive polymer, salt, or treated substrate. The specific resistivity of a hygroscopic material is strongly influenced by the concentration of absorbed water molecules. Its resistance change is typically an inverse exponential relationship to humidity. A typical hygristor (a contraction of *hygro-* and *resistor*) consists of a substrate and two silkscreen-printed conductive electrodes. The substrate surface is coated with a conductive polymer/ceramic binder mixture, and the sensor is installed in a plastic housing with a dust filter (Fig. 2.10).

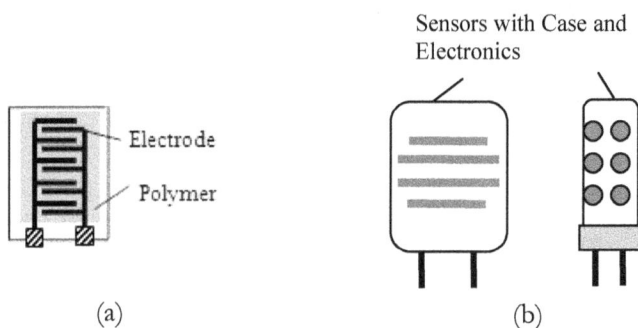

Fig. 2.10. Resistive Humidity Sensors: (a) A Typical Structure; (b) Products.

2.2.2.8. Bioimpedance Sensors

Bioimpedance sensors can extract biomedical information relative to physiology and pathology of the human body based on the electrical properties (resistance and capacitance) of tissue and organs. Usually, a small AC current or voltage is applied to an electrode system placed on the surface of the body to measure the relative impedance of tissue and organs. A German company, *Medizinische Meßstechnik GmbH*, has used an impedance method to measure changes in venous blood volume as well as pulsation of the arteries (see Fig. 2.11 a). As blood volume changes, the electrical impedance also changes proportionally. This impedance can be measured by four electrodes and passing a small amount of high frequency AC current through the body. The two middle electrodes detect a voltage, and their placement defines the measurement segment. The outer electrodes are used to emit a small current required to measure the impedance. The position of these outer electrodes is not critical. This method allows doctors to detect blood flow disorders, early stage arterioscleroses, functional blood flow disturbances, deep venous thromboses, migranes, and general arterial blood flow disturbances. Fig. 2.11b shows a typical measurement result using this setup. This bioimpedance sensor device is safe, non-invasive, inexpensive, and easy to operate.

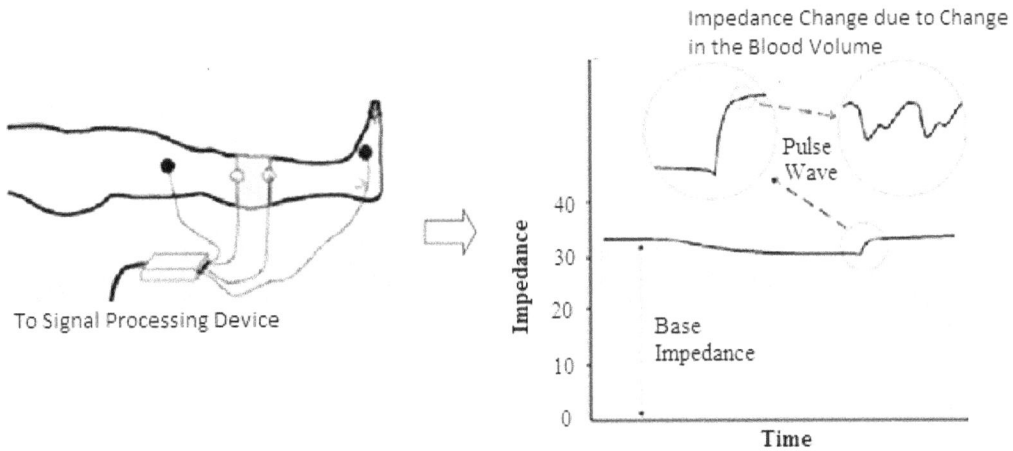

(a) Electrode Placement on Human Subject.   (b) Display of Measurement Results.

Fig. 2.11. Bioimpedance Blood Volume Measurement.

Zetek Inc. (Aurora, Colorado) developed the *CUE Fertility Monitor*. It consists of a hand-held digital monitor and two sensors – the oral and vaginal sensors (see Fig. 2.12) that detect and record the electrical resistance and ionic concentration change of saliva and vaginal secretions, respectively, in response to the cyclical changes in estrogen. The CUE monitor is claimed to both predict and confirm ovulation. The peak electrical resistance in the saliva occurs 5~7 days before ovulation, and the lowest electrical resistance in cervical secretions occurs about a day before ovulation.

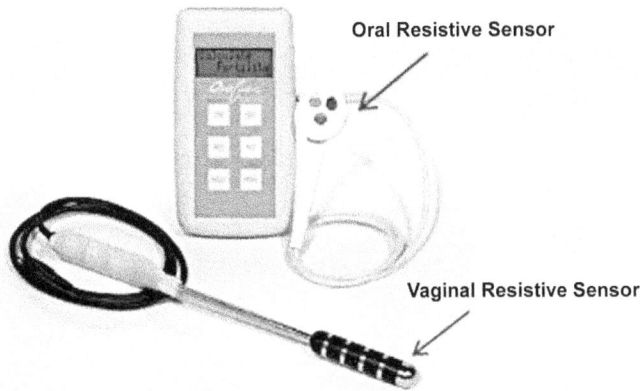

Fig. 2.12. A CUE Fertility Monitor.

2.3. Capacitive Sensing Technologies

2.3.1. Types of Capacitive Sensors

At the heart of any capacitive-sensing system is a capacitor. Capacitors are available in three configurations: flat (parallel), cylindrical (coaxial), and spherical (concentric), as shown in Table 2.3. All capacitive sensors fall into one of these three configurations, with the flat and cylindrical being the most commonly used forms.

Table 2.3. Capacitor Configurations and Their Capacitance.

Flat (Parallel) Capacitor:

$$C = \frac{\varepsilon_0 \varepsilon_r A}{d} \qquad (2.12)$$

Cylindrical (Coaxial) Capacitor:

$(l \gg r_2)$

$$C = \frac{2\pi\varepsilon_0\varepsilon_r l}{\ln(r_2/r_1)} \qquad (2.13)$$

Cylindrical (Coaxial) Capacitor:
Spherical (Concentric) Capacitor:

$$C = \frac{4\pi\varepsilon_0\varepsilon_r r_1 r_2}{r_2 - r_1} \qquad (2.14)$$

In Table 2.3, ε_r is the relative permittivity of the medium between the electrodes: ε_0 is the permittivity of a vacuum. The ratios A/d, $2\pi l/[\ln(r_2/r_1)]$, or $4\pi\ r_1r_2/(r_2-r_1)$ are the *geometry factors* for a parallel-plate, a coaxial, and a spherical capacitor respectively. Eqs. 2.12 to 2.14 describe the relationship between capacitance and the dielectric constants (ε_r, ε_0) as well as the geometric parameters (A, d, l, r_1, r_2). Any change in these components will cause a change in capacitance, which can be accurately measured. The advantages of capacitive sensors are their simple structures, high sensitivity, high resolution, temperature independence, long-term stability, and durability. Some capacitive sensors can achieve sub-nanometer position resolution (< 0.01 nm) and have a bandwidth to 100 kHz. Capacitive sensors can provide contact or noncontact measurement of distance, position, acceleration, separation, proximity, force, pressure, biocells, and chemical concentration, etc.

2.3.2. Design and Applications

2.3.2.1. Parallel Capacitor-based Sensors

The majority of capacitive sensors use the parallel-plate design. The sensing principles of such sensors are shown in Fig. 2.13, where capacitance varies with a change of: (a) the distance between two plates; (b) the overlapped area; (c) the dielectric constant of the media; or (d) the plate conductivity.

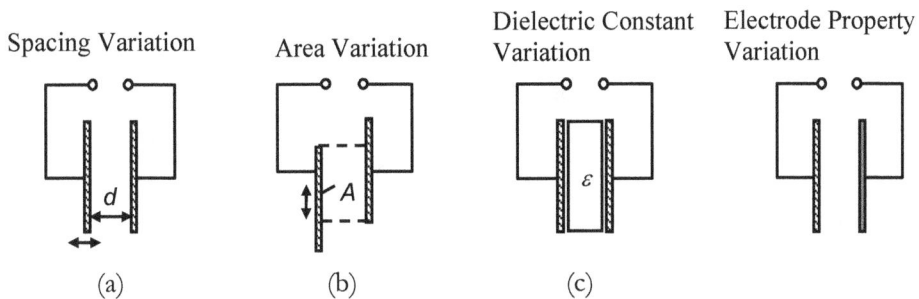

Fig. 2.13. Sensing Principles of Parallel Capacitive Sensors.

Parallel capacitive sensors have either single or dual-plate design. In a single plate (or electrode) design, the target is a conductive object and functions as the second electrode (Fig. 2.14a); while in the dual-plate design (Fig. 2.14b), the target is a non-conductive object and serves to vary the amount of electric flux reaching the second electrode.

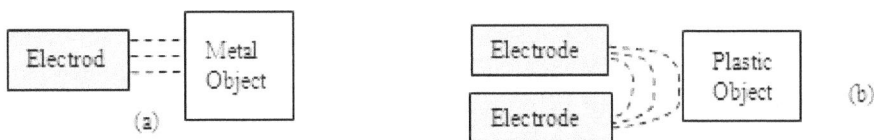

Fig. 2.14. Parallel Capacitive Sensor Designs: (a) Single-Electrode; (b) Dual-Electrode.

Parallel capacitive sensors can detect motion, distance, acceleration, fluid level, biocells, and chemicals. Some of these applications are summarized in Table 2.4. Parallel capacitive sensors can also be designed to measure pressure; such sensors are relatively robust compared to other types of pressure sensors.

Table 2.4. Applications of Parallel Capacitive Sensors.

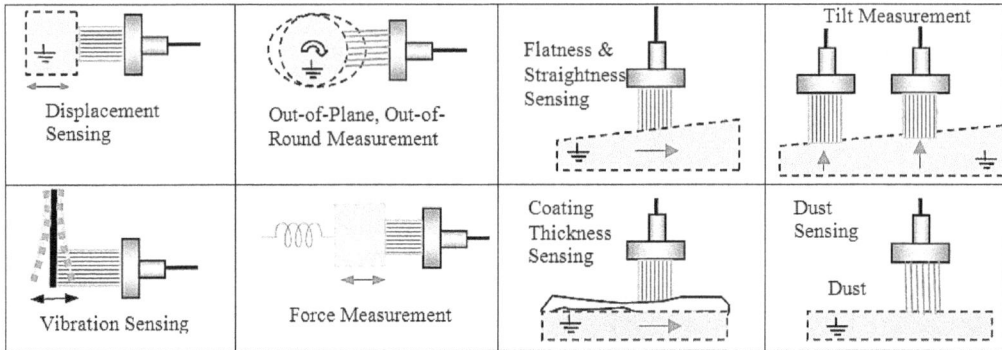

| Displacement Sensing | Out-of-Plane, Out-of-Round Measurement | Flatness & Straightness Sensing | Tilt Measurement |
| Vibration Sensing | Force Measurement | Coating Thickness Sensing | Dust Sensing |

Fig. 2.15a shows a capacitive pressure sensor designed by *VEGA Technique*, France. The unit forms a capacitor whose capacitance variation is caused by the displacement of the diaphragm connected to the electrodes. In this case, the variable parameter of capacitance C is the overlapped area A of the plates; and A itself is a linear function of the displacement l. An alternative design of a pressure sensor is to measure the variation of distance d between two plates (see Fig. 2.15 b).

If one plate of a parallel capacitor moves with respect to acceleration or vibration, a capacitive accelerometer is formed. Fig. 2.16a shows a *PCB Piezotronics*'s capacitive accelerometer. The diaphragm acts as a mass undergoing flexure in the presence of acceleration. Two plates sandwich the diaphragm, creating two capacitors. Each capacitor has its own fixed plate, and they share the movable diaphragm as their common plate. The flexure causes a capacitance change by altering the distance between the two parallel plates. The two capacitance values are sent to a bridge circuit, and the output voltage varies with input acceleration. Such a design can achieve true DC response and high performance for uniform acceleration and low-frequency vibration measurement as shown in Fig. 2.16b.

Capacitive sensors designed based on variation of the dielectric constant are often found in humidity, force/pressure, chemical-substance, and biocell sensing applications. Fig. 2.17 illustrates the mechanism of a pellicular sensor developed by *ONERA French Aerospace Lab*. The sensor detects a change in relative dielectric permittivity, ε_r, between two electrodes when a force or pressure is exerted on the plates. This sensor is very thin (about 80 µm) and can measure micro-pressure. Advantages of such pellicular sensors are their compactness, resistance to vibration, and high bandwidth (50~200 kHz). The primary disadvantage is that they are temperature sensitive.

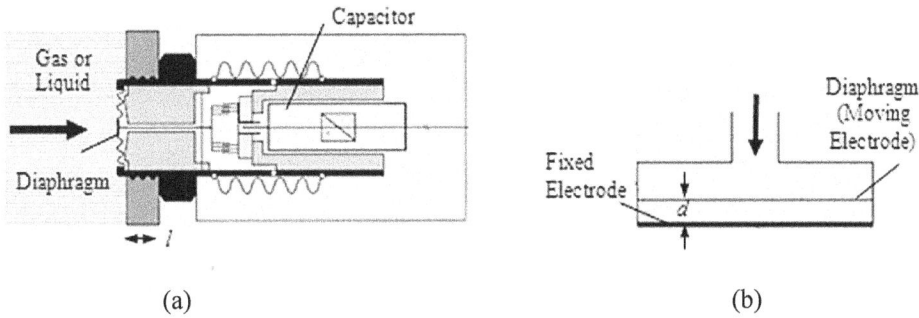

Fig. 2.15. Capacitive Pressure Sensors: (a) with *A* Variation; (b) with *d* Variation.

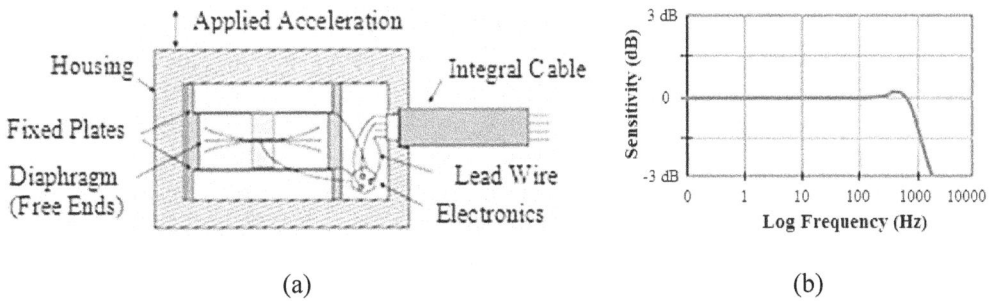

Fig. 2.16. Typical Element Structure of a Capacitive Accelerometer (a) and its Response (b).

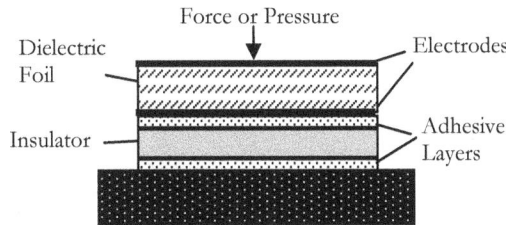

Fig. 2.17. Schematic of a Pellicular Sensor.

A chemical sensor designed at *Biose State University* [10] utilizes a variable capacitor composed of two electrodes separated by a chemically sensitive polymer (Fig. 2.18). The polymer will absorb specific (target) chemicals (analytes). Upon analyte absorption, the polymer swells and increases both the distance between the two electrodes and the polymer's dielectric permittivity, causing changes in capacitance.

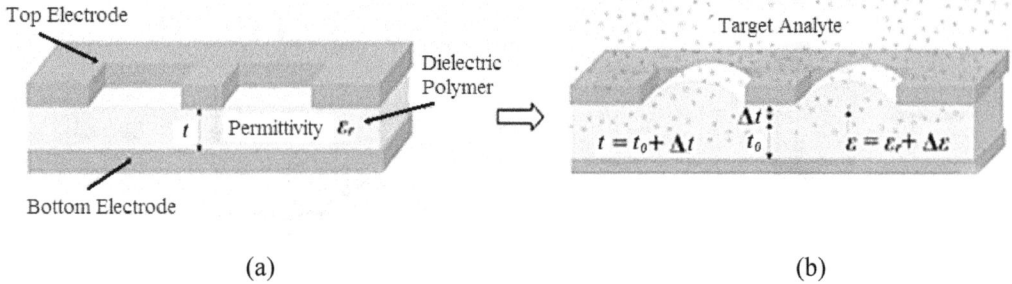

(a)   (b)

Fig. 2.18. A Capacitive Chemical Sensor: (a) Before the Presence of a Target;
(b) After the Presence of A Target.

Researchers at the *University of Illinois at Urbana-Champaign* designed a collapsible capacitive tactile sensor. By constructing the electrodes from a soft conductive elastomer instead of traditional hard materials, the sensor exhibited flexibility and unprecedented robustness [11]. To increase the capacitance of the sensor, a large electrode area and a small electrode gap (2.4 μm) were used. The small gap was achieved by inserting a conductive poly-dimethylsiloxane sheet with a regular array of small pillars between the electrodes (see Fig. 2.19 and Fig. 2.20). The pillars not only define the air gap, but also provide the restoring force to separate the electrodes.

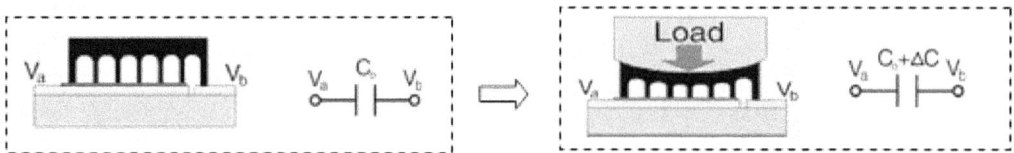

Fig. 2.19. Schematic of the Collapsible Capacitive Sensor Operation.

Fig. 2.20. (a) Flexible Conductor Capacitive Tactile Sensor; (b) Micrograph of Support Pillar Array Built into Flexible Conductor.

Well-designed electrodes of capacitive sensors also allow scientists and researchers to analyze deoxyribonucleic acid (DNA) for applications in medicine and biology. A label-free capacitive DNA sensing algorithm has been developed at the *University of Bologna* in Italy [12]. In this method the electrode-solution interfaces are characterized by capacitance sensitive to the state of the electrode surface. DNA targets in the solution bond with the probe and change the value of interface capacitance. This variation is then measured by accurate instruments. Advantages of this method over the conventional optical marker are: (1) no need for an expensive optical reading device; (2) real-time detection; and (3) improved sensitivity.

2.3.2.2. Coaxial Capacitor-based Sensors

The coaxial configuration is the second most popular design in capacitive sensors. Coaxial capacitive sensors are broadly used in proximity sensing, fluid level gauging, displacement/temperature/humidity measurement, and chemical substance detection. Fig. 2.21 shows the three design principles of cylindrical capacitive sensors.

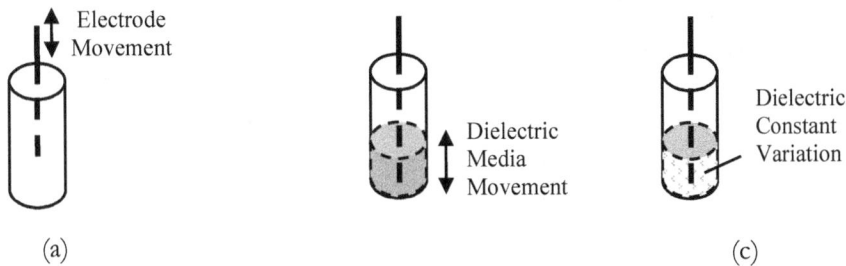

Fig. 2.21. Three Basic Principles Underline the Cylindrical Capacitive Sensor Design:
(a) Movement of Electrode; (b) Movement of Dielectric Media;
(c) Variation of Dielectric Constant.

A simple displacement sensor can be easily built using the cylindrical configuration by moving the inner conductor in and out of the outer conductor (Fig. 2.22). According to Eq. (2.13), the capacitance of a coaxial capacitor is in a linear relationship with the displacement l.

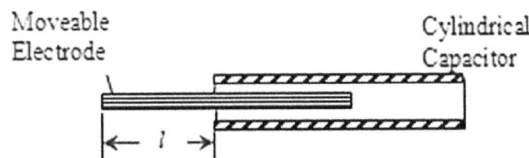

Fig. 2.22. A Capacitive Displacement Sensor.

Coaxial capacitors are commonly used for fluid level measurement with two designs: one for fluids with low dielectric constants (or high conductivity, see Fig. 2.23 a), and the other for fluids with high dielectric constants (or low conductivity, see Fig. 2.23 b). In Design (a), the surface of the metal electrode is coated with a thin isolating layer (e.g., Teflon or Kynar) to prevent an electric short circuit through the liquid. The insulated probe acts as one plate of the capacitor and the conductive liquid acts as the other (it is electrically connected to the ground). The insulating or dielectric medium in this case is the probe's sheath. In Design (b), a bare rod and the metallic vessel wall form the electrodes of a capacitor; the dielectric medium is the liquid. The vessel wall (or reference probe) is grounded in this case. Usually, a potentiometer is built in so that liquids with different densities and different dielectric constants can be measured. A bridge circuit measures the capacitance and provides continuous liquid level monitoring.

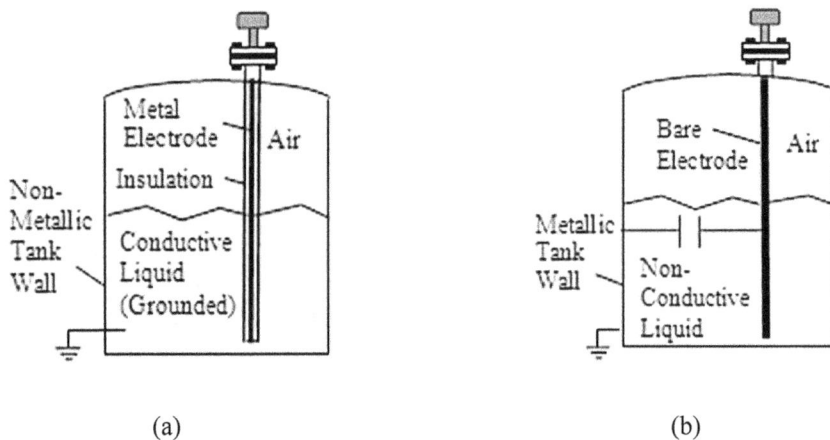

Fig. 2.23. Capacitive Liquid Level Sensors: (a) Design for Fluids with Low Dielectric Constants; (b) Design for Fluids with High Dielectric Constants.

The cross section of a capacitive touch transducer is shown in Fig. 2.24. This sensor uses a coaxial capacitor design and a high dielectric polymer (e.g., polyvinylidene fluoride) to maximize the change in capacitance as force is applied. The movement of one set of the capacitor's plates is used to resolve the displacement and hence applied force, which causes a capacitance change. From an application viewpoint, the coaxial design is better than a flat plate design as it will give a greater capacitance increase for an applied force.

Euro Gulf Group Company developed a*n Oil Analyzer* (EASZ-1) using the capacitance principle to measure moisture content in oil (see Fig. 2.25). The cylindrical sensor and outer barrel form the electrodes of a coaxial capacitor. The oil sample flows between the "plates" as a dielectric fluid, changing the capacitance of the assembly proportionally with the change in dielectric constant of the fluid. The measured capacitance is then converted to a water content output signal by the microprocessor and associated components to deliver stable and accurate readings. It also has a built-in temperature sensor for temperature compensation.

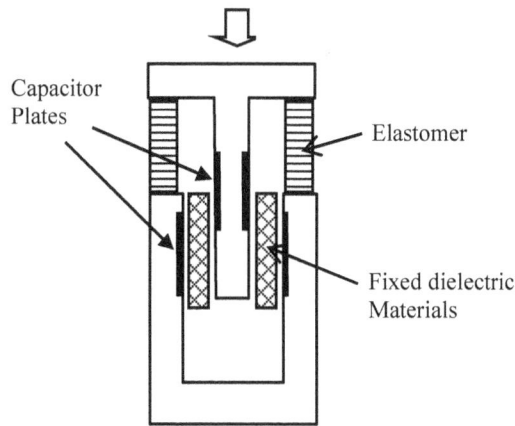

Fig. 2.24. Schematic of a Coaxial Capacitor Touch Sensor.

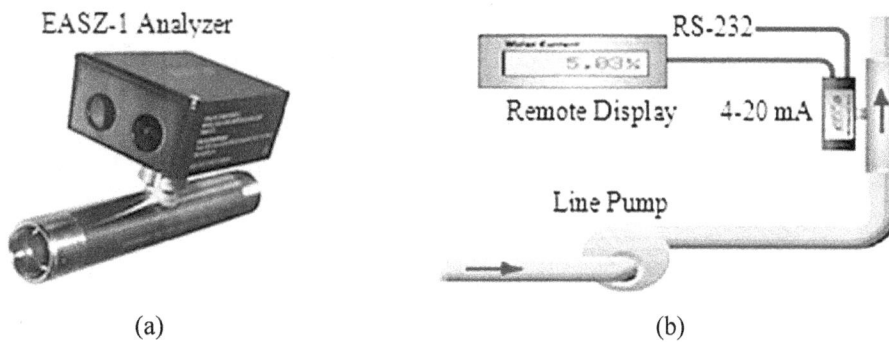

(a)   (b)

Fig. 2.25. EASZ-1 Analyzer (a) and its Typical Installation on a Hydraulic Line (b).

2.3.2.3. Spherical Capacitor-Based Sensors

Spherical capacitive sensors are not as popular as the parallel or cylindrical configurations. This is largely due to the spherical design's complexity and higher manufacturing cost. However, the spherical geometry does provide several unique features, neither flat-plate nor coaxial capacitors have, such as higher capacitance within a limited or compact space, a shape that is more readily adaptable to measure irregular surfaces, spherical equipotentials, and wider bandwidth.

A spherical capacitor provides the ideal shape for generating a nonlinear electric field gradient between its center electrode and its inner surface. This unique feature was utilized by scientists at NASA (National Aeronautics & Space Administration) to create the *Geophysical Fluid Flow Cell* (GFFC, see Fig. 2.26).

This GFFC uses spherical capacitors to simulate gravitational field conditions for studying the behavior of fluids. By applying an electric field across a spherical capacitor filled with a dielectric liquid, a body force analogous to gravity is generated around the fluid. The force acts as a buoyant force with magnitude proportional to the local temperature of the fluid and in a radial direction perpendicular to the spherical surface. In this manner, cooler fluid sinks toward the surface of the inner sphere, while warmer fluid rises toward the outer sphere. Researchers at the University of Shanghai for Science and Technology in China utilized the unique shape of the spherical capacitor to design a probe for measuring the thickness of coatings on metals [13]. This spherical capacitive probe is more accurate in measuring the thickness of non-conducting coatings on metals than the common planar probes. Also, because it is a capacitive sensor, it is not subject to the materials limitations of the magnetic induction method and the eddy current method, both having the disadvantage of being strongly influenced by the electroconductivity and magnetic conductivity of the substrate.

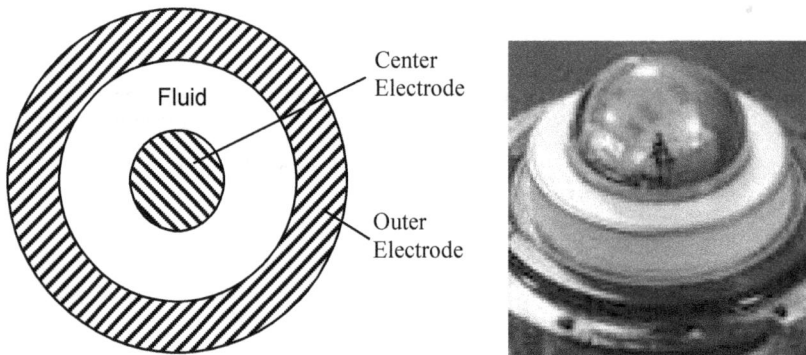

Fig. 2.26. The NASA GFFC Device.

2.3.2.4. Capacitive Sensor Arrays

Capacitive sensors can also be arranged in arrays to perform more sophisticated tasks. Fig. 2.27 shows a spherically folded capacitive pressure sensor array (1 mm thickness) for 3D measurements of pressure distribution in artificial joints [14]. The sensor array consists of 192 sensor elements which are arranged in a 16 × 16 matrix (Fig. 2.27a), then folded spherically (Fig. 2.27 b) and placed in a cavity (60 mm diameter, Fig. 2.27 c), followed by a ball joint (50 mm diameter, Fig. 2.27 d). This unique sensor can be used to measure pressure distributions along curved surfaces such as those in ball joints.

In Fig. 2.28, the *iGuard Security System* [15] and *Fingerprint Cards'* [16] sensor contains tens of thousands of small capacitive plates (functioning as pixels), each with their own electrical circuit embedded in the chip. When a finger is placed on the sensor, extremely weak electrical charges are created. Using these charges the sensor measures the capacitance pattern across the surface. Where there is a ridge or valley, the distance varies, as does the capacitance; building a pattern of the finger's "print". The measured values are digitized by the sensor then sent to the microprocessor.

This capacitance sensing technique is an effective method for acquiring fingerprints. To achieve enough sensitivity, the protective coating must be very thin (a few microns), since an electrical field is measured and the distance between the skin and the pixels is very small. A significant drawback to this design is its vulnerability to strong external electrical fields, the most troublesome being ESD (Electro-Static Discharge).

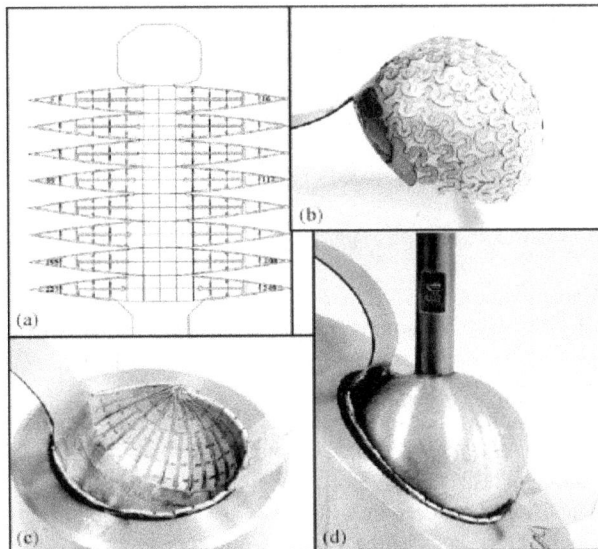

Fig. 2.27. (a) Unfolded Sensor Array; (b) Spherically Folded Sensor Array; (c) Sensor Array Placed in a Cavity; (d) Sensor Array between Ball and Cavity.

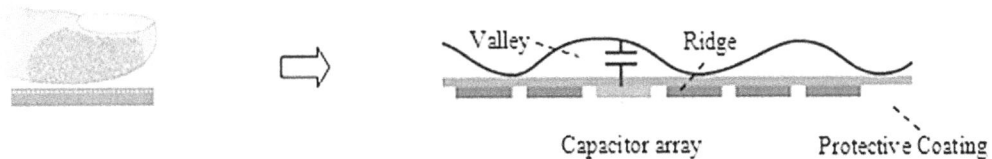

Fig. 2.28. A Capacitive Fingerprint Sensor.

2.4. Summary

Resistive and capacitive sensors are the most broadly used sensors. They can measure or detect a broad range of physical phenomena and parameters, such as position, displacement, acceleration, pressure, force, humidity, temperature, radiation, light, current, flowrate, chemical particles/gases, bioactivity, and more. Electrical resistance is the easiest electrical property to measure and can provide a high degree of precision over a wide range. The physical principles of resistive sensors are governed by several important laws and phenomena such as *Ohm's Law* and *Wiedemann-Franz Law*; *Photoconductive-*, *Piezoresistive-*, *Magnetoresistive-*, *Chemoresistive-*, and *Thermoresistive Effects*, which relate electrical resistance values to the magnitude of the physical or chemical parameters. The operating principles of capacitive sensors are based on the properties of a capacitor: *capacitance variation is a function of changes in dielectric constant, materials, electrode conductivity, electrode movement, and geometric parameters*. Each capacitive sensor is designed to measure one or more of these parameters. Most capacitive sensors fall into one of the three basic capacitor configurations: flat-plate, coaxial, or spherical, depending on their intended use and function. Several typical sensor designs and applications in each configuration, as well as sensor arrays, were described in this chapter.

References

[1]. Introduction to Capacitance Gages, *ADE Technologies,* p. 6.

[2]. W. Du., S. Yelich, Resistive Sensors: Principles, Design, and Applications, in *Proceedings of the 2nd International Conference on Sensing Technology,* Nov. 26-28, 2007, Palmerston North, New Zealand, pp. 326-331.

[3]. W. Du., S. Yelich, Capacitive Sensors: Principles, Design, and Applications, in *Proceedings of the 2nd International Conference on Sensing Technology,* Nov. 26-28, 2007, Palmerston North, New Zealand, pp. 332-337.

[4]. H. P. Schwan, The Bioimpedance Field: Some Historical Observations, in *Proceedings of 9th ICEBI*, Heidelberg, Germany, 1995.

[5]. J. Fraden, Handbook of Modern Sensors: Physics, Designs, and Applications, *Springer-Verlag,* 2003.

[6]. E. Källvesten, P. Melvas, J. Melin, T. Frisk, G. Stemme, Ultra-miniaturized Pressure Sensors for Intravascular Blood Pressure Measurements, *KTH Royal Inst. of Tech., Sweden.*

[7]. T. Montgomery, Industrial Temperature Primer, *Wilkerson Instrument Co.,* 2007, p. 11.

[8]. L. V. King, On the Convention of Heat from Small Cylinders in a Stream of Fluid, *Phil. Trans. Roy. Soc.,* A214, 1914, p. 373.

[9]. G. Man, A. Navabi, T. Raymond, B. Stoeber, K. Walus, Polymer-based Chemical Sensors for Clandestine Lab Detection, *CMC Workshop on MEMS & Microfluidics,* Montreal, Canada, 2007.

[10]. T. J. Plum, V. Saxena, and J. R. Jessing, Design of a MEMS Capacitive Chemical Sensor Based on Polymer Swelling, *IEEE WMED 2006.*

[11]. J. Engel, J. Chen, N. Chen, S. Pandya, and C. Liu, Multi-Walled Carbon Nanotube Filled Conductive Elastomers: Materials & Application to Micro Transducers, in *Proceedings of the 19th IEEE International Conference on Micro Electro Mechanical Systems*, Istanbul, Turkey, January 22-26, 2006, pp. 246-249.

[12]. C. Guiducci, C. Stagni, A. Fischetti, U. Mastro-matter, Microelectrodes on a Silicon Chip for Label-free Capacitive DNA Sensing, *Sensor Journal*, Vol. 6, No. 5, 2006, pp. 1084-1093.

[13]. R. Zhang, S. Dai, and P Mu, A Spherical Capacitive Probe for Measuring the Thickness of Coatings on Metals, *Measurement Science Technology*, 8, 1997, pp. 1028–1033.

[14]. O. Muller, W. J. Parak, M. G. Wiedemann, and F. Martini, Three-Dimensional Measurements of the Pressure Distribution in Artificial Joints with A Capacitive Sensor Array, *Journal of Biomechanics*, 37, 2004, 1623–1625.

[15]. Solid-state, Silicon-based Capacitive Fingerprint Sensor, Retrieved on January 20, 2008 at: http://www.iguardsystem.com

[16]. Fingerprint Sensing Techniques, retrieved on September 20, 2007 at: http://perso.orange.fr/fingerchip/biometrics/types/fingerprint_sensors_physics.htm

Chapter 3
Automated Synthesis of MEMS Sensors

Chenxu Zhao and Tom J. Kazmierski

Part 1: Layout Synthesis of MEMS Component with Distributed Mechanical Dynamics

3.1. Introduction

This two-part chapter presents an effective holistic genetic-based synthesis flow (SystemC-AGNES) applied to automated layout synthesis of mechanical components of Micro-Electro Mechanical Systems (MEMS) and configuration synthesis of associated electronic control system. Part 1 of this chapter focuses on the layout synthesis of mechanical sensing component and automated configuration synthesis of the control system is demonstrated in Part 2.

MEMS are currently used in a wide range of applications due to their significant advantages such as low cost, small form factor and low power consumption. MEMS sensors, for example accelerometers and gyroscopes, are widely used in consumer applications, mainly by the automotive industry, in mixed-technology control designs such as safety air cushion, active suspension or anti-lock brake system. Modern high precision inertial navigation and guidance systems are also based upon MEMS sensors embedded in mixed-technology control loops [1].

The design of a typical MEMS system requires an integration of elements from two or more disparate physical domains: mechanical (translational, rotational, hydraulic), electrical, magnetic, thermal etc. Different parts of a MEMS system are traditionally designed separately using different methodologies and different tools applied to different energy domains. Two engineering teams traditionally collaborate to create a MEMS-based IC: one using 3-D CAD such as CoventorWare to create the MEMS mechanical model, and the other team, meanwhile, using an EDA tool from such companies such as Cadence to create the associated ICs. Although this approach provides accurate

Chenxu Zhao
School of Electronics and Computer Science, University of Southampton, UK

behaviour simulation of MEMS devices with their associated electronics, it requires multiple tools and it is difficult to provide IC designers with an automated synthesis and performance optimization system. This difficulty is primarily caused by disparities between the different tools and the inconvenience of generating new MEMS macromodels, when the MEMS layout changes, for incorporation into the IC simulations performed at the IC design stage.

Analogue and Mixed-Signal(AMS) Hardware Description Languages(HDLs) such as VHDL-AMS which was standardized by the IEEE in 1999 [2] and later equipped with another IEEE standard for multiple energy domain packages [3] or Verilog-AMS [4] are able to integrate components from different energy domains into a single model. However, automated design methodologies for the whole integrated system supporting mixed physical domains are still lagging. This is mainly due to the fact that state-of-the-art tools supporting AMS HDLs. such as the commonly used SystemVision from Mentor Graphics [5] are not designed to support simulation-based synthesis and optimization where users would be able to develop and implement complex numerical algorithms. Wang proposed a methodology to realize a genetic optimization algorithm (GA) in a VHDL-AMS testbench [6], but the software tools used took about 16 hours to fulfill a simple task.

Usually, the design of a MEMS system requires a significant amount of specialist human resources and time in the iterative trial-and-error design process to determine the crucial trade-offs in meeting the performance specifications. Therefore there is an increasing need for automated synthesis techniques that would shorten the development cycle and facilitate the generation of optimal configurations for a given set of performance and constraint guidelines. Some methodologies have already been proposed for automated synthesis of mechanical parts in MEMS systems [7-12]. For example, Tamal presented a method for rapid layout synthesis of a lateral surface-micromachined accelerometer from high-level functional specifications and design constraints [8]. The design problem is regarded as a nonlinear optimization problem. Standard off-the-shelf solvers (NPSOL) and a grid-based numerical optimization algorithm are used to maximize the system's performance. In another approach, Zhun and Wang presented a hierarchical evolutionary synthesis of MEMS device layout [11]. They divided the design into two levels: system-level which uses behavioral macromodels and detailed physical-level based on geometric layout models. At the system level, a combination of genetic programming and bond graphs are used to generate and search for design candidates satisfying design specifications. At the physical layout level, optimizations are carried out to meet more detailed design objectives.

In the above approaches the automated design of MEMS is accomplished either by simulation-based optimization or formulating the design requirements as a numerical nonlinear constrained optimization problem, and solved with powerful optimization techniques. However, these methodologies are constrained to the layout synthesis of a mechanical MEMS device. The salient feature of our approach proposed here is that it realizes an automated design of a whole MEMS system which contains not only layout synthesis of the sensing elements but also optimal configuration synthesis for associated electronics. Synthesis of the electronic control loop is demonstrated in Part 2 of this chapter. The proposed approach integrates a MEMS component library, an electronic control loop library, an efficient fast MEMS simulation engine implemented in SystemC-A [13] and an evolutionary computation method (GA).

SystemC-A is a superset of SystemC developed to extend modeling capabilities of SystemC to the analogue and mixed physical domain [13]. In addition to standard digital modeling capabilities of SystemC, SystemC-A provides constructs to support user-defined ordinary differential and algebraic equations (ODAEs), analogue system variables, and analogue components to enable modeling of analogue and mixed-signal systems from very high levels of abstraction down to the circuit level. Support for digital-analogue interfaces is also provided for smooth integration of digital and analogue parts. The analogue simulator uses efficient linear and nonlinear solvers to assure accurate and fast simulations of the analogue model. Most of the powerful features of VHDL-AMS and Verilog-AMS are provided in SystemC-A in addition to a number of extra advantages such as high simulation speed, support for hardware-software co-design and for high levels of modeling. However, the current SystemC-A can only describe analogue systems by using ODAEs. In modern mixed-domain applications this limits accurate modeling of system blocks, especially in the mechanical domain, which exhibit distributed physical effects described by Partial Differential Equations (PDEs). To ensure accurate modeling of distributed components, Finite Difference Approximation (FDA) approach is applied to convert PDEs to a series of ODAEs which can be solved by SystemC-A. Distributed mechanical modeling is important in MEMS designs with digital control because dynamics of mechanical components may severely affect the system's performance. For example, it has been well documented that sense fingers in lateral capacitive accelerometers may vibrate due to their own dynamics, thus rendering the feedback excitation ineffective, causing an incorrect output and a failure of the system [14, 15]. This scenario cannot be reflected by the conventional mass-damper-spring model based on a 2nd order ordinary differential equation. In our dedicated SystemC-A model, distributed mechanical dynamics models are implemented through FDA to enable accurate performance prediction of critical mechanical components embedded in the mixed-technology control loop.

The synthesis technique presented here is applicable to a wide class of digital MEMS sensors with electronic control. We demonstrate its operation using a capacitive MEMS accelerometer in a Sigma-Delta control loop [16, 17] as a case study. The capacitive digital MEMS accelerometers are notoriously difficult to design using traditional methods because here the mechanical element forms an integral part of the $\Sigma\Delta$ control loop. This feature makes a separation of the two technology domains in the design process very effortful.

3.2. Genetic-based Synthesis of MEMS Sensors with Electronic Control Loop

The proposed automated synthesis approach explores the design according to user defined specifications and optimizes the structural parameters of the mechanical MEMS elements and the associated electronic control loop parameters. The automated optimal synthesis flow is shown in Fig. 3.1. After specifying the design objectives and constraints, such as the die area of the sensing element and feedback voltage in the electronic control loop, available components in the MEMS primitive library and the electronic control loop primitive library are combined automatically to form a valid initial design set. This set of initial designs is loaded into the synthesis module after parameter initialization and encoding phase. The synthesis module uses a genetic algorithm to create new MEMS structures and optimizes their parameters for best performance. Our approach integrates mixed-technology models into a single simulation engine which could be easily invoked from various optimization loops. Unlike traditional MEMS design tool sets, this approach avoids a generation of macromodels in order to realize co-design and co-simulation.

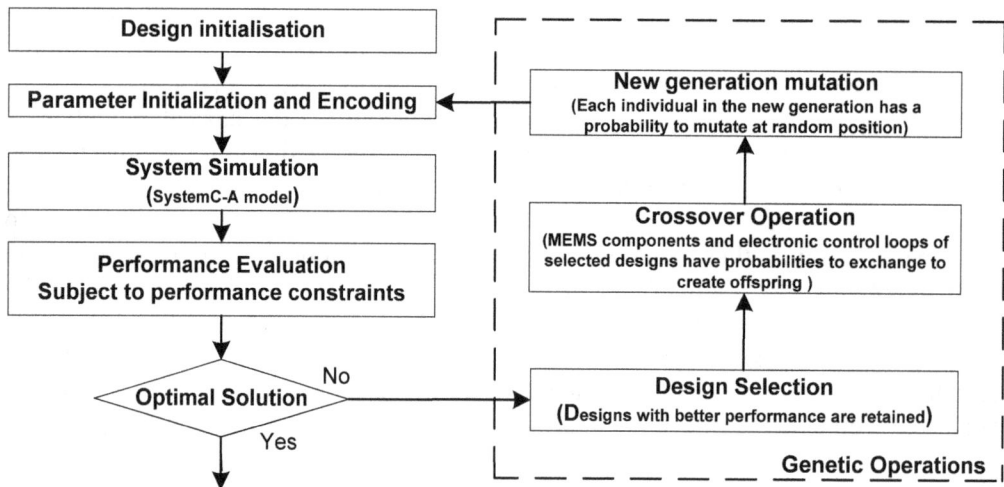

Fig. 3.1. Automated synthesis flow in SystemC-AGNES.

3.2.1. Synthesis Initialization

The two libraries, MEMS primitive library and electronic control loop primitive library, contain typical components that are widely used in practical MEMS designs. Every member in the libraries is a data structure record which includes its type code, geometrical parameters for MEMS primitives, system-level design parameters for electronic primitives and constraints. This Part 1 of the chapter focuses on layout synthesis of the mechanical component while the configuration of the associated electronic control loop is fixed.

3.2.1.1. MEMS Primitive Library

The mechanical part of a lateral capacitive MEMS accelerometer is composed of a proof mass, springs and comb fingers. In the lateral capacitive structure, the proof mass is suspended by springs and it is equipped with sense and force comb fingers which are placed between fixed fingers to form a capacitive bridge. The sense fingers moves with the proof mass resulting in a differential imbalance in capacitance which is measured. The electrostatic force acting on the force fingers is used as the feedback signal to pull the proof mass in the desired direction. The available mechanical components in the MEMS primitive library (Fig. 3.2) are discussed below.

1. *Springs:* 4 typical springs are available in the MEMS components library for this case study: classic serpentine spring, rotated serpentine spring, folded spring and spring beam. The layout and geometrical parameters with constraints are shown in Fig. 3.2.

2. *Proof mass:* The proof mass contains unit squires with etch holes for release. The number of these holes is determined by the size of proof mass and size of holes. There are 4 connecting nodes and 2 connecting sides on the proof mass, 4 connecting nodes is used to connect springs and 2 connecting sides are used for comb sense and force fingers connection.

3. *Comb fingers:* The sensing element dynamics in the sense-direction is normally modeled to reflect only one resonant mode by a lumped mass, spring, and damper, which is represented by a simple 2nd order ordinary differential equation:

$$M \frac{d^2 z(t)}{dt^2} + D \frac{dz(t)}{dt} + Kz(t) = Ma_{in}(t),$$ (3.1)

where M is the total mass of the structure, D and K are the damping and spring coefficients correspondingly. $z(t)$ is the deflecttrion of the proof mass and $a_{in}(t)$ is the input acceleration.

In reality, the sensing fingers in a lateral structure are distributed elements with many resonant modes. As their dynamics affect the performance of a $\Sigma\Delta$ control system, the motion of the sense beam should be distributed, for example using the following partial differential equation [15]:

$$\rho A \frac{\partial^2 y(x,t)}{\partial t^2} + C_D I \frac{\partial^5 y(x,t)}{\partial x^4 \partial t} + EI \frac{\partial^4 y(x,t)}{\partial x^4} = Fe(x,t) + \rho A a_{in}(t),$$ (3.2)

where $y(x, t)$, is the function of time and position, represents the beam deflection, E, I, C_D, ρ, A are the physical properties of the beam: ρ is the material density, A is the cross sectional area ($Wf * T$), where Wf and T are the width and thickness of the beam, E is the Young's modulus and I is the second moment of area and C_D is the internal damping modulus. The product *EI* is usually regarded as the stiffness. *Fe(x, t)* is the distributed electrostatic force along the finger.

Classic serpentine spring Code: 1	Parameters typical range	Beam spring Code: 2	Parameters typical range
	Lo:50 μm~200 μm Wo:2 μm~4 μm Lp:2 μm~5 μm Wp:2 μm~4 μm Lend1=5 μm Lend2=5 μm Wend=4 μm Lroot: 0~ 100 μm Wroot=5 μm Wanchor=10 μm N (number of loop):0~5		Wanchor:10 μm Lo:100 μm~300 μm Wo:2 μm~4 μm
Rotated serpentine spring Code: 3	Parameters typical range	Folded spring Code: 4	Parameters typical range
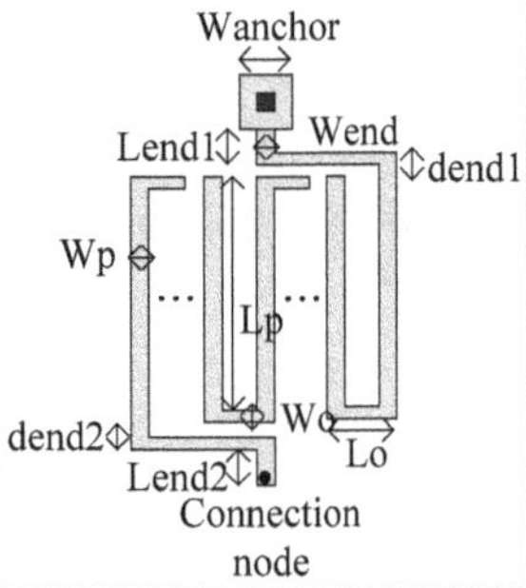	Lo:5 μm ~10 μm Wo:2 μm ~4 μm Lp:50 μm ~200 μm Wp:2 μm ~4 μm dend1:3 μm ~5 μm dend2:3 μm ~5 μm Lend1=5 μm Lend2=5 μm Wend=4 μm Wanchor:10 μm N (number of loop):0~5	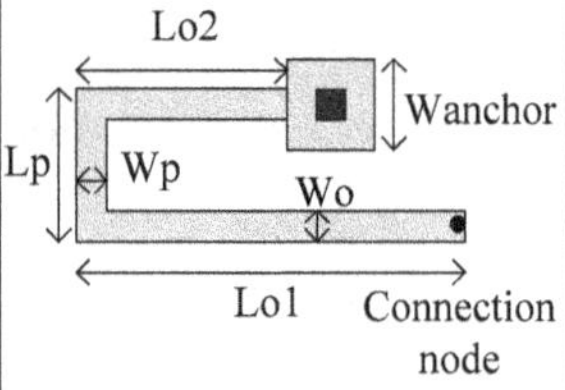	Wanchor:10 μm Lo1:200 μm~300 μm Wo:2 μm ~4 μm Lp:30 μm ~50 μm Wp:2 μm ~4 μm Lo2:determined by Lo1 and width of proof mass
Proof Mass Code: 5	Parameters typical range	Comb fingers Code: 6	Parameters typical range
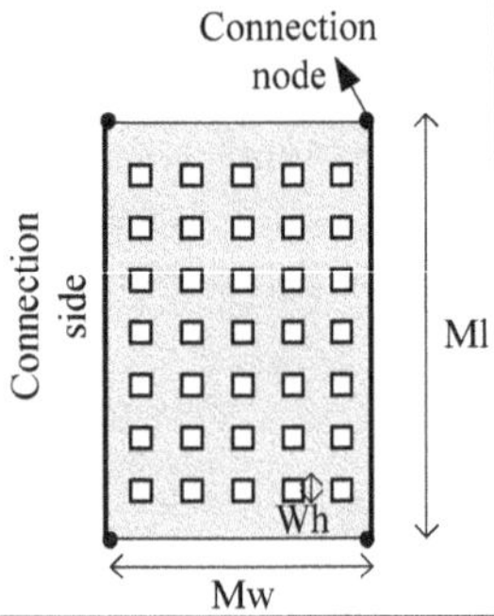	Ml:100 μm~700 μm Mw:50 μm~150 μm Wh:3 μm~6 μm T:2 μm ~3 μm Number of holes is determined by the size of holes and size or proof mass	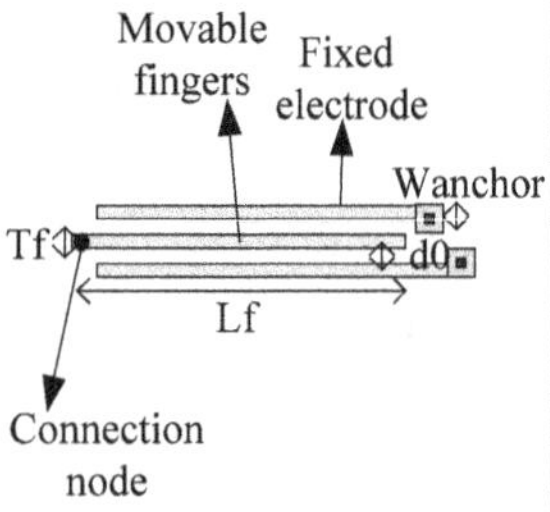	Lf:50 μm~200 μm Tf:2 μm ~3 μm d0:1 μm ~3 μm Ns:10~30 Nf:2~24 Wanchor:4 μm

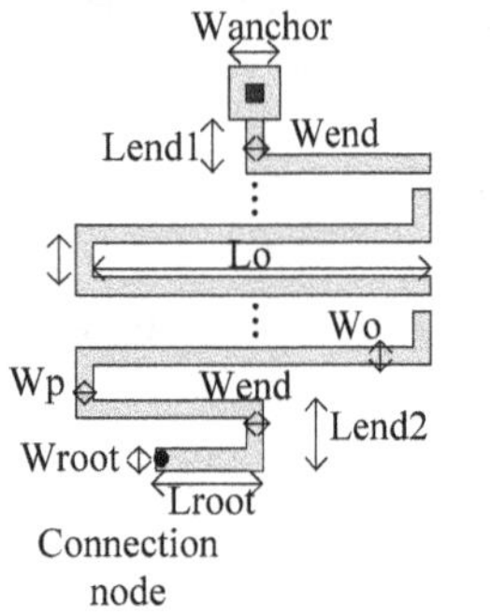

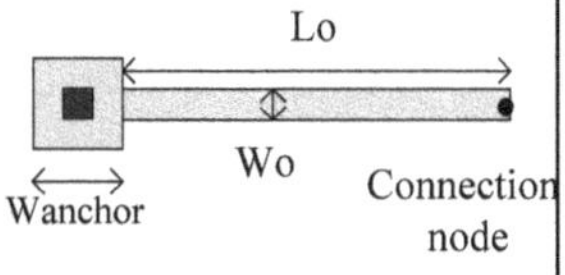

Fig. 3.2. MEMS primitive component library.

The boundary conditions at the clamped end and the free end are shown in the following equations. At the clamped end (x=0):

$$y(0,t) = z(t) \tag{3.3}$$

$$\theta = \frac{\partial y(0,t)}{\partial x} = 0 \tag{3.4}$$

and at the free end (x=l):

$$M = -\frac{\partial^2 y(l,t)}{\partial x^2} = 0 \tag{3.5}$$

$$Q = -\frac{\partial^3 y(l,t)}{\partial x^3} = 0, \tag{3.6}$$

where θ, M and Q denote the slope angle, the bending moment and the shear force respectively and l is the finger length.

The total distributed sense capacitances between the sense fingers and electrodes are:

$$C_1(t) = N_s \varepsilon T \int_0^l \frac{1}{d_0 - y(x,t)} dx \tag{3.7}$$

$$C_2(t) = N_s \varepsilon T \int_0^l \frac{1}{d_0 + y(x,t)} dx, \tag{3.8}$$

where Ns is the number of sense fingers. The output voltage can be calculated as:

$$V_{out}(t) = \frac{C_1 - C_2}{C_1 + C_2} V_m(t), \tag{3.9}$$

where $Vm(t)$ is high frequency carrier voltage applied on the fixed electrode in comb fingers unit.

Here the Finite Difference Approximation (FDA) is applied to convert PDEs to a series of ODAEs. If the beam is divided into N segments and the deflection of the beam is discretized as:

$$y_n(t) = y(n\Delta x, t) \quad n=1,2,3... N \tag{3.10}$$

The first order spatial derivatives can be approximated by finite differences:

$$\frac{\partial y_n(t)}{\partial x} = \frac{y_n(t) - y_{n-1}(t)}{\Delta x} \tag{3.11}$$

Similar approximation can be applied to higher order spatial derivatives. Eq. (3.2) is hence transformed to a system of the following ODAEs:

$$\rho A \frac{d^2 y_n(t)}{\partial t^2} + \frac{C_D I}{(\Delta x)^4} \left(\frac{dy_{n+2}(t)}{\partial t} - 4\frac{dy_{n+1}(t)}{\partial t} + 6\frac{dy_n(t)}{\partial t} - 4\frac{dy_{n-1}(t)}{\partial t} + \frac{dy_{n-2}(t)}{\partial t} \right)$$

$$+ \frac{EI}{(\Delta x)^4} (y_{n+2}(t) - 4y_{n+1}(t) + 6y_n(t) - 4y_{n-1}(t) + y_{n-2}(t)) = \frac{Fe_n(t)}{\Delta x} + \rho A a_{in}(t);$$

$$n=3,4,5...N-2$$ (3.12)

$$y_1(t) = z(t); \quad n=1$$ (3.13)

$$\rho A \frac{d^2 y_2(t)}{\partial t^2} + \frac{C_D I}{(\Delta x)^4} \left(\frac{dy_4(t)}{\partial t} - 4\frac{dy_3(t)}{\partial t} + 6\frac{dy_2(t)}{\partial t} - 3\frac{dy_1(t)}{\partial t} \right)$$

$$+ \frac{EI}{(\Delta x)^4} (y_4(t) - 4y_3(t) + 6y_2(t) - 3y_1(t)) = \frac{Fe_2(t)}{\Delta x} + \rho A a_{in}(t); \quad n=2$$ (3.14)

$$\rho A \frac{d^2 y_{N-1}(t)}{\partial t^2} + \frac{C_D I}{(\Delta x)^4} \left(-2\frac{dy_N(t)}{\partial t} + 5\frac{dy_{N-1}(t)}{\partial t} - 4\frac{dy_{N-2}(t)}{\partial t} + \frac{dy_{N-3}(t)}{\partial t} \right)$$

$$+ \frac{EI}{(\Delta x)^4} (-2y_N(t) + 5y_{N-1}(t) - 4y_{N-2}(t) + y_{N-3}(t)) = \frac{Fe_{N-1}(t)}{\Delta x} + \rho A a_{in}(t); \quad n=N-$$

$$1$$ (3.15)

$$\rho A \frac{d^2 y_N(t)}{\partial t^2} + \frac{C_D I}{(\Delta x)^4} \left(\frac{dy_N(t)}{\partial t} - 2\frac{dy_{N-1}(t)}{\partial t} + \frac{dy_{N-2}(t)}{\partial t} \right)$$

$$+ \frac{EI}{(\Delta x)^4} (y_N(t) - 2y_{N-1}(t) + y_{N-2}(t)) = \frac{Fe_N(t)}{\Delta x} + \rho A a_{in}(t); \quad n=N$$ (3.16)

Boundary conditions provide additional equations. The slope angle at the fixed end is approximated as:

$$\theta = \frac{\partial y_1(t)}{\partial x} = \frac{y_1(t) - y_0(t)}{\Delta x} = 0$$ (3.17)

and the bending moment M and shear force Q at the free end as:

$$M = -\frac{\partial^2 y_N(t)}{\partial x^2} = \frac{y_{N+1}(t) - 2y_N(t) + y_{N-1}(t)}{\Delta x^2} = 0$$ (3.18)

$$Q = -\frac{\partial^3 y_N(t)}{\partial x^3} = \frac{y_{N+2}(t) - 3y_{N+1}(t) + 3y_N(t) - y_{N-1}(t)}{\Delta x^3} = 0 \qquad (3.19)$$

Eq. (3.13) represents the motion of the clamped end of the sense fingers ($y_1(t)$) which moves with the lumped proof mass whose deflection z(t) is obtained from the solution of Eq.(3.1).

3.2.1.2. Electronic Control Loop

High-performance MEMS sensors exploit the advantages of closed-loop control strategy to increase the dynamic range, linearity, and bandwidth of sensor. In particular, digital $\Sigma\Delta$ modulators for closed-loop feedback control schemes, whose output is digital in the form of pulse-density-modulated bitstream, have become very attractive in a number of MEMS applications [16-18]. A conventional 2nd order electromechanical $\Sigma\Delta$ control systems is shown in Fig. 3.3. In this configuration, mechanical sensing element is used as a loop filter to form the 2nd order electromechanical $\Sigma\Delta$ modulator. *Vf1* and *Vf2* are the feedback voltages obtained from the DAC and *Vm(t)* is a high frequency modulation carrier voltage. The gain Kcv represents the signal pick-off from differential change in capacitance to voltage and K is the gain of the voltage booster amplifier following the pick-off stage. The lead compensator is used to ensure the stability of the control loop. It is an optional component in electronic control primitive library depending on whether the sensing element is over damped or under damped. A clocked 1-bit quantizer is used for oversampling and generating a pulse-density modulated digital output signal. However, the equivalent DC gain of the mechanical integrator in the 2nd-order electromechanical $\Sigma\Delta$ modulator is relatively low and this leads to a poor signal-to-noise ratio (SNR). To improve the SNR, the mechanical element can be cascaded with additional electronic integrators to form high order topologies [16]. The example of automated synthesis discussed in this section focuses on the synthesis of MEMS layout and the electromechanical $\Sigma\Delta$ accelerometer is fixed and of 2nd order. Full synthesis which includes both MEMS layout and electronic control loop is presented in Part 2 of this chapter.

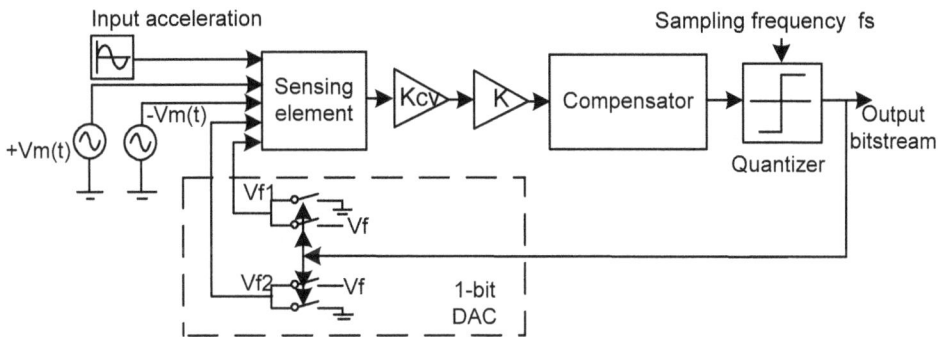

Fig. 3.3. 2nd order electromechanical $\Sigma\Delta$ modulator.

3.2.1.3. Parameter Initialization and Encoding

The automated design process starts with a specification of the design objectives and constraints. Drawing from the MEMS primitive component library and electronic control loop library, a set of configurations is automatically selected (parents of first generation in GA) and loaded into the synthesis module. These feasible configurations not only contain MEMS mechanical layouts but also associated electronic control system topologies. Fig. 3.4 and Table 3.1 show an example of a feasible configuration to illustrate the parameter initialization and encoding phase. This MEMS accelerometer here contains 4 spring beams, 14 force fingers, 20 sense fingers and a proof mass with associated 2nd order $\Sigma\Delta$ control loop. The component code of each component is shown in the Fig. 3.2. Then the geometrical layout parameters of mechanical part and the associated system-level design parameters of electrical control systems are generated to describe the feasible layouts combining with the component code (Fig. 3.4).

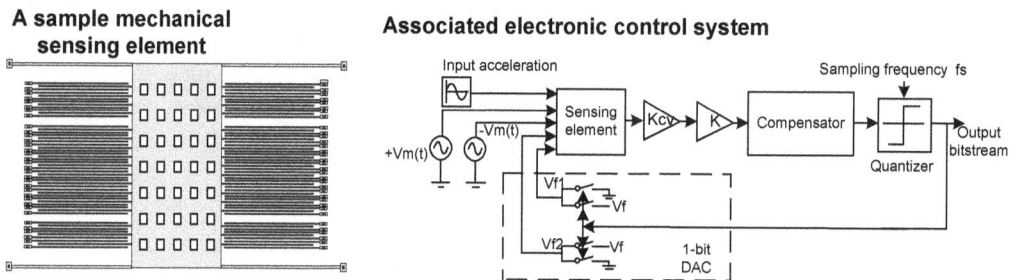

Fig. 3.4. Example of Parameter Initialization and Encoding.

Table 3.1. Representation of a population member in GA for the MEMS accelerometer example.

MEMS component library	Code	Description
Spring	2	Beam spring
Proof mass	5	Proof mass with etching holes
Comb drive	6	Sense and force fingers
Electronic Control loop library	**Code**	**Description**
Sigma-Delta Control system	1(fixed)	2nd order Sigma-Delta Control

3.2.2. Genetic Approach to Synthesis

Genetic Algorithm (GA) has been selected for our case studies as it is a very popular and well tested optimization algorithm which has demonstrated good performance in a wide variety of complex global optimization problems where modeling difficulties arise and there is no obvious way to find optimal solutions [19]. It has already been used for mechanical layout optimization [20].

The optimization problem is considered as a constrained optimization as both of the design and performance parameters are bound by inequality constraints that must be met:

$$\text{Maximize: } F(x) \qquad\qquad (3.20)$$

Subject to:

$$x_n \in [V_{n\_low}, V_{n\_high}], \text{ n=1,2,3...,} \qquad\qquad (3.21)$$

where *F(x)* is the fitness function to be optimized with design parameter vector x, x_n represents the nth design parameter, V_{n_low} and V_{n_high} are the lower and upper constraints of the nth design parameter.

Performance figures of the candidate designs are evaluated by a fitness function that rates the solutions according to their performance parameters. Fitness function is usually constructed in a weighted scalar error form:

$$F(x) = w\frac{R}{R'}, \qquad\qquad (3.22)$$

where w is the weight coefficient, R is the system performance measure obtained from each simulation while R' is the designer specified objective value. In this case study, w is equal to 1 if all user defined performance constraints are met, otherwise w is set to 0.0001. If minimization of a fitness parameter is required, e.g. the sensing element area, w is set to -1 if performance constraints are met or -10 otherwise. In the case study discussed below, a performance evaluation engine is added to the simulator to enable measurements of the power spectrum density (PSD) and signal-to-noise ratio (SNR), as the design objectives, through FFT of the output bitstream. The die area of and static sensitivity of the mechanical sensing element are also used as system performance objectives or constraints.

After the synthesis initialization, the classical genetic operations of selection, crossover and mutation are applied to the current generation parents in order to create a new generation. In the selection operation, designs with better performance (higher fitness) are retained. After the selection, if the crossover operation is triggered, i.e. crossover probability is higher than a fixed threshold, new MEMS layouts are composed from primitives and associated control systems by exchanging elements of randomly selected parents, such as mechanical springs and electronic control blocks. Details of an example of crossover operation in mechanical sensing element synthesis are illustrated in Fig. 3.5. As shown in the figure, in this example only the crossover probability of the spring component is higher than the trigger probability of 70 %, so the spring components of parents A and B exchange leaving the other components unchanged in the creation of new offspring.

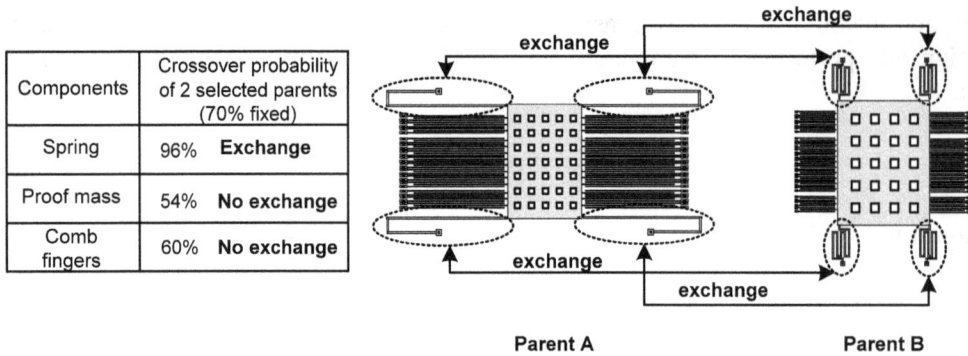

Components	Crossover probability of 2 selected parents (70% fixed)	
Spring	96%	**Exchange**
Proof mass	54%	**No exchange**
Comb fingers	60%	**No exchange**

Fig. 3.5. An example of crossover operation in mechanical layout synthesis.

For each individual in the new generation, the genes in their chromosomes have a fixed probability to mutate at random positions. The mutation operation for the mechanical sensing element is illustrated by the example shown in Fig. 3.6 and Table 3.2. The mutation operation contains two phases: component mutation and component's parameter mutation. In the first phase, if the mutation probability for the components is higher than the fixed trigger (50 % in this example) such as the beam spring and comb fingers, new components are automatically composed using the MEMS primitive library and each parameter of the mutated components gets a random value within its specified range. If there is no mutation in the first phase, the mutation probability of each component parameter will be compared with the trigger (60 % in the example) to decide whether this parameter should mutate. As shown in Fig. 3.6, after the mutation the beam spring mutated to a folded spring and comb fingers mutated to themselves but with different parameters such as a shorter length and a higher number of force fingers. For the proof mass, only the number of holes and width were changed at the second mutation phase.

This evolution process finishes when the generation number exceeds the specified maximum number. The optimal solution within a given generation is that with the highest fitness.

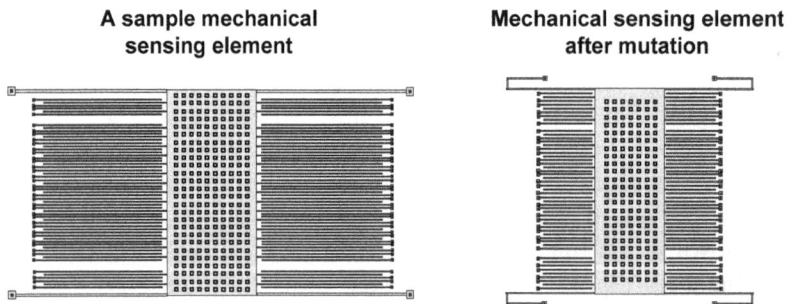

Fig. 3.6. An example of mutation operation in mechanical layout synthesis.

Table 3.2. An example of mutation operation in mechanical layout synthesis.

MEMS component	MEMS component Mutation probability (trigger 50 %)	Component parameters	Component parameters Mutation probability (trigger 60 %)
Beam Spring	56 % Spring mutation		Each parameter of the mutated spring get random value within range
Proof mass with etching holes	30 % No mutation for proof mass	L_m: length of proof mass	55 % No mutation for length
		W_m: Width of proof mass	70 % Mutation
		W_h: Size of holes	23 % No Mutation
		N_s: Number of holes	92 % Mutation
		T: Thickness of proof mass	10 % No mutation
Comb fingers	73 % Comb fingers mutation		Each parameter of the mutated comb sense and force fingers such as L_f, d_0 and N_s get random value within range

3.3. Synthesis Verification to Provide Appropriate Performance Metrics for the Synthesized MEMS Geometries

The practical operation of the proposed synthesis flow for the accelerometer embedded in a conventional $\Sigma\Delta$ control loop is demonstrated below by three experiments listed in Table 3.3. In the first experiment, the system is optimized for maximum SNR with performance constraints, and in the second and third experiments - for maximum static sensitivity and minimum area respectively.

As can be seen from the results presented in Table 3.3 and Fig. 3.7 for each experiment, the genetic synthesis algorithm composed different layout structures and produced different performance parameters. K_x, K_z are the stiffnesses in the axes X and Z correspondingly, and K_y is the stiffness in the sensing axis Y in this case study. The larger the stiffness ratios K_x/K_y, and K_z/K_y the larger the relative movement of the accelerometer along the sensing axis.

The synthesis process was carried out using the following design parameters:
1) Oversampling ratio: OSR=128;
2) Bandwidth: 512 Hz;
3) Oversampling frequency: $fs= 2^{17}$ Hz;
4) Input force: 100 Hz acceleration with 1g amplitude.

Design of MEMS accelerometer in a $\Sigma\Delta$ force feedback control loop contains many crucial trade-offs. For example, in lateral structure, static sensitivity is dependent on the length and number of the sense fingers. However, the performance of $\Sigma\Delta$ modulation may be severely affected by the length of sense fingers to the extent that a complete failure of the $\Sigma\Delta$ control may occur when the fingers are too long. The maximum number of fingers is also limited by the length of proof mass. To maintain the same resonant frequency, the finger width should be reduced if the length of the proof mass increases. This results in a sensitivity decrease. The presented genetic-based synthesis approach deals with these trade-offs effectively for a given choice of the design objectives.

Table 3.3. Summary of synthesis experiments.

	Design objective	Performance constrants	Synthesized layout	SNR (dB)	Static sensitivity	Kzy	Kzy	Area (m^2)
1	Maximum SNR	SNR>30 dB Area<2.0e-7 m^2 Static sensitivity>1fF/G	Fig. 3.7. (a)	39.8	1.8fF/G	202	1.875	1.82e-7
2	Maximum Static sensitivity	SNR>30 dB Static sensitivity>2 fF/G	Fig. 3.7. (b)	32.9	4.77fF/G	4.91	10.72	3.78e-7
3	Minimum area of mechanic al sensing element	SNR>30 dB Area<1.5e-7 m^2	Fig. 3.7. (c)	31.5	0.27fF/G	10346	1.95	1.07e-7

The fitness improvement during the synthesis flow is shown in Fig. 3.8. The synthesized mechanical layouts and parameters of its associated electronic control system are shown in Fig. 3.7 and Table 3.4. The synthesized accelerometer in experiment 1 does not need compensator to assure the stability of the control loop as the mechanical sensing element is over damped system. As expected, the structure optimized for maximum sensitivity has more and longer sense fingers. Area optimized accelerometer in experiment 3 shows a great area improvement over other experiments. The control loop is fixed in this case study to form a conventional 2nd order electromechanical $\Sigma\Delta$ accelerometer. However, the noise floor in higher order $\Sigma\Delta$ accelerometer can be reduced drastically leading to great improvement of the SNR comparing with 2nd order $\Sigma\Delta$ accelerometer. It is discussed in Part 2 of this chapter where the higher order control system is automated optimal synthesized with layout synthesis of mechanical sensing element simultaneously.

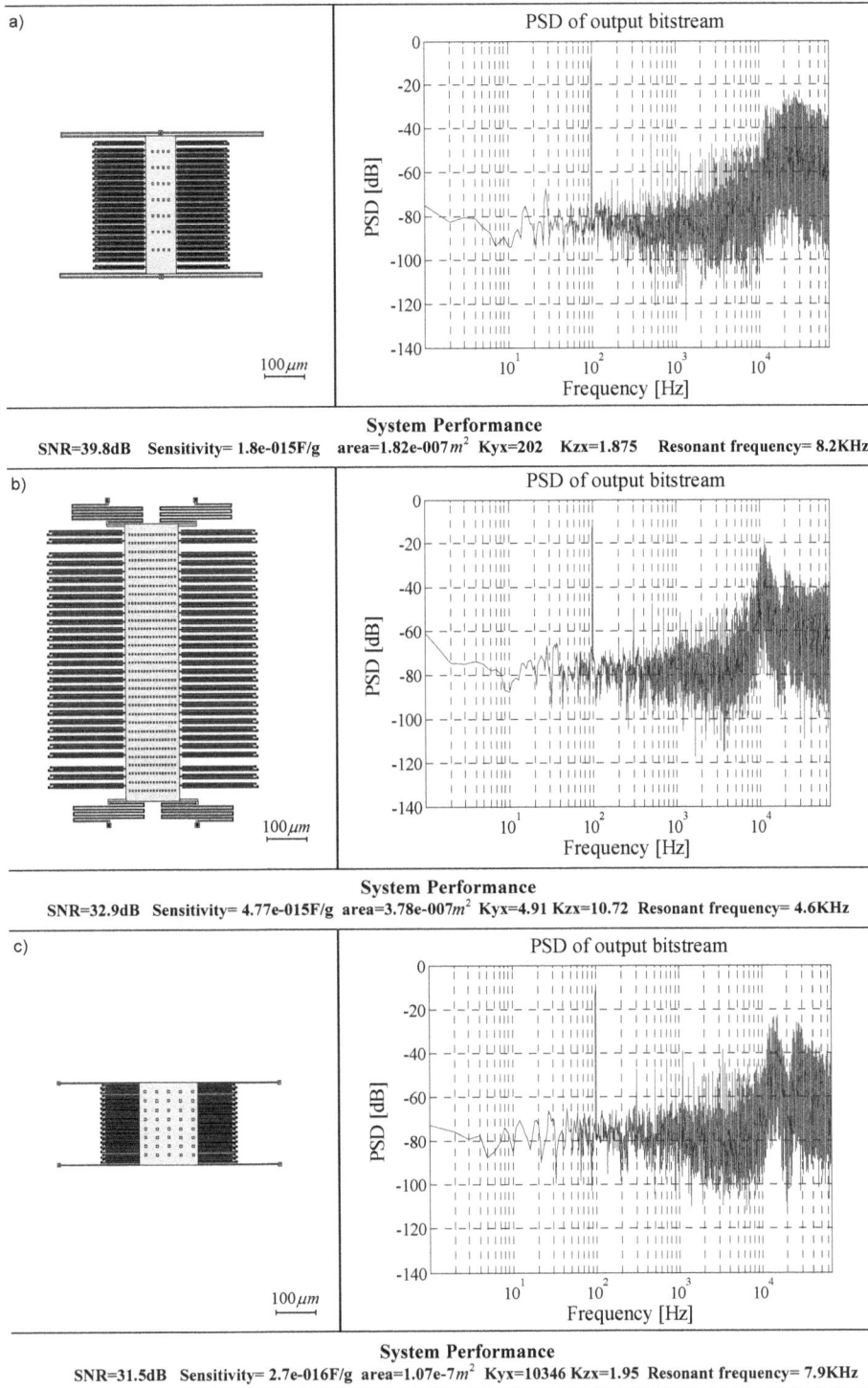

a)

PSD of output bitstream

100 μm

System Performance

SNR=39.8dB   Sensitivity= 1.8e-015F/g   area=1.82e-007m^2 Kyx=202   Kzx=1.875   Resonant frequency= 8.2KHz

b)

PSD of output bitstream

100 μm

System Performance

SNR=32.9dB   Sensitivity= 4.77e-015F/g   area=3.78e-007m^2 Kyx=4.91   Kzx=10.72   Resonant frequency= 4.6KHz

c)

PSD of output bitstream

100 μm

System Performance

SNR=31.5dB   Sensitivity= 2.7e-016F/g   area=1.07e-7m^2 Kyx=10346   Kzx=1.95   Resonant frequency= 7.9KHz

Fig. 3.7. Synthesized results a) and (b: Experiment 1 and 2 (Maximum SNR); c): Experiment 3 (Maximum Static Sensitivity); d): Experiment 4 (Minimum area of mechanical sensing element).

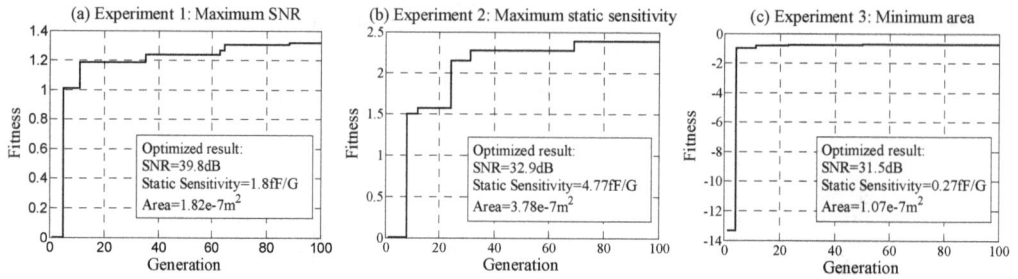

Fig. 3.8. Fitness improvement between generations.

Table 3.4. Summary of synthesized results for Experiments 1, 2 and 3.

MEMS components	Experiment 1	Experiment 2	Experiment 3
Proof mass	Ml = 341 μm Mw = 73 μm T = 2.9 μm Wh = 4.9 μm Nh = 28	Ml = 695 μm Mw = 136 μm T = 2.85 μm Wh = 4.2 μm Nh = 496	Ml = 205 μm Mw = 125 μm T = 2.5 μm Wh = 5.7 μm Nh = 40
Comb fingers	Lf = 122 μm Tf = 2.2 μm d0 = 1.0 μm Ns = 42 Nf = 4 Wanchor=4 μm	Lf = 183 μm Tf = 2.1 μm d0 = 1.5 μm Ns = 50 Nf = 10 Wanchor=4 μm	Lf = 84.6 μm Tf = 2 μm d0 = 1.36 μm Ns = 24 Nf = 8 Wanchor=4 μm
Spring	(Folded spring) Lo1 = 218 μm Lo2=255 μm Wo = 2 μm Lp = 11.5 μm Wp = 2.1 μm	(Classic serpentine spring) N = 2 Lo = 182 μm Wo = 2.0 μm Lp =4.5 μm Wp = 2.6 μm Lroot=45 μm W=5 μm	**(Beam spring)** Lo =200 μm Wo = 2.0 μm
Control system (2nd order Sigma-Delta)	Vf = 0.6 V Vm = 1.2 V K=23.8 Zero=0.1 Pole=12	Vf = 0.94 V Vm = 1.0 V K=4.8 Zero = 0.2 Pole= 13	Vf = 0.72 V Vm = 1.5 V K=9 ZERO = 0.1 POLE= 10

3.4. Conclusions

Part 1 of this two-part chapter presents an effective simulation-based synthesis flow (SystemC-AGNES) for automated layout synthesis of a MEMS component in a mixed-domain electrical-mechanical design. Due to the complex nature of the synthesis process, the synthesis algorithm has been implemented in SystemC-A. This platform is extremely well suited for complex modeling, implementation of post-processing of simulation results and optimization algorithms [13]. A distributed model of the mechanical sensing element is developed to ensure accurate behaviour of the MEMS accelerometer model when embedded in a $\Sigma\Delta$ force feedback control loop. The proposed approach is fully automated and it effectively deals with the trade-offs in complex digital MEMS sensor design to generate the layout of the mechanical sensing element according to user defined performance constraints. Synthesis of a full mixed-technology system which combines the layout synthesis methodology outlined above with the synthesis of the associated electronic control is discussed in Part 2 of this chapter.

Part 2: Synthesis of a MEMS System with Associated Control Loop

3.5. Introduction

This Part 2 of the chapter presents a novel genetic-based methodology for automated optimal synthesis of high order $\Sigma\Delta$ control topology for MEMS sensors. This synthesis approach enables an efficient configuration synthesis coupled with performance optimization. The proposed approach efficiently generates suitable configurations of the $\Sigma\Delta$ control loop by combining primitive components stored in a library and optimizes them according to the user specifications. This methodology can efficiently explore the configuration space and develops new structures with better performance. Compared with a manual designed $\Sigma\Delta$ accelerometer, the synthesized design gets 20 dB improvement for SNR. The approach is combined with the layout synthesis of the mechanical sensing element described in Part 1 of this chapter to realize the automated optimal design of MEMS systems embedded in electronic control circuitry from user defined high-level performance specifications and design constraints.

High performance MEMS sensors usually take advantage of a $\Sigma\Delta$ force feedback control strategy to improve linearity, dynamic range, and bandwidth, and provide direct digital output in the form of pulse density modulated bit stream which can be interface to a digital signal processor [14, 16, 21-25]. This approach has been applied to MEMS accelerometers [21, 22] and gyroscopes [14, 23-25]. Conventionally, the mechanical sensing element of a MEMS sensor is used as a loop filter to form a 2nd order single-loop electromechanical $\Sigma\Delta$ modulator. This is because the sensing element can be approximated by a 2nd order mass-damper-spring transfer function which performs the similar function to that of two cascaded integrators in typical 2nd order electronic $\Sigma\Delta$ modulators. In such a configuration, the dynamics of the mechanical sensing element limit the noise shaping properties. Compared with typical electronic 2nd order $\Sigma\Delta$ modulators, the DC gain of mechanical integrators is quite low which results in a lower signal to noise ratio (SNR) in 2nd order electromechanical $\Sigma\Delta$ modulators [16]. This is considered insufficient in high performance applications.

In order to improve the SNR, higher order $\Sigma\Delta$ control designs are becoming increasingly attractive. There are many results of the research into high order electromechanical $\Sigma\Delta$ control system topologies [16, 17, 26-28]. Dong et al. [16] used a mechanical sensing element and additional cascaded integrators with distributed feedback, based on a 3rd order distributed electronic loop filter, to form a 5th order electromechanical $\Sigma\Delta$ accelerometer. The experiment shows great improvement of the SNR compared with that for a 2nd order structure. Petkov and Boser [27] fabricated a fourth order $\Sigma\Delta$ interface for micromachined inertial sensors based on a chain of integrators with feedforward summation. More available structures, such as a sixth-order multiple-feedback (MF) electromechanical $\Sigma\Delta$ topology, are demonstrated by Dong et al. [28]. These topologies are all based on the idea to insert an additional electronic loop filter between the interface front-end and the quantizer. The additional filter, which provides high gain only in the signal band and rejects the out-of-band electronic noise, increases the order of the $\Sigma\Delta$ modulator [27] and dramatically decreases the noise floor in signal band. Kraft et al. [17] presented a novel multistage noise shaping (MASH) structure in which the electromechanical $\Sigma\Delta$ modulator is cascaded with a purely electronic $\Sigma\Delta$ modulator. Such architecture typically has large fabrication tolerances because accurate cancellation of the quantization noise in this structure relies on the values of mechanical sensor parameters [27].

The performance of the MEMS sensor embedded in a higher order $\Sigma\Delta$ control loop is greatly improved due to the additional purely electronic loop filters. Thus, the design of the higher order electromechanical $\Sigma\Delta$ modulator is focused on the loop filter structure. A general topology of a high order electromechanical $\Sigma\Delta$ accelerometer is shown in Fig. 3.9.

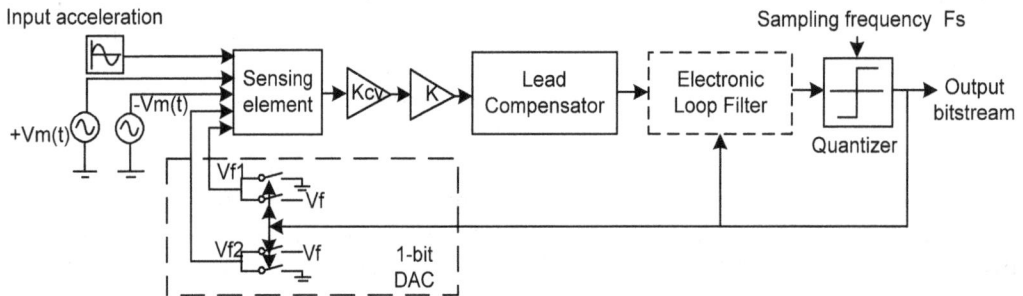

Fig. 3.9. Configuration of high order electromechanical $\Sigma\Delta$ accelerometer.

The sensing element converts the input acceleration signal into a displacement which results in a differential change of capacitance bridges which can be measured easily by the pick-off interface. A lead compensator is required to assure the stability of the feedback control loop when the sensing element is underdamped. The synthesis flow can automatically judge the sensing element properties and decide whether a lead compensator is necessary. A clocked 1-bit quantizer is used for oversampling and generating a pulse-density modulated digital output signal. $Vm(t)$ is a high frequency modulation carrier voltage and DAC is used to provide the feedback voltages Vf_1 and Vf_2. K_{cv} and K represent the signal pick-off gain and the gain of the voltage booster amplifier correspondingly. The block between the compensator and quantizer is an additional electronic loop filter. A high order electronic loop filter can be developed by a series of integrators using different topologies such as multiple feedback topologies and a combination of distributed feedback and feedforward topologies. It is worth noting that not only the topology and order of the electronic integrator loop filter but also the mechanical sensing element determine the noise shaping in a high order electromechanical $\Sigma\Delta$ modulator. This means that both the loop stability and the SNR depend on the sensor as well as loop filter parameters. Dong et al. [28] used a parameter sweep method to explore the optimal coefficients of the loop filters for several fixed topologies. The mechanical sensing element is also fixed. This limits the adaptability of the control system to different types of sensors.

3.6. Genetic-based Synthesis of MEMS Accelerometer with ΣΔ Control Loop

The general automated optimal genetic-based synthesis flow was shown in Fig. 3.1 in Part 1 of this chapter. This section outlines a technique where both the layout synthesis of the MEMS component and the topology synthesis of the associated electronic control loop are combined together to obtain a globally optimized design for the whole MEMS system according to predefined specifications. The detailed topology synthesis approach discussed below adds a genetic-based technique for the synthesis of the associated electronic control loop to the mechanical layout synthesis described in Part 1 of this chapter. The mechanical part in this case study is also a single axis lateral capacitive MEMS accelerometer similar to that synthesized in Part 1.

3.6.1. Synthesis Initialization

In the initialization phase of the synthesis process, a set of configurations is automatically generated from data in the MEMS primitive component and electronic control loop libraries to create the first generation of the GA. The MEMS primitive component library was discussed in Part 1 of this chapter and the primitives stored in the electronic control loop library are explained below. Sample primitive components of the electronic control loop are shown in Fig. 3.10. New loop filter topologies will be automatically generated from these primitives. Some feasible configurations of loop filters are illustrated in Fig. 3.11.

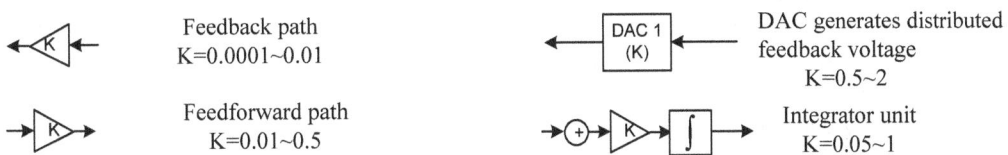

Feedback path K=0.0001~0.01		DAC generates distributed feedback voltage K=0.5~2	
Feedforward path K=0.01~0.5		Integrator unit K=0.05~1	

Fig. 3.10. Primitive components in the electronic control loop library.

The electronic loop filter synthesis is based on a series of integrators (Integrator unit in the library). The max number of the integrator unit is defined by users and the minimum number of integrators is set to zero to form the conventional 2nd order electromechanical ΣΔ accelerometer since the sensor may be used in some applications whose requirement of performance is not crucial. Electromechanical ΣΔ modulator should ideally benefits the advantages of the mature topologies used in ΣΔ AD converters with feedforward or feedback paths or the combination of them both. Thus, feedback and feedforward paths are added to the library. DAC1 is used to generate the distributed feedback voltage to integrator units from output bit stream.

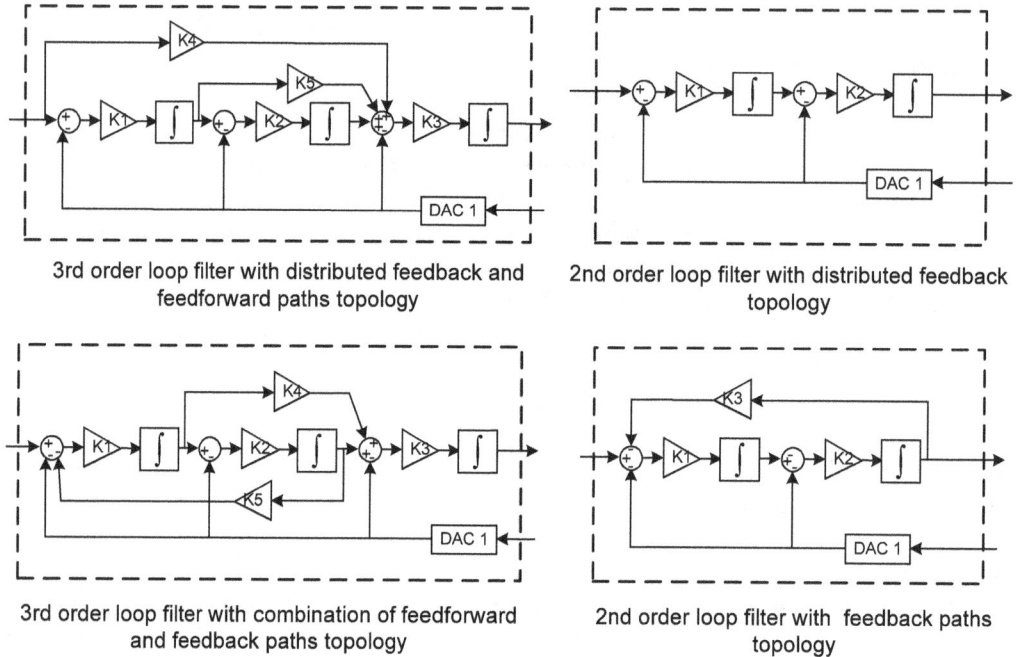

3rd order loop filter with distributed feedback and feedforward paths topology

2nd order loop filter with distributed feedback topology

3rd order loop filter with combination of feedforward and feedback paths topology

2nd order loop filter with feedback paths topology

Fig. 3.11. Examples of feasible configurations.

Typically, the distributed feedback signal from DAC1 is to determine the pole positions and loop stability, the scaling gain in the Integrator unit is to increase the limitation of the integrator output. Feedback paths between integrators form resonators to generate complex pair of zeros in order to further suppress the total noise in signal band while feedforward paths usually provide a larger input range [28].

The automated generation of the loop filter topology in the initialization phase is divided into several steps. Firstly, the system will generate a random number of integrators to determine the order of the loop filter (N). The maximum allowed order of the loop filter (N_{max}) is defined by the user. The number of integrators can be zero such that a conventional 2nd order electromechanical $\Sigma\Delta$ modulator can be generated without an electronic loop filter. Each integrator is randomly connected with DAC1 and other integrators by feedforward and feedback signal paths to produce different topologies of the loop filter.

Fig. 3.12 and Table 3.5 show a sample feasible configuration of a 4th order electromechanical $\Sigma\Delta$ accelerometer to illustrate the parameter initialization and encoding phase. The corresponding parameter initialization and encoding for the mechanical sensing element was discussed in Part 1 of this chapter. The loop filter in this sample MEMS accelerometer configuration is based on a 2nd order distributed feedback and feedforward topology and it contains two integrator units with one feedforward path between them.

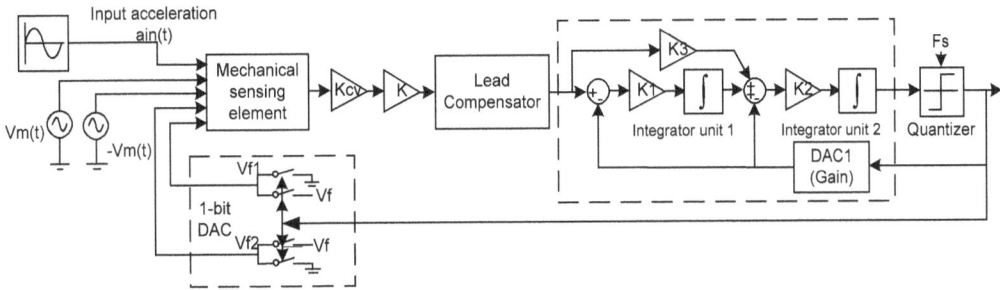

Fig. 3.12. A feasible initial configuration of a 4th order accelerometer.

Table 3.5. Representation of a population member of the MEMS Sigma-Delta electronic loop filter example in GA.

Loop filter topology	Value	Encoding	Description
Order of loop filter	2	1	Integrator 1
		2	Integrator 2
Feedback path	0	0	No feedback path between integrators
Feedforward path	1	1	Feedforward path between integrator 1 and 2
Distributed feedback from DAC1	2	1	Feedback to integrator 1
		2	Feedback to integrator 2

3.6.2. Genetic Synthesis of Electronic Control

In the genetic-based synthesis approach presented here, exploration of the solution is guided by the fitness functions which will be illustrated in next section. To compare the synthesis results with the results reported in Part 1 of this chapter, the Signal to Noise Ratio (SNR), die area of and static sensitivity of the mechanical sensing element are used as system performance constraints or objectives.

The topology synthesis of the loop filter can be divided into two steps: selection and new generation reproduction. In the selection phase, a proportion of designs in the current generation are retained through a fitness-based process (measured by fitness function) to breed the next generation. In the reproduction phase, the standard genetic operations of crossover and mutation are applied to the selected designs to generate the new generation. Firstly, in the crossover operation, the synthesis flow will randomly choose any two topologies as parents to generate offsprings. An example of crossover operation is shown in Fig. 3.13. In this example, crossover probabilities of compensator component and loop filter are higher than the user defined trigger probability 70 % that means these two components of selected parents will exchange to create offsrpings. As long as the crossover operation is triggered, the system will automatically judge the mechanical sensing element whether it is an underdamped system or not after crossover. This operation is used to determine whether the lead compensator is required. The crossover operation will end when a new generation is obtained.

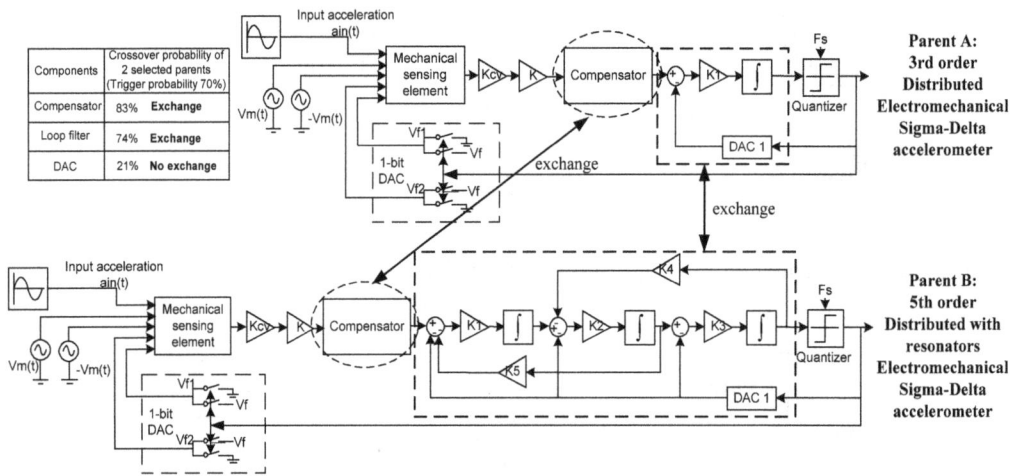

Fig. 3.13. An example of a crossover operation in $\Sigma\Delta$ control loop synthesis.

Subsequently, every individual in the new generation gets a fixed probability to mutate. The mutation operation for the loop filter topology contains two phases: topology mutation and component parameter mutation. In the topology mutation phase, if the topology mutation probability of a selected design is higher than the user defined trigger (50 % as an example), a new topology is generated from the electronic control loop primitive library and each parameter of the mutated component gets a random initial value within the allowed value range as illustrated in Fig. 3.10. Fig. 3.14 and Table 3.6 show an example of the topology mutation process. In this example, the configuration of the randomly selected 3rd order loop filter mutated to a topology with a 2nd order distributed feedback and a feedforward path. Each parameter of the new loop filter is randomly initialized within the constraints following the topology mutation operation. If there is no topology mutation for the selected design, the parameters of each component in the design will have a chance to mutate while keeping the topology unchanged.

Topology before mutation   Topology after mutation

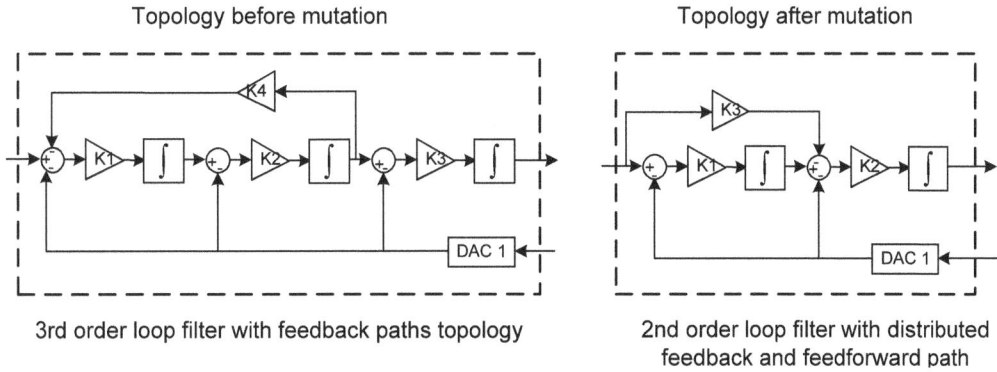

3rd order loop filter with feedback paths topology   2nd order loop filter with distributed
                                                            feedback and feedforward path

Fig. 3.14. An example of mutation operation in mechanical layout synthesis.

Table. 3.6. An example of mutation operation in mechanical layout synthesis.

Control loop component	Mutation probability (trigger 50 %)	Topology Generation Process	Description
Loop filter	71 % Loop filter topology mutation	1. Order of loop filter = 2	Generate random order of loop filter within range (0~Nmax) In this example, it is 2nd order
		2. Integrator units encoding Integrator 1 & Integrator 2	Encode Integrators, 2 integrators in the new topology
		3. Distributed feedback mutation mutation probability: Integrator 1: 61 % > 50 % (trigger) Integrator 2: 89 % > 50 % (trigger)	If mutation probability over trigger, integrator gets feedback from DAC1
		4. Feedback path mutation for integrator 2 mutation probability: 30 % < 50 % (trigger)	If mutation probability over trigger, feedback path is generated between integrator 2 and integrator 1.
		5. Feedforward path mutation for integrator 1 mutation probability: 92 % > 50 % (trigger)	If mutation probability over trigger, feedforward path is generated between integrator 1 and integrator 2.

3.7. Synthesis Experiments

The proposed synthesis flow for high order MEMS Sigma-Delta accelerometer is illustrated by four experiments as shown in Table 3.7. In the first two experiments, the systems are optimized for maximum SNR with different performance constraints, and in the third and forth experiments - for maximum static sensitivity and minimum area respectively. In order to compare the results with the 2nd order $\Sigma\Delta$ accelerometer in Part 1 of this chapter, the same design parameters are applied in the synthesis process:

1) Oversampling ratio: OSR=128;

2) Bandwidth: 512 Hz;

3) Oversampling frequency: $fs= 2^{17}$ Hz;

4) Input acceleration: 100 Hz acceleration with the amplitude of 1 g;

5) Maximum order of electronic loop filter: 3.

The fitness functions for these four experiments are listed below.

Table 3.7. Synthesis experiments.

Design objective	Performance constraints	Synthe-sized layout	Control loop	SNR (dB)	Static sensitivity	Area (m^2)
Maximum SNR	SNR>90 dB, Area<1.5e-7 m^2	Fig. 3.17	5th order	108	0.246 fF/G	1.41e-7
Maximum SNR	SNR>90 dB	Fig. 3.18.	5th order	114	0.76 fF/G	2.7e-7
Maximum Static sensitivity	SNR>75 dB Static sensitivity>2fF/G Area<3.0e-7 m^2	Fig. 3.19.	4th order	88.4	2.27 fF/G	2.1e-7
Minimum area of mechanical sensing element	SNR>75 dB Area<1.5e-7 m^2	Fig. 3.20.	4th order	85	0.11 fF/G	0.85e-7

3.7.1. Experiment Land 2(maximum SNR)

$$Fitness = w\frac{SNR}{SNR'},\qquad(3.23)$$

where SNR' is the designer specified objective value (90 dB in Experiment 1 and 2). SNR is obtained from a performance evaluation engine which is embedded in synthesis flow to enable measurements of the power spectrum density (PSD) and SNR through FFT of the output bit stream after each simulation. w is the weight coefficient which is set to 1 if all user defined performance constraints are met, otherwise w is set to 0.0001. For example, in Experiment 1, if a synthesized design can achieve 90 dB SNR with sensing element area less than 1.5e-7 m^2 , w will equal to 1 that is mean the algorithm finds a feasible solution satisfying specified performance.

3.7.2. Experiment 3 (Maximum Static Sensitivity of Sensing Element)

$$Fitness = w\frac{S}{S'},$$

(3.24)

where S is the static sensitivity of the synthesized sensing element and S' is the user-defined objective value (2 fF/G). Weight coefficient (w) has the same value as illustrated in Experiment 1 and 2.

3.7.3. Experiment 4 (Minimum Area of Mechanical Sensing Element)

$$Fitness = w\frac{Area}{Area'},$$

(3.25)

where $Area$ is the die area of the synthesized mechanical sensing element and $Area'$ is the predefined objective value (1.5e-7 m^2). In order to minimize the fitness parameter, w is set to -1 if performance constraints are met or -10 otherwise.

As the maximum order of electronic loop filter is set to 3 in the experiments, the maximum order of the electromechanical ΣΔ accelerometer is 5. Fitness improvement of synthesis flow is shown in Fig. 3.15. The topology of manual designed 5th order ΣΔ accelerometer with distributed loop filter is illustrated in Fig. 3.16 and Table 3.8. The synthesized mechanical layouts and its associated ΣΔ control system are shown in Fig. 3.17-3.20 and Table 3.8. The system output bit stream is measured by its PSD illustrated in Fig. 3.16-3.19.

The objectives of Experiments 1 and 2 are to maximize the SNR but with different constraints as shown in Table 3.7. It is worth noting that the SNR in Experiment 2 is further improved than that in Experiment 1 because the area is not constrained. Both the synthesized results of experiment 1 and 2 show better performance than the manual designed ΣΔ accelerometer with same order control system. Compared with the manual design shown in Fig. 3.16, the synthesis Experiment 2 improved the SNR figure by nearly 20 dB. As expected, the accelerometer optimized for the area in Experiment 4 shows an almost threefold area improvement over the manual design but the SNR figure is degraded by about 10 dB.

It can be seen from the results of the above synthesis experiments that the proposed synthesis approach efficiently explores the design space to generate suitable configurations of MEMS mechanical layout and its associated ΣΔ control loop by combining primitive components stored in the libraries and optimizes them according to the user-defined specifications.

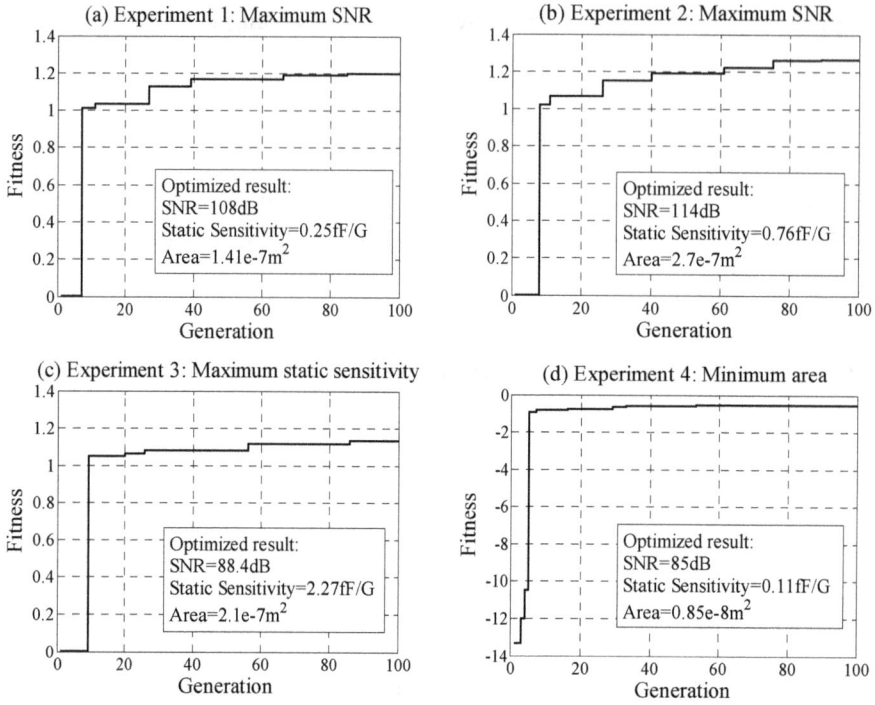

Fig. 3.15. Fitness improvement between generations.

System Performance

SNR=95dB Sensitivity= 0.41e-015F/g area=2.2e-007m^2 Kyx=1089 Kzx=1.95 Resonant frequency= 11.9KHz

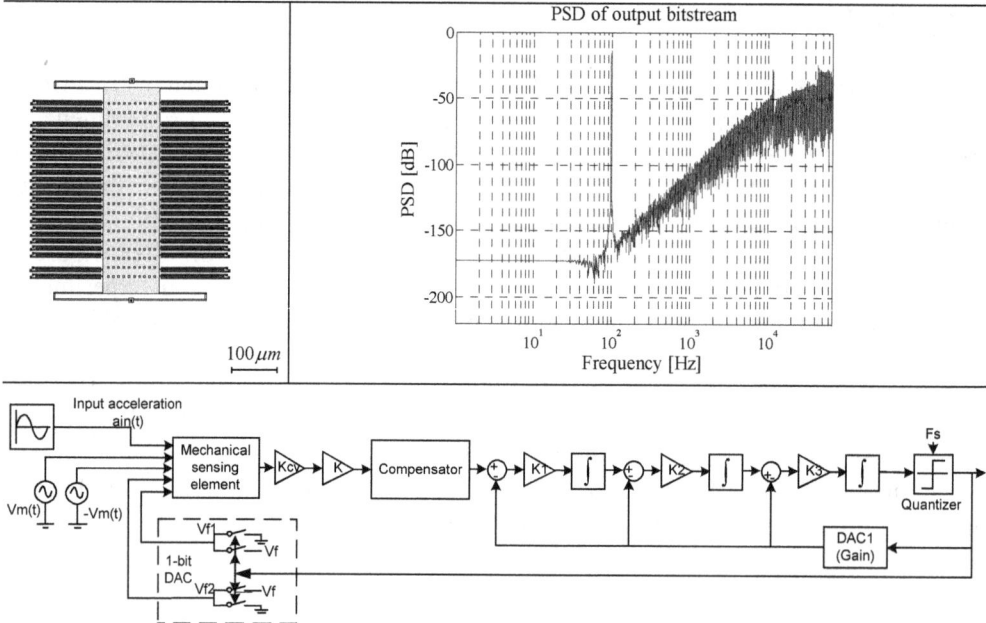

Fig. 3.16. Manual Design (5th order Sigma-Delta accelerometer).

System Performance

SNR=108dB   Static sensitivity= 2.46e-016F/g   area=1.41e-007m^2 Kyx=20802   Kzx=1.16   Resonant frequency= 9.6KHz

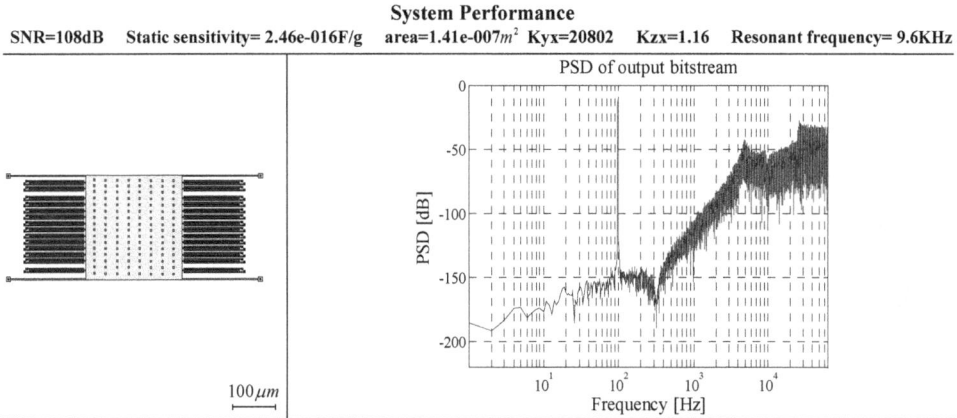

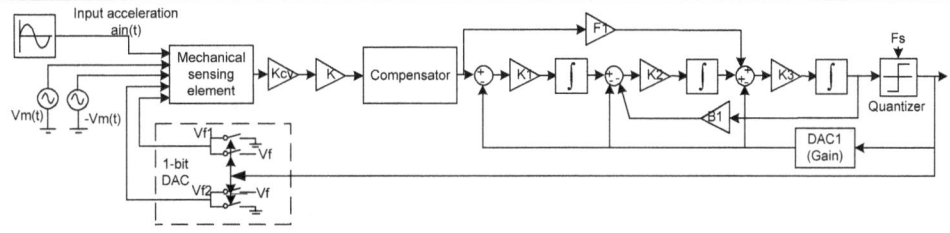

Fig. 3.17. Synthesized result in Experiment 1 (Maximum SNR).

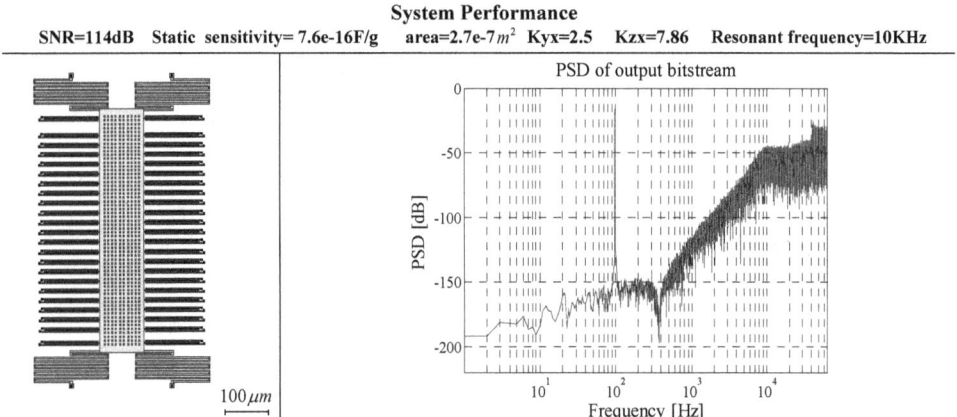

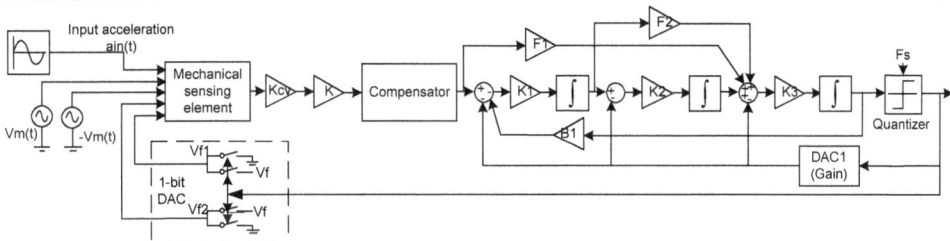

System Performance

SNR=114dB   Static sensitivity= 7.6e-16F/g   area=2.7e-7m^2 Kyx=2.5   Kzx=7.86   Resonant frequency=10KHz

Fig. 3.18. Synthesized result in Experiment 2 (Maximum SNR).

81

System Performance

SNR=88.4dB Static sensitivity= 2.27e-015F/g area=2.1e-007m^2 Kyx=395 Kzx=1.26 Resonant frequency= 9.25KHz

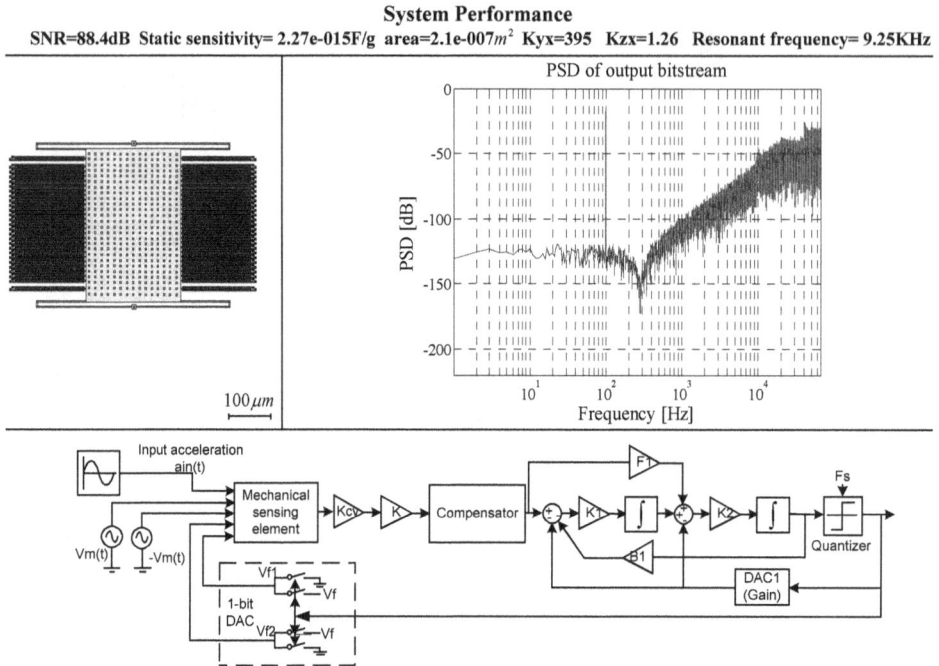

Fig. 3.19. Synthesized result in Experiment 3 (Maximum Static Sensitivity of sensing element).

System Performance

SNR=85dB Static sensitivity= 1.1e-16F/g area=8.5e-8m^2 Kyx=5.35 Kzx=2.17 Resonant frequency= 17.4KHz

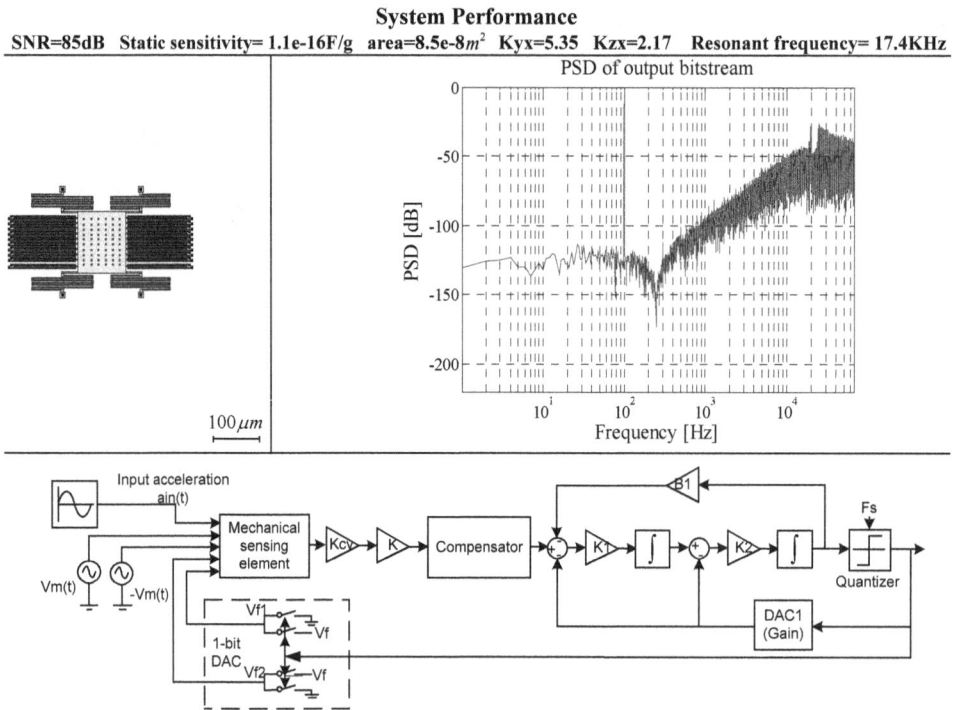

Fig. 3.20. Synthesized result in Experiment 4 (Minimum Area of sensing element).

Table 3.8. Manual design and synthesis results.

MEMS components	Manual design	Experiment 1	Experiment 2	Experiment 3	Experiment 4
Proof mass	Ml = 450 μm Mw = 130 μm T = 2.5 μm Wh = 4 μm Nh = 200	Ml = 236 μm Mw = 237 μm T = 2.7 μm Wh = 3.4 μm Nh = 128	Ml = 563 μm Mw = 102 μm T = 3.0 μm Wh =3.14 μm Nh = 600	Ml = 341 μm Mw = 225 μm T = 2.7 μm Wh = 4.5 μm Nh = 390	Ml = 130 μm Mw = 107 μm T = 2.21 μm Wh =3.45 μm Nh = 60
Comb fingers	Lf = 150 μm Tf = 2 μm d0 = 1.5 μm Ns = 40 Nf = 8 Wanchor=4 μm	Lf = 136 μm Tf = 2 μm d0 = 1.74 μm Ns = 22 Nf = 6 Wanchor=4 μm	Lf = 130 μm Tf = 2.0 μm d0 = 1.4 μm Ns = 42 Nf = 4 Wanchor=4 μm	Lf = 164 μm Tf = 2.3 μm d0 = 1 μm Ns = 50 Nf = 4 Wanchor=4 μm	Lf = 139 μm Tf = 2.0 μm d0 = 1.5 μm Ns = 18 Nf = 2 Wanchor=4 μm
Spring	**(Folded spring)** Lo1 = 180 μm Lo2=115 μm Wo = 2 μm Lp = 16 μm Wp = 2 μm	**(Beam spring)** Lo = 180 μm Wo = 2.5 μm	**(Classic serpentine spring)** N = 3 Lo = 177 μm Wo = 3.0 μm Lp = 3.9 μm Wp = 3.2 μm Lroot=75 μm	**(Folded spring)** Lo1 = 227 μm Lo2=114.5 μm Wo = 2.5 μm Lp = 9.3 μm Wp = 2 μm	**(Classic serpentine spring)** N = 2 Lo = 134 μm Wo = 2.2 μm Lp = 3.2 μm Wp = 3.5 μm Lroot=35 μm
Control system	**(4th order)** Vf = 0.45 V Vm = 1.5 V K=20 ZERO = 0.01 POLE=3000 Gain=0.8 K1=0.078 K2=0.38 K3=0.458	**(5th order)** Vf =0.51 V Vm = 1.1 V K=27.3 ZERO =0.1 POLE= 820 Gain=1.5 K1=0.46 K2=0.076 K3=0.53 F1=0.06 B1=0.006	**(5th order)** Vf =0.35 V Vm = 1.2 V K=33.7 ZERO = 0.1 POLE= 2000 Gain= 1.1 K1= 0.244 K2 = 0.159 K3=0.61 F1=0.1 F2=0.08 B1=0.0073	**(4th order)** Vf = 0.3 V Vm = 1 V K=9 ZERO = 0.001 POLE= 1036 Gain=1.5 K1=0.76 K2=0.61 F1=0.01 B1=0.0004	**(4th order)** Vf =0.35 V Vm = 1.2 V K=48 ZERO = 0.01 POLE= 10 Gain= 1.46 K1= 0.38 K2 = 0.42 B1=0.001

3.8. Conclusion

This chapter presents an effective simulation-based synthesis flow to automated synthesis of MEMS sensors with associated high order electronic $\Sigma\Delta$ control systems. Design of such MEMS systems is notoriously difficult using traditional methods as the mechanical element forms an integral part of the electromechanical $\Sigma\Delta$ control system.

The performance of the system is not only determined by the electronic control system configuration, but also dynamics of the mechanical sensing element. The proposed holistic synthesis approach, implemented in SystemC and named SystemC-AGNES, automates both the layout synthesis of the sensor's mechanical part and the configuration synthesis of the electronic control loop by simultaneously searching for the optimal solution according to user defined constraints. It especially worth noting that the noise floor in synthesized higher order electromechanical $\Sigma\Delta$ modulators can be reduced drastically, by about 20 dB and 74 dB, compared with the high order manual design and 2nd order design respectively (see Part 1 of this chapter).

References

[1]. Kraft M., Micromachined inertial sensors: The state of the art and a look into the future, *IMC Measurement and Control*, 33, 6, 2000, pp. 164-168.

[2]. IEEE standard VHDL analog and mixed-signal extensions, *Design Automation Standards Committee of the IEEE Computer Society,* 1999.

[3]. IEEE Standard VHDL Analog and Mixed-Signal Extensions-Packages for Multiple Energy Domain Support, *Design Automation Standards Committee of the IEEE Computer Society*, 2005.

[4]. Pecheux F., Lallement C., and Vachoux A., VHDL-AMS and Verilog-AMS as alternative hardware description languages for efficient modeling of multidiscipline systems, *IEEE Trans. Computer-Aided Design of Integrated Circuits and Systems*, 24, 2, 2005, pp. 204-225.

[5]. Datasheet, SystemVision for Mechatronic System Modeling, *Mentor Graphics Corporation*, 2006. (http://www.mentor.com/products/sm/system_integration_simulation_ analysis/systemvision/upload/SystemVision datasheet.pdf).

[6]. Wang L. and Kazmierski T. J., VHDL-AMS based genetic optimization of a fuzzy logic controller for automotive active suspension systems, in *Proc. of the Int. Conf. Behavioral Modeling and Simulation Workshop, BMAS*, Sept 2005, pp. 124-127.

[7]. Tran Due Tan, Roy S., Nguyen Phu Thuy, and Huu Tue Huynh, Streamlining the design of MEMS devices: an acceleration sensor, *Circuits and System Magazine, IEEE*, 8, 1, 2008, pp. 18-27.

[8]. Mukherjee T., Zhou Y., and Fedder G. K., Automated optimal synthesis of microaccelerometers, *Proc. Int. Conf. Micro Electro Mechanical Systems, MEMS'99*, Jan. 1999, pp. 326-331.

[9]. Mukherjee T., Zhou Y., and Fedder G. K., Automated optimal synthesis of microresonator, in *Proc. of the Int. Conf. Solid State Sensors and Actuators*, 2, 1997, pp. 1109-1112.

[10]. Mukherjee T. and Fedder G. K., Structured Design of Microelectromechanical Systems, in *Proc. of the Design Automation Conference*, 2, Jun 1997, pp. 680-685.

[11]. Fan Z., Goodman E. D., Wang J., Rosenberg R., Kisung Seo, and Hu J., Hierarchical evolutionary synthesis of MEMS, *Congress on Evolutionary Computation, 2004. CEC2004*, 2, Jun 2004, pp. 2320-2327.

[12]. Gupta V. and Mukherjee T., Layout synthesis of CMOS MEMS accelerometers, *MSM' 2000*, 2000, pp. 150-153.

[13]. H. Al-Junaid and T. Kazmierski, An Analogue and Mixed-Signal Extension to SystemC, *IEE Proc. Circuits, Devices and Systems*, 152, 6, December 2005, pp. 682-690.

[14]. Seeger J. I., Xuesong J., Kraft M., and Boser B. E., Sense finger dynamics in a $\Sigma\Delta$ force feedback gyroscope, *Tech. Digest of Solid State Sensor and Actuator Workshop, Hilton Head Island, USA.*, Jun 2000, pp. 296-299.

[15]. Zhao C., Wang L., and Kazmierski T. J., An efficient and accurate MEMS accelerometer model with sense finger dynamics for applications in mixed-technology control loops, in *Proc. of the Int. Conf. Behavioral Modeling and Simulation Workshop, 2007. BMAS, 2007*, Sept. 2007, pp. 143-147.

[16]. Dong Y., Kraft M., Gollasch C., and Redman-White W., A high-performance accelerometer with a fifth-order Sigma-Delta modulator, *Journal of Micromechanics and Microengineering*, 2, 2005, pp. S22-S29.

[17]. Kraft M., Redman-White W., and Mokhtari M. E., Closed loop micromachined sensors with higher order $\Sigma\Delta$ modulators, in *Proc. of the 4rd Conf. on Modeling and Simulation of Microsystems*, 2001, pp. 104-107.

[18]. Dong Y., Kraft M., Hedenstierna N., and Redman-White W., Microgyroscope control system using a high-order band-pass continuous-time Sigma-Delta modulator, in *Proc. of the Transducers 2007*, 2, June 2007, pp. 2533-2536.

[19]. J. Carnahan and R. Sinha, Nature's algorithm [genetic algorithms], *IEEE Journal of Potential*, 20, 2, Apr-May 2001, pp. 21-24.

[20]. Y. Zhang, R. Kamalian, A. M. Agogino, and H. S. Carlo, Hierarchical MEMS synthesis and optimization. *SPIE Conference on Smart Structures and Materials,* 5763, 12, Mar. 2005.

[21]. Kraft M., Lewis C. P., Hesketh T. G. and Szymkowiak, S., A Novel Micromachined Accelerometer Capacitive Interface, *Sensors & Actuators,* Vol. A68/1-3, 1998, pp. 466-473.

[22]. Kraft, M., Lewis, C. P. and Hesketh, T. G., Closed Loop Silicon Accelerometers, *IEE Proc. Circuits, Devices Syst.,* Vol. 145, No. 5, 1998, pp. 325-331.

[23]. Kraft, M. and Ding, H., Sigma-delta modulator based control systems for MEMS gyroscopes, *Proc. IEEE NEMS*, Jan. 2009, pp. 41-46.

[24]. Damrongsak, B. and Kraft, M., Performance analysis of a micromachined electrostatically suspended gyroscope employing sigma-delta force feedback, in *Proc. MME 2007 Conference,* pp. 269-272, Sept. 2007.

[25]. Dong, Y., Kraft, M., Hedenstierna, N. and Redman-White, W., Microgyroscope control system using a high-order band-pass continuous-time sigma-delta modulator, *Sensors and Actuator, A.,* Vol. 145, 2008, pp. 299-305.

[26]. Dong, Y., Kraft, M. and Redman-White, W., Micromachined vibratory gyroscopes controlled by a high order band-pass sigma delta modulator, *IEEE Sensors Journal,* Vol. 7, 2007, Issue 1, pp. 50-69.

[27]. Petkov V. P. and Boser B. E., A fourth-order $\Sigma\Delta$ interface for micromachined inertial sensors, *IEEE J. of Solid-State Circuit,* Vol. 40, No. 8, 2005.

[28]. Dong Y., Kraft M., and Redman-White W., Higher order noise-shaping filters for high-performance micromachined accelerometers, *IEEE trans. on Instrumentation and Measurement,* Vol. 56, No. 5, 2007.

Modern Sensors, Transducers and Sensor Networks

Chapter 4
Sensors for Food Inspections

Mohd Syaifudin Bin Abdul Rahman, Subhas C. Mukhopadhyay, Pak Lam Yu, Michael J. Haji-Sheikh and Cheng-Hsin Chuang

4.1. Introduction

Food poisoning is a general term used to describe illness from food-borne microorganism. The primary goals for this research work are to detect the natural toxin in seafood which is caused by domoic acid and also the pathogenic bacteria normally from Gram-Negative bacteria which produce endotoxin which is dangerous to human or animal. In order to protect consumers from food poisoning, most countries have a set of regulatory guidelines. All products need to be tested for the toxin, which places a heavy workload on the laboratories and is extremely expensive to the industry. This has motivated us towards the development of a novel low cost sensing system for food inspections without much difficulty.

4.1.1. Seafood Poisoning (Marine Biotoxins)

Illness from seafood poisoning were caused by dangerous contaminated chemicals or marine biotoxins [1-3]. Seafood contaminated by marine biotoxins apparently look, smells or tastes normal but after human or animals eat the seafood, they may suffer a variety of gastrointestinal and neurological illnesses [2]. Shellfish toxins and ciguatoxins are the most dangerous marine biotoxins [1]. Shellfish toxins can cause paralytic shellfish poisoning (PSP), diarrhoeic shellfish poisoning (DSP), amnesic shellfish poisoning (ASP), neurotoxic shellfish poisoning (NSP) and azaspiracid shellfish poisoning (AZP), while ciguatoxins can cause ciguatera fish poisoning (CFP) [1, 4].

In late 19th and early 20th century a large number of illnesses were linked with the consumption of raw oysters, claws and mussels. It was found that these illnesses were related to the ingestion of domoic acid-contaminated mussels which led to ASP [1-4]. ASP is characterized according to both gastrointestinal and neurological symptoms,

Mohd Syaifudin Bin Abdul Rahman
School of Engineering and Advanced Technology, Massey University, Palmerston North, New Zealand

including severe headache, confusion, and either temporary or permanent memory loss. Domoic acid (DA) is a naturally occurring toxin produced by microscopic algae, specifically the diatom species Pseudo-nitzschia [5-7]. DA is a chemical that is produced by algae or plankton when it blooms. Shellfish ingest these algae, where the toxin concentrates and can accumulate this toxin without apparent ill effects [1, 4]. However, for marine mammals and humans, DA is ticarboxylic acid that acts as a neurotoxin. The toxin is not destroyed by cooking or freezing. Fig. 4.1 shows the chemical structure of domoic acid.

The presence of DA in shellfish has been reported in various regions of the world [1, 8]. There have been numerous reports of toxicity in a variety of wildlife species indicating that DA moves up the food chain in marine ecosystems. Studies have proven that certain amount of DA can cause health problems to animals and humans [9-14]. In 1987, DA was identified as the toxin responsible for an outbreak of illness in Prince Edward Island, Canada [15, 16]. It was caused by eating blue mussels. Effects on both the gastrointestinal tract and the nervous system were observed. It was reported that 107 patients (all adult) met the case definition [17]. Dose-related symptoms included nausea, vomiting, abdominal cramps, diarrhoea, headache, memory loss and convulsions and several deaths were attributed to the toxin. As a result of the episode of human illness in Canada, most countries have set a regulatory guideline of 20 µg/g of domoic acid in shellfish meat [2].

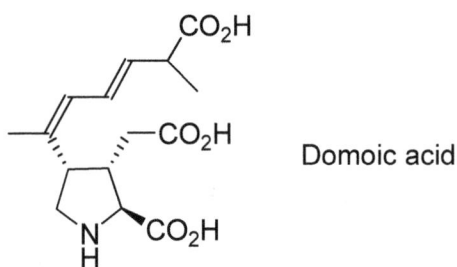

Fig. 4.1. Chemical structure of Domoic Acid.

4.1.2. Food Poisoning (Endotoxin)

Almost 90 % - 95 % of Gram Negative bacteria are pathogens [18]. The difference between Gram Positive and Gram Negative is in the form of their cell wall structure. Endotoxins (Lipopolysaccharide, LPS) are dangerous toxins which are associated with Gram-negative bacteria [19]. Endotoxins or LPS are part of the outer membrane of the Gram-negative bacteria cell wall which comprises of O antigen, Core and Lipid A [20]. The detection of endotoxins which are caused by the pathogens such as Escherichia coli and Salmonella is the main concerns the food industries [21]. These bacteria are commonly found on many raw foods. High risk foods which are commonly affected by these bacteria are meat, poultry, seafood, dairy products, and eggs. Bacteria capable of causing food poisoning are often found on meat and on several occasions meat has been implicated in food poisoning outbreaks which have caused human illness and death [22]. Food may be contaminated because of poor food processing, handling and storage, poor hygiene, food poisoning bacteria in the soil, water, on animals and people [23]. Fig. 4.2 shows the cell wall of Gram-Negative bacteria.

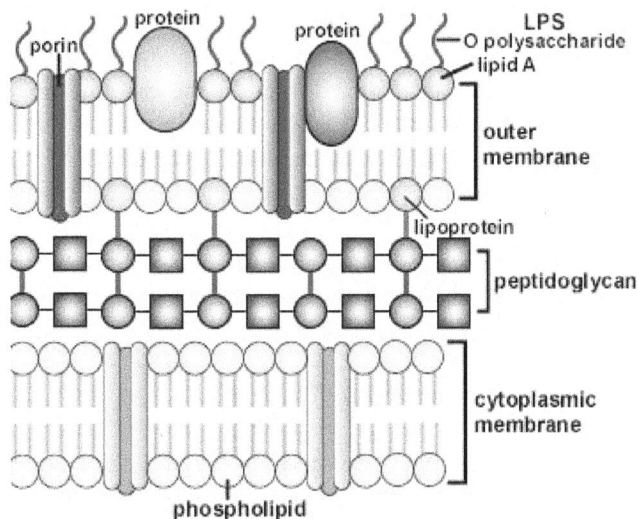

Fig. 4.2. Cell wall of Gram-Negative bacteria.

4.2. Existing Method of Domoic Acid and Pathogens Detection

4.2.1. Domoic Acid Detection

The existing method of DA detection is based on using chromatography technique, surface plasmon resonance (SPR), and immunoassay technique using enzyme-linked Immunosorbent assay (ELISA). Chromatographic techniques have been widely used for the detection of marine toxins. Overview of different chromatographic techniques for marine toxins detection has been reported by Quilliam [24]. Liquid chromatography (LC) has been used as Association of Analytical Communities (AOAC) official method for DA in mussels [25]. Identification of DA in mussels using LC technique has been reported in [6, 26, 27]. A new sensitive determination method of DA using high-performance liquid chromatography (HPLC) has been reported in [5, 10, 28, 29]. Although chromatographic technique is one of the best methods for the detection of DA but they require expensive equipments, trained personnel, sophisticated method of sampling preparation and also it is a time consuming method.

Surface Plasmon Resonance (SPR) has been widely used as detection technique in biosensing system [3]. A rapid and sensitive immuno-based screening method was reported to detect DA present in extracts of shellfish species using a surface plasmon resonance-based optical biosensor [30]. An immunosensor based on surface plasmon resonance (SPR) was used for the detection of DA [31]. A portable SPR biosensor has been used to detect domoic acid in clam extracts [32]. The detection method based on SPR is suitable for laboratory analysis and not suitable for in-situ monitoring since the samples need to be prepared accordingly, analysis may take longer time and the equipment (SPR) is expensive.

Immunoassay techniques are based on the affinity recognition between antibodies and antigens, and the most commonly found format is the enzyme-linked immunosorbent assay (ELISA) [3]. Research of using ELISA to determine DA has been reported in [7, 33-36]. ELISA method normally can be used to detect only one particular toxin. Only one research work reported by Garthwaite et. al [35], which integrates ELISA for screening of DSP, PSP, ASP and NSP toxins. To develop the ELISA strip and to prepare the samples will need to follow some laboratory procedures and tedious work. Also there is no guarantee that the ELISA strip will respond very well to all samples.

4.2.2. Pathogens Detection (Endotoxin)

The endotoxic is caused by pathogens from Gram-Negative bacteria [18]. Conventional method is conducted by enriching the food sample and performing various media-based metabolic testing (agar plates or slants) [37, 38]. Conventional techniques are capable of giving accurate result because of high selectivity and sensitivity. The conventional method of detection of pathogens is by culturing techniques which are tedious, time-consuming and expensive [39]. Polymerase chain reaction (PCR) is one of the most popular methods of pathogens detection [37, 40]. PCR is based on the isolation, amplification and quantification of a short DNA sequence. Recent advancement in PCR technology has developed a real-time PCR [41, 42] and a multiplex PCR [43] which can obtain results in a few hours.

Enzyme linked immunosorbent assays (ELISA) and culture techniques for determining and quantifying pathogens in food have been established [22]. ELISA is combined with a specific antibodies and the sensitivity of simple enzyme assays by using antibodies or antigens coupled to an easily assayed enzyme [21]. The technique is developed to replace the detection using culture and colony counting. However, the ELISA also needs to be confirmed using conventional test. Biosensor has been widely used to detect pathogens in food [21, 39]. Most biosensors have a sensitivity of 10 – 10 000 colony from units (CFU)/ml. Biosensor is one of the rapid methods, based on immunochemical or nuclei acid technologies [21]. The transduction methods used in biosensor are optical fibre [44, 45], electrochemical [46, 47], piezoelectric, interdigitated array microelectrodes [48], surface plasmon resonance (SPR) [49] and others.

4.3. Development of Novel Planar Interdigital Sensor

4.3.1. Introduction to Planar Interdigital Sensors

Planar interdigital sensor is based on the interdigital electrode structures which have been widely used in photosensitive detectors [50-53], surface acoustic wave (SAW) [54-58], humidity sensors [59-62], sensors for chemicals and gasses [63-68], measurement of electrolyte conductivity (EC) [69], determination of components contained in aqueous solutions [70], determination of moisture, fibre and titanium dioxide in paper pulp [71, 72], for the complex permittivity characterization of materials [73], measuring tablet hardness and coating thickness (pharmaceutical products) [74], the detection of small amount of magnetic beads over the surface of micro sensors [75]. Currently, the numbers of research works in developing planar interdigital sensors are increasing. Planar interdigital sensors have been used for estimation of properties of dielectric material such as milk, saxophone reeds, meat and leather [76-83] . Evaluation of diffusion-driven material property profiles using three-wavelength interdigital sensor has been reported by Mamishev et. al. [84]. Interdigital type of micro impedance biosensor has been developed for bacterial detection and foodborne bioterrorism agents [85, 86]. The dielectric behaviour of nematic liquid crystals was studied with interdigital capacitors [87]. Design and applications of interdigital capacitive sensors were discussed in [88].

The operating principle of planar interdigital sensor basically follows the rule of two parallel plate capacitors, where electrodes open up to provide a one sided access to the material under test (MUT) [88]. The electric field lines of parallel plate capacitor and an interdigital sensor are shown in Fig. 4.3 (a) and (b). The electric field lines generated by the sensor penetrate into the MUT and will change the impedance of the sensor. The sensor behaves as a capacitor in which the capacitive reactance becomes a function of system properties. Therefore by measuring the capacitive reactance, the system properties can be evaluated.

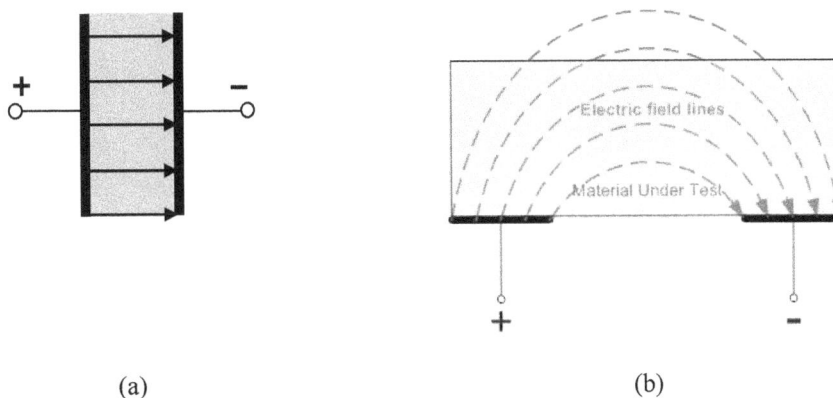

(a)   (b)

Fig. 4.3. Electric field lines of (a) Parallel plate capacitors; (b) Planar interdigital sensor.

Since the electrodes of an interdigital sensor are coplanar, therefore the measured capacitance will have a high signal-to-noise ratio. In order to get a strong signal, the electrode pattern of the interdigital sensor can be repeated many times. The term "interdigital" refer to a digit-like or finger-like periodic pattern of parallel in-plane electrodes, used to build up the capacitance associated with the electric fields that penetrate into a material sample [88]. The conventional interdigital sensor is shown in Fig. 4.4. AC voltage source will be applied as an excitation voltage between the positive terminal and the negative terminal. An electric field is formed from positive terminal to negative terminal. Mamishev et. al. [59], also stated that for a semi-infinite homogeneous medium placed on the surface of the sensor, the periodic variation of the electric potential along the X-axis creates an exponentially decaying electric field along the Z-axis, which penetrates the medium. Fig. 4.5 illustrates Mamishev statement.

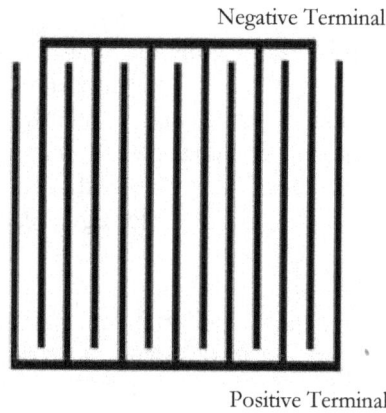

Fig. 4.4. Configuration of conventional planar interdigital sensor.

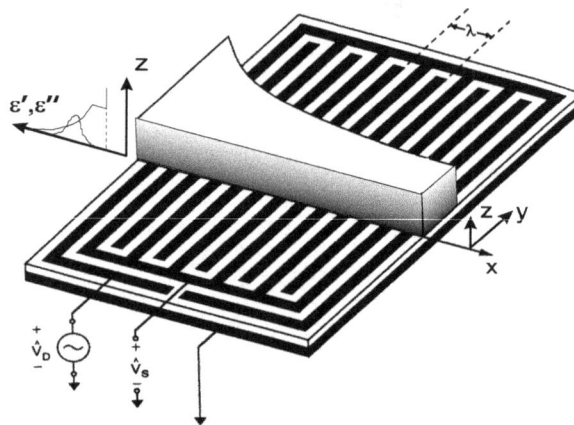

Fig. 4.5. Fringing electric field of interdigital sensor [59].

Fig. 4.6 shows the side view of the interdigital sensor showing how electric field was formed between positive and negative electrodes. It is shown clearly in Fig. 4.6 that the penetration depths of the electric field lines vary for different pitch length. The pitch length of the interdigital sensors is the distance between two consecutive electrodes. Also in Fig. 4.6, there are three pitch length (l_1, l_2 and l_3) showing the different penetration depths with respect to the pitch length of the sensor. The penetration depth can be increased by increasing the pitch length, but the electric field strength generated at the neighbouring electrodes will be weak. Planar interdigital sensors can be used for different sensing application. Fig. 4.7 illustrates on the sensing possibilities of planar interdigital sensors. These sensing possibilities for various characteristic of samples have given us the opportunity to design and fabricate a miniature type of planar interdigital sensor to detect the contaminated food with dangerous toxins.

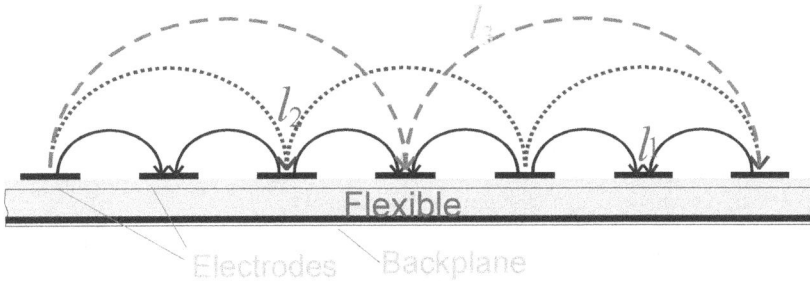

Fig. 4.6. Electric field formed between positive and negative electrodes for different pitch lengths, (l_1, l_2 and l_3) [88].

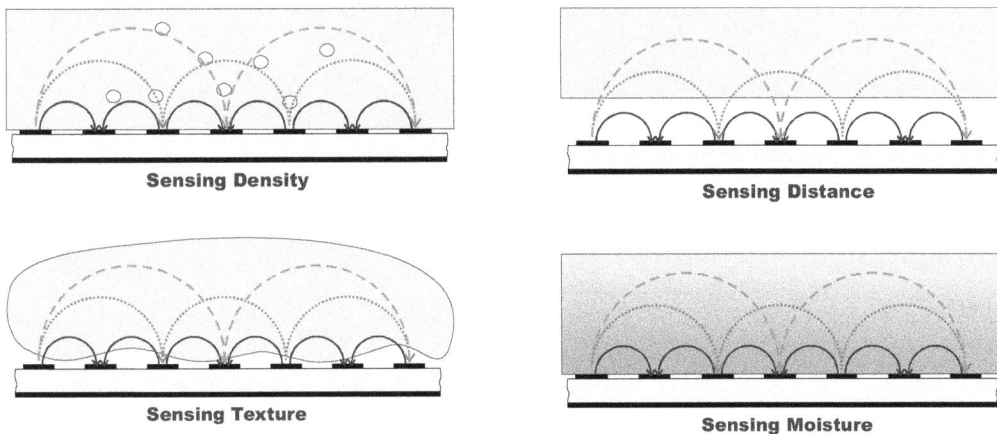

Fig. 4.7. Sensing possibilities to detect various characteristic of samples [59].

4.3.2. Analytical Analysis and Modeling

4.3.2.1. Calculation of Capacitance using Circuit Analysis

Analytical analysis to calculate the equivalent capacitance for each sensor's geometry has been carried out. The equivalent capacitive circuit for each sensor's geometry is shown in Fig. 4.8. All sensors have the same area and electrodes spacing between two consecutive electrodes. The capacitors are formed between positive and negative electrodes. Therefore, all capacitors are effectively connected in parallel. The capacitance is a function of distance between two positive and negative electrodes, d and is given by:

$$C = \frac{\varepsilon_0 \varepsilon_r A}{d}$$

(4.1)

where

C = capacitance in farads, F

ε_0 = the permittivity of free space ($\varepsilon_0 = 8.854 \times 10^{-12}$ F/m)

ε_r = the relative static permittivity or dielectric constant (vacuum = 1)

A = effective area, square meters

d = effective spacing between positive and negative electrode, meters

With references to Fig. 4.8, we can write

$$C_{12} = \frac{1}{2} C_{11}; ... C_{1n} = \frac{1}{n} C_{11}$$

and

$$C_{22} = \frac{1}{2} C_{21}; ... C_{2n} = \frac{1}{n} C_{21}$$

(4.2)

where $C_{11} = C_{21}$.

Therefore the total equivalent capacitance (C_{eq}) for each sensor's geometry is given by:

$$C_{eq(1)} = C_{11} + C_{12} + ... + C_{1n}$$

(4.3)

where $C_{eq(1)}$ is for equivalent capacitance in the first geometry of each sensor.

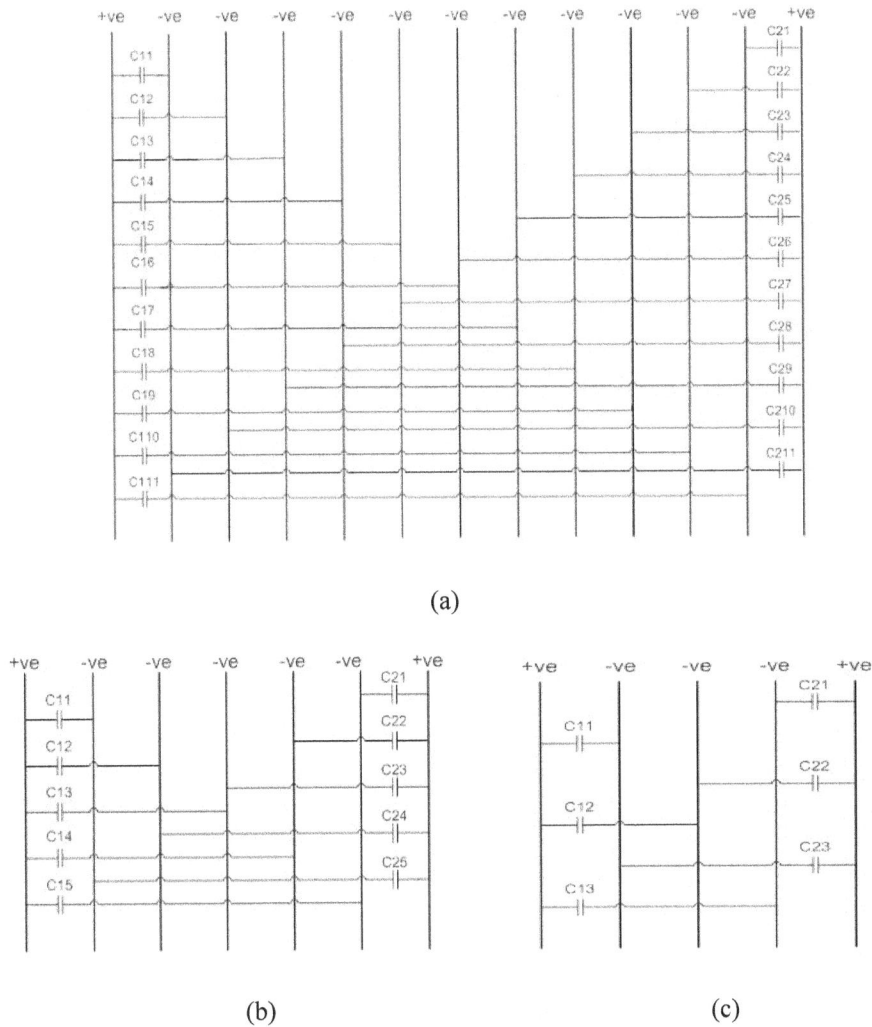

(a)

(b)   (c)

Fig. 4.8. The capacitive circuit of different sensors; (a) Sensor 1_11 with electrodes configuration 1-11-1 (b) Sensor 1_5 with electrodes configuration 1-5-1-5-1 and (c) Sensor 1_3 with electrodes configuration 1-3-1-3-1-3-1.

The estimated capacitance within each sensor's geometry is shown in Fig. 4.9. It is shown that Sensor 1_11 has better uniformity in term of distribution of effective capacitance though out the sensors geometry compared to Sensor 1_5 and Sensor 1_3, but has lower total capacitance value. The result from analytical analysis is in line with the result shown in the simulation using COMSOL. It is important to achieve better sensitivity in term of uniform distribution of total capacitance value and the uniformity of electric field distributions throughout the sensor geometry. The performance of the sensors depends on the characteristic of the sensor and also the characteristic of the material under test (MUT). Therefore the experiments of sensors with MUT will determine the best sensing performance of different sensor configurations.

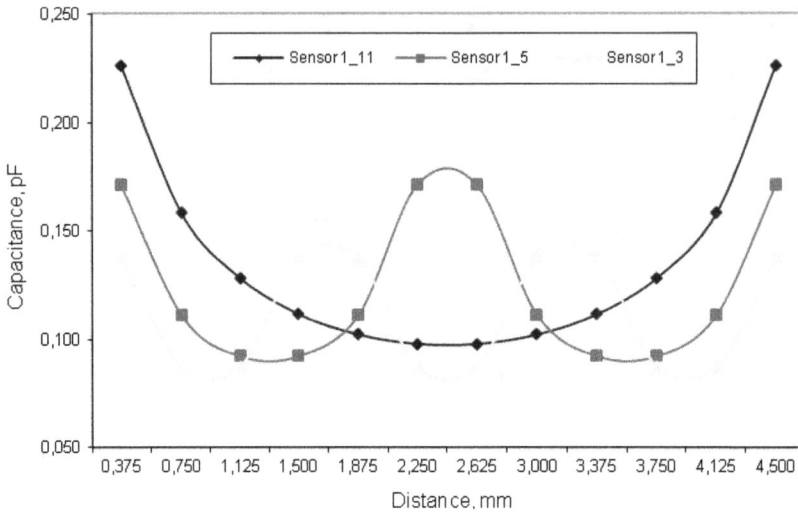

Fig. 4.9. Equivalent capacitance within each sensor geometry

4.3.2.2. Modeling using COMSOL Multiphysics

The COMSOL Multiphysics® (formerly FEMLAB) is a finite element analysis and solver software package for various physics and engineering applications, especially coupled phenomena, or multiphysics. The software applications are based on partial differential equations (PDEs). The sensors were modelled to derive the optimum structure of sensor for better performance. The Electromagnetic module of COMSOL was used for analysis of novel interdigital sensors. The application was in 3D and AC/DC module for electrostatics mode for dielectric material was selected. Fig. 4.10 shows the arrangement of positive electrodes and negatives electrodes of the sensor design in 3D view. The sensors were modelled using the actual size of the fabricated sensors.

The value for electrostatic energy density, We can be obtained from the simulation. The energy required to charge a capacitor should equal that of the electrostatic field, which is given by

$$W_e = \frac{Q^2}{2C} \qquad (4.4)$$

We is available in the Electrostatics application mode; the software calculates it by integrating across the domain

$$W_e = \int_\Omega (D.E)\,d\Omega \quad , \qquad (4.5)$$

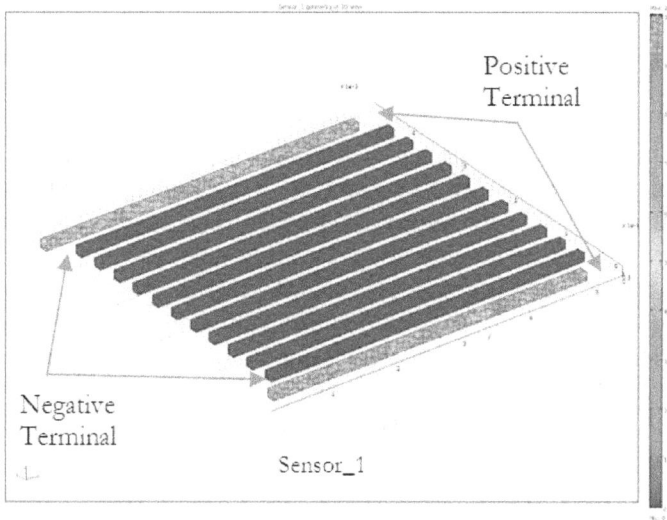

Fig. 4.10. The geometry of interdigital sensor in 3D view.

where, D is the electric displacement, and E is the electric field intensity. The capacitance, C, is related to the charge on the two conductive plates, Q, and the voltage difference across those plates, ΔV, by:

$$C = \frac{Q}{\Delta V} \tag{4.6}$$

The calculation of C is carried out from the stored electric energy, We and the voltage across the two plates, ΔV, and is given by:

$$C = \frac{Q^2}{2W_e} = \frac{C^2 \Delta V^2}{2W_e} \Rightarrow C = \frac{2W_e}{\Delta V^2} \tag{4.7}$$

So,

$$C = \frac{2W_e}{V_0^2} \tag{4.8}$$

where: W_e is the stored electrostatic energy density and V_0 is the applied voltage to the positive electrode. The negative electrode is kept at 0 V.

The analysis using COMSOL was carried out to observe how optimum number of negative electrodes between two positive electrodes of interdigitated configuration contributes to highest sensitivity measurement. Analysis was conducted for different number of negative electrodes between two positive electrodes. All electrodes were modelled to have the same pitch (250 μm), length (4750 μm) and width (125 μm). The calculated capacitance and reactance as a function of number of negative electrodes between two positive electrodes obtained from simulation are shown in Fig. 4.11 and Fig. 4.12 respectively. It was observed number of negative electrodes between five to thirteen has the highest capacitance and consequently the lowest reactance value. Further analysis was conducted to evaluate the distribution of electric field and to measure the electric field intensity for all three sensors. Fig. 4.13 shows the simulation results of electric field distribution of each sensor. As for Sensor 1_11, the magnitude of electric field strength is stronger at neighbouring electrodes but as it goes to the middle it becomes weak. It is also shown in Fig. 4.13 that the electric field strength becomes weak as it goes up. The same results were observed from simulation of Sensor 1_5 and Sensor 1_3 but with different pattern. It is shown that optimum numbers of negative electrodes between positive electrodes will give better distribution of electric field but comparatively weak in field strength. The electric field strength is sufficient to interact with material under test.

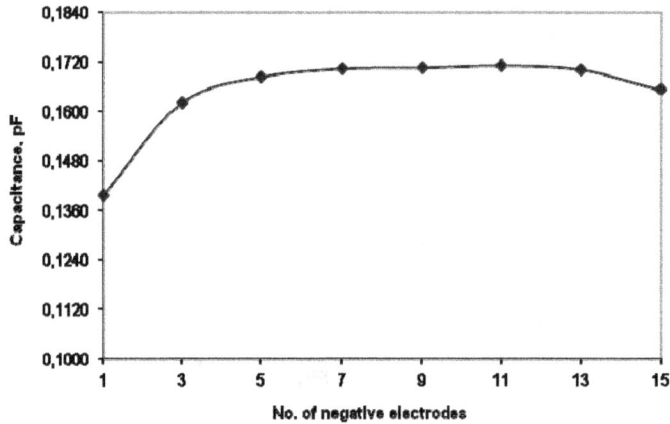

Fig. 4.11. The relationship between Capacitance and no. of negative electrodes.

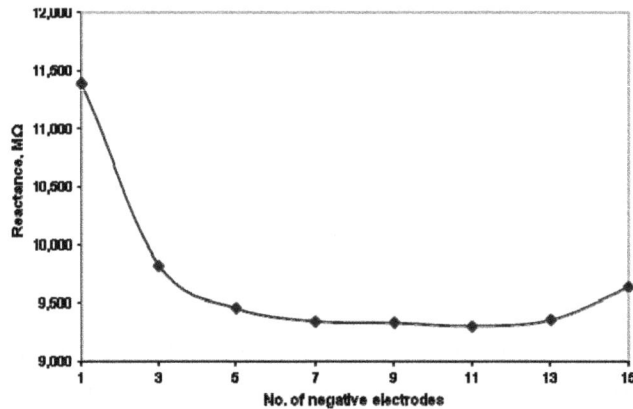

Fig. 4.12. The relationship between Reactance and no. of negative electrodes.

(a)

(b)

(c)

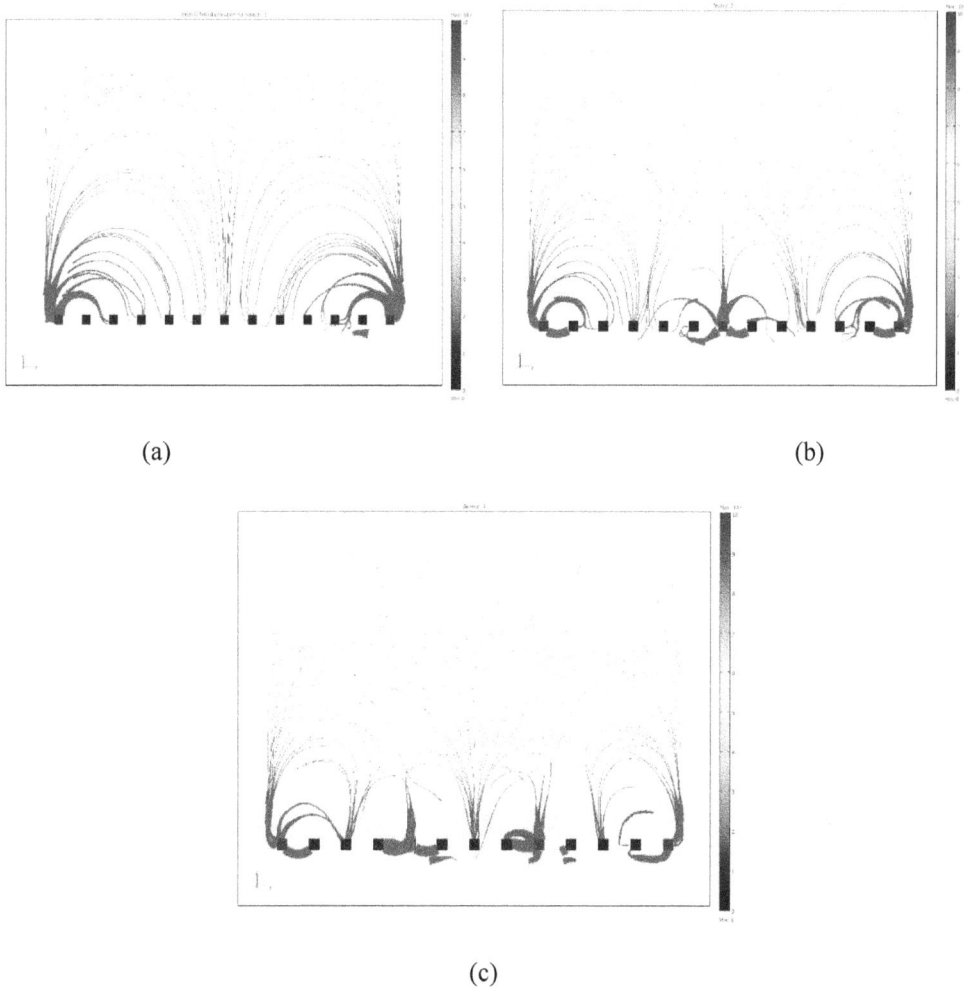

Fig. 4.13. The electric field distribution of each sensors; (a) Sensor 1_11, (b) Sensor 1_5 and (c) Sensor 1_3.

Variation of electric field intensity was then analyzed for each sensor designed. Electric field intensity is a measurement of electric field (V/m) along the z-axis. The cross-section line data were set starting from coordinates X0=2.75 mm, Y0= 0 mm, Z0= 0.125 mm and X1=2.75 mm, Y1=4.75 mm, Z1=0.125 mm. The coordinate of Z0 and Z1 were replaced from 0.125 mm to 3.50 mm and electric field intensity data were recorded. The average of electric field intensity is shown in Fig. 4.14. Sensor 1_11 has been chosen for analysis. Table 4.1 shows the calculated capacitance value of each sensor measured from the simulation.

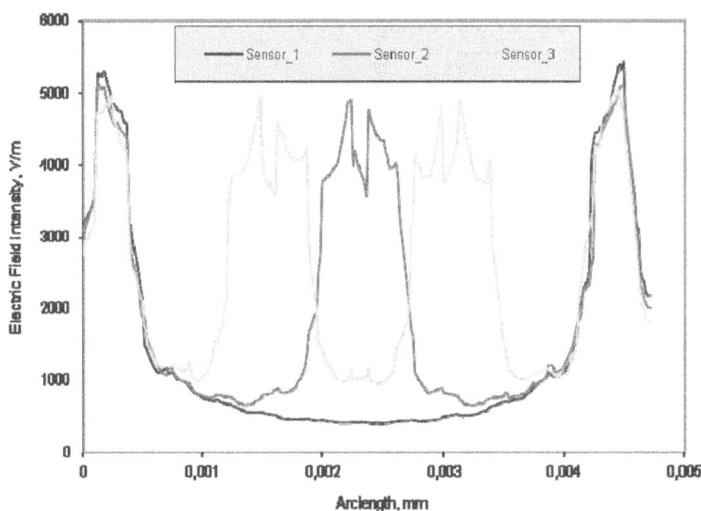

Fig. 4.14. Electric field intensity of each sensor.

Table 4.1. Capacitance Calculated from the Modeling using COMSOL Multiphysics for Interdigital Sensors.

Sensor Type	Simulation Parameters			Calculated Parameters
	Ve (V)	Freq (kHz)	Electrostatic energy density (J)	Capacitance (pF)
Sensor 1_11	10.0	10.0	8.5421×10^{-11}	1.709
Sensor 1_5	10.0	10.0	1.3823×10^{-11}	2.765
Sensor 1_3	10.0	10.0	1.7091×10^{-10}	3.418

The analysis was conducted by placing a sample into the model. Two analyses have been conducted. The first analysis was to observe the relationship of capacitance for different values of effective permittivity (ε_r) of sample while maintaining the same thickness (1 mm). The second analysis was to evaluate the capacitance produced by the sensor with different sample thickness while keeping the permittivity value constant (air gap, $\varepsilon_r = 1$ and sample, $\varepsilon_r = 3.6$). The 3D model for the analysis is shown in Fig. 4.15. The model consists of Sensor 1_11, a sample and a box with constant height of 3 mm (enclosed system). The electric field passing through the sample and the effect of permittivity of sample and air are shown in Fig. 4.16. From Fig. 4.17, it is seen that the capacitance has increased with the increase in permittivity of the sample. The increase of capacitance is not proportional to sample permittivity due to significant influence of air. Result in Fig. 4.18 shows that the capacitance value decrease as the sample thickness increases.

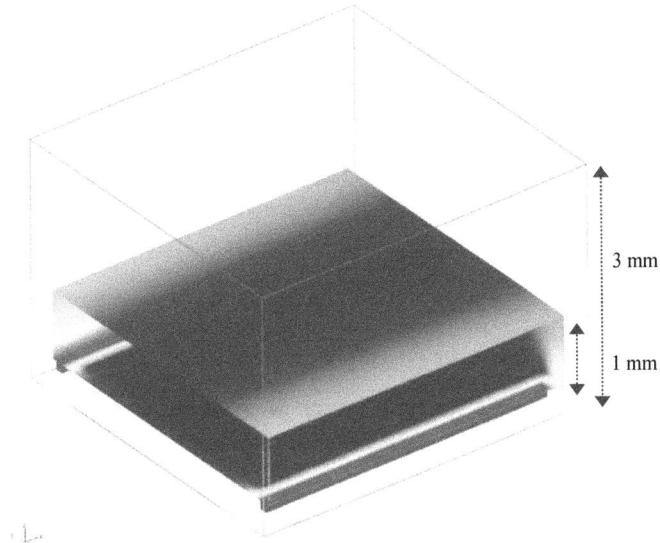

Fig. 4.15. 3D view of Sensor 1_11 with a known sample.

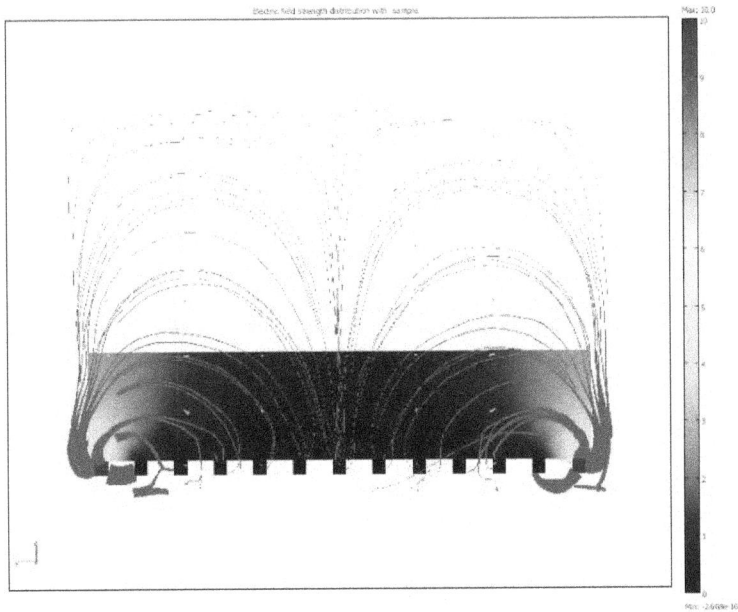

Fig. 4.16. Electric field distribution of Sensor 1_11 for a known sample.

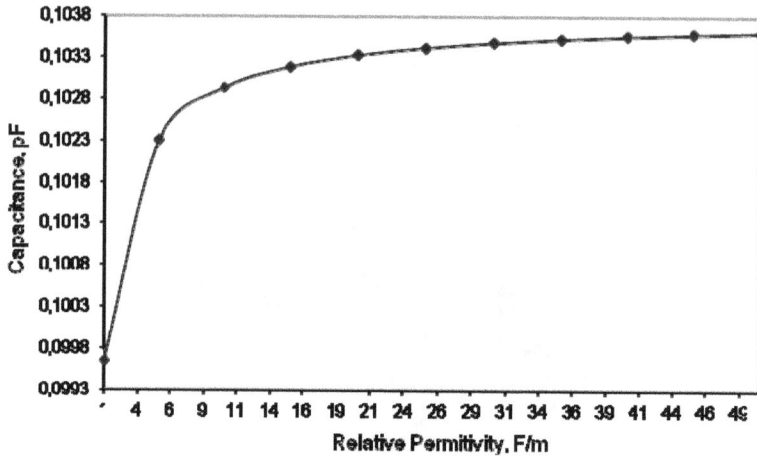

Fig. 4.17. Calculated capacitance for different values of sample permittivity.

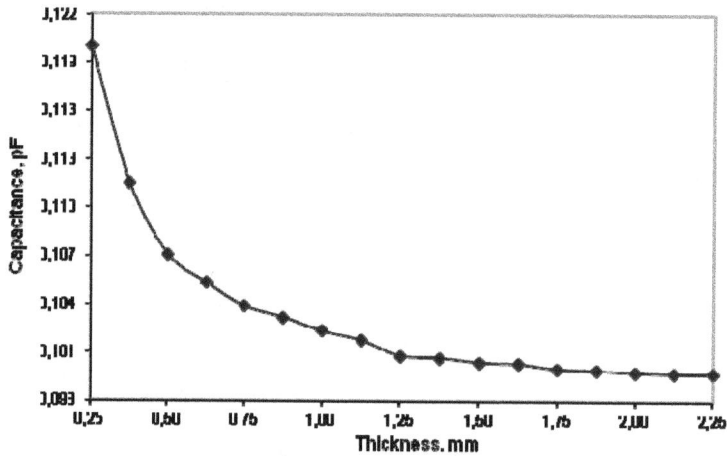

Fig. 4.18. Calculated capacitance for different values of sample thickness.

4.3.3. Sensor Design and Fabrication

4.3.3.1. Design and Fabrication Process

The first sensors were design using Altium Designer 6 software. The fabrication process of the sensors was done in Massey Electronics Laboratory. The final design of each sensor from the Altium Designer was printed in negative and then was transferred into a film. The conducting layers of the board are typically made of thin copper foil. The insulating layers (dielectric) are typically laminated together with epoxy resin pre-impregnated. The film together with the board was exposed to the UV light. This process will impress and burn the desired sensor design onto the board. The printed circuit was developed. The printed circuit board was immersed into a special chemical for etching process to remove the unwanted copper, leaving only the desired copper trace. The sensor was cut to a desired design to make it suitable for testing.

4.3.3.2. Conventional Interdigital Sensors

Initial part of the research works were involving designed and fabricated of three conventional planar interdigital sensors. The initial goal is to understand how conventional interdigital sensor works and respond to materials. All three sensors were designed with different configurations, so that the difference of their response can be evaluated. Each sensor was designed to have same effective area of 5000 µm by 5000 µm but with different pitch lengths of 250 µm, 510 µm and 1020 µm. The negative and positive electrodes have the same length of 4750 µm and width of 250 µm. Table 4.2 shows the parameters for conventional interdigital sensors. Fig. 4.19 shows the representation of conventional interdigital sensor with configuration #1 (Din_10mil) and Fig. 4.20 shows the representation of conventional interdigital sensor with configuration #2 (Din_20mil) and configuration #3 (Din_40mil). Fig. 4.21 shows the fabricated conventional interdigital sensors.

Table 4.2. Conventional interdigital sensor parameters.

Sensor	Sensing area, (mm²)	Pitch length, (mm)	Number of electrodes	
			Positive	Negative
Din_10mil	25	0.25	5	6
Din_20mil	25	0.51	4	4
Din_40mil	25	1.02	2	3

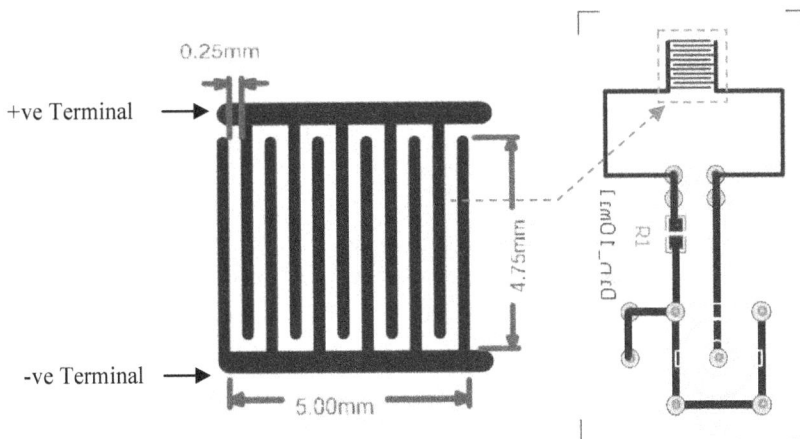

Fig. 4.19. Representation of conventional interdigital sensor with configuration #1.

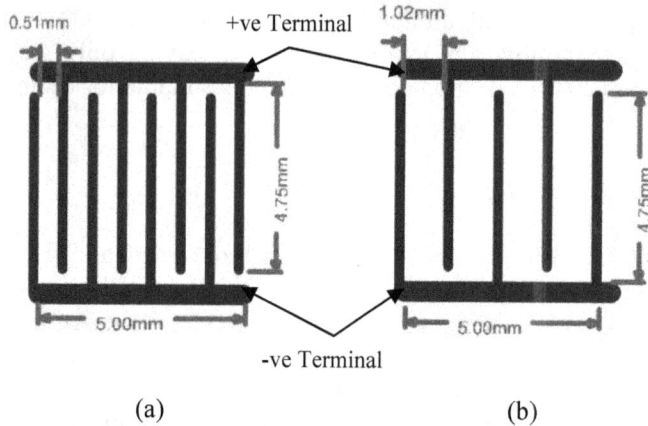

Fig. 4.20. Conventional sensor with (a) Configuration #2, and (b) Configuration #3.

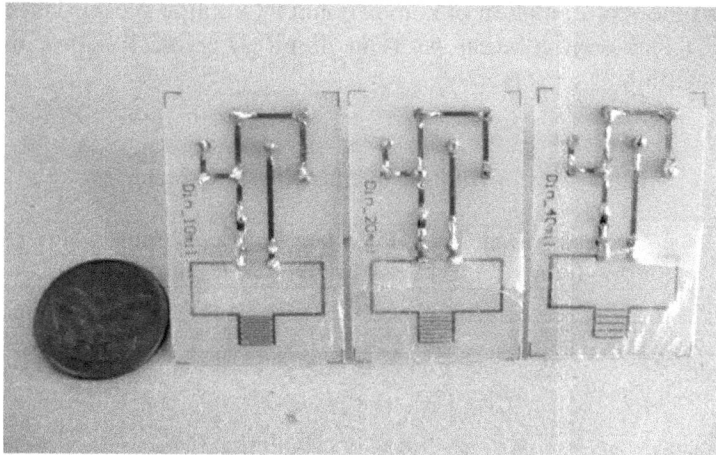

Fig. 4.21. The fabricated conventional interdigital sensors compared to 20¢ New Zealand coin.

4.3.3.3. Novel Planar Interdigital Sensors

All sensors were designed to have same effective area of 4750 μm by 5000 μm and having pitches of 250 μm. The positive and negative electrodes have the same length and width of 4750 μm and 125 μm respectively. The sensors were fabricated on three different materials with different fabrication process. The initial (first) sensors were fabricated on FR4 at Massey University using the normal printed circuit board technology. All sensors have equal numbers of electrodes (thirteen). The only parameter changing in the sensor design is the d, spacing between two adjacent positive and negative electrodes between which the electric field-lines exist. Sensor 1_11 was designed to have two positive electrodes at each end separated by eleven negative electrodes. Sensor 1_5 and Sensor 1_3 were designed with the same dimensions but with different configurations. Sensor 1_5 has five negative electrodes between two positive electrodes and has the same pitch like Sensor 1_11. Sensor 1_3 has three negative electrodes between two positive electrodes. Table 4.3 shows the parameters for novel interdigital sensors. Fig. 4.22 shows the representation of sensor configuration #1 (Sensor 1_11). Fig. 4.23 shows sensor configuration #2 (Sensor 1_5) and sensor configuration #3 (Sensor 1_3). The fabricated novel interdigital sensors are shown in Fig. 4.24.

Table 4.3. Novel interdigital sensor parameters.

Sensor	Sensing area, (mm²)	Pitch length, (mm)	Number of electrodes	
			Positive	Negative
Sensor 1_11	23.75	0.125	2	11
Sensor 1_5	23.75	0.125	3	10
Sensor 1_3	23.75	0.125	4	9

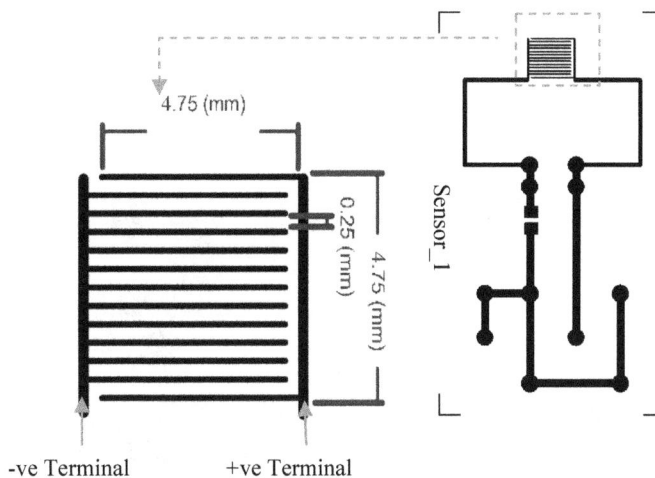

Fig. 4.22. Representation of Sensor 1_11 configuration.

The second sensors were fabricated at Northern Illinois University, United States using alumina as it platform. The sensors were constructed using standard thick film printing methodologies. The pattern was drawn in AutoCad and then printed on a Fire 9500 photoplotter. The patterns were then transferred to three 325 mesh screens. The printer used in the fabrication was a Presco 435, which is capable of printing up to a 100 mm × 100 mm substrate. The top trace layer was printed using a PdAg 850 °C firing alloy while the ground plane was made using an 850 °C silver material. The solder dam was made using a 600 °C low k dielectric. The wet paste was dried and fired after each individual layer was printed. The paste was dried at 150 °C then inspected. The dried substrate was then placed on the belt of a BTU firing furnace profiled to deliver

850 °C $\pm$ 5°C for 10 minutes. The entire firing cycle is approximately 30 s. The substrates were then patterned for the next layer and the cycle was repeated. The final layer is a low k dielectric to limit the solder flow during the mounting of a chip resistor. Fig. 4.25 shows the design of fabricated sensors which were fabricated at Northern Illinois University, United States. Fig. 4.26 shows the fabricated sensors.

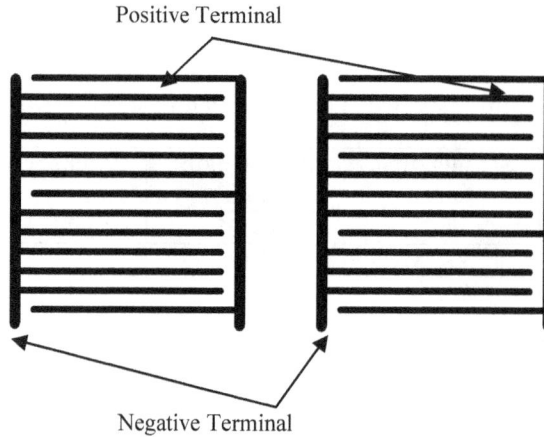

Fig. 4.23. Sensor 1_5 and Sensor 1_3 with different configurations.

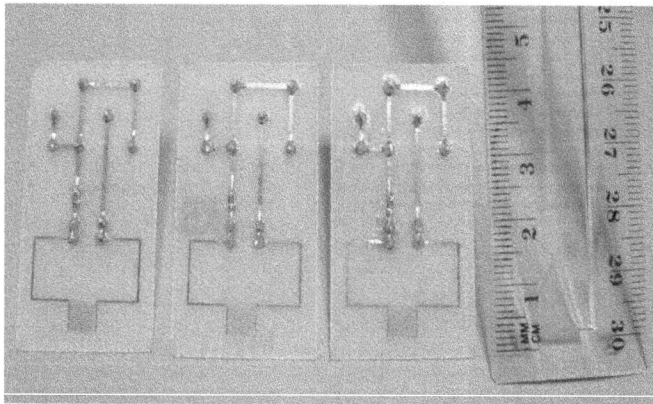

Fig. 4.24. The fabricated novel interdigital sensors (FR4).

The third sensors were fabricated on special glass using MEMS technology and were fabricated in Taiwan at Southern Taiwan University. The plastic photo mask design was used because of low cost and easy to prepare. The glass slip (sensor platform) was first cleaned with acetone. Then, an ultrasonic cleaning machine was used to clean the glass in methanol solution. N_2 gun was used to dry the glass slide. The IDT electrode fabrication process starts by depositing the Cr or Au on the glass slide using E-beam evaporated machine. Then, the positive photo-resist (PR) was coated on the metal layer by using spin coater. The coated metal layer was exposed to the UV light. The Developer machine was used to remove the unexposed PR. The etching process was done to remove the surplus metal and also to remove the remaining PR leaving only the patterned electrode. Fig. 4.27 shows the IDT electrode fabrication process. Fig. 4.28 shows the representation of interdigital sensor which were using MEMS and have been fabricated at Southern Taiwan University, Taiwan. Fig. 4.29 shows the fabricated interdigital sensors using MEMS technology. Fig. 4.30 shows the three different designs of fabricated novel interdigital sensors.

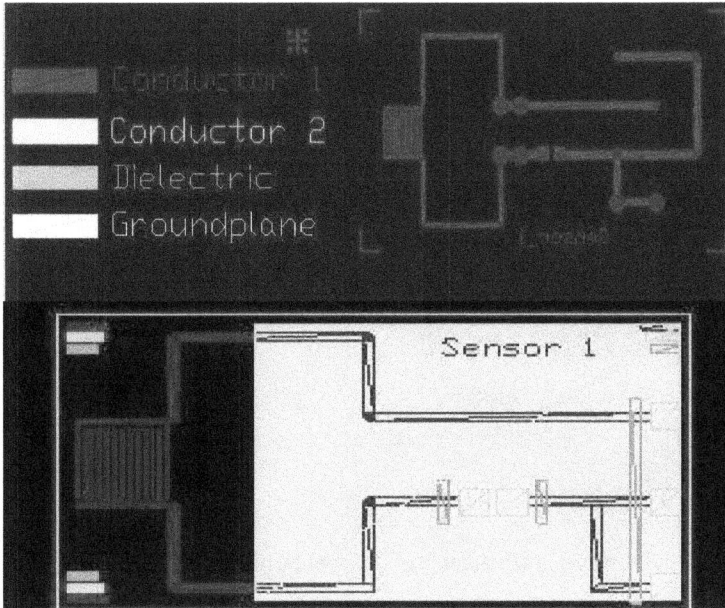

Fig. 4.25. The representation of interdigital sensor (Sensor 1_11) fabricated in USA.

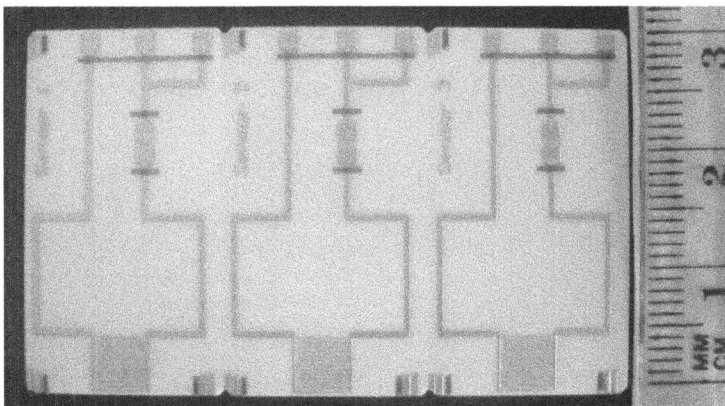

Fig. 4.26. The fabricated interdigital sensors (Alumina) in USA.

107

(a) Deposit Au/Cr layer

(b) Coat PR - S1813

Glass

Au

Cr

PR-S1813

(c) Expose and Develop

(d) Etching Electrode

(e) Remove PR

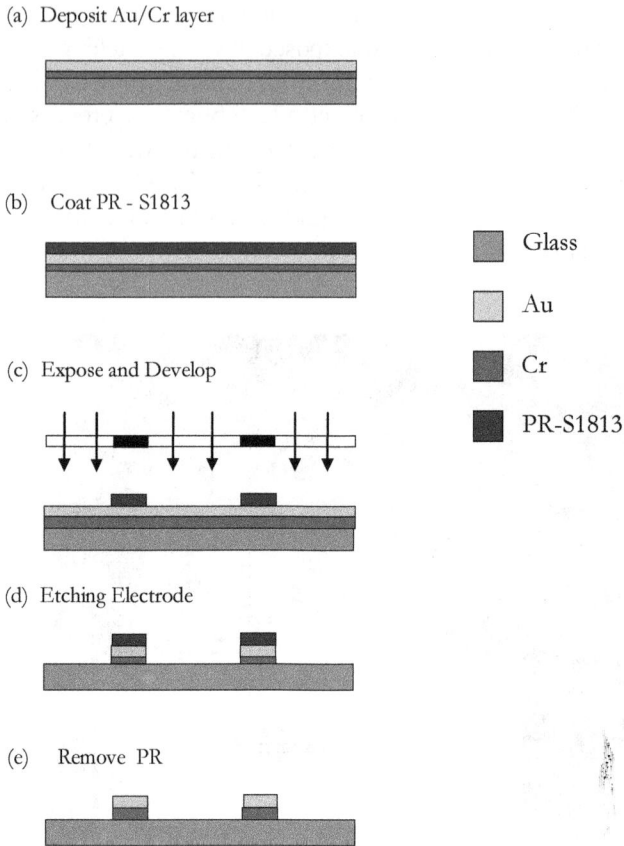

Fig. 4.27. The IDT electrode fabrication process at Southern Taiwan University, Taiwan.

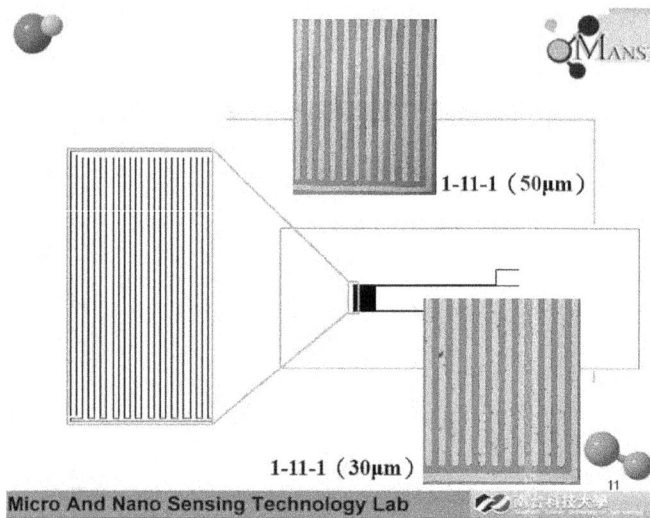

1-11-1（50μm）

1-11-1（30μm）

Micro And Nano Sensing Technology Lab

Fig. 4.28. The representation of interdigital sensor (Glass) fabricated in Taiwan.

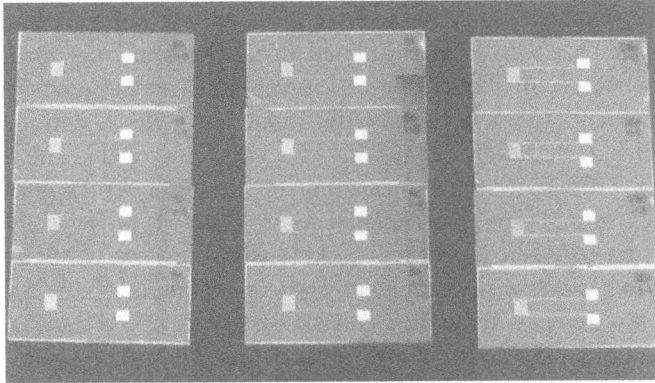

Fig. 4.29. The fabricated interdigital sensors using MEMS technology.

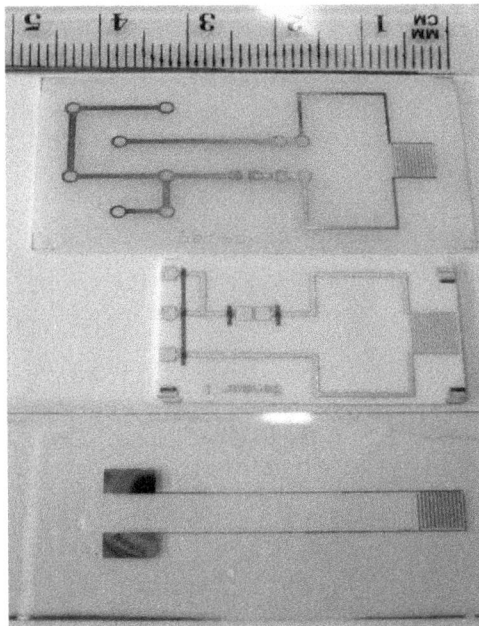

Fig. 4.30. The representation of three different designs of novel interdigital sensors.

4.4. Experimental Results and Discussions

4.4.1. Characterization of Sensors only without Material under Test (MUT)

All different fabricated sensors were characterized based on Impedance Spectroscopy (IS) method. The IS is a powerful technique used to evaluate electrical properties of materials and their interfaces with surface-modified electrodes [89]. Recent method of detection in pathogen sensing has been discussed by Heo et. al, [90] which utilized the applications of interdigital sensors with impedance spectroscopy measurement. The first experiments were conducted to characterize the sensors alone without MUT. Each developed sensors has difference impedance characteristic because of their configurations, substrate, different fabrication methods and others. The best sensors which have higher sensitivity will have higher capacitance value, uniform distribution of electric field, higher penetration depth with low impedance [91]. The Instek LCR-821 was linked to a LabVIEW auto measurement program to capture the electrical parameters of each sensor. The experimental setup is shown in Fig. 4.31. The operating frequencies were set between 12 Hz to 100 kHz with constant voltage of 1 Vrms. The LCR meter is set to have slow measurement for better accuracy (0.05 %) with average of three measurements for each frequency and all data were saved into *.csv format. The initial test is to calibrate the sensors with respect to air (without any chemicals). Since the sensors will response differently with different chemicals, it is important to establish the sensor characteristic with respect to air. The impedance characteristic of all sensors varies with frequency. With the increase in operating frequency of exciting voltage, the impedance decreases. The interdigital sensor connected in AC circuit is frequency dependent.

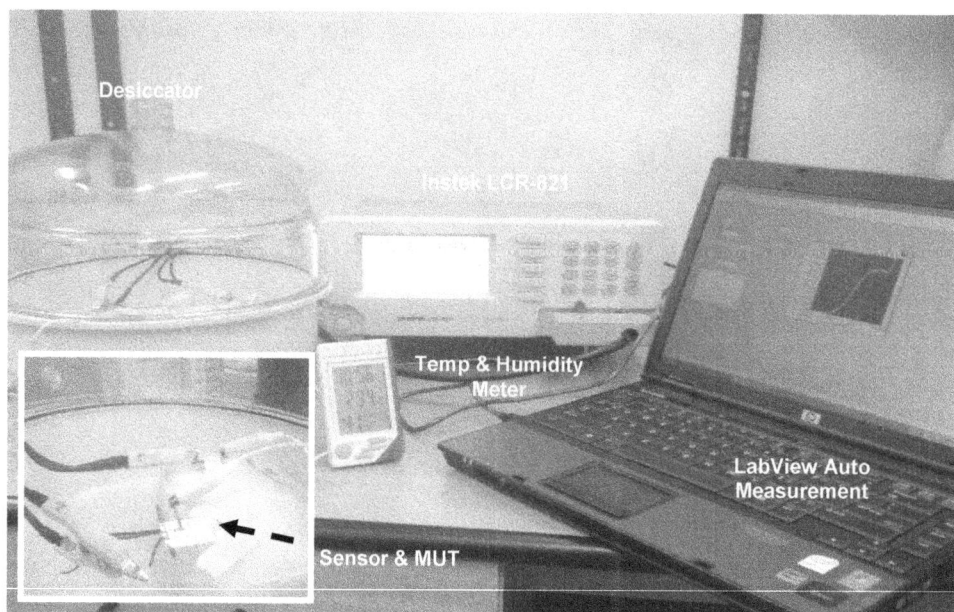

Fig. 4.31. Experimental setup for Impedance Spectroscopy (IS) measurement.

4.4.1.1. Characterization of FR4 Sensors

The impedance characteristics of FR4 sensors with three different configurations are shown in Fig. 4.32. Fig. 4.33 (a) and (b) show the bode plot of real part and the imaginary part of the impedance. The Nyqusit plot of the complex impedance plane for FR4 sensors is shown in Fig. 4.34. The impedance characteristics with respect to air were observed to have higher changed of imaginary part compared to real part. The phase angle measured were closed to 90 degree which indicates that the sensor behave as capacitive sensors. The Nyquist plot shows the real part change smaller compared to the imaginary part at certain frequency range. The sensor configuration 1_11 has higher impedance compared to sensor 1_5 and 1_3 for frequency less than 10 kHz.

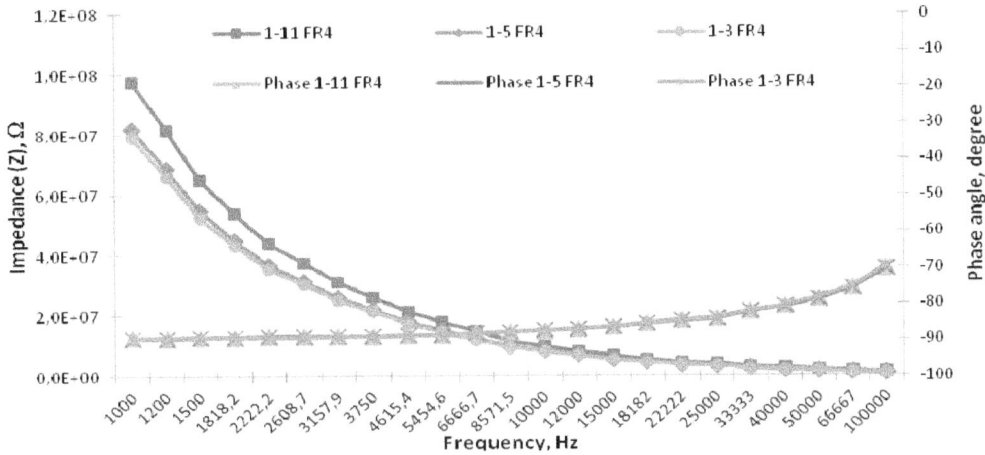

Fig. 4.32. Impedance and phase angle of FR4 sensors.

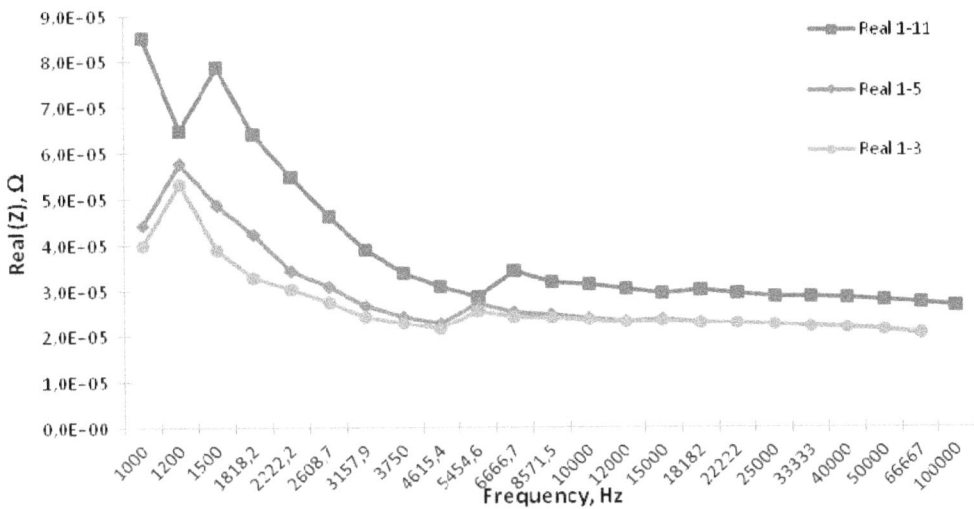

(a) Real part of FR4 sensor.

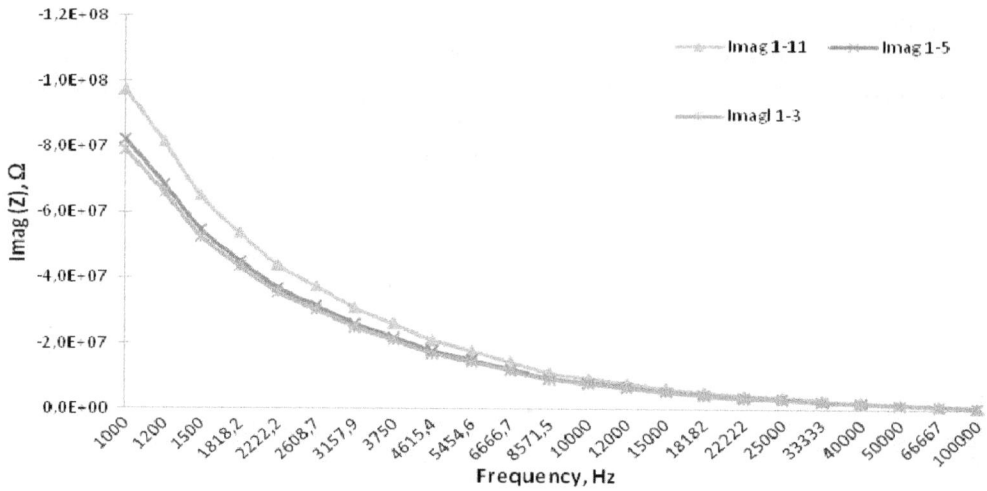

(b) Imaginary part FR4 sensor

Fig. 4.33. Bode plot of FR4 sensors showing (a) Real part, and (b) Imaginary part.

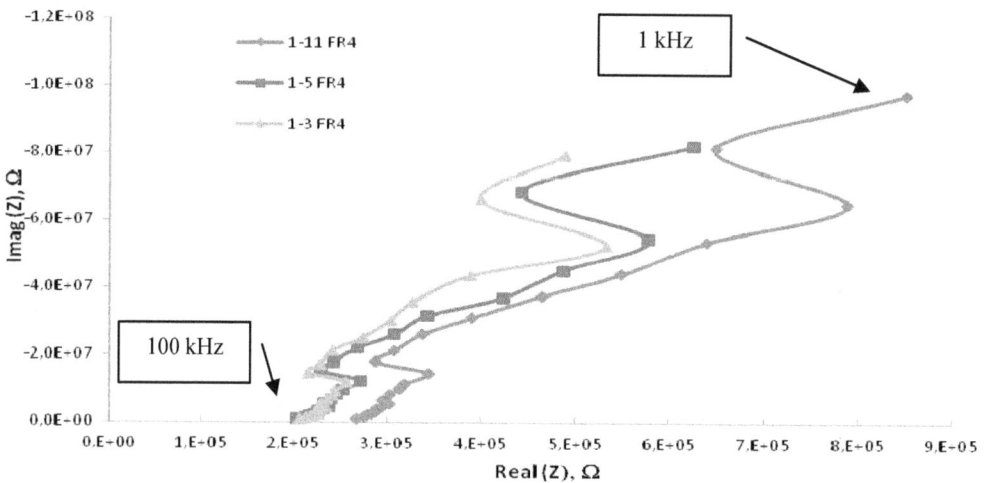

Fig. 4.34. Nyquist plot of FR4 sensors with respect to air.

4.4.1.2. Characterization of Alumina Sensors

The impedance characteristics of Alumina sensors are shown in Fig. 4.35. Fig. 4.36 (a) and (b) show the real part and the imaginary part of the impedance. The Nyquist plot is shown in Fig. 4.37. Alumina sensors also shown that the changed of imaginary part is higher compared to the real part. The phase angle measured were closed to 90 degree which indicates that the sensor behave as capacitive sensors. The sensor configuration 1_11 has higher impedance compared to sensor 1_5 and 1_3 for frequency less than

10 kHz. The impedance characteristic for Alumina sensors discriminate clearly for different configurations compared to FR4 sensors. The impedance for all configurations also lower compared to FR4 sensors which indicated that the different substrates have significant effect on the sensing performance.

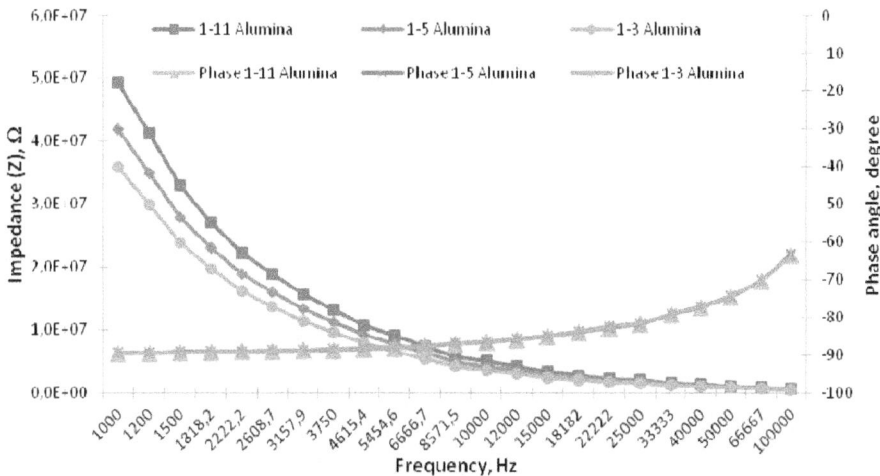

Fig. 4.35. Impedance and phase angle of Alumina sensors.

4.4.1.3. Characterization of Glass Sensors

The impedance characteristics of Glass sensors with three different configurations are shown in Fig. 4.38. Fig. 4.39 (a) shows the real part of the impedance while Fig. 4.39 (b) shows the imaginary part of the impedance. Fig. 4.40 shows the Nyqusit plot of Glass sensors with respect to air. The same characteristic were observed for Glass sensors for different configurations. The results for Glass sensors were observed to have less discrimination between configurations for imaginary part compared to results shown for Alumina sensors but has lower impedance compared to FR4 sensors.

Analyses of results from sensors characterization have shown that the change of resistive part of all sensors is small compared to the imaginary part of the sensors. From the figures, it can be said that the interdigital sensors will have better sensing performance as it produces a large change of the imaginary part. The phase measurement at different frequencies shown that all sensors have the phase angle closed to 90° at a particular low frequency and as the frequency increase the phase angle decreases. As for the interdigital sensor, at low frequencies the sensor will behave better as a capacitive sensor. Fig. 4.41 shows the impedance characteristic of all fabricated sensors. The sensors fabricated on FR4 materials create very high impedance as compared to the sensors fabricated on Alumina and Glass. Based on the observation results from impedance characteristic, the Alumina sensors has lower impedance value (higher capacitance) and better phase angle measurement compared sensors fabricated on Glass and on FR4. Analyses of characteristic impedance of fabricated sensors with air have shown that Alumina sensors will have better sensing performance. Therefore, only Alumina sensors were chosen for further experiments.

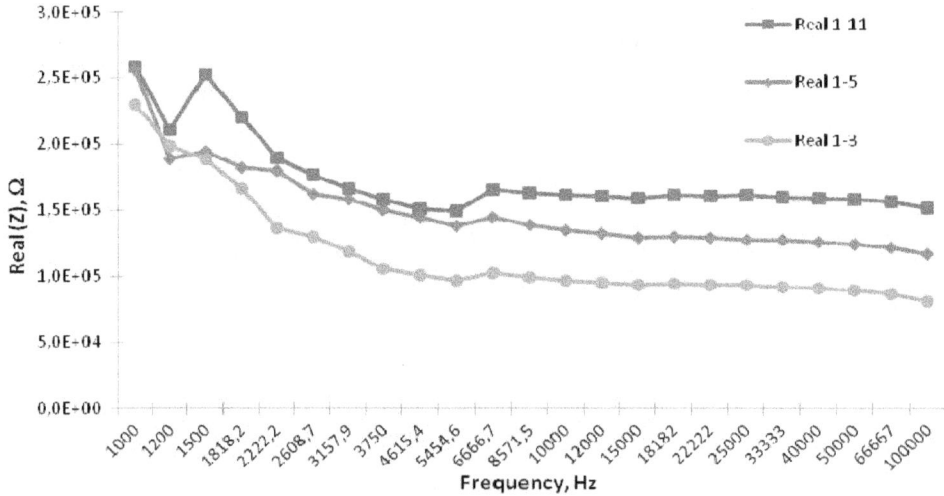

(a) Real part of Alumina sensors.

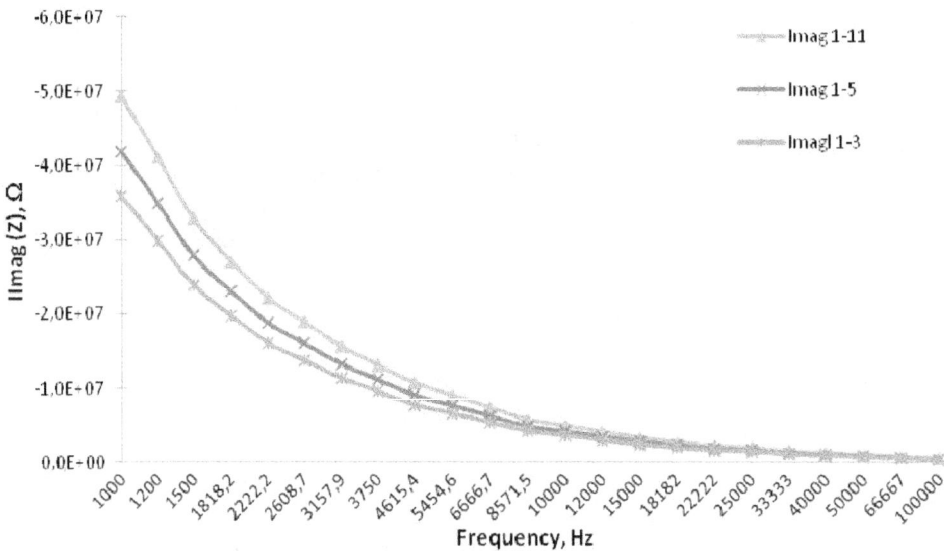

(b) Imaginary part of Alumina sensors.

Fig. 4.36. Bode plot of Alumina sensors showing (a) Real part, and (b) Imaginary part.

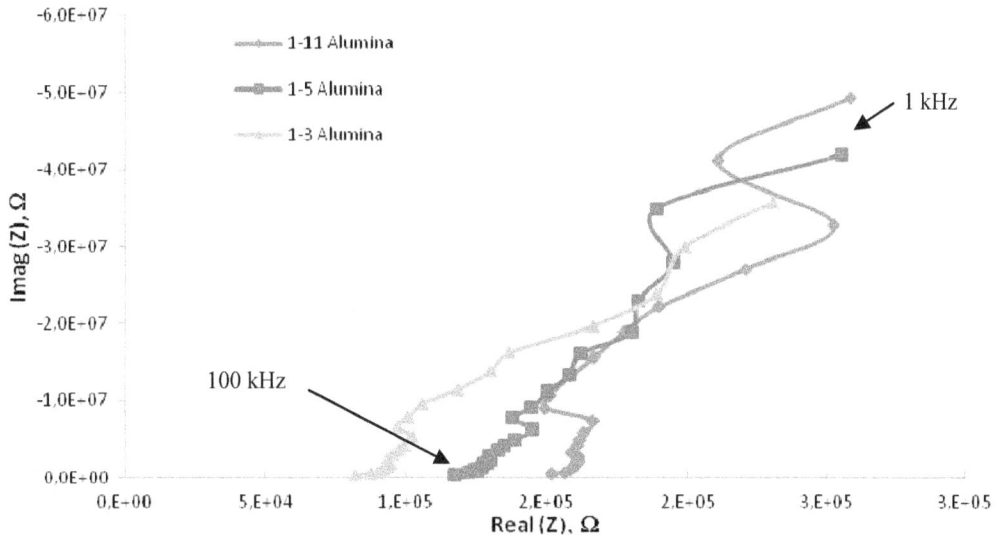

Fig. 4.37. Nyquist plot of Alumina sensors with respect to air.

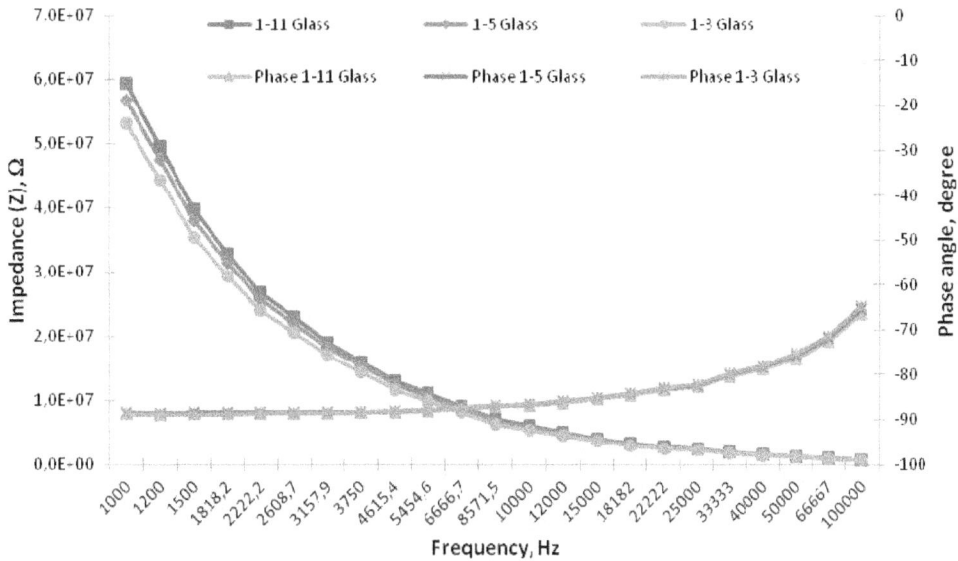

Fig. 4.38. Impedance and phase angle of Glass sensors.

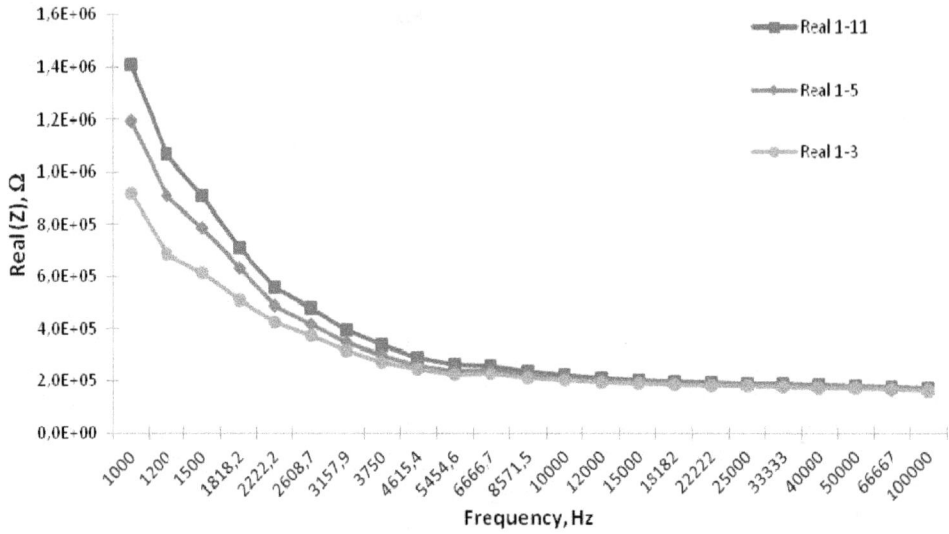

(a) Real part

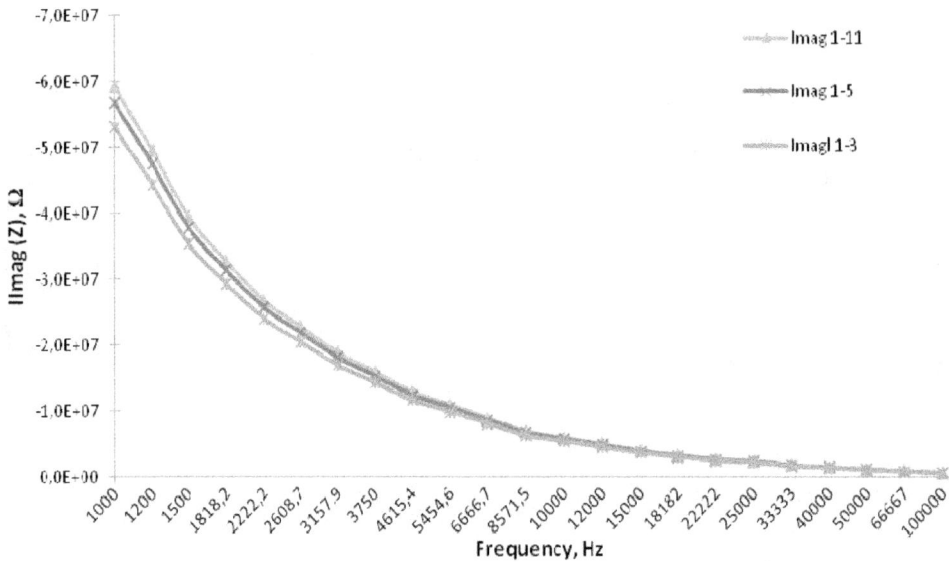

(b) Imaginary part

Fig. 4.39. Bode plot of Glass sensors showing (a) Real part, and (b) Imaginary part.

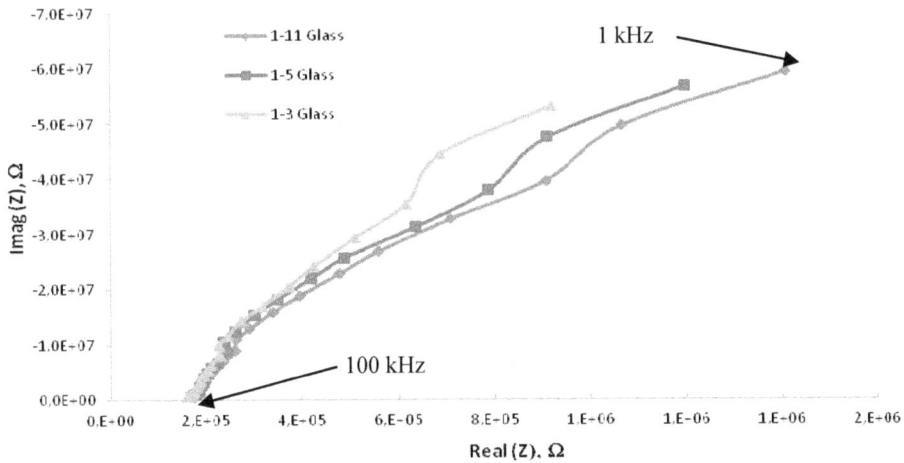

Fig. 4.40. Nyquist plot of Glass sensors with respect to air.

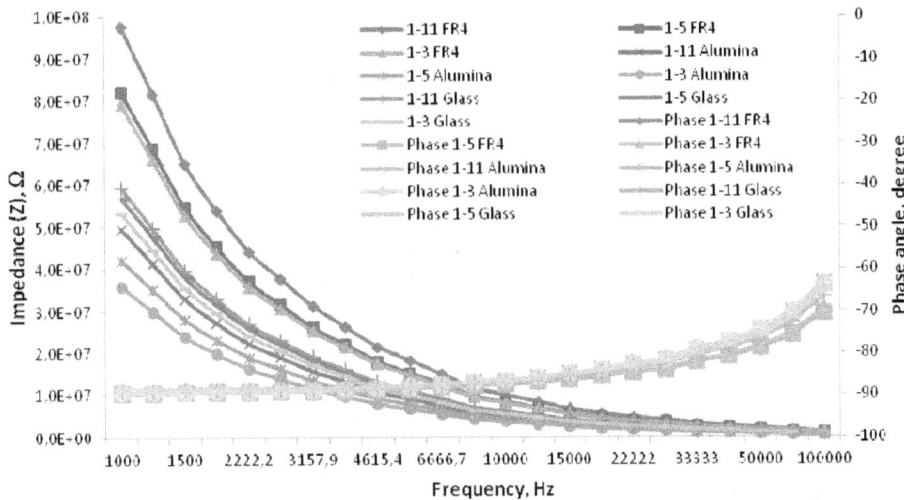

Fig. 4.41. Comparison of all fabricated sensors showing impedance characteristic and phase angle.

4.4.2. Characterization of sensors with ionic solution, Sodium Chloride (NaCl)

Characterization of Alumina sensors with a known ionic solution has been conducted. The LCR-821 interfaced with LabVIEW program was used for the experiment setup. The measurement was set to slow for 0.05 % accuracy with input voltage of 1 V_{rms}. The frequency range is set to 12 Hz – 100 kHz. The exposure time of the solution to sensor's surface is 7 minutes. The experiments were replicate three times and mean values of three runs were taken. The experiments were conducted to study the behaviour of different configuration of Alumina sensors response to the different molarity of ionic solutions. Sodium chloride (NaCl) of different molarity has been used. Sodium chloride is a salt, it is a chemical compound that consists of a positively charged cation, the sodium ion (Na^+), and a negatively charged anion, the chloride ion (Cl^-). Dissolve the sodium chloride in MilliQ water (control) for required molarity. Bring the volume to 1 litre with MilliQ water using volumetric plus. Finally, sterilize the solutions. The prepared NaCl solutions have different concentrations between 0.01 M to 1 M.

All sensors were coated with thin Incralac (5 µm – 10 µm) and cured overnight at room temperature. Incralac has been widely used as a coating material to protect metallic substrates and studies have been conducted using EIS (electrochemical impedance spectroscopy) to analyze its performance [92]. Fig. 4.42 (a) – (c) show the impedance characteristics of Alumina 1_11 sensor with NaCl solution. Impedance characteristics of Alumina 1_5 sensor with NaCl solution are shown in Fig. 4.43 (a) – (c). Fig. 4.44 (a) – (c) show the impedance characteristics of Alumina 1_3 sensor. Different configurations of Alumina sensors show different impedance characteristic. All sensors were observed could discriminate NaCl of different concentrations. From the analyses Alumina sensor with configuration 1_5 has shown better results compared to other configurations.

4.4.3. Experiments with Peptides Related to Domoic Acid

Experiments were conducted to analyze the sensor performance with pure chemicals which are related to food poisoning. The experiments were conducted in a desiccator to control the temperature and humidity. The experiments were conducted at room temperature between 23 °C – 26 °C with humidity between 40 % - 50 %. Two peptide derivatives namely sarcosine and proline were used for the initial studies, which are structurally and closely related to the target molecule. N-methyl glycine or sarcosine represents the simplest structure and the proline molecule is arguably the most important amino acid in peptide conformation, containing the basic structural similarity to the domoic acid. Fig. 4.45 shows the chemical structure of the target molecule. These peptides were readily available at laboratory and since the domoic acid is too expensive for reproducibility of the experiments, therefore, these two chemicals were chosen. Each chemical was diluted in MilliQ water (control) and the solutions were prepared for 100 mg/ml each. Small amount of 40 µl of each solution was taken for sensor measurement. The exposure time of 12 minutes was established after a few experiments with the same chemicals of same concentrations. These measurements were observed frequently for any changes in measurement signals. The sensitivity measurement can be calculated by

$$\%Sensitivity = \frac{|V_{Target}| - |V_{Control}|}{|V_{Control}|} \times 100, \tag{4.9}$$

where V_{Target} is the voltage sensing value of peptides on the sensor. $V_{Control}$ is the resulting voltage of control solution (MilliQ water) on the sensor.

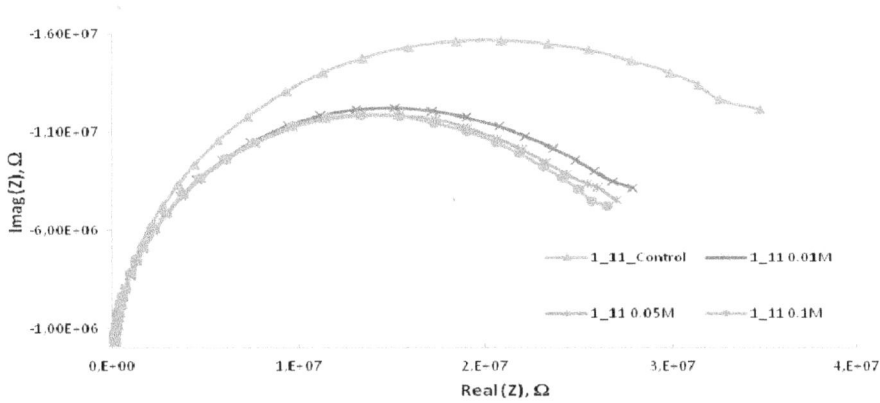

(a) 1_11: Nyquist plot.

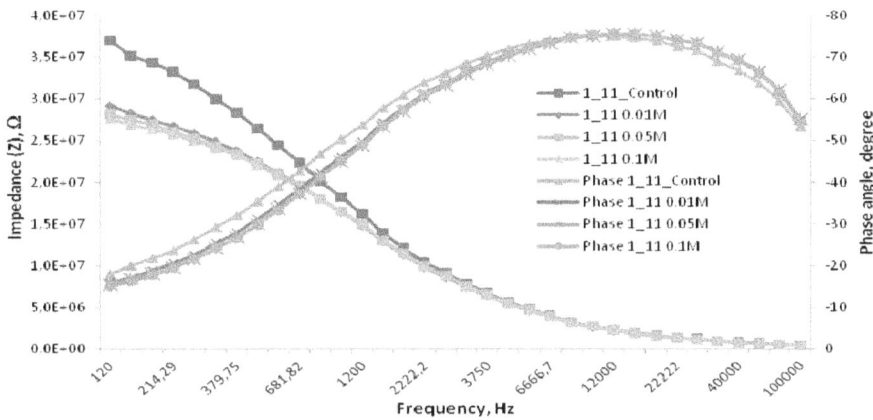

(b) 1_11: Impedance and phase angle.

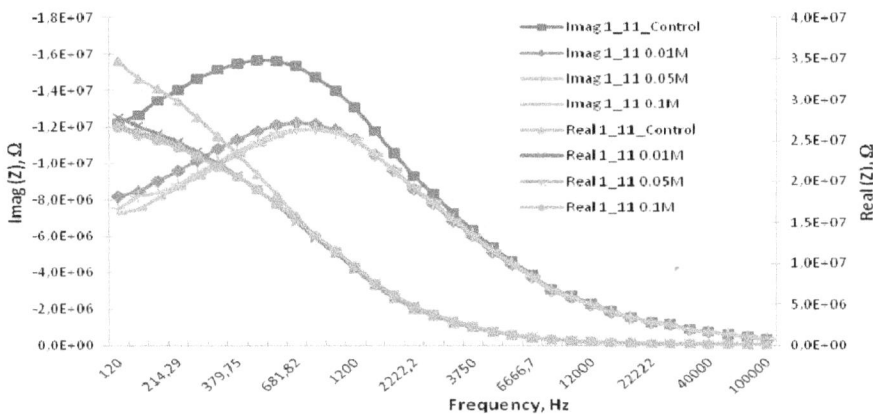

(c) 1_11: Real and Imaginary part.

Fig. 4.42. Characterization of Alumina1_11 sensor: (a) Nyqusit plot showing complex impedance plane; (b) Impedance and phase measurement plot; (c) Impedance behaviour in showing the Real, Z and the Imaginary, Z.

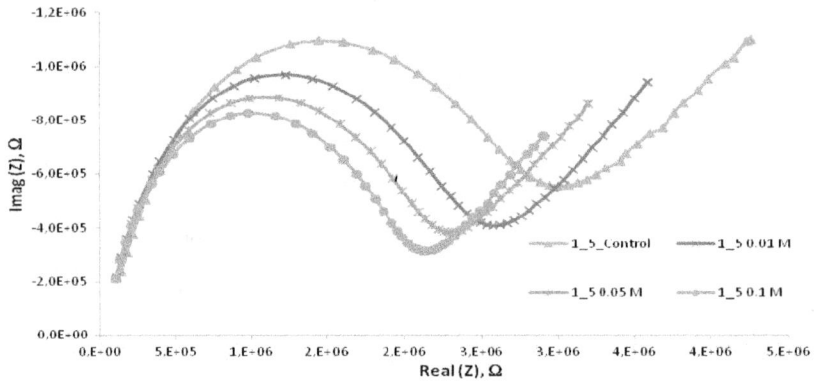

(a) 1_5: Nyquist plot.

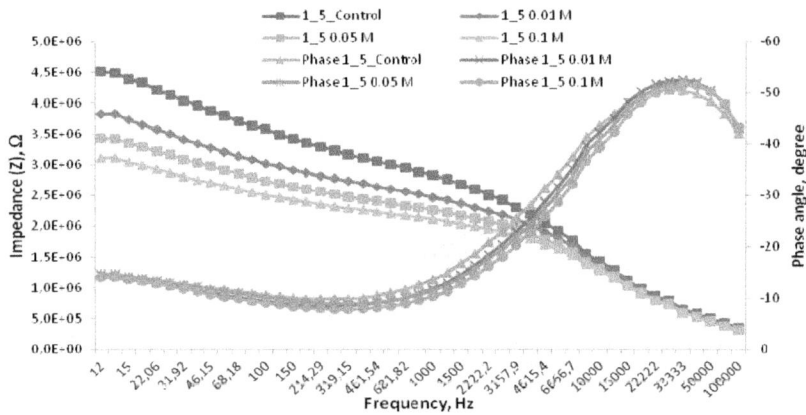

(b) 1_5: Impedance and phase angle.

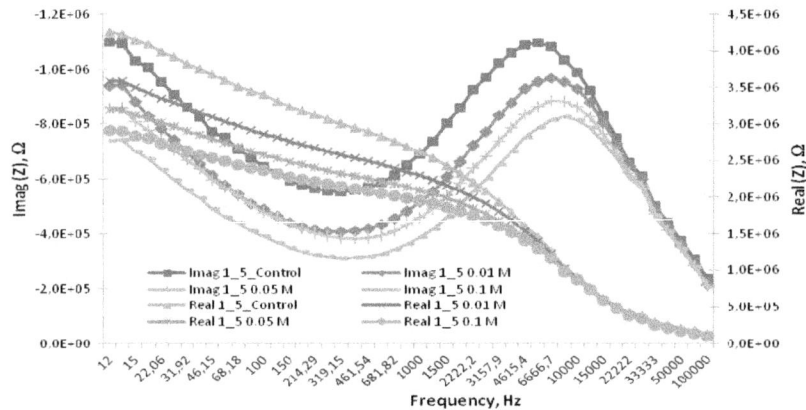

(c) 1_5: Real and Imaginary part.

Fig. 4.43. Characterization of Alumina1_5 sensor: (a) Nyqusit plot showing complex impedance plane; (b) Impedance and phase measurement plot; (c) Impedance behaviour in showing the Real, Z and the Imaginary, Z.

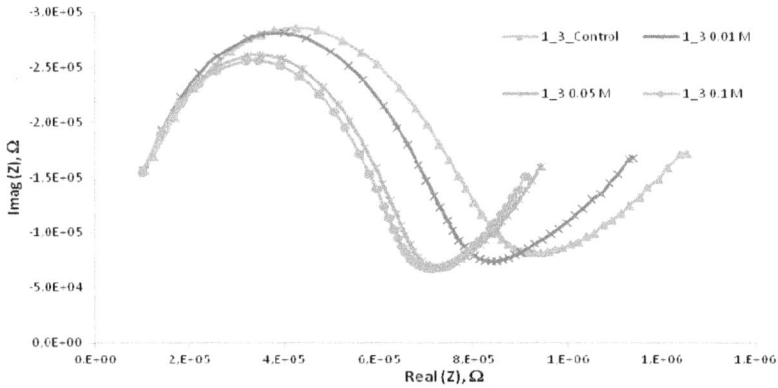

(a) 1_3: Nyquist plot.

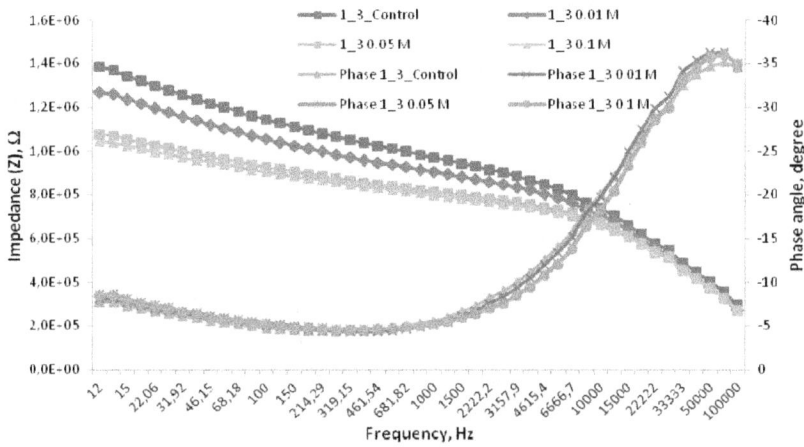

(b) 1_3: Impedance and phase angle.

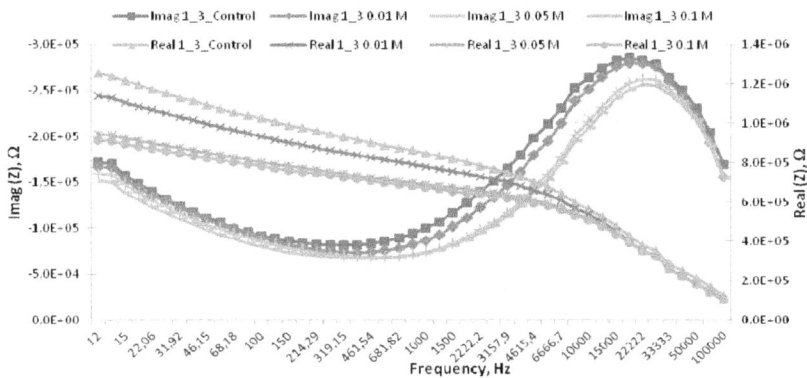

(c) 1_3: Real and Imaginary part.

Fig. 4.44. Characterization of Alumina1_3 sensor; (a) Nyqusit plot showing complex impedance plane (b) Impedance and phase measurement plot (c) Impedance behaviour in showing the Real, Z and the Imaginary, Z.

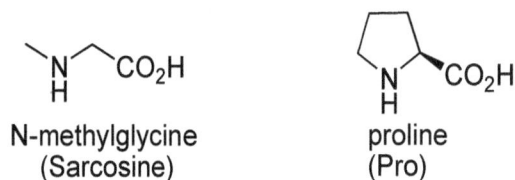

Fig. 4.45. The target molecules of two different peptides.

It was observed from the experimental results that Alumina sensors give a good response to the target molecules. Fig. 4.46 and Fig. 4.47 show the Nyquist plots of each Alumina sensor for different peptides. Only sensor with configurations 1_5 and 1_3 response very well to the peptides and it is possible to discriminate the different peptides from the output of the sensor. The Nyquist plot of 1_5 sensor has shown better result. Sensor 1_11 has shown less response to the peptides since it has higher impedance value which limits the mobility of electron flow through the system. The impedance behaviours of both configurations (1_5 and 1_3) are shown in Fig. 4.48 and Fig. 4.49. It is clearly observed that at a particular frequency range (0.3 kHz – 2 kHz) there is no change of real part whereas there is a significant change of the imaginary part of the system. The change of imaginary part is higher for 1_5 configuration as compared to 1_3, which indicates that 1_5 configuration has better sensing performance. The impedance characteristics and phase angle for different peptides and control for both sensors are shown in Fig. 4.50 and Fig. 4.51. The change of phase angle with respect to control solution is particularly small for different peptides shown for 1_5 sensors as compared to 1_3 sensors. Fig. 4.52 shows the percentage of sensitivity measurement for different peptides with respect to control. It is observed that the percentage difference of target molecules is higher for Alumina 1_5 sensor compared to 1_3 sensor which indicates that 1_5 configurations can distinguish better between different peptides. As for 1_11 sensor, since the measurement only valid after 120 Hz and above (because of very high impedance to be captured by the LCR meter between 12 Hz – 120 Hz), that is why in the figure it shows no change until the frequency reach 120 Hz and is it noticed that there is very small change of this percentage for 1_11 sensor. The Alumina 1_5 sensor has been chosen for further analysis for endotoxin detection.

4.4.4. Experiments with LPS O111:B4

Experiments were conducted to evaluate the Lipopolysaccharides (LPS) or endotoxins. The commercial product of LPS O111:B4 was purchased from Sigma-Aldrich, USA. This pure LPS product was extracted using phenol extraction which contains less than 3 % of protein [93]. 1 mg of O111:B4 is diluted into 1 ml of MilliQ water (control). The pH of each solution was measured using pH meter model 420A from Orion Research Inc with the new pHoenix Tuff Tip® Combination pH electrode model 5733534-003B. It was observer that the pH reading of MilliQ water was 6.50 and pH of LPS O111:B4 solution is 6.59. It can be said that the signal measurements from the experiments conducted using these pure LPS will not be affected by protein or pH of each samples.

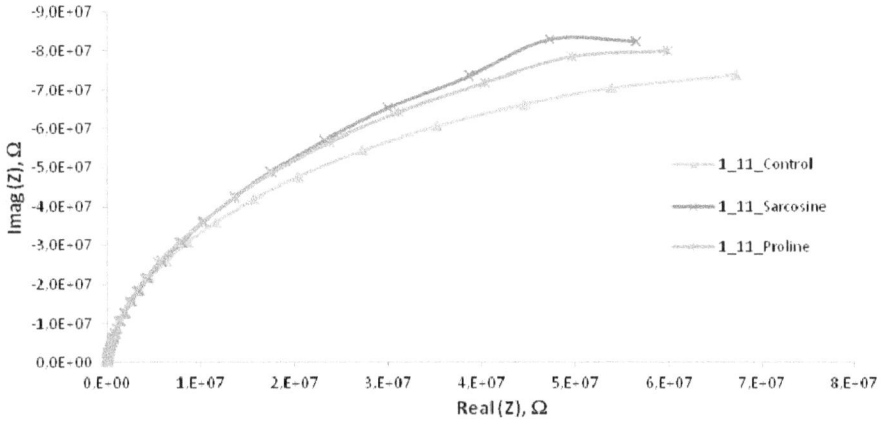

Fig. 4.46. Nyqusit plot of two different peptides and MilliQ (control) for Alumina, 1_11 configuration.

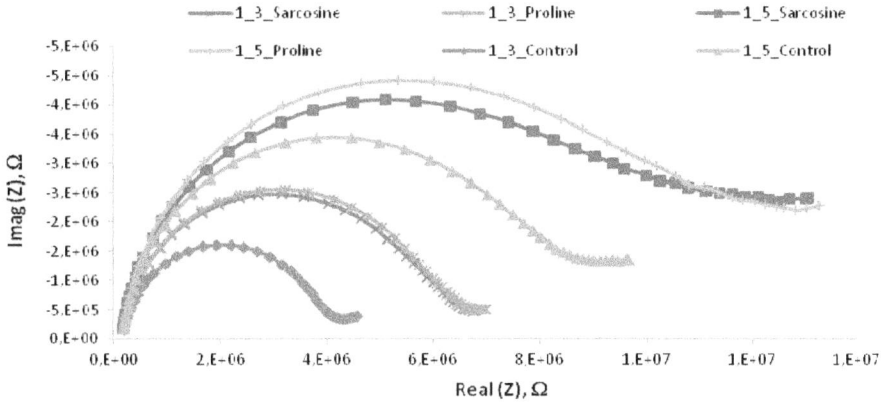

Fig. 4.47. Nyqusit plot of two different peptides and MilliQ (control) for Alumina, 1_5 and 1_3 configurations.

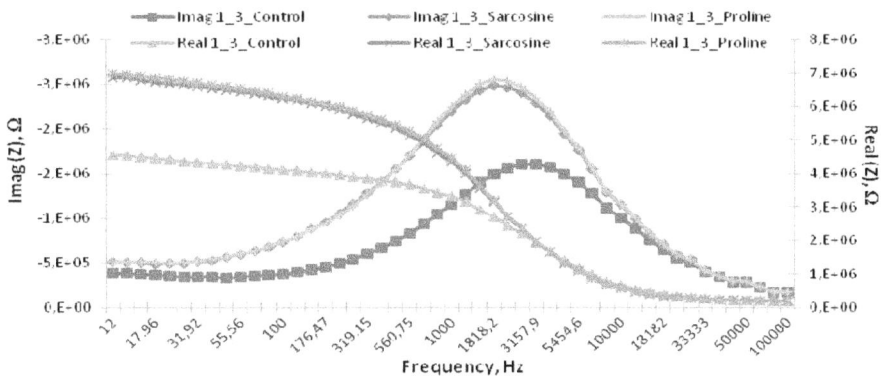

Fig. 4.48. Impedance behavior of Alumina 1_3 for Sarcosine, Proline and MilliQ (control).

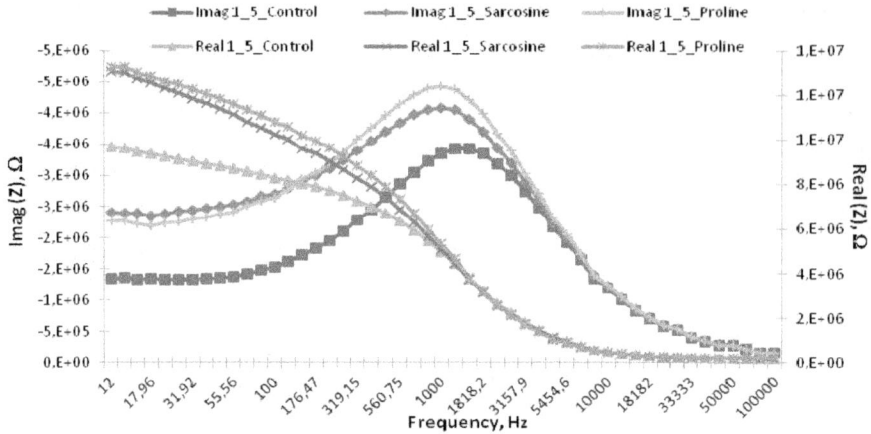

Fig. 4.49. Impedance behavior of Alumina 1_5 for Sarcosine, Proline and MilliQ (control).

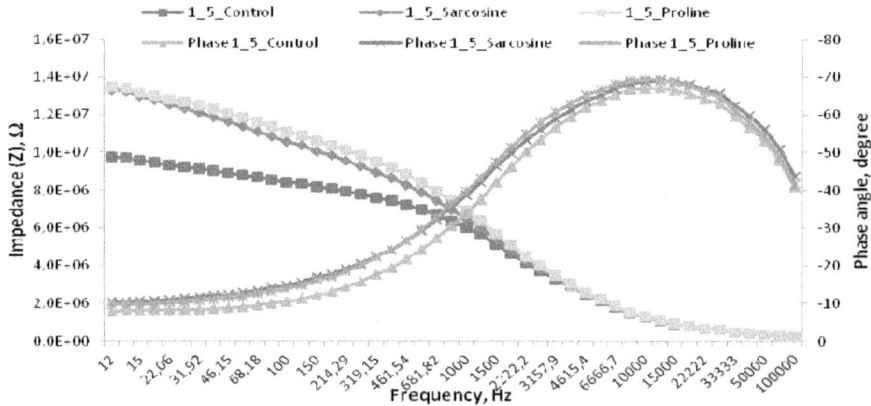

Fig. 4.50. Impedance and phase angle of different peptides and control for 1_5 sensor.

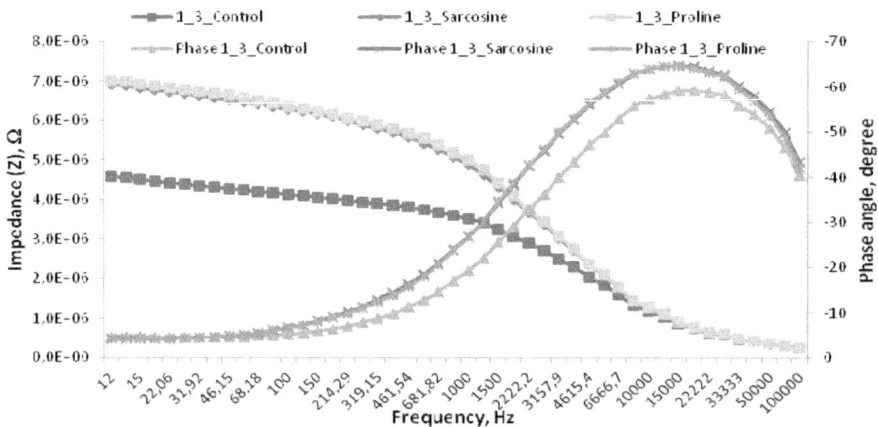

Fig. 4.51. Impedance and phase angle of different peptides and control for 1_3 sensor.

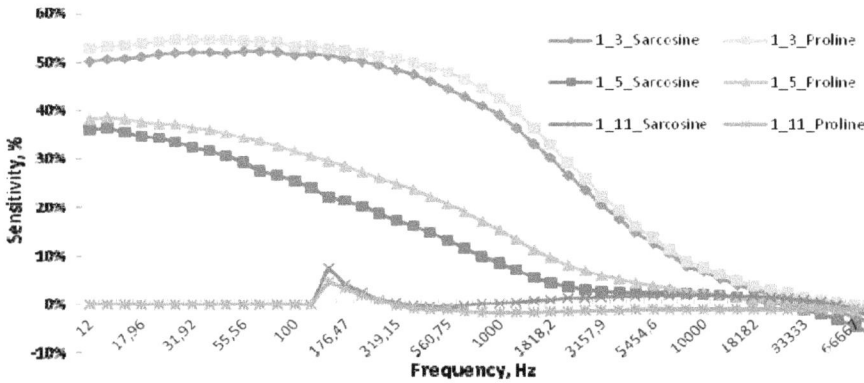

Fig. 4.52. Sensitivity measurement for different peptides with respect to MilliQ.

A small amount (40 µl) of each solution was pipetted using 10 µl pipettor onto the sensor. The change of impedance and the phase angle were recorded. Fig. 4.53 shows the Nyquist plot of the LPS O111:B4 and control (MilliQ). The sensor shows a significant result to discriminate between the target LPS with the control. The impedance behavior of the LPS O111:B4 is shown in Fig. 4.54. It is noticed that any change in absolute value of the real part of the signal is negligible at certain frequency between 1 kHz – 5 kHz. So the effect of conductivity does not have any influence on the signal. Moreover to eliminate the effect of water on the results, the amount of water used in the experiments has been kept constant. Therefore, the change of measured signal is due to the change of the effective permittivity of the solutions. Results from the experiments have shown that Alumina sensor with configuration 1_5 can be used to detect LPS or endotoxins. This sensor can be used to sense the contaminated food with dangerous toxins.

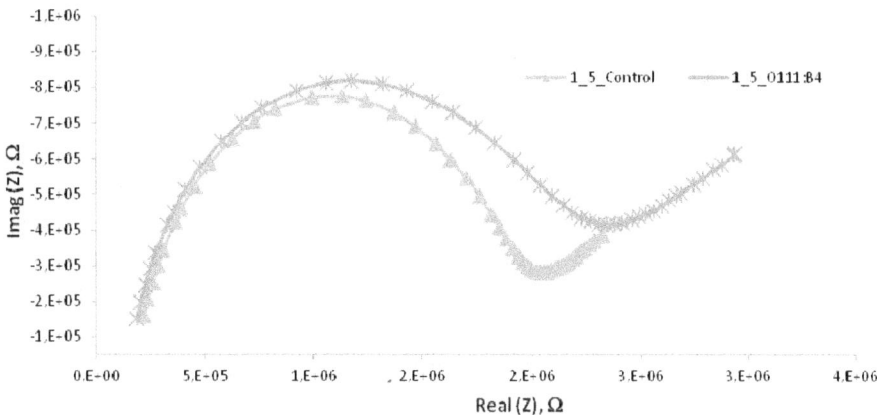

Fig. 4.53. The Nyquist plot of LPS O111:B4 and Control for Alumina 1_5 sensor.

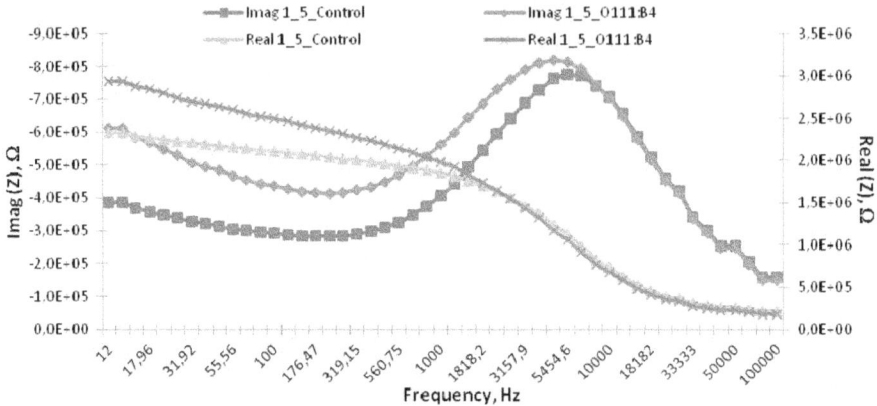

Fig. 4.54. The impedance behavior of LPS O111:B4 and Control for Alumina 1_5 sensor.

4.5. Conclusion and Future Work

Three novel interdigital sensors have been analyzed and tested. The optimum number of negative electrodes is important in the design consideration of interdigital sensors. Uniformity of electric field distribution with highest capacitance value within the sensor's geometry contributes to the highest sensitivity. Modeling using COMSOL Multiphysics and analytical methods have been used to evaluate the sensors performance. The sensors have been fabricated using different method of manufacturing. Each manufacturing technique used different substrates. Due to different values of permittivity of different substrate, the capacitance values and corresponding impedance values are influenced by these substrates. Analyses of characteristic impedance have been conducted using LCR meter and auto measurement program from LabVIEW. Analyses from sensors characterization have shown that Alumina sensors have better sensing performance. Experiments were conducted to investigate how sensors behave with pure chemicals which are related to dangerous toxins. Results have shown that Alumina sensor with 1_5 configuration has better sensitivity measurement.

More experiments are currently being conducted to evaluate new type of silicon based interdigital sensors. Further experiments are planned to be conducted to investigate the significance of any molecular selectivity or resolution as well as the effect of residual adsorption. The experiments with different coating materials are currently conducted. Carboxyl-functional polymer, 3-Aminopropyltriethoxysilane (APTES) and Thionine (organic dye) have been used as a coating layer on the sensor electrodes as these materials have effective binding properties to certain biomolecules. The Polymyxin B (PmB) has been immobilized on the coated sensors because of it has a specific binding properties to LPS. PmB is an antimicrobial peptide produced by the Gram-positive bacterium-*Bacillus*. PmB has been widely used for detection of endotoxins. Equivalent circuit modeling and statistical analysis using principal component analysis (PCA) are implemented for better analyses. The novel sensors need to be selective with higher sensitivity measurement in order to detect the toxins effectively.

Acknowledgement

The authors would like to acknowledge Massey University, New Zealand, Northern Illinois University, United States, Southern Taiwan University, Taiwan, and Malaysian Agricultural Research and Development Institute (MARDI), researchers referenced throughout the paper and also to whom that had fruitful discussions and collaboration with the authors.

References

[1]. FAO, Marine Biotoxins: Food and Nutrition Paper 80, *Food and Agriculture Organization (FAO)*, Rome, Report 2004.

[2]. J. Osek, K. Wieczorek, and M. Tatarczak, Seafood as potential source of poisoning by marine biotoxins, *Medycyna Weterynaryjna,* Vol. 62, April 2006, pp. 370-373.

[3]. M. Campas, B. Prieto-Simon, and J. L. Marty, Biosensors to detect marine toxins: Assessing seafood safety, *Talanta,* Vol. 72, May 15 2007, pp. 884-895.

[4]. I. Garthwaite, Keeping shellfish safe to eat: a brief review of shellfish toxins, and methods for their detection, *Trends in Food Science & Technology,* Vol. 11, July 2000, pp. 235-244.

[5]. P. R. Costa, R. Rosa, J. Pereira, and M. A. M. Sampayo, Detection of domoic acid, the amnesic shellfish toxin, in the digestive gland of Eledone cirrhosa and E-moschata (Cephalopoda, Octopoda) from the Portuguese coast, *Aquatic Living Resources,* Vol. 18, October-December 2005, pp. 395-400.

[6]. C. K. Cusack, S. S. Bates, M. A. Quilliam, J. W. Patching, and R. Raine, Confirmation of domoic acid production by Pseudo-nitzschia australis (Bacillariophyceae) isolated from Irish waters, *Journal of Phycology,* Vol. 38, December 2002, pp. 1106-1112.

[7]. J. A. Maucher and J. S. Ramsdell, Ultrasensitive detection of domoic acid in mouse blood by competitive ELISA using blood collection cards, *Toxicon,* Vol. 45, April 2005, pp. 607-613.

[8]. F. M. V. Dolah, Marine Algal Toxins: Origins, Health Effect, and Their Increased Ocurrence, *Environmental Health Perspectives,* Vol. 108, March 2000, pp. 133-141.

[9]. G. J. Doucette, K. L. King, A. E. Thessen, and Q. Dortch, The effect of salinity on domoic acid production by the diatom Pseudo-nitzschia multiseries, *Nova Hedwigia,* 2008, pp. 31-46.

[10]. P. McCarron, S. Burrell, and P. Hess, Effect of addition of antibiotics and an antioxidant on the stability of tissue reference materials for domoic acid, the amnesic shellfish poison, *Analytical and Bioanalytical Chemistry,* Vol. 387, April 2007, pp. 2495-2502.

[11]. S. F. Qiu, C. W. Pak, and M. C. Curras-Collazo, Sequential involvement of distinct glutamate receptors in domoic acid-induced neurotoxicity in rat mixed cortical cultures: Effect of multiple dose/duration paradigms, chronological age, and repeated exposure, *Toxicological Sciences,* Vol. 89, January 2006, pp. 243-256.

[12]. R. Duran, M. C. Arufe, B. Arias, and M. Alfonso, Effect of Domoic Acid on Brain Amino-Acid Levels, *Revista Espanola De Fisiologia,* Vol. 51, March 1995, pp. 23-27.

[13]. B. Arias, M. Arufe, M. Alfonso, and R. Duran, Effect of Domoic Acid on Metabolism of 5-Hydroxytryptamine in Rat-Brain, *Neurochemical Research,* Vol. 20, April 1995, pp. 401-404.

[14]. G. Debonnel, L. Beauchesne, and C. Demontigny, Domoic Acid, the Alleged Mussel Toxin, Might Produce Its Neurotoxic Effect through Kainate Receptor Activation - an Electrophysiological Study in the Rat Dorsal Hippocampus, *Canadian Journal of Physiology and Pharmacology,* Vol. 67, January 1989, pp. 29-33.

[15]. J. L. C. Wright, R. K. Boyd, A. S. W. Defreitas, M. Falk, R. A. Foxall, W. D. Jamieson, M. V. Laycock, A. W. Mcculloch, A. G. Mcinnes, P. Odense, V. P. Pathak, M. A. Quilliam, M. A. Ragan, P. G. Sim, P. Thibault, J. A. Walter, M. Gilgan, D. J. A. Richard, and D. Dewar, Identification of Domoic Acid, a Neuroexcitatory Amino-Acid, in Toxic Mussels from Eastern Prince-Edward-Island, *Canadian Journal of Chemistry-Revue Canadienne De Chimie,* Vol. 67, March 1989, pp. 481-490.

[16]. T. M. Perl, L. Bedard, T. Kosatsky, J. C. Hockin, E. C. D. Todd, and R. S. Remis, An Outbreak of Toxic Encephalopathy Caused by Eating Mussels Contaminated with Domoic Acid, *New England Journal of Medicine,* Vol. 322, June 21 1990, pp. 1775-1780.

[17]. FAO/IOC/WHO, Report of the Joint FAO/IOC/WHO ad hoc Expert Consultation on Biotoxins in Bivalve Molluscs, FAO, IOC, WHO, Oslo, Norway, Report, 26-30 September 2004.

[18]. A. B. Wagner, Bacterial Food Poisoning, *Texas Agric. Ext. Publication,* 1989, pp. 1-6.

[19]. M. Khafaji, S. Shahrokhian, and M. Ghalkhani, Electrochemistry of Levo-Thyroxin on Edge-Plane Pyrolytic Graphite Electrode: Application to Sensitive Analytical Determinations, *Electroanalysis,* Vol. 23, August 2011, pp. 1875-1880.

[20]. C. Xhoffer, K. Van den Bergh, and H. Dillen, Electrochemistry: a powerful analytical tool in steel research, *Electrochimica Acta,* Vol. 49, July 30 2004, pp. 2825-2831.

[21]. O. Lazcka, F. J. Del Campo, and F. X. Munoz, Pathogen detection: A perspective of traditional methods and biosensors, *Biosensors & Bioelectronics,* Vol. 22, February 15 2007, pp. 1205-1217.

[22]. T. A. McMeekin, Ed., Detecting Pathogens in Food, *Woodhead Publishing Limited,* Cambridge, England, 2003.

[23]. FAO/WHO, Microbiological Risk Assessment Series: Hazard Characterization for Pathogens in Food and Water, Vol. 3, *W.H.O.W. Food and Agriculture Organization (FAO),* Ed., WHO Library Cataloguing, 2003, pp. 1-76.

[24]. M. A. Quilliam, The role of chromatography in the hunt for red tide toxins, *Journal of Chromatography A,* Vol. 1000, June 6 2003, pp. 527-548.

[25]. Domoic Acid in Mussels, *AOAC,* Gaithersburg Official Method 991.26, 2005.

[26]. P. Hess, S. Gallacher, L. A. Bates, N. Brown, and M. A. Quilliam, Determination and confirmation of the amnesic shellfish poisoning toxin, domoic acid, in shellfish from Scotland by liquid chromatography and mass spectrometry, *Journal of Aoac International,* Vol. 84, Sep-Oct 2001, pp. 1657-1667.

[27]. J. F. Lawrence, C. F. Charbonneau, and C. Menard, Liquid-Chromatographic Determination of Domoic Acid in Mussels, Using Aoac Paralytic Shellfish Poison Extraction Procedure - Collaborative Study, *Journal of the Association of Official Analytical Chemists,* Vol. 74, January-February 1991, pp. 68-72.

[28]. H. Kodamatani, K. Saito, N. Niina, S. Yamazaki, A. Muromatsu, and I. Sakurada, Sensitive determination of domoic acid using high-performance liquid chromatography with electrogenerated tris(2,2'-bipyridine)ruthenium(III) chemiluminescence detection, *Analytical Sciences,* Vol. 20, July 2004, pp. 1065-1068.

[29]. C. Hummert, M. Reichelt, and B. Luckas, Automatic HPLC-UV determination of domoic acid in mussels and algae, *Chromatographia,* Vol. 45, 1997, pp. 284-288.

[30]. I. M. Traynor, L. Plumpton, T. L. Fodey, C. Higgins, and C. T. Elliott, Immunobiosensor detection of domoic acid as a screening test in bivalve molluscs: Comparison with liquid chromatography-based analysis, *Journal of Aoac International,* Vol. 89, May-June 2006, pp. 868-872.

[31]. Q. M. Yu, S. F. Chen, A. D. Taylor, J. Homola, B. Hock, and S. Y. Jiang, Detection of low-molecular-weight domoic acid using surface plasmon resonance sensor, *Sensors and Actuators B-Chemical,* Vol. 107, May 27 2005, pp. 193-201.

[32]. R. C. Stevens, S. D. Soelberg, B. T. L. Eberhart, S. Spencer, J. C. Wekell, T. M. Chinowsky, V. L. Trainer, and C. E. Furlong, Detection of the toxin domoic acid from clam extracts using a portable surface plasmon resonance biosensor, *Harmful Algae,* Vol. 6, February 2007, pp. 166-174.

[33]. B. R. Hesp, J. C. Harrison, A. I. Selwood, P. T. Holland, and D. S. Kerr, Detection of domoic acid in rat serum and brain by direct competitive enzyme-linked immunosorbent assay (cELISA), *Analytical and Bioanalytical Chemistry,* Vol. 383, November 2005, pp. 783-786.

[34]. M. Kania and B. Hock, Development of monoclonal antibodies to domoic acid for the detection of domoic acid in blue mussel (mytilus edulis) tissue by ELISA, *Analytical Letters,* Vol. 35, 2002, pp. 855-868.

[35]. I. Garthwaite, K. M. Ross, C. O. Miles, L. R. Briggs, N. R. Towers, T. Borrell, and P. Busby, Integrated enzyme-linked immunosorbent assay screening system for amnesic, neurotoxic, diarrhetic, and paralytic shellfish poisoning toxins found in New Zealand, *Journal of Aoac International,* Vol. 84, Sep-Oct 2001, pp. 1643-1648.

[36]. Z. Tsao, Y. Liao, B. Liu, C. Su, and F. Yu, Development of a Monoclonal Antibody against Domoic Acid and Its Application in Enzyme-Linked Immunosorbent Assay and Colloidal Gold Immunostrip, *Journal of Agriculture and Food Chemistry,* Vol. 55, 2007, pp. 4921-4927.

[37]. P. M. Fratamico, Comparison of culture, polymerase chain reaction (PCR), TaqMan Salmonella, and Transia Card Salmonella assays for detection of Salmonella spp. in naturally-contaminated ground chicken, ground turkey, and ground beef, *Molecular and Cellular Probes,* Vol. 17, Oct 2003, pp. 215-221.

[38]. E. Leoni and P. P. Legnani, Comparison of selective procedures for isolation and enumeration of Legionella species from hot water systems, *Journal of Applied Microbiology,* Vol. 90, Jan 2001, pp. 27-33.

[39]. E. Alocilja and S. Radke., Market Analysis of Biosensor for Food Safety, *Biosensors & Bioelectronics,* Vol. 18, 2003, pp. 841-846.

[40]. A. K. Bej, M. H. Mahbubani, J. L. Dicesare, and R. M. Atlas, Polymerase Chain Reaction-Gene Probe Detection of Microorganisms by Using Filter-Concentrated Samples, *Applied and Environmental Microbiology,* Vol. 57, Dec 1991, pp. 3529-3534.

[41]. K. Levi, J. Smedley, and K. J. Towner, Evaluation of a real-time PCR hybridization assay for rapid detection of Legionella pneumophila in hospital and environmental water samples, *Clinical Microbiology and Infection,* Vol. 9, Jul 2003, pp. 754-758.

[42]. D. Rodriguez-Lazaro, M. D'Agostino, A. Herrewegh, M. Pla, N. Cook, and J. Ikonomopoulos, Real-time PCR-based methods for detection of Mycobacterium avium Subsp paratuberculosis in water and milk, *International Journal of Food Microbiology,* Vol. 101, May 1 2005, pp. 93-104.

[43]. A. Jofre, B. Martin, M. Garriga, M. Hugas, M. Pla, D. Rodriguez-Lazaro, and T. Aymerich, Simultaneous detection of Listeria monocytogenes and Salmonella by multiplex PCR in cooked ham, *Food Microbiology,* Vol. 22, February 2005, pp. 109-115.

[44]. A. J. Baeumner, Biosensors for environmental pollutants and food contaminants, *Analytical and Bioanalytical Chemistry,* Vol. 377, October 2003, pp. 434-445.

[45]. M. A. Cooper, Label-free screening of bio-molecular interactions, *Analytical and Bioanalytical Chemistry,* Vol. 377, November 2003, pp. 834-842.

[46]. Z. Mubammad-Tahir and E. C. Alocilja, A conductometric biosensor for biosecurity, *Biosensors & Bioelectronics,* Vol. 18, May 2003, pp. 813-819.

[47]. E. Barsoukov and J. R. Macdonald, Eds., Impedance Spectroscopy Theory, Experiment and Applications, *John Wiley & Sons Ltd.,* Hoboken, New Jersey, 2005.

[48]. M. Varshney and Y. B. Li, Interdigitated array microelectrodes based impedance biosensors for detection of bacterial cells, *Biosensors & Bioelectronics,* Vol. 24, June 15 2009, pp. 2951-2960.

[49]. B. A. Morris and A. Sadana, A fractal analysis for the binding of riboflavin binding protein to riboflavin immobilized on a SPR biosensor, *Sensors and Actuators B-Chemical,* Vol. 106, May 13 2005, pp. 498-505.

[50]. S. Yoshida, T. Gyoji, and K. Toda, Position-Sensitive Photodetector Using Interdigital Electrodes on Pb_2CrO_5 Thin-Films, *International Journal of Electronics,* Vol. 68, April 1990, pp. 525-531.

[51]. S. Averin, R. Sachot, J. Hugi, M. deFays, and M. Ilegems, Two-dimensional device modeling and analysis of GaInAs metal-semiconductor-metal photodiode structures, *Journal of Applied Physics,* Vol. 80, August 1 1996, pp. 1553-1558.

[52]. S. H. Lee, S. L. Lee, H. Y. Kim, and T. Y. Eom, Analysis of light efficiency in homogeneously aligned nematic liquid crystal display with interdigital electrodes, *Journal of the Korean Physical Society,* Vol. 35, December 1999, pp. S1111-S1114.

[53]. Y. J. Chen, C. L. Zhu, M. S. Cao, and T. H. Wang, Photoresponse of SnO2 nanobelts grown in situ on interdigital electrodes, *Nanotechnology,* Vol. 18, July 18 2007.

[54]. T. M. A. Gronewold, Surface acoustic wave sensors in the bioanalytical field: Recent trends and challenges, *Analytica Chimica Acta,* Vol. 603, November 12 2007, pp. 119-128.

[55]. R. Weigel and R. Hauser, Introduction to the special issue on acoustic wave sensors and applications, *IEEE Transactions on Ultrasonics Ferroelectrics and Frequency Control,* Vol. 51, Nov 2004, pp. 1365-1366.

[56]. S. Shiokawa and J. Kondoh, Surface acoustic wave sensors, *Japanese Journal of Applied Physics Part 1-Regular Papers Brief Communications & Review Papers,* Vol. 43, May 2004, pp. 2799-2802.

[57]. B. Drafts, Acoustic wave technology sensors, *IEEE Transactions on Microwave Theory and Techniques,* Vol. 49, April 2001, pp. 795-802.

[58]. M. Hoummady, A. Campitelli, and W. Wlodarski, Acoustic wave sensors: design, sensing mechanisms and applications, *Smart Materials & Structures,* Vol. 6, December 1997, pp. 647-657.

[59]. J. J. Steele, M. T. Taschuk, and M. J. Brett, Nanostructured metal oxide thin films for humidity sensors, *IEEE Sensors Journal,* Vol. 8, July-August 2008, pp. 1422-1429.

[60]. J. J. Steele, G. A. Fitzpatrick, and M. J. Brett, Capacitive humidity sensors with high sensitivity and subsecond response times, *IEEE Sensors Journal,* Vol. 7, May-June 2007, pp. 955-956.

[61]. C. Y. Lee and G. B. Lee, Humidity sensors: A review, *Sensor Letters,* Vol. 3, March 2005, pp. 1-15.

[62]. P. Furjes, A. Kovacs, C. Ducso, M. Adam, B. Muller, and U. Mescheder, Porous silicon-based humidity sensor with interdigital electrodes and internal heaters, *Sensors and Actuators B-Chemical,* Vol. 95, October 15 2003, pp. 140-144.

[63]. M. Kitsara, D. Goustouridis, S. Chatzandroulis, M. Chatzichristidi, I. Raptis, T. Ganetsos, R. Igreja, and C. J. Dias, Single chip interdigitated electrode capacitive chemical sensor arrays, *Sensors and Actuators B-Chemical,* Vol. 127, October 20 2007, pp. 186-192.

[64]. R. Igreja and C. J. Dias, Dielectric response of interdigital chemocapacitors: The role of the sensitive layer thickness, *Sensors and Actuators B-Chemical,* Vol. 115, May 23 2006, pp. 69-78.

[65]. R. Igreja and C. J. Dias, Organic vapour discrimination using sorption sensitive chemocapacitor arrays, *Advanced Materials Forum Iii, Pts 1 and 2,* Vol. 514-516, 2006, pp. 1064-1067.

[66]. R. Igreja and C. J. Dias, Capacitance response of polysiloxane films with interdigital electrodes to volatile organic compounds, *Advanced Materials Forum Ii,* Vol. 455-456, 2004, pp. 420-424.

[67]. C. Ziegler, J. Maier, and H. D. Wiemhofer, Chemical sensors and sensor technology, *Physical Chemistry Chemical Physics,* Vol. 5, December 1 2003, pp. Vii-Vii.

[68]. C. Hagleitner, A. Hierlemann, D. Lange, A. Kummer, N. Kerness, O. Brand, and H. Baltes, Smart single-chip gas sensor microsystem, *Nature,* Vol. 414, November 15 2001, pp. 293-296.

[69]. W. Olthuis, A. J. Sprenkels, J. G. Bomer, and P. Bergveld, Planar interdigitated electrolyte-conductivity sensors on an insulating substrate covered with Ta2O5, *Sensors and Actuators B- Chemical,* Vol. 43, 1997, pp. 211-216.

[70]. K. Toda, Y. Komatsu, S. Oguni, S. Hashiguchi, and I. Sanemasa, A planar gas sensor combined with interdigitated array electrodes, *Analytical Sciences,* Vol. 15, January 1999, pp. 87-89.

[71]. K. Sundara-Rajan, L. Byrd, and A. V. Mamishev, Moisture content estimation in paper pulp using fringing field impedance Spectroscopy, *IEEE Sensors Journal,* Vol. 4, June 2004, pp. 378-383.

[72]. K. Sundara-Rajan, L. Byrd, and A. V. Mamishev, Measuring moisture, fiber, and titanium dioxide in pulp with impedance spectroscopy, *Tappi Journal,* Vol. 4, February 2005, pp. 23-27.

[73]. E. Fratticcioli, M. Dionigi, and R. Sorrentino, A simple and low-cost measurement system for the complex permittivity characterization of materials, *IEEE Transactions on Instrumentation and Measurement,* Vol. 53, August 2004, pp. 1071-1077.

[74]. X. Li, Impedance Spectroscopy for Manufacturing Control of Material Physical Properties, Master of Science in Electrical Engineering, Department of Electrical Engineering, University of Washington, Washington, 2003.

[75]. S. Baglio, S. Castorina, and N. Savalli, Integrated inductive sensors for the detection of magnetic microparticles, *IEEE Sensors Journal,* Vol. 5, June 2005, pp. 372-384.

[76]. V. Kasturi and S. C. Mukhopadhyay, Estimation of Property of Sheep Skin to Modify the Tanning Process Using Interdigital Sensors, in Sensors: Advancements in Modeling, Design Issues, Fabrication and Practical Applications, Vol. 21, S. C. M. a. R. Y. M. Huang, Ed., *Springer Berlin Heidelberg,* 2008, pp. 91-110.

[77]. S. C. Mukhopadhyay, S. D. Choudhury, T. Allsop, V. Kasturi, and G. E. Norris, Assessment of pelt quality in leather making using a novel non-invasive sensing approach, *Journal of Biochemical and Biophysical Methods,* Vol. 70, April 24 2008, pp. 809-815.

[78]. S. C. Mukhopadhyay, G. Sen Gupta, and S. N. Demidenko, Special issue on intelligent sensors - Guest Editorial, *IEEE Sensors Journal,* Vol. 7, May-June 2007, pp. 608-610.

[79]. S. C. Mukhopadhyay, Planar electromagnetic sensors: Characterization, applications and experimental results, *Tm-Technisches Messen,* Vol. 74, 2007, pp. 290-297.

[80]. S. C. Mukhopadhyay, Novel planar electromagnetic sensors: Modeling and performance evaluation, *Sensors,* Vol. 5, December 2005, pp. 546-579.

[81]. S. C. Mukhopadhyay, J. D. M. Woolley, G. S. Gupta, and S. Deidenko, Saxophone Reed Inspection Employing Planar Electromagnetic Sensors, *IEEE Transactions on Instrumentation and Measurements,* Vol. 56, December 2007, pp. 2492-2503.

[82]. S. C. Mukhopadhyay, C. Goonerate, G. S. Gupta, and S. Demidenko, A Low Cost Sensing System for Quality of Dairy Products, *IEEE Transactions on Instrumentation and Measurements,* Vol. 55, August 2006, pp. 1331-1338.

[83]. S. C. Mukhopadhyay and C. P. Gooneratne, A Novel Planar-Type Biosensor for Noninvasive Meat Inspection, *IEEE Sensors Journal,* Vol. 7, September 2007, pp. 1340-1346.

[84]. A. V. Mamishev, Y. Du, J. H. Bau, B. C. Lesieutre, and M. Zahn, Evaluation of diffusion-driven material property profiles using three-wavelength interdigital sensor, *IEEE Transactions on Dielectrics and Electrical Insulation,* Vol. 8, October 2001, pp. 785-798.

[85]. S. Radke and E. Alocilja., A Microfabricated Biosensor for Detecting Foodborne Bioterrorism Agents, *IEEE Sensors Journal,* Vol. 5, August 2005, pp. 744-750.

[86]. S. Radke and E. Alocilja, Design and Fabrication of a Microimpedance Biosensor for Bacterial Detection, *IEEE Sensors Journal,* Vol. 4, August 2004, pp. 434-440.

[87]. F. L. Dickert, G. K. Zwissler, and E. Obermeier, Liquid-Crystals on Interdigital Structures - Applications as Capacitive Chemical Sensors, *Berichte Der Bunsen-Gesellschaft-Physical Chemistry Chemical Physics,* Vol. 97, Feb 1993, pp. 184-188.

[88]. A. V. Mamishev, K. Sundara-Rajan, F. Yang, Y. Q. Du, and M. Zahn, Interdigital sensors and transducers, *Proceedings of the IEEE,* Vol. 92, May 2004, pp. 808-845.

[89]. E. J. Olson and P. Buhlmann, Minimizing Hazardous Waste in the Undergraduate Analytical Laboratory: A Microcell for Electrochemistry, *Journal of Chemical Education,* Vol. 87, November 2010, pp. 1260-1261.

[90]. L. A. Avaca and C. Comninellis, Electrochemistry for a Healthy Planetenvironmental Analytical and Engineering Aspects, Selection of papers from the 6th ISE Spring Meeting, 16-19 March 2008, Foz do Iguacu, Brazil Foreword, *Electrochimica Acta,* Vol. 54, February 28 2009, pp. 1717-1718.

[91]. A. R. M. Syaifudin, K. P. Jayasundera, and S. C. Mukhopadhyay, A low cost novel sensing system for detection of dangerous marine biotoxins in seafood, *Sensors and Actuators B-Chemical,* Vol. 137, March 28 2009, pp. 67-75.

[92]. E. Cano, D. Lafuente, and D. M. Bastidas, Use of EIS for the evaluation of the protective properties of coatings for metallic cultural heritage: a review, *Journal of Solid State Electrochemistry,* Vol. 14, Mar 2010, pp. 381-391.

[93]. Sigma-Aldrich Catalogue, Biochemical, Reagents and Kits for Life Science Research, 2009-2010, pp. 1437-1440.

Chapter 5
Concept of Force Sensing and Measurements: from Past to Present

Ebtisam H. Hasan

5.1. Introduction

It is imperative to define the difference between force, mass, weight and load. In summary, force is a measure of the interaction between bodies. It takes a number of forms including short-range atomic force, electromagnetic, and gravitational forces. The SI unit of force is the Newton (N). Newton is defined as the force which would give to a mass of one kilogram an acceleration of one meter per square second. Mass is a measure of the amount of material in an object, the SI unit of mass is the kilogram (kg). Kilogram is defined to be equal to the mass of the international prototype kilogram mass held at the International Bureau of Weight and Measures (BIPM). Weight is the gravitational force acting on a body, it is taken to mean the same as mass and measured in kilograms .Load usually means the force exerted on a surface or body. It is a term frequently used in engineering to mean the force exerted on a surface or body [1].

5.2. Type of Forces

5.2.1. Force Concept

In physics, the concept of force is used to describe an influence that causes a free body to undergo acceleration. Force can also be described by intuitive concepts such as a push or pull that can cause an object with mass to change its velocity or which can cause a flexible object to deform. Forces which do not act uniformly on all parts of a body will also cause mechanical stresses, a technical term for influences which cause deformation of matter. Newton's second law can be formulated to state that an object with a constant mass will accelerate in proportion to the net force acting upon [2].

Ebtisam H. Hasan
National Institute for Standards (NIS), Giza, Egypt

Sergey Y. Yurish (ed.), *Modern Sensors, Transducers and Sensor Networks*
© International Frequency Sensor Association Publishing, 2012

Before Newton, the tendency for objects to fall towards the Earth was not understood to be related to the motions of celestial objects. Today, this acceleration due to gravity towards the surface of the Earth is usually designated as $\vec{g}$ and has a magnitude of about 9.81 meters per second squared (this measurement is taken from sea level and may vary depending on location), and points toward the center of the Earth. This observation means that the force of gravity on an object at the Earth's surface is directly proportional to the object's mass. Thus an object that has a mass of m will experience a force:

$$\vec{F} = m\vec{g}$$

5.2.2. Force Models

5.2.2.1. Electromagnetic Force

The electrostatic force was first described in 1784 by Coulomb as a force which existed intrinsically between two charges. Meanwhile, the Lorentz force of magnetism was discovered to exist between two electric currents. The connection between electricity and magnetism allows for the description of a unified electromagnetic force that acts on a charge. This force can be written as a sum of the electrostatic force (due to the electric field) and the magnetic force (due to the magnetic field). Fully stated, this is the law:

$$\vec{F} = q(\vec{E} + \vec{v} \times \vec{B}),$$

where $\vec{F}$ is the electromagnetic force, q is the magnitude of the charge of the particle, $\vec{E}$ is the electric field, $\vec{v}$ is the velocity of the particle which is crossed with the magnetic field ($\vec{B}$).

5.2.2.2. Nuclear Force

There are two "nuclear forces" which today are usually described as interactions that take place in quantum theories of particle physics. The strong nuclear force is the force responsible for the structural integrity of atomic nuclei while the weak nuclear force is responsible for the decay of certain nucleons into leptons and other types of hadrons.

5.2.2.3. Non-fundamental Forces

5.2.2.3.1. Normal Force

The normal force is the repulsive force of interaction between atoms at close contact. The normal force, for example, is responsible for the structural integrity of tables and floors as well as being the force that responds whenever an external force pushes on a solid object.

5.2.2.3.2. Friction

Friction is a surface force that opposes relative motion. The frictional force is directly related to the normal force which acts to keep two solid objects separated at the point of contact.

5.2.2.3.3. Elastic Force and Hooke's Law

An elastic force acts to return a spring to its natural length as shown in Fig. 5.1. An ideal spring is taken to be massless, frictionless, unbreakable, and infinitely stretchable. Such springs exert forces that push when contracted, or pull when extended, in proportion to the displacement of the spring from its equilibrium position. This linear relationship was described by Robert Hooke in 1676, for whom Hooke's law is named. If Δx is the displacement, the force exerted by an ideal spring equals:

$$\vec{F} = -k\Delta\vec{x},$$

where k is the spring constant (or force constant), which is particular to the spring. The minus sign accounts for the tendency of the elastic force to act in opposition to the applied load.

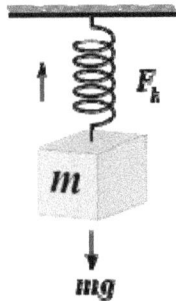

Fig. 5.1. Spring model for elastic force.

5.2.2.3.4. Rotations and Torque

Forces that cause extended objects to rotate are associated with torques as indicated in Fig. 5.2.

Mathematically, the torque on a particle is defined as the cross-product:

$$\vec{\tau} = \vec{r} \times \vec{F},$$

where

$\vec{r}$ is the particle's position vector relative to a pivot;
$\vec{F}$ is the force acting on the particle.

Torque is the rotation equivalent of force in the same way that angle is the rotational equivalent for position, angular velocity for velocity, and angular momentum for momentum [2].

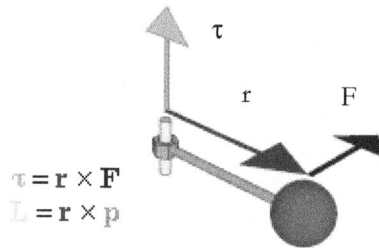

Fig. 5.2. Rotating force model.

5.3. Force-measurement System

The SI unit of force is the Newton (symbol N), which is the force required to accelerate a one kilogram mass at a rate of one meter per second squared, or $kg \cdot m \cdot s^{-2}$.

A complete force-measurement system has two parts; a sensor or signal-producing system and an indicating system. The sensor system contains elements which have measurable attributes, the magnitudes of which have a correspondence to the magnitude of the applied force, e.g., the deflection of a ring, the state of strain at one or more locations on the element, etc. With the exception of elastic proving rings, commercial sensing systems are usually sealed units, the details of which are proprietary to the manufacturer. Ordinarily, the indicating system, which quantitatively displays the signal from the sensing system, is also essentially a 'black box'.

There are a variety of electronic indicating systems which can be used with strain-gage-type sensing systems. Such systems ordinarily have provisions for a 'zero' adjusts (A_z) and for an amplifier gain adjust (G) relative to a span reference (S) or relative to some particular load. The span reference (S) may be a bypass resistor inserted across one arm of the sensing bridge, an internal calibration network or a ratio device which is external to the system. Most proving rings use a fixed-gain mechanical amplifier with no 'zero' adjust. Since no two sensing systems have exactly the same characteristics, and since amplifier-gain settings are arbitrary, all force measurement systems must be calibrated, i.e., subjected to known forces in order to establish the correspondence between the applied load, x, and the numerical output of the indicating system, y, for a particular amplifier-gain setting. In addition to the 'known' applied forces, there are a number of perturbing effects which can cause observable changes in the response of a force-measurement system [3].

5.3.1. Force Transducers

There are many types of force transducer/load cell and they are used with instrumentation of varying complexity. In designing or specifying a force measurement system for an application, it is useful to understand the basic operation of the transducer to be used and also their broad operating characteristics. Load cell designs can be distinguished according to the type of output signal generated (pneumatic, hydraulic, electric) or according to the way they detect weight (bending, shear, compression, tension, etc.)

5.3.1.1. Strain Gauge Load Cells

These are the most common type of force transducer, and a clear example of an elastic device. Each cell is based on an elastic element to which a number of electrical resistance strain gauges are bonded. The geometric shape and modulus of elasticity of the element determine the magnitude of the strain field produced by the action of the force. Each strain gauge responds to the local strain at its location, and the measurement of force is determined from the integration of these individual measurements of strain. The rated capacities of strain gauge load cells range from 5 N to more than 50 MN. They have become the most widespread of all force measurement systems and can be used with high resolution digital indicators as force transfer standards [4, 5]. The point of weakness for some strain gauge load cell types are that the strain gauges are exposed and require protection.

5.3.1.2. Piezoelectric Crystal

When a force is exerted on certain crystalline materials, electric charges are formed on the crystal surface in proportion to the rate of change of that force. To make use of the device, a charge amplifier required to integrate the electric charges to give a signal that is proportional to the applied force and big enough to measure. These piezoelectric crystal sensors are different from most other sensing techniques in that they are active sensing elements. No power supply is needed and the deformation to generate a signal is very small which has the advantage of a high frequency response of the measuring system without introducing geometric changes to the force measuring path. Extremely fast events such as shock waves in solids, or impact printer and punch press forces can be measured with these devices when otherwise such measurements might not be achievable. Piezoelectric sensors operate with small electric charge and require high impedance cable for the electrical interface. It is important to use the matched cabling supplied with a transducer.

5.3.1.3. Measuring Force through Pressure

The hydraulic load cell is a device filled with a liquid (usually oil) which has a pre-load pressure. Application of the force to the loading member increases the fluid pressure which is measured by a pressure transducer or displayed on a pressure gauge dial via a Bourdon tube. Measurement uncertainties of around 0.25 % can be achieved with careful design and favorable application conditions. Uncertainties for total systems are more realistically 0.5 % to 1 %. The cells are sensitive to temperature changes and usually have facilities to adjust the zero output reading, the temperature coefficients are of the order of 0.02 % to 0.1 % per °C.

5.3.1.4. Other Types of Force Measuring System

5.3.1.4.1. Elastic Devices

The loading column is probably the simplest elastic device, being simply a metal cylinder subjected to a force along its axis. In this case the length of the cylinder is measured directly by a dial gauge or other technique, and an estimate of the force can be made by interpolating between the lengths measured for previously applied known forces. The proving ring is functionally very similar except that the element is a circular ring, and the deformation is usually measured across the inside diameter. These transducers have the advantage of being simple and robust, but the main disadvantage is the strong effect of temperature on the output. Such methods find use in monitoring the forces in building foundations and other similar applications.

5.3.1.4.2. Magneto-elastic Devices

The magneto-elastic force transducer is based on the effect that when a ferromagnetic material is subjected to mechanical stress, the magnetic properties of the material are altered and the change is proportional to the applied stress. The rated capacities of these devices are in the range from 2 kN to 5 MN [6].

5.3.2. Selection Criteria for Force Transducers

There are different styles of Force transducers/Load cells. Load cell selection in the context of trouble free operation concerns itself primary with the right capacity, accuracy class and environmental protection.

5.3.2.1. Capacity Selection

Overload is still the primary reason for load cell failure, although the process of selection the right load cell capacity looks easy and straight forward on first sight.

Capacity selection requires a fundamental understanding of the load related terms for load cells as well as the load related factors associated with systems. The load related terms for load cells are: load cell measuring range, safe load limit, ultimate overload and safe side load. A load cell will perform within specifications until the safe load limit(the maximum load that can be applied without producing a permanent shift in the performance characteristics beyond those specified; specified as a percentage of the measuring range i.e. 150 %) or safe side load limit (the maximum load that can act 90° to the axis along which the load cell is designed to be loaded at the point of axial load application without producing a permanent shift in the performance beyond those specified; specified as a percentage of the measuring range i.e. 100 %) is passed. Beyond this point, even for a very short period of time, the load cell will be permanently damaged. The load cell may physically break at the ultimate load limit.

5.3.2.2. Accuracy

load cells are ranked, according to their overall performance capabilities into differing accuracy classes. Some of these accuracy classes are related to standards which are used in legal for trade weighing instruments, while others accuracy classes are defined by the individual load cell manufacturer.

5.3.2.3. Environmental Protection

No area of load cell operation causes more confusion and contention than that of environmental protection and sealing standards. In the absence of such standards, most manufacturers have adopted the International Protection system (IP/IEC 529 or EN 40.050) or National Electrical Manufacturers Association Standards (NEMA publication 250) [7].

5.3.3. Development of the Load Cell Design Technology, Force Application Systems and New Force Measurement Techniques

5.3.3.1. Load Cells

In the past, the load cells are constructed using electric resistance metal foil strain gauges bonded to an elastic flexure element. The load cell is a passive analog device with continuous resolution limited ultimately by noise, due to electron motion on the order of 10^{-9} Volts (nano Volt). Therefore, practically speaking, sensitivity and stability of the electronic instrumentation used is critical when high resolution is required. High electronic gain alone will not achieve good results if the zero stability or gain stability is poor because the readings will drift with time or temperature changes. Generally, it is desired to read physical units instead of counts [8]. Modern industrial processes are digitally controlled but the majority of sensors for measuring force and weight are still transmitting analog signals (voltage or current). There are obvious benefits in generating digital signals directly from the sensors in relation to ease of integration, implementation, use, and maintenance.

For this reason, the strain gauge load cell (as mentioned in 3.1.1) with digital readout to receive the load cell output is considered the development in force measurement area. These cells convert the load acting on them into electrical signals. The gauges themselves are bonded onto a beam or structural member that deforms when weight is applied. In most cases, four strain gages are used to obtain maximum sensitivity and temperature compensation. Two of the gauges are usually in tension, and two in compression, and are wired with compensation adjustments. When load is applied, the strain changes the electrical resistance of the gauges in proportion to the load.

Other load cells are fading into obscurity, as strain gage load cells continue to increase their accuracy and lower their unit costs. Piezoresistive is Similar in operation to strain gages. Piezoresistive sensors generate a high level output signal, making them ideal for simple weighing systems because they can be connected directly to a readout meter. An added drawback of piezoresistive devices is their nonlinear output [9].

Adding the LCB500 tension and compression load cell to USB sensor is one solution for the elimination of the need for an analog amplifier, power supply and display equipment making usage that much easier. The LCB500 is machined from a single block of metal, so the primary sensing element, the mountings and the case contain no welds allowing smaller dimensions and an enhanced grade of protection. The configuration of the point of measurement reduces errors caused by imperfect application of the load. The stainless steel construction is suitable for use in aggressive environments in the chemical and petrochemical industries. The standard LCB500 tension and compression load cell model has male/female threads, is a welded until and comes with Bendix receptacle. This model is well suited for both low ranges to high range capacity applications (100-5000 lb) [10].

In the area of new sensor development, fiber optic load cells are gaining attention because of their immunity to electromagnetic and radio frequency interference (EMI/RFI), suitability for use at elevated temperatures, and intrinsically safe nature. In this work developing, there are two techniques showing promise: measuring the micro-bending loss effect of single-mode optical fiber and measuring forces using the Fiber Bragg Grating (FBG) effect. Optical sensors based on both technologies are undergoing field trials in Japan. But, a few fiber optic load sensors are commercially available [9].

S. Mäuselein et all investigated a new type of load cell, Silicon type, with thin-film strain gauges as shown in Fig. 5.3. The results offer the usability of the Si LC in the range of precision measurement if temperature behavior of sensitivity and linearity are compensated [11].

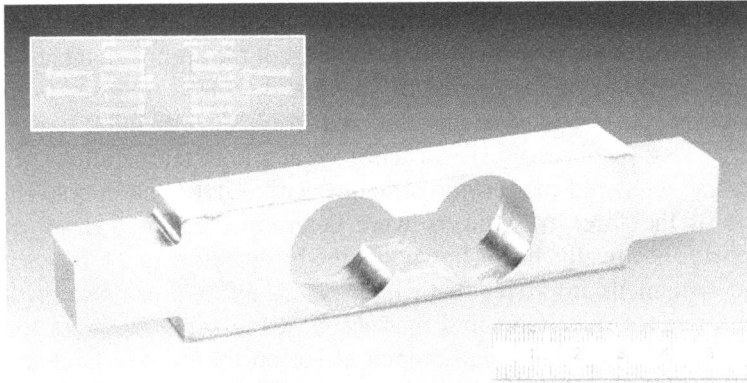

Fig. 5.3. Silicon cell model.

The thin-film SGs were applied on the upper surface of the Si Spring in the area of the thin places of the LC body by using the sputtering technology. The material of the SGs is NiCr and the thickness of the SGs amount to 250 nm. The small application area of one SG with only 0.8 mm times 1.8 mm leads to a higher sensitivity compared to glue SGs with greater application areas. The investigations result in a high reproducibility and a low hysteresis which are about one magnitude better than for conventional strain gauge load cells.

Not only the load cell development is related to the new design, but the improvement of the load cell function in force measurement is also required. Surasith Piyasin is a researcher who concentrated to improve the hollow clevis-pin type load cells. These cells are widely used in heavy-duty machinery such as crams, hoists and conveyors. He tried to determine the best geometric proportion for the load cell. He proved that the certain positions for strain gauges offer some advantage over the proportions and gauge positions. Various kinds of geometric parameters for the clevis-pin have been created. These parameters were inner diameter, outside diameter, length and height. By using the ANSYS version 5.4 finite element code for the stress analysis, he concluded that: when the hollow-bore clevis-pin is used as a strain gauge load cell, the proportions are likely to be of length to diameter ratio varying from about 2.5:1 to about 3.5:1 or perhaps 4.0:1. Also, the best results are likely to be obtained with bore to outside diameter ratio giving thick-wall proportions and, typically, of from 1/2 to 1/3 [12].

Continuing the improvement of force transducers, some researches focused on the application of heat treatments on spring elements of transducers, since this is a very effective method for attaining good performance in force measurements. Heat treatments change the microstructure of the spring material, which plays a major role in the improvement of performance characteristics of force transducers, particularly in terms of hysteresis error. Sinan Fank and Mehmet Demirkol studied the changing of the microstructure of AISI 4340 steel through the application of different heat treatments, and the subsequent measurement of the hysteresis performance of force transducers in relation to the change in the structural characteristics of the spring elements. Some of the specimens were quenched and tempered to 35, 45 and 55 HRC (Hardness Rockwell C). Some of the other specimens were austempered to obtain a bainitic structure with 45 HRC. The remaining specimens were austenitized at a high temperature for a long time to obtain a coarse tempered martensitic structure with 45HRC hardness. The hysteresis characteristics of the force transducers were determined using a dead weight, force standard machine. The results have shown that the hysteresis characteristics of quenched and tempered specimens improved with increasing hardness. Bainite exhibited better hysteresis performance over tempered martensite at the same hardness level, while a coarse martensite structure has a detrimental effect on the hysteresis characteristics of force transducers [13].

5.3.3.2. Force Application Systems

In the area of using Dead Weight concept as a primary method for applying the forces, the Electronic Deadweight Tester is a modern replacement for the Conventional Deadweight Tester (CDT) which has some disadvantages by usage. In this dead weight concept, the forces applied to the sensor are generated by suspending weights of known mass in a known gravity field. Some of the CDT disadvantages are: The minimum pressure increment is limited by the minimum mass value in the mass set, so if the test gauge is of high resolution, it may not be possible to position its needle directly on a nominal test point. The second disadvantage are the alternative to setting nominal pressures on the gauge under test by loading incremental masses is to interpolate the gauge's reading, creating an opportunity for errors. Another disadvantage is, since the deadweight tester is inherently a mechanical device, there is no convenient way to automate the data acquisition. Once a test point is defined, the operator must write the data on a data sheet or manually enter it into a customized computer application [14].

Force Standard Machines (FSM's) operating with dead weights in the earth's gravitation field are the most accurate way to generate forces because the realization is traceable to the base units of mass, length and time. On the other hand, cost limitations for high capacity machines lead to technical solutions using force amplification of the direct loading by hydraulic or by mechanical lever system. The best measurement uncertainty achieved by dead weight machines is 20 ppm, while the best measurement uncertainties reached with hydraulic amplification are in the range from 100 ppm up to 500 ppm, and 50 ppm up to 200 ppm for lever system.

In Germany, in year 2001, G. Navrozidis et al evaluated new state of the art force standard machine. The fully automated force standard facility is a combined dead weight – lever amplification machine with 110 kN direct load capacity and a 10:1 lever multiplication part. The lever system is affected by disturbing moments that are caused by eccentric coupling of the masses to the lever and eccentric moment of the test transducer [15].

In the case of force standard machines with hydraulic transmission, a defined weight force acts on a piston/cylinder system in which a constant oil pressure builds up. By coupling this system with a second piston/cylinder system having a larger sectional area (pressure balance), a transmission of force by a factor $Q=A_2/A_1$ is achieved, where A_1 and A_2 are the effective areas of the two piston/cylinder systems.

Indeed the advantages of the modern automatic machines are not limited to the possibility of checking the time response of the devices under test, but even the accuracy can be generated with a small number of masses, with the relevant advantage of lower costs of mass production and calibration.

The possibility of self calibration is also important. Self calibration consists of internal mutual comparison of the forces generated by the different masses. Beside the great practical advantage of avoiding the separation of the masses from the machine for calibration, it produces a great metrological advantage. The metrological advantage of self calibration is due to the fact that forces produced by different groups of masses are compared directly at their point of application to the force transducer. It takes, therefore, into account eventual systematic effects due to the force transfer and cancels them.

Besides force standard machines with deadweight, with lever transmission and with hydraulic transmission, there are some others procedures used. Build-up procedure is one of them. Fig. 5.4 indicates the base plate of buildup setup.

Fig. 5.4. Base plate of the Build-up prototype.

For very large capacities, in general above 1 MN, the "build up" procedure may be employed, consisting of comparing the dynamometer to be calibrated with one or more dynamometers linked in parallel. With the "build up" method the field of force measurement may be extended to values that would not appear possible with the use of deadweight or multiplication primary standards, at a much lower cost [16]. Incorrect positioning of a force transducer at the build-up base plate, can be a source of systematic errors for the behavior evaluation in this type of standardization system.

5.3.3.3. Techniques

In the field of dynamic force measurements, there are some improved techniques. On of them is discussed by Yusaku Fujii and J. D. R. Valera. They studied a novel method for accurately measuring small and steep impact force. In this method, the inertial force of a moving mass is used as the reference force applied to the material under test. A pneumatic linear bearing is used to obtain linear motion with negligible friction acting on the mass, i.e., the piston-shaped moving part of the bearing. The impact force is generated and applied to the material by collision with the mass. An initial velocity is manually given to the moving part. A corner-cube prism (CC), that forms part of the interferometer, and a metal block with a round-shaped tip (for adjusting the collision position) are attached to the moving part (made of aluminium with square pole shape); its total mass M is approximately 21.17 g. The inertial force acting on the mass is measured highly accurately using an optical interferometer. The high performance of the proposed method is shown by evaluating the viscoelasticity of a small rubber block under small and steep impact load [17].

Due to the progress achieved in micro system technology and materials science, e.g., medical technology and hardness measurement, there is an increasing need for traceable measurements in the mN range. For small force measurements, there are some difficulties raised from: (1) No National Measurement Institute supports a direct force realization linked to the International System of Units (SI) below 1 N even for static force. (2) Small force to be generated and/or measured is usually a varying force and any dynamic calibration technique for force sensors has not been established yet. In other words, this fact means that the uncertainty evaluations both for the measured value of the small force and for the time of the measurement are very difficult. The proposed optical method is demonstrated in Fig. 5.5 for Micro Force Measurement according to Levitated-Mass Method, is free from these problems [18].

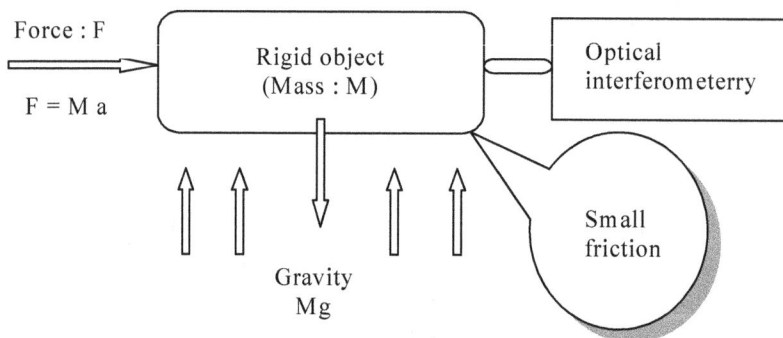

Fig. 5.5. Levitated-Mass Method.

Another trend for measuring the small force is described in Fig. 5.6 by Stefan Niehe at 2003. He has evaluated metrological characteristics for a new force measuring facility for the range from 10 mN to 10 N consisting of a piezoelectric adjustment unit and a precision compensation balance [19].

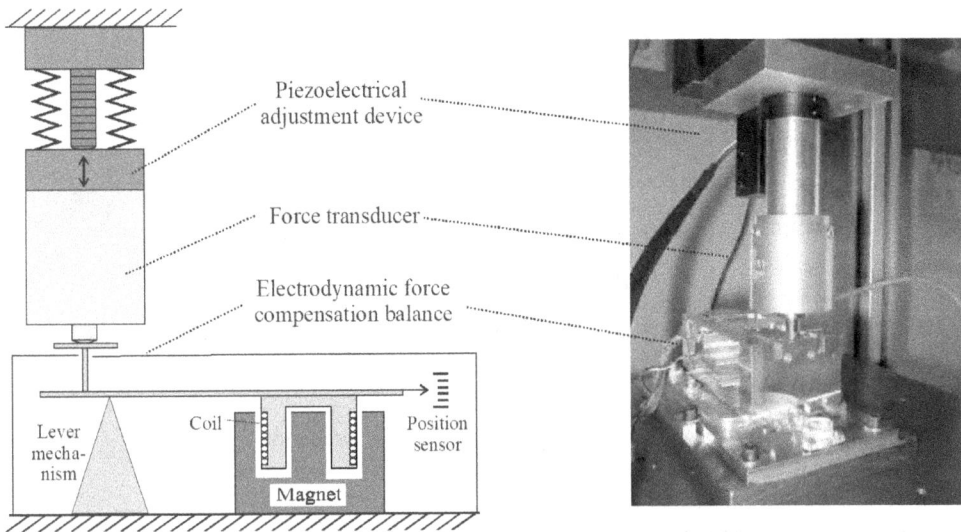

Fig. 5.6. Measuring facility.

It essentially consists of a piezoelectric ID fine adjustment unit with an integrated capacitive feedback sensor and a precise electrodynamic compensation balance with lever mechanism. The force transducer to be investigated is screwed overhead, together with the fine adjustment device, and traced by the coarse adjustment unit to contact the balance. When the fine adjustment device travels downwards, the force transducer presses the load receptor of the balance downwards. A piston sensor records the movement of the lever arm, and the balance automatically changes the current through a coil rigidly connected to the lever arm. The action of force on this current – carrying coil in a magnetic field procedures a counterforce and thus ensures resulting to the initial position by the piston sensor. The force – proportional coil current is a measured of the balance. By variation of the piezoelectrically produced translation, the force transducer can be loaded in compression with different forces which are recorded by the balance.

Fig. 5.7 indicates a new trend for force measurement by hand force measurement system. B. D. Lowe et. al are studying this point [20]. The system incorporates 16 to 20 thin profile conductive polymer pressure sensitive resistors attached to a thin athletic grip glove to cover the pulpar regions of the phalangeal segments and palm that contact the grasped object. The electrical responses of these sensors have been found to be linear when calibrated against applied force distributed over the sensor surface area. The active area of each sensor is circular with a diameter of 9.53 mm. The thickness of each sensor is 0.127 mm. Raw data are sampled via a pc and digitized through a 12-bit data acquisition board set to a ±5 V range. The raw voltage data from the sensors are converted to calibrated force units and displayed in real time.

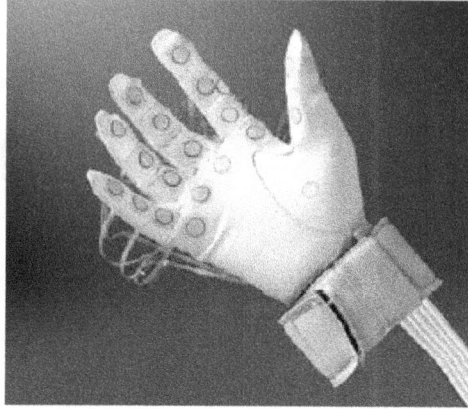

Fig. 5.7. Sensor placement on the force glove.

Attaching thin profile pressure/force sensors to a glove has the advantage of facilitating measurements of hand force on any grasped object, such as a hand tool, without the need to instrument the tool itself. The disadvantage of this measurement approach is that the material properties of the glove and sensor may alter the frictional conditions between the hand and the tool and may degrade tactile representation of the coupling between the hand and the tool. If these limitations are acceptable the glove-based system is a useful measurement device for characterizing hand grip contact forces in the evaluation of many hand tools. The glove system is also advantageous because of its low cost, and replaceable sensors. Through repeated use in industrial activities that involve high grip force levels we have observed damage and failure of the thin profile force sensors. These sensors are relatively inexpensive and can be replaced individually. Other commercial systems have been developed with multiple force sensing elements embedded within a single printed medium. If one of the sensing elements in this medium is damaged the entire sensor medium must be replaced – at significant cost.

5.4. Summary

Developments of force measurement are passed by several steps in the past and still continue in future. The classic applications, e.g. material testing and safety engineering, require force measurements extend from 1 N to 100 MN. In the aerospace industry, off-shore industry and in opencast mining, applications in the MN range dominate. In the medium force range from 1 kN to 1 MN, applications are found in the automotive industry, in materials handling and in aviation, whereas in the textile industry and in automation and medical engineering, forces in the lower force range of up to a few kN are measured. Whilst in the past, traceability was especially required for forces larger than 1 Newton, the need for traceability of smaller forces, in the milli-, micro- and nanonewton range, is constantly increasing today [21]. Also the traceability for large force measurements (e.g. build up system) is needed.

Nowadays, the self check is one of the modern techniques in dead weight machine design. It is solving the problem for mass recalibrating, without release the masses far from the machine [22]. The smallest relative measurement uncertainties are 0.002 % for calibrations with deadweight.

The force sensors are usually mechanical elastic bodies which deform under the action of a force. Most of the time, this deformation is measured electrically, e.g., according to the metrological principle of strain gauges. There are, however, other principles (e.g. piezo-electric force transducers) which generate a charge proportional to the force. For the precision measurement of static forces, e.g. within the scope of international comparison measurements, especially strain gauge force transducers have established themselves, whereas dynamic measurements often require the use of piezo-electric force transducers.

Also, this study review reveals the advantages of some force measurement tools and disadvantages of others. Silicon load cell, for example, could be used for precision measurements which are normally the field of the electromagnetic force compensation technology. Beyond, in spite of the electromagnetic force compensation technology, the Si LCs are not limited to low and medium loads and thus they are suitable to be used as transfer standards [11]. Investigations are performed with mechanical spring made of single-crystalline silicon (S-spring) for the load cell with sputtered-on thin film strain gauges. The measurements revealed that the Silicon load cell behavior is better than the conventional metallic material load cell [23, 24]. But, it is costly use and still under development.

Forces in the nanonewton and piconewton range have already been for several years in scanning force microscopy for the high resolution measurement of surfaces. The increasing industrial use of plastic microparts which get scrathed when they are measured by means of contact measurements where the measuring force is too high, has led to new challenges for quality assurance for both scanning force microscopes (SFMs) and usual contact stylus instruments. A new field of application for scanning force microscopes has resulted from the development of automated force spectroscopy devices for molecular analysis. These devices can determine where active ingredients of medicaments bind to the target molecule and how strong the bond is. It is thus possible to measure forces in the piconewton range. All these measuring procedures are based on the use of soft bending test beams (cantilevers) with integrated stylus tip.

The force measurement concept is focused on the usage of the modern technique not only to measure the small forces but also to characterize the materials. Atomic force microscope on of this method to characterize the nanotribological properties [25].

References

[1]. The UK's National Measurement Institute, National Physical Laboratory (NPL): http://www.npl.co.uk/commercial-services/measurement-services/force/

[2]. Wikipedia Encyclopedia: http://en.wikipedia.org/wiki/force

[3]. P. E. Pontius and R. A. Mitchell, Inherent Problems in Force Measurement, *Experimental Mechanics*, March 1982, pp. 81-88.

[4]. The UK's National Measurement Institute, National Physical Laboratory (NPL): http://www.npl.co.uk/reference/faqs

[5]. Load cell technology in practice, *RT Revere Transducers. Application Notes,* 07/6-13/01, pp. 1-13.

[6]. OMEGA Engineering Technical Reference: http://www.omega.com/prodinfo/loadCells

[7]. http://www.mikesearch.com/tags/1/load-cell

[8]. Interface advanced force measurement: http://www.loadcelltheory.com

[9]. Omega company, Load Cell Designs: http://www.omega.com/literature/transactions/volume3/load.html

[10]. Futek Advanced Sensor Technology: http://www.globalspec.com/FaturedPoducts

[11]. S. Mäuselein, O. Mack, R. Schwartz and G. Jäger, Investigations of new Silicon Load Cells with thin film atrain gauges, *XIX IMEKO World Congress,* Lisbon, Portugal, September 6−11, 2009, pp. 379-382.

[12]. Surasith Piyasin, The detail design of Hollow Clevis-Pin type load cells, Ph. D., Department of Mechanical Engineering, *Khon Kaen University.*

[13]. Sinan Fank and Mehmet Demirkol, Effect of microstructure on the hysteresis performance of force transducers using AISI 4340 steel spring material, *Sensors and Actuators, A,* 126, 2006, pp. 25-32.

[14]. Jeff Grossman, The Electronic Deadweight Tester — A Modern Replacement for the Conventional Deadweight Tester, *The International Journal of Metrology*, Jul.-Aug.-Sep. 2009, pp. 36-39.

[15]. G. Navrozidis, F. Strehle, D. Schwind and H. Gassmann, Operation of a new force standard machine at Hellenic Institute of Metrology, *Proceeding of the IMEKO TC3 Conference,* Istanbul, Turkey, September 17-21, 2001, pp. 119-128.

[16]. C. Ferrero and C. Marinari, Long-Term stability of the IMGC reference force transducers up to 9 MN, in *Proceedings of the IMEKO 17th International Conference on Force, Mass, Torque and Pressure Measurements, TC3,* Istanbul, Turkey, Sept. 17-21, 2001.

[17]. Yusaku Fujii and J D R Valera, Impact force measurement using an inertial mass and a digitizer, *Measurement Science and Technology,* 17, March 2006, pp. 863–868.

[18]. Yusaku FUJII, Optical Method for Micro Force Measurement, *Gunma University.*

[19]. Stefan Niehe, A new force measuring facility for the range of 10 mN to 10 N, in *Proceedings of the XVII IMEKO World Congress,* Dubrovnik, Croatia, June 22-27, 2003, pp. 335-340.

[20]. B. D. Lowea, Y. Kongb, and J. Hanc, Development and application of a hand force measurement system, in *Proceedings of the 16th World Congress on Ergonomics of the International Ergonomics Association*, Maastricht the Netherlands, 10-14 July 2006.

[21]. Rolf Kumme, Jens IIIemann, VIadimir Nesterov and Uwe Brand, Kraftmessung von Mega- bis Nanonewton, *PTB-Mitteilungen,* 118, No. 2 and No. 3, 2008, pp. 33-41.

[22]. Prof. A. A. El-Sayed, Prof. H. M. El-Hakeem, Mr. B. Gloeckner and Dr. T. Allgeier, Performance evaluation and metrological characteristics of deadweight force standard machine with substitute load control system, *IMEKO-TC3,* September 24-26, 2002, Celle, Germany.

[23]. S. Mäuselein, O. Mack, R. Schwartz and G. Jäger, Investigations of load cells made of single-crystalline silicon with sputrered-on strain gauges, *IMEKO 20th, 3rd TC16 and 1th*

TC22 International Conference Cultivating metrological knowledge, Merida, Mexico, November 27–30, 2007.

[24]. S. Mäuselein, O. Mack, R. Schwartz and G. Jäger, Investigations of new silicon load cells with thin-film strain gauges, *XIX IMEKO World Congress, Fundamental and Applied Metrology,* Lisbon, Portugal, September 6–11, 2009, pp. 379-382.

[25]. B. Stegemann, H. Backhaus, H. Kloss, and E. Santner, Spherical AFM Probes for Adhesion Force Measurements on Metal Single Crystals, Modern Research and Educational Topics in Microscopy, A. Méndez-Vilas and J. Díaz (Eds.), 2007, pp. 820-827.

Chapter 6
Gas Sensors Based on Inorganic Materials

K. R. Nemade

6.1. Introduction

In 1962, Seiyama *et al* discovered that the electrical conductivity of ZnO changed dramatically by the presence of reactive gases in the air. In same year his *et al* had written paper on this phenomenon of title "A new detector for gaseous components using semi conductive thin films". Consequently 1962 is birth year of gas sensor technology. After that plenty of work published on this phenomenon by using different inorganic and organic materials throughout the world. Literature survey of metal oxides suggest that they are the most significant class of materials with such diverse properties that cover almost all aspects of material science in the field of superconductivity, ferroelectricity, magnetism and particularly gas sensors. Effective metal oxide catalysts are characterized with large specific surface area and high selectivity. Surface can be greatly enhanced by utilizing new developments in nano technology, especially those related to the synthesis of nano particles [1].

Generally solid state gas sensors can be classified mainly into two main categories that are organic and inorganic materials. Metal oxide thin films are giving their poor performances regarding at lower concentration of gas. Although most commercially available sensors are based on metal oxide gas sensors due to their eco friendliness, easy synthesis methods and they can operated at high temperature. Solid state gas sensors mainly are composed of metal oxide. Metal oxides are most suited gas sensors due to their high sensitivity and their ability to maintain structural integrity in harsh conditions, namely high temperature and pressure. In recent years, the development of inorganic-organic hybrid materials has grown. An electronic sensor system is highly desirable to provide online monitoring of the chemical composition of gas emitted from combustion facilities, in order to minimize air pollution, maintain the concentration of dangerous gaseous species within the limit stipulated by regulations. Semiconductor metal oxide gas sensors of a wide variety of advantage over traditional analytical instruments such as low cost, short response time, easy manufacturing and small size. Despite these qualities

K. R. Nemade
Department of Physics, Sant Gadge Baba Amravati University, Amravati 444 602, India

Semiconductor metal oxide sensors suffers a lack of selectivity. The metal oxide investigated to date are non selective, that is they are sensitive simultaneously to wide range of reducing and oxidizing gases. Some methods to improve Semiconductor metal oxide sensors selectivity that is optimization of temperature, bulk/surface doping and use of molecular filters have been successfully employed. It has been shown by many research teams that a sensor matrix based on several technologies or several Semiconductor metal oxide sensors materials could provide a specific and unique response patterns for different individual chemical species or mixture of species [2].

6.2. Experimental Methodology

Experimental Methodology adapted for study of gas sensors consist of following methods:
1) DC resistance measurements are the 'normal' way for sensor measurements and allow for the screening of a large number of sensors;
2) Work function change measurements by the Kelvin method are also performed in normal operation conditions. They additionally provide insight about surface reactions where free charge carriers are not involved.

6.2.1. DC Measurements

The typical measurement technique for metal oxide sensors is the measurement of conductance or resistance. In this work dc measurements were performed by using electronic circuitry which ensures a constant voltage drop over the sensing layer. With such a technique, a defined polarization and a known and constant measurement potential, were ensured. In this way, effects of the possible influence of the measuring potential, like the ones already reported for very similar samples, were avoided [3].The measuring potential was adjusted in a range in which its influence on the measurement is minimal.

6.2.2. Work Function Change Measurements

The technique used here for measuring differences in work functions is the well established Kelvin oscillator method [4, 5]. In our case, the Kelvin oscillator consists of a metallic grid in electrical contact with the sample, which oscillates at a mean distance d over the sample. Due to the electrical contact, a contact potential VC, which is equal to the difference in work function of the two materials $\Delta\Phi$, occurs. This results in an electrical field between the two plates of the capacitor, the sample and the grid. Changes in distance due to the oscillation result in changes of the capacity C and therefore in a current i(t) according to

$$C(t) = \varepsilon\varepsilon_0 *A / (d_0 + \delta \sin \omega t) = Q(t)/VC \qquad (6.1)$$

$$i(t) = Q = -VC * \varepsilon\varepsilon_0 * A\delta\omega * \cos\omega t / (d_0^2 (1 + \delta/d0) \sin \omega t)^2 \qquad (6.2)$$

Experimentally, a counter-voltage VG is adjusted until i(t) disappear and the contact potential

$$VC = -VG = \Delta\Phi$$

is measured. Consequently, if the work function of the metallic grid is unaffected by changes in the ambient atmosphere it is possible to determine the work function changes of the sample dependent on gas adsorption. By performing simultaneous conductance/resistance and work function measurements one can get information about the changes in surface concentration species that are not carrying a net charge, such as dipoles. This is of special interest when one studies the effect of water vapors because their surface interaction can lead to such species. This happens because work functions Φ of semiconductors contains three contributions: the energy difference between the Fermi level and conduction band in the bulk $(E_C - E_F)_b$, bandbending qV_S and electron affinity χ. All three contributions may, in principle, change upon gas exposure

$$\Phi = (E_C - E_F)_b + q*V_S + \chi \qquad (6.3)$$

The corresponding conductance G may formally be described by equation (6.4) [6] here we assume that the work function changes measured by the Kelvin probe at the outer layers are identical with the electrostatic potential changes involving the inner grains. The latter control the conduction changes upon gas exposure; one also assumes a narrow size distribution of the electrically active grains in the sensing layer and hence homogeneous electrical properties, percolation paths of the conductivity independent of the work function changes, and constant mobility of the charge carriers in the nanoparticles

$$G = G_0 \exp\{(E_F - E_C)_b - q*V_S / kT\} = G_0 \exp\{\chi - \Phi / k*T\} \qquad (6.4)$$

It is worth mentioning that changes in the electron affinity χ influence the work function Φ but not the electrical conductance G. By combining equations (6.3) and (6.4), it gives that

$$k*T \ln (G/G_0) = \chi - \Phi \qquad (6.5)$$

If the sensors are exposed to two different gas atmospheres (initial gas atmosphere I and final gas atmosphere F) and the related resistances (R_I, R_F) and work functions (Φ_I, Φ_F) are monitored, the different contributions to the work function change, $\Delta\Phi$, can be determined according to

$$\Phi_{I,F} = -k*T \ln(G_{I,F}/G_0) + \chi_{I,F} \qquad (6.6)$$

Hence,

$$\Delta\Phi = \Phi_F - \Phi_I = k*T \ln(G_I/G_F) + \chi_F - \chi_I = k*T \ln(R_F/R_I) + \Delta\chi, \qquad (6.7)$$

where, Φ_I, R_I and χ_I are the work function, the resistance and the electron affinity in the initial gas atmosphere, respectively, and Φ_F, R_F and χ_F are the corresponding values in the final gas atmosphere. In all experiments $\Delta\Phi$ and the electrical resistance are measured and $\Delta\chi$ is calculated. An increase in humidity results in a decreased resistance and an increased electron affinity χ.

6.3. Sensing Mechanism

Physics, chemistry, and technology of semiconductor sensors require a better understanding both the bulk and surface properties of the sensing materials. The sensing material interact the gas and changed it fully or partially. The adsorption of the gas or vapor by the sensor surface can be associated with decomposition or dissociation. A reducing molecule (CO, H_2 etc.) adsorbed on the sensor surface acts as a surface donor, injecting electrons into it. The opposite phenomenon occurs during exposure to oxidizing gases like that of NO.

Physical-chemical processes can be rather complicate, which lead to changes of many parameters of a sensing device. For example, most often the sensor element gets covered with decomposition products like carbon, CO_2, and H_2O causing a gradual decrease in the sensitivity at the operating temperature. A periodic change of the sensor temperature removes all the adsorbed species and unburnt organic contaminants from the surface. Oxidation of hydrocarbons generally proceeds through partially oxidized intermediates. The hydrocarbons (CnHm) are adsorbed on the sensor surface and react with the surface oxygen species.

Sensing Mechanism of metal oxide includes two functions. The first is the receptor which represents the chemical interaction on the surface of metal oxide layer between chemisorbed molecules, mostly oxygen ions and ambient gas molecules such as CO and H_2. The second is the transduction function which represents the dependence of the electrical conductivity on the occupied surface trap density [7-10]. The occupied surface trap density determines the depletion layer length of metal oxide. Therefore the ratio between the average grain size of metal oxide and the depletion layer length is important if high sensitivity is to be achieved [11].

Sensing Mechanism considering the influence factors on gas sensing properties of metal oxides, it is necessary to reveal the sensing mechanism of metal oxide gas sensor. The exact fundamental mechanisms that cause a gas sensors response are still controversial, but essentially trapping of electrons at adsorbed molecules and band bending induced by these charged molecules are responsible for a change in conductivity. The negative charge trapped in these oxygen species causes an upward band bending and thus a reduced conductivity compared to the flat band situation .When O_2 molecules are adsorbed on the surface of metal oxides, they would extract electrons from the conduction band E_c and trap the electrons at the surface in the form of ions. This will lead a band bending and an electron depleted region. The electron-depleted region is so called space-charge layer, of which thickness is the length of band bending region. Reaction of these oxygen species with reducing gases or a competitive adsorption and replacement of the adsorbed oxygen by other molecules decreases and can reverse the band bending, resulting in an increased conductivity. O^- is believed to be dominant at the operating temperature of 300–450 °C [12] which is the work temperature for most metal oxide gas sensors. When gas sensors exposure to the reference gas with CO, CO is oxidized by O^- and released electrons to the bulk materials. Together with the decrease of the number of surface O^-, the thickness of space-charge layer decreases. Then the Schottky barrier between two grains is lowered and it would be easy for electrons to conduct in sensing layers through different grains. However, the mechanism in n-type semiconducting metal oxides of which depletion regions is smaller than grain size.

6.4. Relation between Resistance and Sensitivity

According to the thermionic emission theory the current density across the grain boundary [13] is

$$J_{th} = e^2 n_b (1/2\pi m_n k^*T)^{1/2} \exp(-e^*V_s/k^*T)V_a , \qquad (6.8)$$

where, k is the Boltzmann constant, T is the absolute temperature, m_n is the effective mass of electron, and V_a is the applied voltage across the grain boundary, which is approximately equal to the electric field across the grain boundary multiplied by the grain size L. Current density $J_{th} = \sigma E$ where σ is the conductivity, and the grain boundary current

$$I = A_{GB} * \sigma * (Va/L),$$

where A_{GB} is the cross section of grain boundary. Then grain boundary resistance can be derived as

$$R_{GB} = (2\pi\alpha*k*T)^{1/2}/A_{GB}e^2 n_b \exp(e^*V_s/k^*T) , \qquad (6.9)$$

where α is the ratio of effective electron mass to electron mass. All quantities are measure in SI system. Supposing the free electron density in depleted layer is n_d, the total current through the neck is

$$I_N = \pi e[\mu_b * n_b * r_0^2 + n_d * \mu_d * (r_m^2 - r_0^2)] * V_N/L , \qquad (6.10)$$

where, μ_b is the electron mobility in the neutral grain body, μ_d is the electron mobility in the depleted layer, and V_N is the applied voltage across the neck. The neck resistance is

$$R_N = L/\pi e[\mu_b * n_b * r_0^2 + n_d * \mu_d * (r_m^2 - r_0^2)] , \qquad (6.11)$$

where, $n_d = n_b \exp(-e^*V_s/k^*T)$.

Actually there is no straight forward uniform definition for gas sensor sensitivity now. Usually, sensitivity defined as

$$S = (R_g - R_a)/R_a$$

where, R_a stands for the resistance of gas sensors in the reference gas (usually the air) and R_g stands for the resistance in the target gases. Both R_a and R_g have a significant relationship with the surface reaction.

6.5. Factors Affecting Sensitivity

6.5.1. Doping

The sensitivity of metal-oxide gas sensors can be substantially improved by dispersing a low concentration of additives, such as Pd [14], Pt [15], Au [16, 17], Ag [18, 19], Cu [20], Co [21], and F [22] on oxide surface or in its volume. Although doping has been used for a long time now in preparation of commercial gas sensors, the working principle of additive-modified metal-oxide materials is still not completely understood.

6.5.2. Surface Area of Grain

One of the common ways to tune the sensitivity and selectivity of the materials is to dope them with various noble metals. A simple, yet promising alternative to enhance the gas-sensing performance is to control the morphological features of the materials during the chemical synthesis. In particular, the generation of high specific surface areas and uniform systems of large (meso) pores will result in a higher probability for a gas to interact with the semiconductor, which is likely to increase the sensitivity of the material. The concept of utilizing self-assembled arrays of amphiphiles to synthesize well-ordered mesoporous materials [23] has recently been applied to tin(IV) oxides (SnO_2) [24-28].

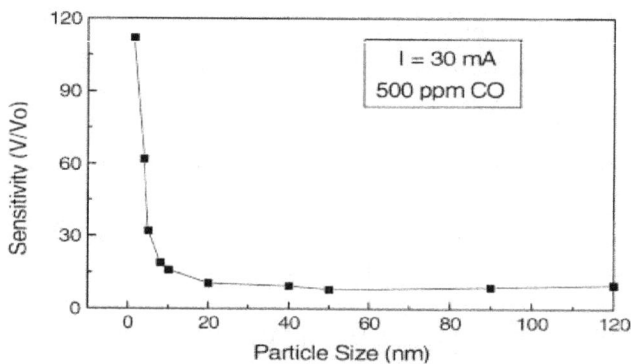

Fig. 6.1. Effect of particle size on gas sensitivity for CO (adapted from [29]).

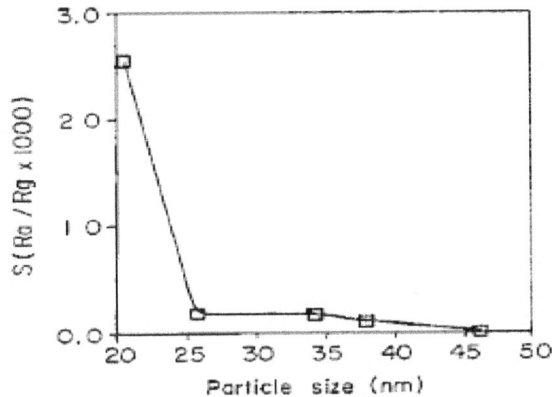

Fig. 6.2. Variation in sensitivity with average particle size (adapted from [30]).

6.5.3. Working Temperature

Working temperature is usually the temperature that corresponds to maximum sensitivity. Semiconductor Metal Oxide such as SnO_2, ZnO, TiO_2, MoO_3 and Fe_2O_3, cannot be used at high temperature due to the sensitivity decreasing with increasing operating temperature. For instance, the sensitivity of SnO_2 based sensors vanished when temperature exceeds 450 °C [31]. For harsh high temperature environment, semiconductor sensors are less attractive.

6.5.4. Dielectric Constant

The Maxwell–Garnett theory is an Effective Medium Theory (EMA) with which the complex dielectric function of the composite material can be calculated as a weighted average dielectric constant of two components. The composite dielectric function only depends on the volume action and shape of the metal particles [32]. The general equation that describes the EMA model is given by,

$$(\varepsilon_t\text{-}\varepsilon_h)/(\varepsilon_t\text{-}Y\varepsilon_h) = \sum f_j(\varepsilon_t\text{-}\varepsilon_h)/(\varepsilon_t\text{-}Y\varepsilon_h) , \qquad (6.12)$$

where value of j vary from 1 to m, m is the number of materials included in the basic material matrix, ε_t and ε_h are the dielectric functions of the EMA (total system) and the host materials, respectively, ε_j is the dielectric constant of the included material, f_j is the volume fraction of the material included and Y is a screening factor used to describe the microstructure of the mixed materials. The screening factor describes the shape of microstructure through the equation:

$$Y + 1 = 1/ \text{ screening factor} \qquad (6.13)$$

The screening factor of 1/3 describes a spherical microstructure, the screening factor of 1 describes a flat disc, and a 0 value represents a columnar microstructure [33]. Recently, a theoretical model suggested that the sensitivity is inversely proportional to average grain size of metal oxide. Since the depletion layer is proportional to dielectric constant, metal oxide with higher dielectric constant such as TiO_2 ($\varepsilon=100$) will result in better sensitivity.

6.6. Selectivity of Metal Oxide Gas Sensors

One of the main challenges to the developers of metal-oxide gas sensors is high selectivity. Currently, two general approaches exist for enhancing the selective properties of sensors. The first one is aimed at preparing a material that is specifically sensitive to one compound and has low or zero cross-sensitivity to other compounds that may be present in the working atmosphere. To do this, the optimal temperature, doping elements, and their concentrations are investigated [34-36].

Nonetheless, it is usually very difficult to achieve an absolutely selective metal oxide gas sensor in practice, and most of the materials possess cross-sensitivity at least to humidity and other vapors or gases. Another approach is based on the preparation of materials for discrimination between several analytes in a mixture. It is impossible to do this by using one sensor signal; therefore, it is usually done either by modulation of sensor temperature or by using sensor arrays [37].

6.7. Stability of Metal Oxide Gas Sensors

Another issue of metal-oxide gas sensor materials is their low stability and long-range signal drift. This problem leads to in uncertain results, false alarms and the need to frequently recalibrate or replace sensors. Little attention is paid in the literature to the problems of stability. Only a few papers, cited in this chapter, report the stability of sensor response in a period of several days. Generally, nanostructured oxides with small grains as well as nanotubes, nanorods etc. are subject to degradation because of their high reactivity. There is no unified approach to increasing the stability of metal-oxide gas sensors. To some extent, stability can be increased by calcination and annealing as the post-processing treatment and by reducing the working temperature of the sensor element. Doping metal oxides with metal particles or carbon nanotubes as well as synthesis of mixed oxides have been also reported to increase the stability of sensor elements. One can distinguish between the two types of sensor stability. One is connected with reproclucibility of sensor characteristics during a certain period of time at working conditions, which may include high temperature and the presence of a known analyte. Such stability may be referred to as active stability. Another type of sensor stability, which can be called conservative stability, is connected with retaining the sensitivity and selectivity during a period of time at normal storage conditions, such as room temperature and ambient humidity [38].

6.8. Metal Oxides for Gas Sensors

Many metal oxides are suitable for detecting combustible, reducing, or oxidizing gases by conductive measurements. The following oxides show a gas response in their conductivity: Cr_2O_3, Mn_2O_3, Co_3O_4, NiO, CuO, SrO, In_2O_3, WO_3, TiO_2, V_2O_3, Fe_2O_3, GeO_2, Nb_2O_5, MoO_3, Ta_2O_5, La_2O_3, CeO_2, Nd_2O_3. Metal oxides selected for gas sensors can be determined from their electronic structure. The range of electronic structures of oxides is so wide that metal oxides were divided into two the following categories:

1) Transition-metal oxides (Fe_2O_3, NiO, Cr_2O_3, etc.).

2) Non-transition-metal oxides, which include (a) pre-transition-metal oxides (Al_2O_3, etc.) and (b) post-transition-metal oxides (ZnO, SnO_2, etc.). Pre-transition-metal oxides (MgO, etc.) are expected to be quite inert, because they have large band gaps. Neither electrons nor holes can easily be formed. They are seldom selected as gas sensor materials due to their difficulties in electrical conductivity measurements [39]. Responses expected for metal oxides (n-type and p-type) to reducing and oxidizing gases are shown in Table 6.1.

Table 6.1. Responses expected for semiconducting metal oxide to reducing and oxidizing gases are as follows.

Material	Reducing Gas	Oxidizing Gas
n-type	Resistance falls	Resistance rises
p-type	Resistance rises	Resistance falls

Today, most of the commercial metal-oxide gas sensors are manufactured by screen printing on small and thin ceramic substrates. The advantage of this preparation technique is that the thick films of metal oxide semiconductor can be deposited in batch processing, leading to small deviations of characteristics for different sensor elements. Although this fabrication technology is well-established, it possesses a number of drawbacks and needs to be improved. Primarily, the drawbacks are connected with the necessity to keep the thick metal oxide film at high temperature. Due to this reason the power consumption of screen-printed sensors can be as high as 1 W, which makes them unable to be used in battery-driven devices. These problems have promoted the development of substrate technology and strong research in preparation of the sensitive layer. One promising solution is the integration of a sensing layer in standard microelectronic processing, which overcomes the difficulties of the screen-printed sensors. In this case, an oxide layer is deposited onto a thin dielectric membrane of low thermal conductivity, which provides good thermal isolation between the substrate and the heated area on the membrane. Such a construction allows the power consumption to be kept at very low levels [38].

6.8.1. Hydrogen Gas Sensors

Hydrogen is the most attractive and ultimate candidate for a future fuel and an energy carrier. Its generation can be realized by a variety of methods, including reforming of natural gas and alcohols (methanol etc.).Metal oxides are extremely studied with hydrogen gas. Sensing response of metal oxides towards hydrogen gas shown in Table 6.2.

6.8.2. Oxygen Gas Sensor

Oxygen sensors are producing in a large scale exceeding several ten million sets yearly. Oxygen sensors are widely used as sensors for automotive applications. To decrease the exhaust emissions from gasoline internal combustion engine automobiles, the air/fuel ratio is monitored with oxygen sensors which usually made of a metal oxide. Sensing response of metal oxides towards oxygen gas shown in Table 6.3.

Table 6.2. Some recent papers on detection of hydrogen.

Sensing material	Operating temperature range (°C)	Range of detection limit	Sensor physical parameter	Response time	Reference
ZnO	250-350	200-5000 ppm	Electrical resistance	2 min	40
Pt-ZnO	300-400	8000 ppm	Electrical resistance	30 min	41
ZnO-Ru	100-450	200 ppm	Electrical resistance	-	42
ZnO(Al, Ag)	200-400	2-1000 ppm	Electrical resistance	2-5 min	43
ZnO	300	200-5000 ppm	WorkFunction	-	44
CdO	450	2.1 %	Electrical resistance	5-90 s	45
CeO	500	2.1 %	Electrical resistance	10-60 s	45
CuO	450	9 %	Electrical resistance	<1 min	45
TiO_2	420	0.5 %	Electrical resistance	30 s	45
In_2O_3	350	1000 ppm	Electrical resistance	-	46
SnO_2	25-575	50-1000 ppm	Electrical resistance	12-25 s	47
SnO_2-Sn	150-250	100-5000 ppm	Electrical resistance	-	48
SnO_2-Pd	200-450	100-5000 ppm	Electrical resistance	5-7 s	49
TiO_2-Nb, Pd	200-250	1000 ppm	Electrical resistance	-	50
Ga_2O_3	600	500 ppm	Electrical resistance	-	51
In_2O_3	25-250	1000 ppm	Electrical resistance	30 s	51
Ga_2O_3-SiO_2	700	12.5-500 ppm	Electrical resistance	30 min	51
MoO_3-V_2O_5	150	1000-10000 ppm	Electrical resistance	20 s	51
TiO_2-WO_3	200	500 ppm	Electrical resistance	1-20 min	51
TiO_2 nanotube	25-250	1000 ppm	Electrical resistance	30 s	51

Table 6.3. Some recent papers on detection of oxygen.

Sensing material	Operating temperature range (°C)	Range of detection limit	Sensor physical parameter	Response time	Reference
CuO	450	2.1 %	Electrical resistance	< 1 min	45
CdO	450	2.1 %	Electrical resistance	< 1 min	45
CeO_2	700-1100	-	Electrical resistance	5-10 ms	52
SnO_2	27-650	2-10 ppm	Electrical resistance	10-20 min	53
TiO_2	200-800	Up 100 %	Electrical resistance	-	51
TiO_2 (Pd/Pt)	225	-	Electrical resistance	-	51
ZrO_2	253	-	EMF	-	51

6.8.3. Nitrogen Oxides Gas Sensor

Large amounts of nitrogen oxides are produced by motor vehicles. Other sources are electric Utilities and other industrial, commercial, and residential sources that burn fuels. Nitrogen Oxides gas can cause various problems such as smog and acid rain. Thus, sensors are needed for environmental monitoring and for use in cars to control the combustion process. Sensing response of metal oxides towards nitrogen oxide gas shown in Table 6.4.

Table 6.4. Some recent papers on detection of Nitrogen Oxides.

Sensing material	Operating temperature range (°C)	Range of detection limit	Sensor physical parameter	Response time	Reference
ZnO	200	1000 ppm	Electrical resistance	-	38
$SnO(WO_3)$	150	500 ppm	Electrical resistance	-	38
ZnO:In	275	5 ppm	Electrical resistance	-	38
ZnO	300-350	1 ppm	Electrical resistance	-	38
SnO_2/SWCNT	150	5-60 ppm	Electrical resistance	-	38
SnO_2:Zn	100-500	0.1-5 ppm	Electrical resistance	-	38
ZnO:Al	100	20 ppm	Electrical resistance	-	38

6.8.4. Carbon Monoxide Gas Sensor

It is well known that CO is produced due to the incomplete combustion of fuels, it is commonly found in the emission of automobile exhaust. Such toxic and dangerous is colorless and odorless. The demand for better environmental control and safety has increased research activities of solid-state gas sensors. Sensing response of metal oxides towards carbon monoxide gas shown in Table 6.5.

Table 6.5. Some recent papers on detection of carbon monoxide.

Sensing material	Operating temperature range (°C)	Range of detection limit	Sensor physical parameter	Response time	Reference
CuO	350	1-1000 ppm	Electrical resistance	-	51
Nb_2O_5	400-500	100-1000 ppm	Electrical resistance	-	51
$SnO_2(ZnO)$	27-800	50-10000 ppm	Electrical resistance	5 min	51
TiO_2 (CuO)	267-600	20-100 ppm	Electrical resistance	1 min	51
$SnO_2(Pt)$	300	100 ppm	Electrical resistance	2 s	51
$SnO_2(Bi_2O_3)$	20-300	1500 ppm	Electrical resistance	1 s	51
SnO_2-Pd/Pt	523	1000 ppm50-	Electrical resistance	13 s	51

6.9. Conclusions

Gas sensors based on metal oxide are complex devices and that it is not possible to understand them in the absence of systematic approach. Inorganic material has very poor sensitivity and selectivity at lower concentration. Much effort is being made to extend the working temperature range of metal-oxide gas sensors and lower the optimal working temperature to 20-25 °C. The goal of these investigations is to decrease the power consumption of sensor elements. Metal oxide have virtue can maintain their integrity in harsh condition. Present review suggests that sensitivity, selectivity and stability of inorganic material can improve by various methods like optimization of temperature, doping with novel element. Surface area of grains also play important role in sensitivity. Surface can be greatly enhanced 3S parameters of metal oxide gas sensors, consequently new developments in nano technology, especially those related to the synthesis of nano particles are needed.

Acknowledgement

Author is very much thankful to Head, Department of Physics, Sant Gadage Baba Amravati University, Amravati, for providing the necessary facilities.

References

[1]. M. Hosseinpour, S. J. Ahmadi, M. Outokesh, T. Mousavand, A. Charkhi, A. Tayebi, A Novel granulated copper oxide nano-catalyst with high porousity, *Nanotechnology*, Vol. 15, 2002, pp. 37–42.

[2]. Alexey A. Tomchenko, Gregory P. Harmer, Brent T. Marquis, John W. Allen, Semiconducting metal oxide sensor array for selective detection of combustion gases, *Sensors and Actuators B: Chemical*, Vol. 93, 2003, pp. 126–134.

[3]. M. Bauer, N. Barsan, K. Ingrisch, A. Zeppenfeld, I. Denk, B. Schuman, U. Weimar, W. Gopel, Geometry and electrodes on the characteristics of thick films SnO_2 gas sensors, *Sensors and Actuators B: Chemical*, Vol. 43, 1997, pp. 45–51.

[4]. J. C. Riviere, Work Function: Measurements and Results in Solid State Surface Science, Ed. M. Green, Vol. 1, *Dekker*, New York, 1969.

[5]. J. C. Tracy, J. M. Blakely, A study of faceting of tungsten single crystal surfaces, *Surface Science*, Vol. 13, Issues 2, 1968, pp. 313-336.

[6]. N. Barsan, U. Weimar, Conduction model of metal oxide gas sensors, *Jr. of Electroceramics*, Vol. 7, Issues 3, 2001, pp. 143-167.

[7]. M. E. Frank, T. J. Koplin, U. Simon, Nanogaps for Sensing, *Procedia Chemistry, Proceedings of the Eurosensors XXIII Conference*, Vol. 1, 2006, pp. 746-749.

[8]. C. Xu, J. Tamaki, N. Miura, N. Yamazoe, Grain size effects on gas sensitivity of porous SnO_2-based elements, *Sensors and Actuators B: Chemical*, Vol. 3, 1991, pp. 147-155.

[9]. N. Yamazoe, Humidity sensors: Principles and applications, *Sensors and Actuators B: Chemical*, Vol. 10, 1986, pp. 379 – 398.

[10]. A. Rothschild, F. Edelman, Y. Komem, F. Cosandey, Fabrication of Anatase Thin Films from Peroxo- polytitanic Acid by Spray Pyrolysis, *J. of Electrochemical Society*, Vol. 143, 2001, pp. 191-193.

[11]. S. R. Morrison, Chemical sensing with solid state devices, *Academic Press*, 1989.

[12]. G. Korotcenkov, Metal Oxides for Solid-State Gas Sensors: What Determines Our Choice?, *Mater. Sci. Eng.* B, Vol. 139, 2007, pp. 1-23.

[13]. Y. Ma, W. L. Wang, K. J. Liao, C. Y. Kong, Study on Sensitivity of Nano-Grain ZnO Gas Sensors, *Wide Bandgap Materials*, Vol. 10, 2002, pp. 113-120.

[14]. T. B. Fryberger, S. Semancik, Conductance response of $Pd/SnO_2(110)$ model gas sensors to H_2 and O_2, *Sensors and Actuators B: Chemical*, Vol. 2, 1990, pp. 305-310.

[15]. Mädler L, Sahm T, Gurlo A, Grunwaldt J. D., Barsan N, Weimar U., Pratsinis S. E., Sensing low concentrations of CO using flame-spray-made Pt/SnO_2 nanoparticles, *Nanoparticle Research*, 2006, Vol. 8, pp. 783-796.

[16]. U.-S. Choi, G. Sakai, K. Shimanoe, and N. Yamazoe, Sensing properties of Au-loaded SnO_2-Co_3O_4 composites to CO and H_2, *Sensors and Actuators B: Chemical*, Vol. 107, 2006, pp. 397-401.

[17]. O. Wurzinger and G. Reinhardt, CO-sensing properties of doped SnO_2 titania thin films, *Sensors and Actuators B: Chemical*, Vol. 115, 2006, pp. 403-411.

[18]. R. K. Joshi, F. E. Kruis and O. Dmitrieva, Gas sensing behavior of $SnO_{1.8}$:Ag films composed of size- selected nanoparticles, *Journal of Nanoparticle Research*, Vol. 8, 2006, pp. 797-801.

[19]. J. Gong, Q. Chen, M. R. Lian, N. C. Liu, R. G. Stevenson, and F. Adami, Micromachined nanocrystalline silver doped SnO_2 H_2S sensor, *Sensors and Actuators B: Chemical, Vol.* 32, 2006, pp. 114-119.

[20]. A. Galdikas, A. Mironas, and A. Setkus, Cu-doping level controlled sensitivity and selectivity of tin oxide based thin films, *Sensors and Actuators B: Chemical*, Vol. 26, 1995, pp. 29-33.

[21]. S. B. Patil, M. A. Patil, and P. P. More, Acetone vapour sensing characteristics of cobalt-doped SnO2 thin films, *Sensors and Actuators B: Chemical*, Vol. 125, 2007, pp. 126-130.

[22]. C. H. Han, D. U. Hong, J. Gwak, and S. D. Han, Mo/Schottky barrier diodes on 4H-silicon carbide, Vol. 24, *Sensors and Actuators B: Chemical* 2007, pp. 927-930.

[23]. M. L. Madou, S. R. Morrison, Chemical Sensing with Solid State Devices, *Academic Press*, 1989.

[24]. W. Chu, D. H. Olson, E. W Sheppard, S. B McCullen, J. B. Higgins, J. L. Schlenker, A 28-year-old synthesis of micelle-templated mesoporous silica, *Microporus Materials*, Vol. 10, 1992, pp. 283-286.

[25]. K. G. Severin, T. M. Abdel-Fattah, T. J. Pinnavaia, Supramolecular Assembly of Mesostructured Tin Oxide, *Chemical Communication*, 1998, pp. 1471–1472.

[26]. F. Chen, M. Liu, Metal oxide semiconductor gas sensors present interesting advantages in relation to other gas-sensor, *Solid State Ionics*, Vol. 166, 2004, pp. 241-250.

[27]. (a) Y. Wang, C. Ma, X. Sun, H. Li, Synthesis of mesoporous structured material based on tin oxide, Microporous and Mesoporous Materials, *Sensors and Actuators B: Chemical*, Vol. 49, 2001, pp. 171-178. (b) Y. D. Wang, C. L. Ma, X. H. Wu, X. D. Sun, H. D. Li, Electrical and gas-sensing properties of Mesostructured tin oxide-based H_2 sensor, *Sensors and Actuators B: Chemical*, Vol. 85, 2002, pp. 270-276.

[28]. D. N. Srivastava, S. Chappel, O. Palchik, A. Zaban, A. Gedanken, Sonochemical synthesis of mesoporous tin oxide, *Mater. Res. Bull.*, Vol. 37, 2002, pp. 1721-1735.

[29]. Lu, F., Liu, Y., Dong, M., Wang, X. P., Nanosized Tin Oxide as the Novel Material with Simultaneous Detection towards CO, H_2 and CH_4, *Sensors and Actuators B: Chemical*, 2000, Vol. 66, pp. 225- 227.

[30]Ansari, S. G., Boroojerdian, P., Sainkar, S. R., Karekar, R. N., Aiyer, R. C., Kulkarni, S. K., Grain Size Effects on H2 Gas Sensitivity of Thick Film Resistor Using SnO_2 Nanoparticles, *Thin Solid Films,* 1997, Vol. 295, pp. 271-276.

[31]. T. Hyodo, N. Nishida, Y. Shimizu, Y. Egishara, Preparation and gas-sensing properties of thermally stable mesoporous SnO_2, *Sensors and Actuators B: Chemical*, Vol. 83, 2002, pp. 209-215.

[32]. F. Menil, V. Coillard, C. Lucat, Critical review of nitrogen monoxide sensors for exhaust gases of lean burn engines, *Sensors and Actuators B: Chemical*, Vol. 67, 2000, pp. 1-23.

[33]A. Rothschild, Y. Komem, Si substrates on which the TiO_2 sensors were prepared, *J. Appl. Phys.*, Vol. 95, 2004, pp. 374-377.

[34] S. Chnkraborty, Sen A., and H. S. Maiti, Selective detection of methane and butane by temperature modulation in iron doped tin oxide sensors, *Sensors and Actuators B: Chemical*, Vol. 115, 2006, pp. 610-615.

[35]. F. Parret, Ph. Ménini, A. Martinez, K. Soulantica, A. Maisonnat and B. Chaudret, Improvement of micromachined SnO2 gas sensors selectivity by optimised dynamic temperature operating mode, *Sensors and Actuators B: Chemical*, Vol. 118, 2006, pp. 276-281.

[36]. S. Nakata, H. Okunishi, and Y. Nakashimn, Distinction of gases with a semiconductor sensor under a cyclic temperature modulation with second-harmonic heating, *Sensors and Actuators B: Chemical*, Vol. 119, 2006, pp. 556-561.

[37]. S. Nakata, H. Okunishi, and Y. Nakashima, Distinction of gases with a semiconductor sensor depending on the scanning profile of a cyclic temperature, *Analyst.*, 2006, Vol. 131, Issue 1, pp. 148-154.

[38]. V. E. Bochenkov, G. B. Sergeev, Sensitivity, Selectivity, and Stability of Gas-Sensitive Metal-Oxide Nanostructures, *American Scientific Publishers*, 2010, Vol. 3, pp. 31-52.

[39]. E. Kanazawa, G. Sakai, K. Shimanoe, Y. Kanmura, Y. Teraoka, N. Miura, N. Yamazoe, Metal Oxide Semiconductor N_2O Sensor for Medical Use, *Sensors and Actuators B: Chemical*, Vol. 77, 2001, pp. 72-77.

[40]. D. Davazoglou, K. Georgouleas., Low pressure chemically vapor deposited WO_3 thin films for integrated gas sensor applications, *J. Electrochem. Soc.*, 1998, Vol. 145, pp. 1346-1350.

[41]. B. Bott, T. A. Jones and B. Mann, The detection and measurement of CO using single crystals, *Sensors and Actuators B: Chemical,* 1984, Vol. 5, pp. 65-73.

[42]. P. Van Geloven, J. Moons, M. Honore and J. Roggen, Tin (IV) oxide gas sensors: thick-film versus metallo-organic based sensors, *Sensors and Actuators B: Chemical*, Vol. 17, 1989, pp. 361-368.

[43]. H. Nanto, T. Minami, S. Takata, Zinc oxide thin film as ammonia gas sensor, *J. Appl. Phys.*, 1986, Vol. 60, pp. 482-484.

[44]. Y. Shimizu, Microstructure of TiO_2 and ZnO films fabricated by the sol-gel method., *Sensors and Actuators B: Chemical*, 2002, Vol. 83, pp. 195-201.

[45]. T. Seiyama and S. Kagawa, Study on a detector for gaseous components using semiconductive thin films, *Analytical Chemistry*, 1966, Vol. 38, pp. 1069-1073.

[46]. W. Y. Chung, Semiconductor metaloxide hydrocarbon gas sensors, *Sensors and Actuators B: Chemical*, 1998, Vol. 46, pp. 139-145.

[47]. K. Katsuki and K. Fukui, H_2 selective gas sensor based on SnO_2, *Sensors and Actuators B: Chemical,* Vol. 52, 1998, pp. 30-37.

[48]. W. K. Choi, S. K. Song, J. S. Cho, Y. S. Yoon, H_2 gas-sensing characteristics by an activator layer, *Sensors and Actuators B: Chemical,* 1997, Vol. 40, pp. 21-27.

[49]. G. S. Korotchenkov, Semiconductor metaloxide hydrocarbon gas sensors, *Sensors and Actuators B: Chemical*, 1999, Vol. 54, pp. 202-209.

[50]. Y. Li, W. Wlodarski, K. Galatsis, S. H. Moshli, J. Cole, S. Russo, N. Rockelmann, Gas Sensing Properties of p-type Semiconducting Cr-doped TiO_2 Thin Films, *Sensors and Actuators B: Chemical*, Vol. 83, 2002, pp. 160-163.

[51]. V. M. Aroutiounian, Metal oxide hydrogen, oxygen and carbon Monooxide sensors, *International Scientific Journal for Alternative Energy*, 2006, Vol. 11, pp. 12-22.

[52]. Saito, S., Miyayama, M., Koumoto, K., Yanagida, H., Gas sensing with metal oxide ceramics, *J. Am. Ceram. Soc.*, Vol. 68, 1985, pp. 40-43.

[53]. N. Yamazoe, J. Fuchigami, M. Kiskawa, T. Seiyama, Interaction of tin oxide surface with O_2, H_2O and H_2, *Surface Science*, Vol. 86, 1979, pp. 402-410.

[54]. L. Zhang, TiO_2 as oxygen sensor, *Int. Conf. Solid-State Sensors & Actuators*, 2001, pp. 153-158.

Chapter 7
Microcantilever-based Sensors for Biological and Chemical Sensing Applications

Qing Zhu

7.1. Introduction

Biological and chemical sensors have recently become a major scientific interest with a wide variety of biomedical, environmental and homeland security applications. Especially after 9-11, there is an urgent need for highly sensitive sensors for rapid, real time, and *in situ* biological and chemical warfare agent (CWA) detections in both civilian and military environments.

After microcantilever was introduced as a novel sensing paradigm in the 90s, [1, 2] numerous sensing applications ranging from chemical, [3-15] physical, [2, 16-21] and biological and biomedical [22-57] areas have been demonstrated. For the chemical and biological detection, receptors were immobilized on the microcantilever surface to bind target chemical molecules, DNA, protein molecules, or bacteria. For example, Fig. 7.1 showed a Scanning Electron Micrograph (SEM) picture of a single *Escherichia coli (E. coli)* cell captured on an antibody immobilized silicon microcantilever after detection [39]. Binding of target species to the receptors on the microcantilever surface is detected by monitoring the tip bending displacement or the resonance frequency shift of the microcantilever.

7.2. Detection Schemes

Typically, for microcantilever sensors, two approaches are used to perform sensing and detection: 1) Dynamic (Resonance) method, and 2) Static (deflection) method.

Qing Zhu
FUJIFILM Dimatix, Inc.

Sergey Y. Yurish (ed.), *Modern Sensors, Transducers and Sensor Networks*
© International Frequency Sensor Association Publishing, 2012

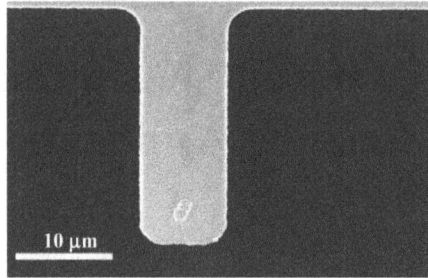

Fig. 7.1. A Scanning electron micrograph (SEM) micrograph of a single E. coli O157:H7 cell bound to the immobilized antibody layer on top of a silicon microcantilever [39].

7.2.1. Dynamic Sensing Method

The flexural resonance frequency (*f*) of an oscillating cantilever beam can be expressed as [58]:

$$f = \frac{\upsilon_n'^2}{2\pi}\sqrt{\frac{k}{m_e}},$$

(7.1)

where $\upsilon_n'^2 = \upsilon_n^2\sqrt{0.236/3}$ (ν_n is *n-th* mode eigen value), *k* is the effective spring constant and m_e is the effective mass at the tip of the cantilever. Assuming the effective spring constant (*k*) of the cantilever remains constant during the detection, and the effective mass m_e can be obtained by approximating the cantilever as a point mass at the free end and for a rectangular cantilever is $m_e=0.236\ m$ where *m* is the mass of the cantilever. [58-60] the adsorbed target species (*Δm*) can cause the resonance frequency shift Δf as

$$\frac{\Delta f}{f} = \frac{1}{2}\frac{\Delta m}{m_e}$$

(7.2)

Therefore, for example, when an *E. coli* cell was adsorbed on the cantilever as shown in Fig. 7.1, the resonance frequency shifted correspondingly (Fig. 7.2 (a)). Equation 7.2 implies that the frequency shift is linear to the loaded mass, and Fig. 7.2 (b) experimentally confirmed that the frequency shift is proportional to the captured cells.

7.2.2. Scaling Law Based on Mass Loading Model

As a sensing platform, mass sensitivity is one of the most important criteria of the microcantilever. From Eq. 7.2, the resonance frequency shift of the cantilever will change if foreign mass attaches to the sensor. The mass sensitivity (Δm/Δf) can be defined as mass change (Δm) during detection over the measured resonance frequency shift (Δf). It depicts the capability of the response of the microcantilever to the loaded mass (target). The smaller the value, the more sensitive the sensor is. In 2002, Yi et al. reported both experimental and theoretical investigations of the resonance frequency change and the mass sensitivity of a piezoelectric unimorph cantilever due to the mass loaded at the tip of the cantilever [60]. Theoretically, based on the mass loading model, i.e., assuming the spring constant is constant, and once the effective Young's modulus and effective density of the cantilever is fixed, i.e., by maintaining the same layer thickness fractions, the mass sensitivity of a cantilever follows the scaling law [60]:

$$\frac{\Delta m}{\Delta f_n} \propto \frac{L^3 w}{v_n^2}, \tag{7.3}$$

where Δm, Δf_n, L, w, and v_n are the loaded mass, resonance frequency shift of the *n-th* flexural mode, length, width, and *n-th* mode eigen value of the cantilever, respectively. This indicated that the mass sensitivity of the cantilever can be improved via miniaturization, i.e., if the length and width were shrunk 10 times, the mass sensitivity can be increased by 10,000 times.

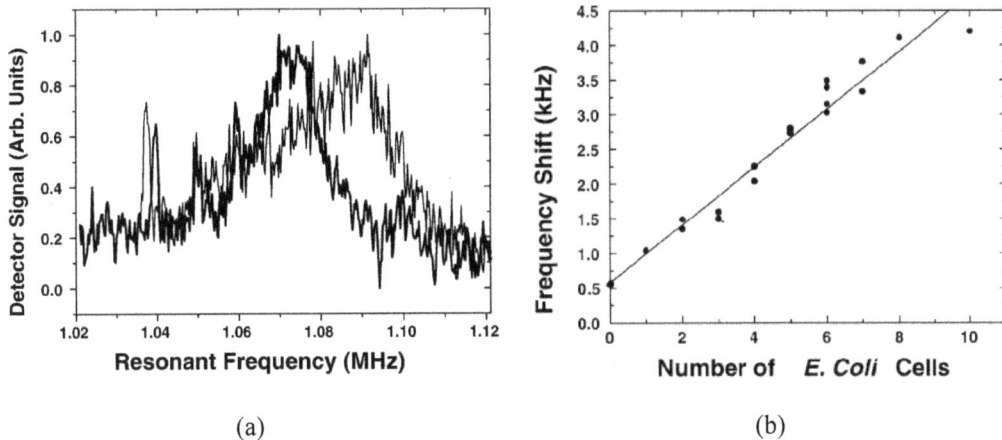

(a)   (b)

Fig. 7.2. The resonance frequency spectra of the cantilever before and after antibody immobilization and single cell attachment (a), and Measured frequency shift versus the number of bound E. coli cells (b) [39].

Experimentally, they examined the cantilevers composed of a lead zirconate titanate (PZT) layer and a stainless steel layer, and the length of the cantilevers varied from 4.4 mm to 13.7 mm. In Fig. 7.3, the result of the first-mode resonance frequency shifts, Δf, versus the loaded mass, Δm, for the cantilevers were shown with various lengths. The slopes are the mass sensitivities of the cantilevers. It's clear that the shorter cantilever has higher mass detection sensitivity.

The normalization in Fig. 7.4 validated the scaling law depicted by Eq. 7.3. This model was validated by numerous publications on biological or chemical detection using microcantilevers [34, 38-40, 61, 62].

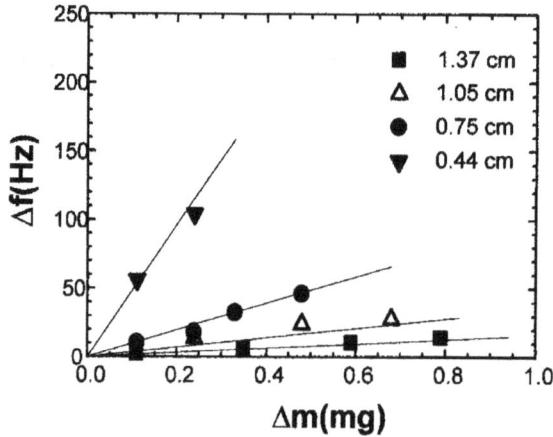

Fig. 7.3. Δf versus Δm of a cantilever 0.4 cm in width and 1.37 cm, 1.05 cm, 0.75 cm, and 0.44 cm in length [60].

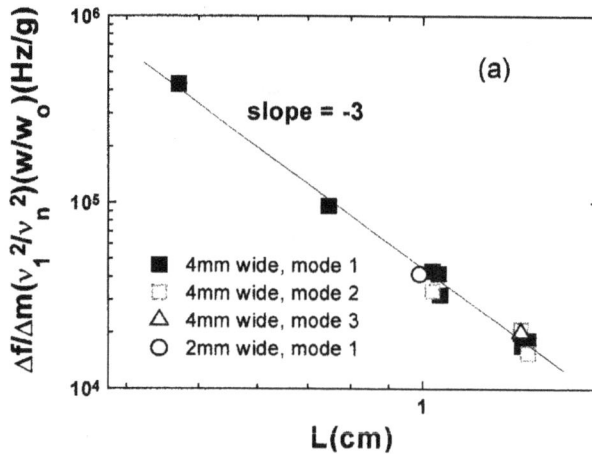

Fig. 7.4. $(\Delta f_n /\Delta m)(v_1^2 / v_n^2)(w/w_0)$ versus L on a double logarithmic plot [60].

7.2.3. Adsorption Induced Surface Stress

Adsorption-induced surface stress has been reported and characterized in a wide range of adsorption systems both in air and in liquid such as adsorption of molecule [11, 54, 63-66], polymer [67], ions [4, 21] biological antigen, [43] protein [44, 45], and DNA [31, 68]. The adsorption induced surface stress arises from the molecular interactions which are universal and mainly originate from the following sources:

1) Intermolecular forces of attraction and repulsion. An example was illustrated in Fig. 7.5 - a schematic of the adsorption process of gas molecule to adsorbent layer from the vapor phase. The gas molecules adsorb to the adsorbent immobilized on cantilever surface, and upon adsorption the molecules pack with an average spacing d_a which is less than the equilibrium intermolecular spacing (d_0) of the gas molecules in the ambient. Hence, once the gas molecules absorb on the cantilever, the cantilever surface will want to expand to increase the intermolecular spacing and a tensile surface stress is present as a result of adsorption.

2) The electrostatic force. Typically, this kind of force is present for biological species because normally antigen, cell, protein, or DNA is charged in liquid. Once they adsorb on the sensor surface, electrostatic force is present because of the charges they carry.

3) Steric force. Steric force between polymer-covered surfaces or in solutions containing non-adsorbing polymer can modulate interparticle forces, producing an additional steric repulsive force or an attractive depletion force between them [69].

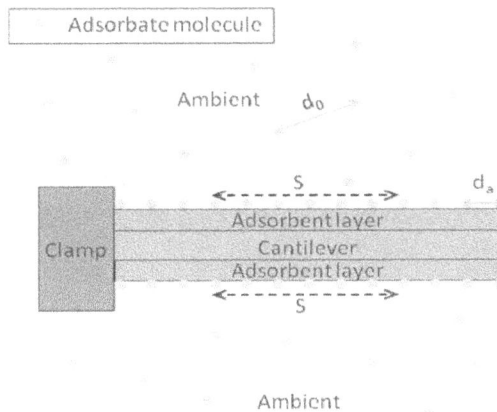

Fig. 7.5. A schematic of adsorption induced surface stress on cantilever.

It was observed that when surface stress is present, the resonance frequency of the microcantilever also shifted and this frequency shift could not be explained by mass loading model described in Section 7.2.2. The stress effect in cantilever was first reported in 1975. Lagowski et al. from Massachusetts Institute of Technology first observed the natural frequency shift of a fresh etched GaAs cantilever plate due to surface stress induced by adsorption/desorption of gas molecules [70]. Since 90 s, the silicon-based microcantilever had become a hot research topic as a sensing platform for a wide range of chemical and biological detections. Observations of the stress effect of the silicon microcantilever in both gaseous and in liquid detections were reported [6, 21, 63]. It was found that for silicon-based microcantilever, the enhancement in frequency shift due to stress effect could be 10 times larger than predicted by the mass loading model [21, 64]. Some researchers argued [63, 71] that the spring constant of the cantilever could change when adsorption induced stress is present. Therefore, the resonance frequency shift would follow:

$$\frac{\Delta f}{f} = \frac{1}{2}\left(\frac{\Delta k}{k} - \frac{\Delta m}{m_e} \right),$$

(7.4)

where Δf, Δk, and Δm are the frequency shift, spring constant change, and mass change due to adsorption.

7.2.4. Static Sensing Method

Because of the universality of the adsorption induced stress, another way of detecting molecular adsorption is by measuring the tip deflection of the cantilever due to adsorption stress. In order to generate bending or deflection, only one side of the cantilever is functionalized or two sides are functionalized differently [4, 6, 11, 21, 31, 64, 65]. Therefore, one of its sides is relatively passive, whereas the other exhibits high affinity to the targeted analyte. As a result, upon adsorption, depending on the nature of the bonding of the target species, the cantilever can bend up or down because the induced stresses are unequal on two sides. For example, Fig. 7.6 illustrated the deflection of a cantilever due to the hybridization (adsorption) of DNAs. In this case, only one side of the cantilever was functionalized with receptors. Upon adsorption, the deflection (Δx) of the cantilever can be measured. The adsorption induced surface stress can be quantified using Stoney's Equation [72]:

$$\Delta x = \frac{3L^2(1 - v)}{Yt^2}\Delta S,$$

where L, t, v, and Y are the length, thickness, Poisson's ratio, and Young's modulus of the cantilever, and ΔS is the differential surface stress of the two surfaces.

In general, sophisticated optical instruments are needed to measure the tip displacement while the resonance frequency measurement can be done by electrical means and thereby can be easily deployable.

7.3. Applications of Microcantilever Sensors in Biological and Chemical Detections

Based on the operation mechanisms and materials, microcantilevers can be classified into silicon-based, piezoresistive, capacitive, magnetoresistive, and piezoelectric microcantilevers.

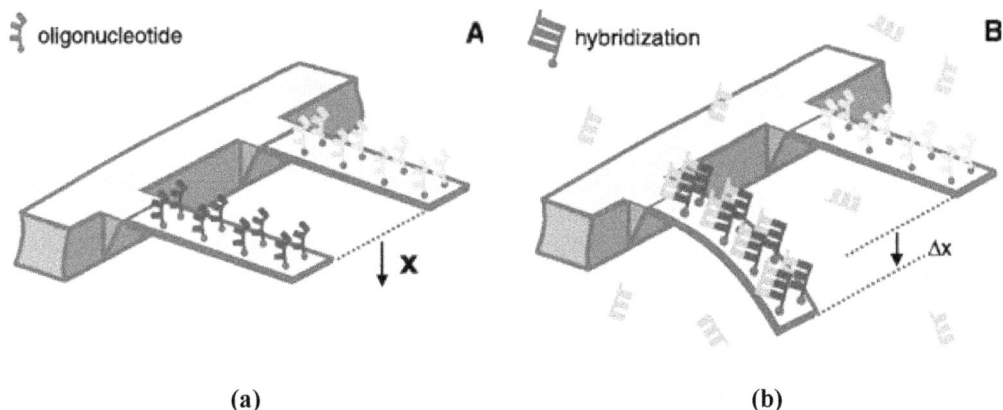

(a)   (b)

Fig. 7.6. Scheme illustrating the hybridization experiment. Each cantilever is functionalized on one side with a different oligonucleotide base sequence (red or blue). (a) The differential signal is set to zero (before detection); (b) After injection of the first complementary oligonucleotide (green), hybridization occurs on the cantilever that provides the matching sequence (red), increasing the differential signal Δx [65].

7.3.1. Silicon-based Microcantilever Sensor

Silicon-based microcantilever has been widely used in Atomic Force Microscopy (AFM) and has been demonstrated as a versatile sensing platform in a wide range of areas [2, 4, 6-11, 13, 27, 30, 35, 37-42, 51, 52, 54, 57, 64, 73-77]. It has several advantages over the conventional analytical techniques in terms of high sensitivity, label-free, quick response, and array capability. With the development of silicon microfabrication and most recently developed nanofabrication, the mass sensitivity of the cantilever has been boosted significantly.

In 2001, Illic, et al from Cornell University reported single *Escherichia coli* (*E. coli*) cell detection [39]. The cantilevers they used are 5 or 10 μm wide, and 15 or 25 μm long (one of the cantilevers was shown in Fig. 7.1). The binding of a single *E. coli* cell caused 4.7 kHz shift in resonance frequency (see Fig. 7.2 (a)) and the detection was performed in air which was not *in situ*. The mass sensitivity of the cantilevers was 1.4×10^{-16} g/Hz.

In 2004, Gupta, et al from Purdue University reported single vaccinia virus detection using arrays of silicon microcantilevers with nanoscale thickness (Fig. 7.7) [34]. The dimensions of the fabricated cantilever were in the range of 4-5 μm in length, 1-2 μm in width and 20-30 nm in thickness. The resonance frequency of the cantilever was in the 1-2 MHz range with quality factor of around 5-7. After loading of a single virus particle, there was 60 kHz decease in the resonance frequency (Fig. 7.8) and the mass sensitivity was 1.6×10^{-19} g/Hz.

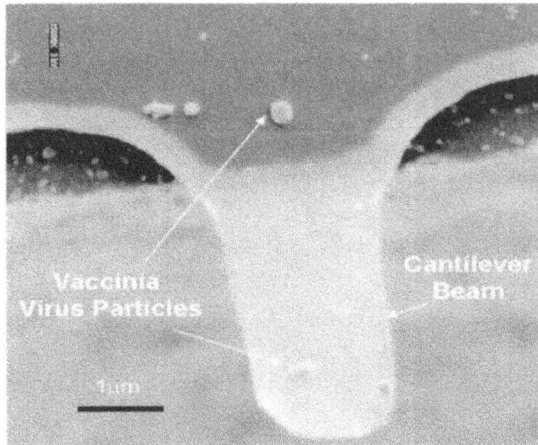

Fig. 7.7. Scanning electron micrograph showing a cantilever beam with a single vaccinia virus particle. The cantilever beam has planar dimensions of length, L=4 μm, and width, W=1.8 μm [34].

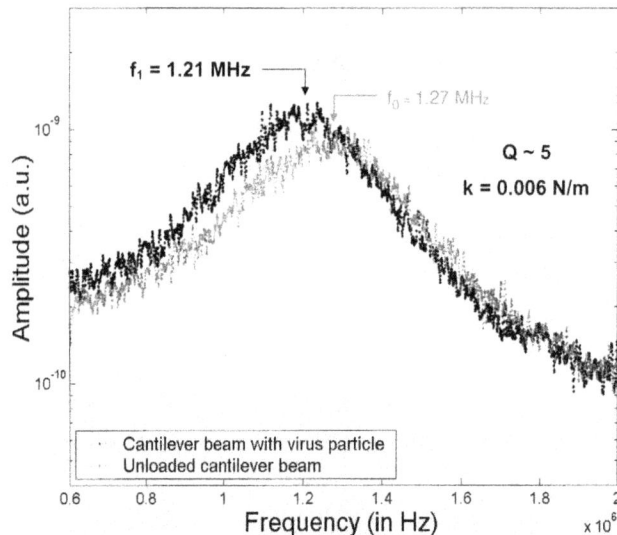

Fig. 7.8. Plot of resonant frequency shift after loading of a single virus particle [34].

In 2005, Ilic et al. from Cornell University demonstrated single DNA detection in vacuum using a cantilever fabricated from 90 nm thick low-pressure chemical vapor deposited low-stress silicon nitride [40]. The lengths of the cantilevers were between 3.5-5 μm (Fig. 7.9 (a) and (b)). To localize binding site, they formed the cantilever with nanoscale gold dots (see Fig. 7.9 (c)) at precise locations to act as spatially and chemically discriminant binding sites to selectively capture disulfide modified 1578 base pair long double-stranded deoxyribonucleic acid (dsDNA) molecules

(m_{DNA}=999 kDaltons). Fig. 7.10 showed the frequency shift corresponding to a single dsDNA molecule bonding. The mass sensitivity of the cantilever was as high as 5×10^{-21} g/Hz and better than attogram (10^{-18} g) mass detection was demonstrated.

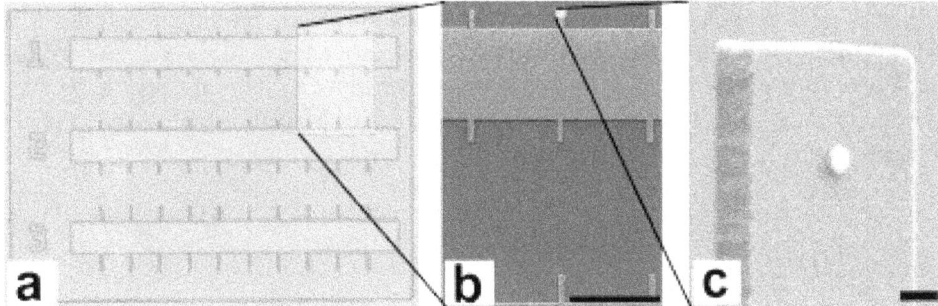

Fig. 7.9. (a) Optical micrograph showing arrays of cantilevers of varying lengths; (b) Zoomed-in scanning electron microscope (SEM) image; (c) Oblique angle SEM image of the 90 nm thick silicon nitride cantilever with a 40 nm circular Au aperture centered 300 nm away from the free end. Scale bar corresponds to 100 nm [40].

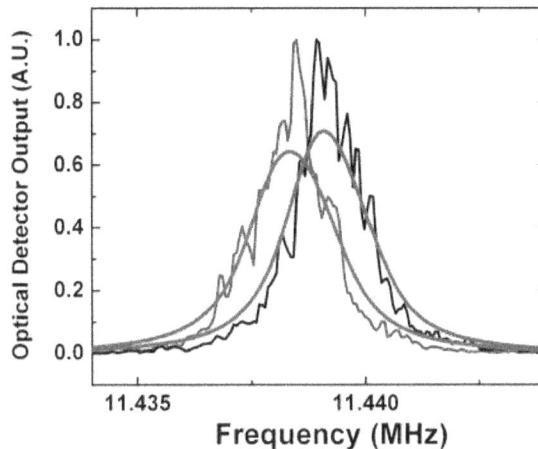

Fig. 7.10. Frequency spectra before (black) and after (red) the binding events show a frequency shift due to a single dsDNA molecule bound to the Au surface of the cantilever [40].

In 2006, Yang et al from California Institute of Technology demonstrated zeptogram (10^{-21} g) scale mass sensing in high vacuum using a doubly clamped micocantilver beams (Fig. 7.11). The dimensions of the beam were 2.3 μm long, 150 nm wide, and 70 nm thick and the resonance frequency was around 190 MHz. As shown in Fig. 7.11, a minute, calibrated, highly controlled flux of Xe atoms or N_2 molecules is delivered to the device surface by a mechanically shuttered gas nozzle within the apparatus. Fig. 7.12 showed the frequency shift upon deposition of the N_2 molecules. The mass detection sensitivity of the device was as high as 8.6×10^{-22} g/Hz and the best mass resolution corresponds to 7 zg, equivalent to an individual 4 kDa molecule.

In general, the advantages of the silicon based microcantilever over the conventional analytical techniques are high sensitivity, low cost, label free, low analyte requirement (in μl), and quick response [78]. However, the main challenge for silicon cantilever is in liquid detection capability, i.e., the quality of the resonance peak is too low to use in liquid. For the recently developed ultrasensitive cantilevers, they are unable to sustain the damping in air, therefore, high vacuum (10^{-6} Torr or higher) is necessary to perform sensing [37, 40, 57, 61].

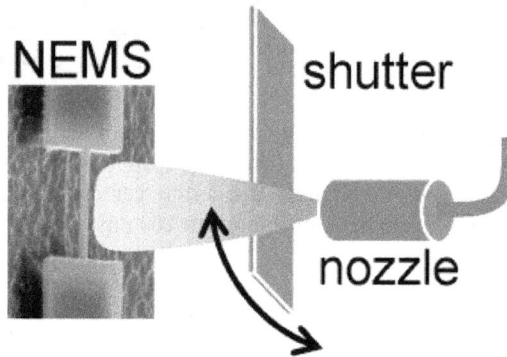

Fig. 7.11. Experimental configuration. A gas nozzle with a 100 μm aperture provides a controlled flux of atoms or molecules. The flux is gated by a mechanical shutter to provide calibrated, pulsed mass accretions upon the NEMS device [57].

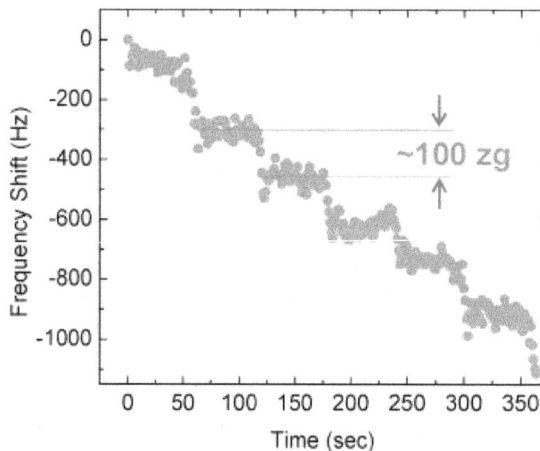

Fig. 7.12. *Real time* zeptogram-scale mass sensing experiment. Sequential mass depositions are executed *in situ* upon the 190 MHz device within a cryogenic ultrahigh vacuum apparatus [57].

7.3.2. Piezoresistive Microcantilever Sensor

Typically, optical method is adopted to detect the static deflection or dynamic vibration of silicon based microcantilever. Alternatively, piezoresistive read-out method can be used. Piezoresistive method is based on the changes observed in the electrical resistance of the material of the cantilever as a consequence of a surface-stress change [79-82]. This method involves the embedding of a piezoresistive material near the top surface of the cantilever during fabrication to record the stress change occurring at the surface of the cantilever. Fig. 7.13 shows a schematic of a piezoresistive cantilever. Piezoresistive elements fabricated onto or into cantilevers comprise either semiconductor or metallic strain gauges [83].

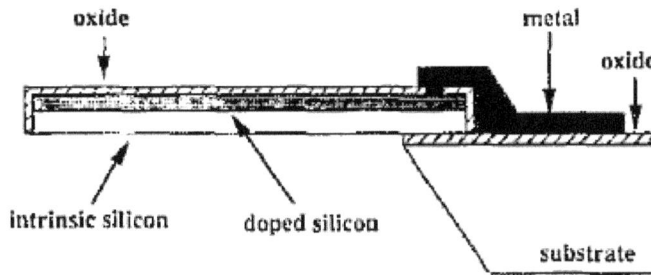

Fig. 7.13. Schematic drawing of the cross section of a piezoresistive cantilever (doped silicon is piezoresistive layer) [79].

As the microcantilever deflects, it undergoes a stress change that will create strain to the piezoresistor, thereby causing a change in resistance. The relative change in resistance as function of applied strain can be written as: [80, 81].

$$\frac{\Delta R}{R} = K_l \delta_l K_t \delta_t$$

where K denotes the Gage Factor, which is a material parameter, and δ is the strain in the material and R is the resistance. The subscripts l and t refer to the longitudinal and the transversal part of the Gage Factor. The piezoresistor material in the beam must be localized as close to one surface of the cantilever as possible for maximum strain/sensitivity.

The piezoresistive cantilever beam can be measured by Wheatstone Bridge circuit as shown in Fig. 7.14 [78, 80, 81].

The resistance of the variable resistance arm ($R_0 + \Delta R$) in the above figure can be determined by using the common Voltage divider formula and is shown as below:

$$\Delta V = V_0 \left\{ \frac{R_2}{R_1 + R_2} - \frac{R_3}{R_0 + \Delta R + R_3} \right\} \Rightarrow R_0 + \Delta R = R_3 \left\{ \frac{V_0 (R_1 + R_2)}{R_2 V_0 - \Delta V (R_1 + R_2)} - 1 \right\}$$

Numerous physical, chemical and biomedical applications have been demonstrated using piezoresistive microcantilevers including calorimetry [84], humidity sensing [73], TNT detection [85], C-reactive protein detection [86], *Salmonella enterica* detection [87], *Vaccinia* virus detection [88] and allergy check [89].

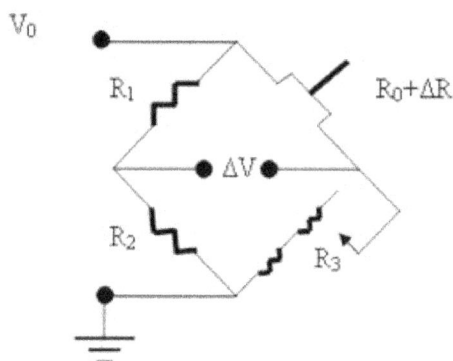

Fig. 7.14. The Wheatstone Bridge Circuit used for the piezoresistive microcantilever [78].

In general, the advantage of piezoresistive cantilever is that the read-out electronics can be integrated onto the chip containing the cantilever array and it is unaffected by light-absorbing or scattering components in the analyte stream. Because current is flowing through the cantilevers while measurements are being made, local heating can occur and it is a major problem for practical applications, although it can be managed by changing the amount of current flowing through the resistive layer [55]. Other drawbacks to this technique are thermal, electronic, and conductance fluctuation noise, thermal drifts, nonlinearity in piezoresponse, and poor sensitivity [90]. In addition, a piezoresistor has to be embedded in the cantilever, therefore, fabrication of such a cantilever with a composite structure is more complicated.

7.3.3. Capacitive Microcantilever Sensor

Capacitive cantilever acts as one of the parallel plates of a capacitor [91, 92]. Fig. 7.15 showed a schematic of a capacitive microcantilever system [93]. As the cantilever deflection takes place due to the adsorption of the analyte, the distance between the two plates changes and this changes the capacitance of the system.

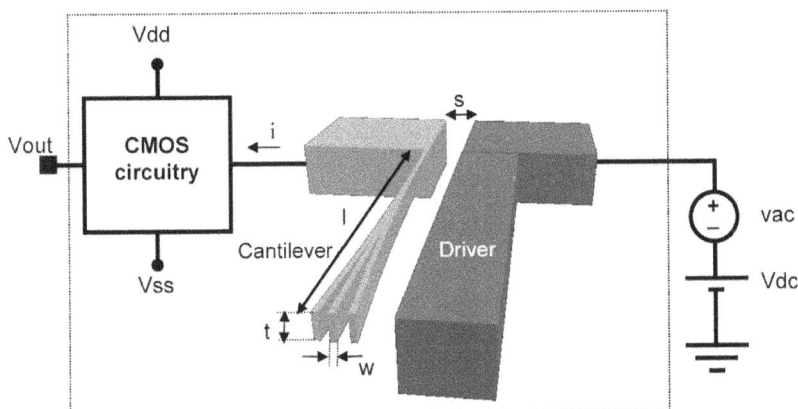

Fig. 7.15. Schematic drawing of the capacitive microcantilever system based on a laterally vibrating cantilever (s direction) electrostatically excited and with capacitive readout [93].

The advantage of capacitive detection is the simplicity of the associated electronics [94] and it is highly sensitive and provides absolute displacement. However, this technique is not one of the most commonly used because of a number of limitations [83]. To accurately record cantilever deflection, the dielectric material between the conductive plates must be constant throughout the experiment. The presence of analyte within the gap often changes its effective dielectric constant. In addition, although the capacitive cantilevers can be integrated onto a microchip [93-95] scaling down the size of the capacitive cantilever will lower its overall sensitivity because the capacitance of a capacitor is directly proportional to its surface area.

Capacitive cantilever is mainly used for gaseous phase chemical sensing since it does not work in electrolyte solutions due to the faradic currents between the capacitive plates. For gas sensing, Amirola and co-workers [96, 97] used capacitive detection of gaseous molecules and found the limit of detection to be 50 ppm for toluene and 10 ppm for octane. Britton Jr. and co-workers demonstrated hydrogen detection using their capacitive cantilever array and the detection limit of the hydrogen was as low as 100 ppm [94] Verd and co-workers report mass resolution for their specific capacitive cantilever system is on the order of 10^{-18} g [93].

7.3.4. Magnetostrictive Microcantilever Sensor

Recently, a research group from Auburn University developed magnetostrictive microcantilever (MSMC) as a sensing platform [98-101]. MSMS consists of two layers - one is active (magnetostrictive) and the other is inactive. MSMC is actuated remotely using a magnetic field. Due to the magnetic nature, the vibration of the microcantilever results in an emission of a magnetic signal, which is sensed remotely using a pickup coil. Figs. 7.16 (a)-(d) illustrated the operation mechanism of the MSMC: [101] the length of the active layer in the MSMC would be changed with a magnetic field due to the magnetostrictive effect. Therefore, an applied magnetic field on the MSMC (Fig. 7.16 (a)) would lead to a length difference between the active and inactive layers, which would bend the MSMC since the active and inactive layers are bonded together. Therefore, a time-varying magnetic field would make the MSMC bending vibration as shown in Fig. 7.16 (b). Due to the magnetic nature of the magnetostrictive alloy, the bending vibration of MSMC would emit a magnetic signal, which can be measured using a pick-up coil (see Fig. 7.16 (c)). If the time-varying magnetic field is a sine wave, the bending vibration of the MSMC would also be a sine function of time. The amplitude of the bending vibration of the MSMC changes with the amplitude and frequency of the magnetic field. Additionally, there would be a phase difference between the driving magnetic field and the bending vibration. If an ac magnetic field is swept over a frequency range with constant amplitude, as shown in Fig. 7.16 (a), the amplitude of the bending vibration of the MSMC would change with the frequency as shown in Fig. 7.16 (d).

Since there is no physical connection between the MSMC and the integration device, this is the principal advantage of MSMCs over other microcantilevers [99, 100]. Another advantage of MSMC is the capability of operation in liquid. Real time in liquid biological detections of yeast cells [99], *Salmonella typhimurium* [101], and *Bacillus anthracis* spores [98] were demonstrated.

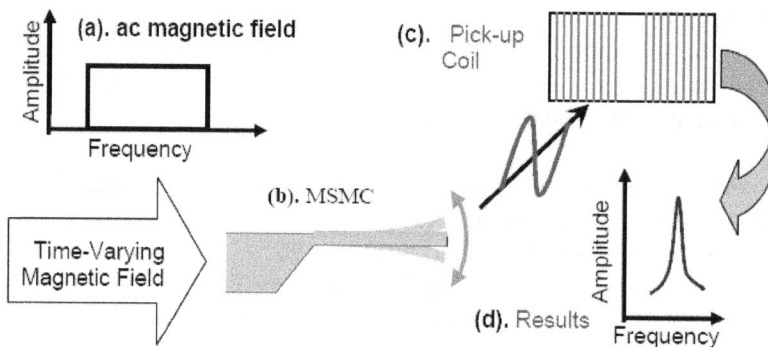

Fig. 7.16. Schematic illustration of the principle of MSMC as a transducer for biosensors [101].

7.3.5. Piezoelectric Microcantilever Sensor

A typical Piezoelectric Microcantilever sensor (PEMS) consists of a piezoelectric layer bonded to a nonpiezoelectric layer (Fig. 7.17). Because of the converse piezoelectric effect, when an AC voltage is applied to the thickness direction (also poling direction) of the piezoelectric layer, it will elongate or shrink along the length and width directions due to its piezoelectric characteristics. However, the nonpiezoelectric layer does not deform thereby constraining the movement of the piezoelectric layer and resulting in the alternative bending (vibration) of the cantilever structure. PEMS was originally developed to overcome the complexity of the force detector of the conventional noncontact AFM [102, 103]. Comparing to the silicon-based microcantilever, PEMS has several advantages:

1. PEMS can self-excite and self-sense, i.e., the exciting can be performed by applying an AC field on the piezoelectric layer and the sensing can be achieved by monitoring the phase angle change;
2. PEMS can withstand high environmental damping and it can operate in liquid;
3. The detection scheme is all electric and the system can be easily made portable (Lab-On-A-Chip).

Fig. 7.17. A SEM micrograph of a ZnO/SiO_2 piezoelectric microcantilever [102].

Recently, PEMS has become a very hot research focus and various physical [16, 104-106] chemical [3, 66, 107-111] and biological [22-26, 43-45, 47, 48, 68, 112-117] sensing applications have been demonstrated using PEMS.

In 2001, Shih et al. performed simultaneous liquid viscosity and density determination using lead zirconate titanate (PZT)/stainless steel cantilever. Their study indicated that the viscosity and the density of a liquid can be determined simultaneously by measuring resonance frequency and peak width. In 2002, Yi et al. demonstrated real time *in situ* in liquid yeast cells detection using PZT/stainless steel microcantilever [22]. The peak height of the PEMS was not reduced much when immersed in water and the quality factor (Q) almost remained the same as that in air (Fig. 7.18). This study demonstrated PEMS as a power sensing platform for real time in liquid detection which is a big hurdle for silicon microcantilevers.

Since 2004, Lee et al. fabricated PZT/SiN_x PEMS via microfabrication (Fig. 7.19) and various biological detections of prostate-specific antigen (PSA) [43, 112], C-reactive protein [44, 45, 116], myoglobin [113], protein kinase [115], and aptamer [68] were demonstrated. However, because the ions in the liquid would cause 'short-circuit' between the two electrodes of the piezoelectric layer, most of the above detections were performed *ex situ* [43-45, 68, 113].

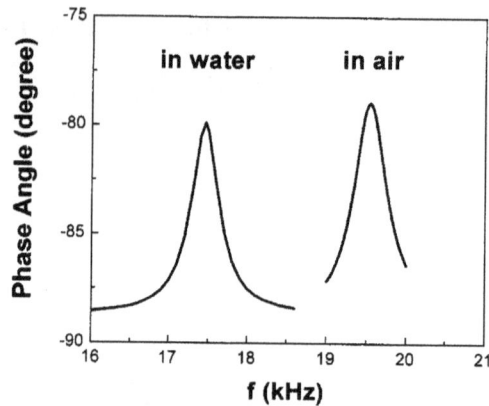

Fig. 7.18. Phase angle vs. frequency of cantilever both in air (right) and in water (left) [60].

Fig. 7.19. A SEM photograph of the micromachined PZT cantilever arrays designed for simultaneous self-actuating and sensing [44].

In 2005, selective nerve gas simulant Dimethyl methylphosphonate (DMMP) detection has been demonstrated using PZT/stainless steel PEMS array [3]. In this study, three PEMSs were coated with different adsorbents and the response of each PEMS/adsorbent generated a unique pattern to DMMP. Meanwhile, *in situ* real time *Salmonella typhimurium* detection in liquid was demonstrated using PZT/Gold coated glass PEMS [23, 24]. The detection was achieved without insulation by partially dipping the sensor in the liquid at nodal point [23, 59] or at controlled humidity [24]. In 2006, Shen et al. fabricated PZT/SiO$_2$ PEMS with a 60×25 μm PZT/SiO$_2$ section and a 24×20 μm SiO$_2$ extension (Fig. 7.20) and mass sensitivity was $1×10^{-15}$ g/Hz calibrated by quartz crystal microbalance (QCM) in humidity detection.

Fig. 7.20. A SEM micrograph of the PEMS with a 60×25 μm PZT/SiO$_2$ section and a 24×20 μm SiO$_2$ extension [66].

In 2006, Capobianco et al. invented a novel insulation scheme [25, 26] for PEMS which enabled the PEMS to be fully immersed in conductive solution. Since then, various *in situ* in liquid biological detections of *Escherichia coli* [25], single chain variable fragment (scFv) protein [26], and *Bacillus anthracis* spores [47, 48] were demonstrated using insulated lead magnesium niobate-lead titanate (PMN-PT)/Metal PEMS.

Interestingly, stress effect was observed and reported in a wide range of biological and chemical detections using PEMS [3, 16, 22, 24-26, 43, 44, 47, 48, 66, 104, 111]. Quantitatively, the enhancement observed in PEMS in both gaseous and aqueous detection was 100-200 times larger than prediction by the mass loading model, and was 10-50 times larger than the enhancement in the silicon-based microcantilever (see Table 7.1).

Table 7.1. Enhancement due to stress effect comparing to the mass loading model of the microcantilever sensors reported in the literature.

Cantilever	Detection system	Enhancement*	Reference
Silicon Nitride	*E. coli* detection	No	[39]
Silicon	Virus detection	No	[34]
Silicon	Mercury detection	~ 4 times	[64]
Silicon Nitride	Na+ adsorption	10 times	[21]
PZT/SiO$_2$/SiN$_x$	C-protein detection	100-120 times	[44, 45]
PZT/SiO$_2$/SiN$_x$	PSA detection	100-200 times	[43, 112]
PZT/SiO$_2$	Humidity detection	100 times	[66]
PZT/Glass	Salmonella cells detection	100-200 times	[23, 24]

*Enhancement in frequency shift in comparison to prediction by mass loading model.

It could be concluded that the two-order-of-magnitude enhancement would be a unique feature and advantage of the PEMS and a wide range of biological and chemical detections using PEMS showed dominant stress effect. The most recent study [111] showed this enhancement in frequency shift of a PEMS is a result of Young's modulus change in piezoelectric layer induced by surface stress. Furthermore, it is shown that the resonance frequency shift can be further enhanced by applying a DC bias electric field to the piezoelectric layer during detection and 1000 times enhancement has been demonstrated for humidity detection [118].

7.4. Conclusions

Microcantilevers have several advantages over the conventional sensing techniques such as optical or fluorescence-based sensors, chemiresistive sensors, Quartz crystal microbalance (QCM), or surface acoustic wave (SAW) devices in terms of high sensitivity, label free, versatility, array capability, and quick response. Single cell [39], single virus [34] and single DNA [40] detection have been demonstrated and single molecule [57] detection could be possible in the near future.

References

[1]. T. Thundat, R. J. Warmack, G. Y. Chen et al., Thermal and ambient-induced deflections of scanning force microscope cantilevers, *Applied Physics Letters,* Vol. 64, No. 21, 1994, pp. 2894-2896.

[2]. J. R. Barnes, R. J. Stephenson, M. E. Welland *et al.,* Photothermal spectroscopy with femtojoule sensitivity using a micromechanical device, *Nature,* Vol. 372, No. 6501, 1994, pp. 79-81.

[3]. Q. Zhao, Q. Zhu, W. Y. Shih et al., Array adsorbent-coated lead zirconate titanate (PZT)/stainless steel cantilevers for dimethyl methylphosphonate (DMMP) detection, *Sensors and Actuators B: Chemical,* Vol. 117, No. 1, 2006, pp. 74-79.

[4]. Y. Zhang, H. F. Ji, G. M. Brown *et al.,* Detection of CrO42- Using a Hydrogel Swelling Microcantilever Sensor, *Analytical Chemistry,* Vol. 75, No. 18, 2003, pp. 4773-4777.

[5]. X. Yan, X. K. Xu, and H. F. Ji, Glucose Oxidase Multilayer Modified Microcantilevers for Glucose Measurement, *Analytical Chemistry,* Vol. 77, No. 19, 2005, pp. 6197-6204.

[6]. T. Thundat, G. Y. Chen, R. J. Warmack et al., Vapor Detection Using Resonating Microcantilevers, *Analytical Chemistry,* Vol. 67, No. 3, 1995, pp. 519-521.

[7]. H. P. Lang, M. K. Baller, R. Berger et al., An artificial nose based on a micromechanical cantilever array, *Analytica Chimica Acta,* Vol. 393, 1999, No. 1-3, pp. 59-65.

[8]. B. C. Fagan, C. A. Tipple, Z. Xue et al., Modification of micro-cantilever sensors with sol-gels to enhance performance and immobilize chemically selective phases, *Talanta,* Vol. 53, No. 3, 2000, pp. 599-608.

[9]. R. Bashir, J. Z. Hilt, O. Elibol et al., Micromechanical cantilever as an ultrasensitive pH microsensor, *Applied Physics Letters,* Vol. 81, No. 16, 2002, pp. 3091-3093.

[10]. F. Lochon, L. Fadel, I. Dufour et al., Silicon made resonant microcantilever: Dependence of the chemical sensing performances on the sensitive coating thickness, *Materials Science and Engineering: C,* Vol. 26, No. 2-3, 2006, pp. 348-353.

[11]. Y. Yang, H. F. Ji, and T. Thundat, Nerve Agents Detection Using a Cu2+/L-Cysteine Bilayer-Coated Microcantilever, *Journal of the American Chemical Society,* Vol. 125, No. 5, 2003, pp. 1124-1125.

[12]. X. Xu, T. G. Thundat, G. M. Brown et al., Detection of Hg2+ Using Microcantilever Sensors, *Analytical Chemistry,* Vol. 74, No. 15, 2002, pp. 3611-3615.

[13]. M. K. Baller, H. P. Lang, J. Fritz et al., A cantilever array-based artificial nose, *Ultramicroscopy,* Vol. 82, No. 1-4, 2000, pp. 1-9.

[14]. S.-H. S. Lim, D. Raorane, S. Satyanarayana et al., Nano-chemo-mechanical sensor array platform for high-throughput chemical analysis, *Sensors and Actuators B: Chemical,* Vol. 119, No. 2, 2006, pp. 466-474.

[15]. D. Then, A. Vidic, and C. Ziegler, A highly sensitive self-oscillating cantilever array for the quantitative and qualitative analysis of organic vapor mixtures, *Sensors and Actuators B: Chemical,* Vol. 117, No. 1, 2006, pp. 1-9.

[16]. W. Y. Shih, X. Li, H. Gu et al., Simultaneous liquid viscosity and density determination with piezoelectric unimorph cantilevers, *Journal of Applied Physics,* Vol. 89, No. 2, 2001, pp. 1497-1505.

[17]. A. Markidou, W. Y. Shih, and W.-H. Shih, Soft-materials elastic and shear moduli measurement using piezoelectric cantilevers, *Review of Scientific Instruments,* Vol. 76, No. 6, 2005, pp. 064302-7.

[18]. P. I. Oden, G. Y. Chen, R. A. Steele et al., Viscous drag measurements utilizing microfabricated cantilevers, *Applied Physics Letters,* Vol. 68, No. 26, 1996, pp. 3814-3816.

[19]. R. Berger, C. Gerber, J. K. Gimzewski et al., Thermal analysis using a micromechanical calorimeter, *Applied Physics Letters,* Vol. 69, No. 1, 1996, pp. 40-42.

[20]. E. T. Arakawa, N. V. Lavrik, S. Rajic et al., Detection and differentiation of biological species using microcalorimetric spectroscopy, *Ultramicroscopy,* Vol. 97, No. 1-4, 2003, pp. 459-465.

[21]. S. Cherian, and T. Thundat, Determination of adsorption-induced variation in the spring constant of a microcantilever, *Applied Physics Letters,* Vol. 80, No. 12, 2002, pp. 2219-2221.

[22]. J. W. Yi, W. Y. Shih, R. Mutharasan et al., In situ cell detection using piezoelectric lead zirconate titanate-stainless steel cantilevers, *Journal of Applied Physics,* Vol. 93, No. 1, 2003, pp. 619-625.

[23]. Q. Zhu, W. Y. Shih, and W.-H. Shih, Real-time, label-free, all-electrical detection of Salmonella typhimurium using lead titanate zirconate/gold-coated glass cantilevers at any relative humidity, *Sensors and Actuators B: Chemical,* Vol. 125, No. 2, 2007, pp. 379-388.

[24]. Q. Zhu, W. Y. Shih, and W.-H. Shih, In situ, in-liquid, all-electrical detection of Salmonella typhimurium using lead titanate zirconate/gold-coated glass cantilevers at any dipping depth, *Biosensors and Bioelectronics,* Vol. 22, No. 12, 2007, pp. 3132-3138.

[25]. J. A. Capobianco, W. Y. Shih, and W.-H. Shih, Methyltrimethoxysilane-insulated piezoelectric microcantilevers for direct, all-electrical biodetection in buffered aqueous solutions, *Review of Scientific Instruments,* Vol. 77, No. 12, 2006, pp. 125105-6.

[26]. J. A. Capobianco, W. Y. Shih, and W.-H. Shih, 3-mercaptopropyltrimethoxysilane as insulating coating and surface for protein immobilization for piezoelectric microcantilever sensors, *Review of Scientific Instruments,* Vol. 78, No. 4, 2007, pp. 046106-3.

[27]. Y. Arntz, J. D. Seelig, H. P. Lang et al., Label-free protein assay based on a nanomechanical cantilever array, *Nanotechnology,* Vol. 14, No. 1, 2003, pp. 86-90.

[28]. D. R. Baselt, G. U. Lee, and R. J. Colton, Biosensor based on force microscope technology, *Journal of Vacuum Science and Technology B,* Vol. 14, 1996, pp. 789-793.

[29]. M. Calleja, M. Nordstrom, M. Alvarez et al., Highly sensitive polymer-based cantilever-sensors for DNA detection, *Ultramicroscopy,* Vol. 105, No. 1-4, 2005, pp. 215-222.

[30]. B. Dhayal, W. A. Henne, D. D. Doorneweerd et al., Detection of Bacillus subtilis Spores Using Peptide-Functionalized Cantilever Arrays, *Journal of the American Chemical Society,* Vol. 128, No. 11, 2006, pp. 3716-3721.

[31]. J. Fritz, M. K. Baller, H. P. Lang et al., Translating Biomolecular Recognition into Nanomechanics, *Science,* Vol. 288, No. 5464, 2000, pp. 316-318.

[32]. H. J. G. Wu, K. M. Hansen, T. Thundat, R. Datar, R. Cote, M. F. Hagan, A. Chakraborty, A. Majumdar, Origin of nanomechanical cantilever motion generated from biomolecular interactions, *Proc. Natl. Acad. Sci. USA,* Vol. 98, No. 4, 2001, pp. 1555-1559.

[33]. K. Y. Gfeller, N. Nugaeva, and M. Hegner, Rapid Biosensor for Detection of Antibiotic-Selective Growth of Escherichia coli, *Applied and Environmental Microbiology,* Vol. 71, No. 5, 2005, pp. 2626-2631.

[34]. A. Gupta, D. Akin, and R. Bashir, Single virus particle mass detection using microresonators with nanoscale thickness, *Applied Physics Letters,* Vol. 84, No. 11, 2004, pp. 1976-1978.

[35]. K. M. Hansen, H. F. Ji, G. Wu et al., Cantilever-Based Optical Deflection Assay for Discrimination of DNA Single-Nucleotide Mismatches, *Analytical Chemistry,* Vol. 73, No. 7, 2001, pp. 1567-1571.

[36]. S.-J. Hyun, H.-S. Kim, Y.-J. Kim et al., Mechanical detection of liposomes using piezoresistive cantilever, *Sensors and Actuators B: Chemical,* Vol. 117, No. 2, 2006, pp. 415-419.

[37]. B. Ilic, H. G. Craighead, S. Krylov et al., Attogram detection using nanoelectromechanical oscillators, *Journal of Applied Physics,* Vol. 95, No. 7, 2004, pp. 3694-3703.

[38]. B. Ilic, D. Czaplewski, H. G. Craighead et al., Mechanical resonant immunospecific biological detector, *Applied Physics Letters,* Vol. 77, No. 3, 2000, pp. 450-452.

[39]. B. Ilic, D. Czaplewski, M. Zalalutdinov et al., Single cell detection with micromechanical oscillators, *Journal of Vacuum Science and Technology B,* Vol. 19, No. 6, 2001, pp. 2825-2828.

[40]. B. Ilic, Y. Yang, K. Aubin et al., Enumeration of DNA Molecules Bound to a Nanomechanical Oscillator, *Nano Letters,* Vol. 5, No. 5, 2005, pp. 925-929.

[41]. Y. Lam, N. I. Abu-Lail, M. S. Alam et al., Using microcantilever deflection to detect HIV-1 envelope glycoprotein gp120, *Nanomedicine: Nanotechnology, Biology and Medicine,* Vol. 2, No. 4, 2006, pp. 222-229.

[42]. H. P. Lang, R. Berger, C. Andreoli et al., Sequential position readout from arrays of micromechanical cantilever sensors, *Applied Physics Letters,* Vol. 72, No. 3, 1998, pp. 383-385.

[43]. J. H. Lee, K. S. Hwang, J. Park et al., Immunoassay of prostate-specific antigen (PSA) using resonant frequency shift of piezoelectric nanomechanical microcantilever, *Biosensors and Bioelectronics,* Vol. 20, No. 10, 2005, pp. 2157-2162.

[44]. J. H. Lee, T. S. Kim, and K. H. Yoon, Effect of mass and stress on resonant frequency shift of functionalized $Pb(Zr_{0.52}Ti_{0.48})O_3$ thin film microcantilever for the detection of C-reactive protein, *Applied Physics Letters,* Vol. 84, No. 16, 2004, pp. 3187-3189.

[45]. J. H. Lee, K. H. Yoon, K. S. Hwang et al., Label free novel electrical detection using micromachined PZT monolithic thin film cantilever for the detection of C-reactive protein, *Biosensors and Bioelectronics,* Vol. 20, No. 2, 2004, pp. 269-275.

[46]. G. V. Lubarsky, M. R. Davidson, and R. H. Bradley, Hydration-dehydration of adsorbed protein films studied by AFM and QCM-D, *Biosensors and Bioelectronics,* Vol. 22, No. 7, 2007, pp. 1275-1281.

[47]. J.-P. McGovern, W. Y. Shih, R. Rest et al., Label-free flow-enhanced specific detection of Bacillus anthracis using a piezoelectric microcantilever sensor, *The Analyst*, Vol. 133, No. 5, 2008, pp. 649-654.

[48]. J.-P. McGovern, W. Y. Shih, and W.-H. Shih, In situ detection of Bacillus anthracis spores using fully submersible, self-exciting, self-sensing PMN-PT/Sn piezoelectric microcantilevers, *The Analyst*, Vol. 132, No. 8, 2007, pp. 777-783.

[49]. C. Milburn, J. Zhou, O. Bravo et al., Sensing Interactions Between Vimentin Antibodies and Antigens for Early Cancer Detection, *Journal of Biomedical Nanotechnology*, Vol. 1, 2005, pp. 30-38.

[50]. A. M. Moulin, S. J. O'Shea, R. A. Badley et al., Measuring Surface-Induced Conformational Changes in Proteins, *Langmuir*, Vol. 15, No. 26, 1999, pp. 8776-8779.

[51]. A. M. Moulin, S. J. O'Shea, and M. E. Welland, Microcantilever-based biosensors, *Ultramicroscopy*, Vol. 82, No. 1-4, 2000, pp. 23-31.

[52]. R. Mukhopadhyay, V. V. Sumbayev, M. Lorentzen et al., Cantilever Sensor for Nanomechanical Detection of Specific Protein Conformations, *Nano Letters*, Vol. 5, No. 12, 2005, pp. 2385-2388.

[53]. C. Z. N. Backmann, F. Huber, A. Bietsch, A. Pluckthun, H.-P. Lang, H.-J. Guntherodt, M. Hegner, C. Gerber, A label-free immunosensor array using single-chain antibody fragments, *Proc. Natl. Acad. Sci. USA*, Vol. 102, No. 41, 2005, pp. 14587-14592.

[54]. L. A. Pinnaduwage, J. E. Hawk, V. Boiadjiev et al., Use of Microcantilevers for the Monitoring of Molecular Binding to Self-Assembled Monolayers, *Langmuir*, Vol. 19, No. 19, 2003, pp. 7841-7844.

[55]. G. N. Roberto Raiteri, Hans-Jürgen Butt, Wolfgang Knoll, Petr Skladal, Sensing of biological substances based on the bending of microfabricated cantilevers, *Sensors and Actuators B: Chemical*, Vol. 61, No. 1-3, 1999, pp. 213-217.

[56]. C. A. Savran, S. M. Knudsen, A. D. Ellington et al., Micromechanical Detection of Proteins Using Aptamer-Based Receptor Molecules, *Analytical Chemistry*, Vol. 76, No. 11, 2004, pp. 3194-3198.

[57]. Y. T. Yang, C. Callegari, X. L. Feng et al., Zeptogram-Scale Nanomechanical Mass Sensing, *Nano Letters*, Vol. 6, No. 4, 2006, pp. 583-586.

[58]. J. Merhaut, Theory of Electroacoustics, *McGraw-Hill*, New York, 1981, p. 100.

[59]. Z. Shen, W. Y. Shih, and W.-H. Shih, Mass detection sensitivity of piezoelectric cantilevers with a nonpiezoelectric extension, *Review of Scientific Instruments*, Vol. 77, No. 6, 2006, pp. 065101-10.

[60]. J. W. Yi, W. Y. Shih, and W.-H. Shih, Effect of length, width, and mode on the mass detection sensitivity of piezoelectric unimorph cantilevers, *Journal of Applied Physics*, Vol. 91, No. 3, 2002, pp. 1680-1686.

[61]. B. Ilic, Y. Yang, and H. G. Craighead, Virus detection using nanoelectromechanical devices, *Applied Physics Letters*, Vol. 85, No. 13, 2004, pp. 2604-2606.

[62]. M. Maute, S. Raible, F. E. Prins et al., Detection of volatile organic compounds (VOCs) with polymer-coated cantilevers, *Sensors and Actuators B: Chemical*, Vol. 58, No. 1-3, 1999, pp. 505-511.

[63]. G. Y. Chen, T. Thundat, E. A. Wachter et al., Adsorption-induced surface stress and its effects on resonance frequency of microcantilevers, *Journal of Applied Physics*, Vol. 77, No. 8, 1995, pp. 3618-3622.

[64]. T. Thundat, E. A. Wachter, S. L. Sharp et al., Detection of mercury vapor using resonating microcantilevers, *Applied Physics Letters*, Vol. 66, No. 13, 1995, pp. 1695-1697.

[65]. R. Berger, E. Delamarche, H. P. Lang et al., Surface Stress in the Self-Assembly of Alkanethiols on Gold, *Science*, Vol. 276, No. 5321, 1997, pp. 2021-2024.

[66]. Z. Shen, W. Y. Shih, and W.-H. Shih, Self-exciting, self-sensing $PbZr_{0.53}Ti_{0.47}O_3/SiO_2$ piezoelectric microcantilevers with femtogram/Hertz sensitivity, *Applied Physics Letters,* Vol. 89, No. 2, 2006, pp. 023506-3.

[67]. M. Toda, A. N. Itakura, S. Igarashi et al., Surface Stress, Thickness, and Mass of the First Few Layers of Polyelectrolyte, *Langmuir,* Vol. 24, No. 7, 2008, pp. 3191-3198.

[68]. K. S. Hwang, S.-M. Lee, K. Eom et al., Nanomechanical microcantilever operated in vibration modes with use of RNA aptamer as receptor molecules for label-free detection of HCV helicase, *Biosensors and Bioelectronics,* Vol. 23, No. 4, 2007, pp. 459-465.

[69]. H.-J. Butt, M. Kappl, H. Mueller et al., Steric Forces Measured with the Atomic Force Microscope at Various Temperatures, *Langmuir,* Vol. 15, No. 7, 1999, pp. 2559-2565.

[70]. J. Lagowski, H. C. Gatos, and J. E. S. Sproles, Surface stress and the normal mode of vibration of thin crystals :GaAs, *Applied Physics Letters,* Vol. 26, No. 9, 1975, pp. 493-495.

[71]. W. M. Andrew, A. P. Mark, J. D. Margaret et al., Influence of surface stress on the resonance behavior of microcantilevers, 5, AIP, 2005, p. 053505.

[72]. G. G. Stoney, *Proc. R. Soc. London Ser. A,* Vol. 82, 1909, pp. 172-177.

[73]. A. Boisen, J. Thaysen, H. Jensenius et al., Environmental sensors based on micromachined cantilevers with integrated read-out, *Ultramicroscopy,* Vol. 82, No. 1-4, 2000, pp. 11-16.

[74]. L. G. Carrascosa, M. Moreno, M. Alvarez et al., Nanomechanical biosensors: a new sensing tool, *TrAC Trends in Analytical Chemistry,* Vol. 25, No. 3, 2006, pp. 196-206.

[75]. J. Drelich, C. L. White, and Z. H. Xu, Laboratory tests on mercury emission monitoring with resonating gold-coated silicon cantilevers, *Environmental Science & Technology,* Vol. 42, No. 6, Mar, 2008, pp. 2072-2078.

[76]. B. H. Kim, F. E. Prins, D. P. Kern et al., Multicomponent analysis and prediction with a cantilever array based gas sensor, *Sensors and Actuators B: Chemical,* Vol. 78, No. 1-3, 2001, pp. 12-18.

[77]. C. Y. Wang, D. Y. Wang, Y. D. Mao et al., Ultrasensitive biochemical sensors based on microcantilevers of atomic force microscope, *Analytical Biochemistry,* Vol. 363, No. 1, Apr. 2007, pp. 1-11.

[78]. S. K. Vashist, A Review of Microcantilevers for Sensing Applications, *AZojono - Journal of Nanotechnology Online,* 2007.

[79]. M. Tortonese, R. C. Barrett, and C. F. Quate, Atomic resolution with an atomic force microscope using piezoresistive detection, *Applied Physics Letters,* Vol. 62, No. 8, 1993, pp. 834-836.

[80]. T. G. R. Linnemann, L. Hadjiiski and I. W. Rangelow, Characterization of a cantilever with an integrated deflection sensor, *Thin Solid Films,* Vol. 264, No. 2, 1995, pp. 159-164.

[81]. R. Linnemann, T. Gotszalk, I. W. Rangelow et al., Atomic force microscopy and lateral force microscopy using piezoresistive cantilevers, pp. 856-860.

[82]. B. W. Chui, T. D. Stowe, T. W. Kenny et al., Low-stiffness silicon cantilevers for thermal writing and piezoresistive readback with the atomic force microscope, *Applied Physics Letters,* Vol. 69, No. 18, 1996, pp. 2767-2769.

[83]. K. M. Goeders, J. S. Colton, and L. A. Bottomley, Microcantilevers: Sensing Chemical Interactions via Mechanical Motion, *Chemical Reviews,* Vol. 108, No. 2, 2008, pp. 522-542.

[84]. N. Abedinov, P. Grabiec, T. Gotszalk et al., Micromachined piezoresistive cantilever array with integrated resistive microheater for calorimetry and mass detection, *Journal of Vacuum Science & Technology A: Vacuum, Surfaces, and Films,* Vol. 19, No. 6, 2001, pp. 2884-2888.

[85]. L. A. Pinnaduwage, A. Gehl, D. L. Hedden et al., Explosives: A microsensor for trinitrotoluene vapour, *Nature,* Vol. 425, No. 6957, 2003, pp. 474-474.

[86]. G. Y. K. Kyung Wook Wee, Jaebum Park, Ji Yoon Kang, Dae Sung Yoon, Jung Ho Park and Tae Song Kim, Novel electrical detection of label-free disease marker proteins using piezoresistive self-sensing micro-cantilevers, *Biosensors and Bioelectronics,* Vol. 20, No. 10, 2005, pp. 1932-1938.

[87]. J. C. B. L. Weeks, A. Noy, A. E. Miller, L. Stanker and J. J. De Yoreo, A Microcantilever-Based Pathogen Detector, *Scanning,* Vol. 25, No. 6, pp. 297-299, 2003, November 2003.

[88]. W. G. D. R. L. Gunter, K. Manygoats, A. Kooser and T. L. Porter, Viral detection using an embedded piezoresistive microcantilever sensor, *Sensors and Actuators A: Physical,* Vol. 107, No. 3, 2003, pp. 219-224.

[89]. A. I. Hayato Sone, Takashi Izumi, Haruki Okano and Sumio Hosaka, Femtogram Mass Biosensor Using Self-Sensing Cantilever for Allergy Check, *Jpn. J. Appl. Phys.,* Vol. 45, 2006, pp. 2301-2304.

[90]. G. Shekhawat, S.-H. Tark, and V. P. Dravid, MOSFET-Embedded Microcantilevers for Measuring Deflection in Biomolecular Sensors, *Science,* Vol. 311, No. 5767, 2006, pp. 1592-1595.

[91]. G. Neubauer, S. R. Cohen, G. M. McClelland et al., Force microscopy with a bidirectional capacitance sensor, *Review of Scientific Instruments,* Vol. 61, No. 9, 1990, pp. 2296-2308.

[92]. N. Blanc, J. Brugger, N. F. de Rooij et al., Scanning force microscopy in the dynamic mode using microfabricated capacitive sensors, *Journal of Vacuum Science & Technology A: Vacuum, Surfaces, and Films,* Vol. 14, No. 2, 1996, pp. 901-905.

[93]. J. Verd, G. Abadal, J. Teva et al., Design, fabrication, and characterization of a submicroelectromechanical resonator with monolithically integrated CMOS readout circuit, *Microelectromechanical Systems, Journal of,* Vol. 14, No. 3, 2005, pp. 508-519.

[94]. C. L. Britton, R. L. Jones, P. I. Oden et al., Multiple-input microcantilever sensors, *Ultramicroscopy,* Vol. 82, No. 1-4, 2000, pp. 17-21.

[95]. E. Forsen, G. Abadal, S. Ghatnekar-Nilsson et al., Ultrasensitive mass sensor fully integrated with complementary metal-oxide-semiconductor circuitry, *Applied Physics Letters,* Vol. 87, No. 4, 2005, pp. 043507-3.

[96]. J. Amírola, A. Rodiguez, and L. Castaner, Design fabrication and test of micromachined-silicon capacitive gas sensors with integrated readout, in *Proceedings of the SPIE on Smart Sensors, Actuators, and MEMS,* Vol. 5116, pp. 2003. pp. 92-99.

[97]. A. R. Jorge Amírola, Luis Castañer, J. P. Santos, J. Gutiérrez and M. C. Horrillo, Micromachined silicon microcantilevers for gas sensing applications with capacitive read-out, *Sensors and Actuators B: Chemical,* Vol. 111-112, 2005, pp. 247-253.

[98]. M. L. Johnson, J. Wan, S. Huang et al., A wireless biosensor using microfabricated phage-interfaced magnetoelastic particles, *Sensors and Actuators A: Physical,* Vol. 144, No. 1, 2008, pp. 38-47.

[99]. S. Li, L. Orona, Z. Li et al., Biosensor based on magnetostrictive microcantilever, *Applied Physics Letters,* Vol. 88, No. 7, 2006, pp. 073507-3.

[100]. Z. Li, S. Li, and Z.-Y. Cheng, Microbiosensor based on magnetostrictive microcantilever, pp. 333-343.

[101]. S. L. Liling Fu, Kewei Zhang, I-Hsuan Chen, Valery. A. Petrenko, and Zhongyang Cheng, Magnetostrictive Microcantilever as an Advanced Transducer for Biosensors, *Sensors,* Vol. 7, pp. 2929-2941, 2007.

[102]. T. Itoh, and T. Suga, Development of a force sensor for atomic force microscopy using piezoelectric thin films, *Nanotechnology,* Vol. 4, No. 4, pp. 218-224, 1993.

[103]. C. Lee, T. Itoh, T. Ohashi et al., Development of a piezoelectric self-excitation and self-detection mechanism in PZT microcantilevers for dynamic scanning force microscopy in liquid, pp. 1559-1563.

[104]. X. Li, W. Y. Shih, J. Vartuli et al., Detection of water-ice transition using a lead zirconate titanate/brass transducer, *Journal of Applied Physics,* Vol. 92, No. 1, 2002, pp. 106-111.

[105]. V. Mortet, R. Petersen, K. Haenen et al., Wide range pressure sensor based on a piezoelectric bimorph microcantilever *Applied Physics Letters,* Vol. 88, No. 13, 2006, pp. 133511-3.

[106]. D. W. Chun, K. S. Hwang, K. Eom et al., Detection of the Au thin-layer in the Hz per picogram regime based on the microcantilevers, *Sensors and Actuators A: Physical,* Vol. 135, No. 2, 2007, pp. 857-862.

[107]. J. D. Adams, G. Parrott, C. Bauer et al., Nanowatt chemical vapor detection with a self-sensing, piezoelectric microcantilever array, *Applied Physics Letters,* Vol. 83, No. 16, 2003, pp. 3428-3430.

[108]. B. Rogers, L. Manning, M. Jones et al., Mercury vapor detection with a self-sensing, resonating piezoelectric cantilever, *Review of Scientific Instruments,* Vol. 74, No. 11, 2003, pp. 4899-4901.

[109]. J. Zhou, P. Li, S. Zhang et al., Self-excited piezoelectric microcantilever for gas detection, *Microelectronic Engineering,* Vol. 69, No. 1, 2003, pp. 37-46.

[110]. N. Ledermann, P. Muralt, J. Baborowski et al., Piezoelectric $Pb(Zr_x, Ti_{1-x})O_3$ thin film cantilever and bridge acoustic sensors for miniaturized photoacoustic gas detectors, *Journal of Micromechanics and Microengineering,* Vol. 14, No. 12, 2004, pp. 1650-1658.

[111]. Q. Zhu, W. Y. Shih, and W.-H. Shih, Mechanism of flexural resonance frequency shift of a piezoelectric microcantilever sensor during humidity detection, *Applied Physics Letters,* Vol. 92, No. 18, 2008, pp. 183505-3.

[112]. K. S. Hwang, J. H. Lee, J. Park et al., In-situ quantitative analysis of a prostate-specific antigen (PSA) using a nanomechanical PZT cantilever, *Lab on a Chip,* Vol. 4, No. 6, 2004, pp. 547-552.

[113]. K. S. Hwang, K. Eom, J. H. Lee et al., Dominant surface stress driven by biomolecular interactions in the dynamical response of nanomechanical microcantilevers, *Applied Physics Letters,* Vol. 89, No. 17, 2006, pp. 173905-3.

[114]. G. Y. Kang, G. Y. Han, J. Y. Kang et al., Label-free protein assay with site-directly immobilized antibody using self-actuating PZT cantilever, *Sensors and Actuators B: Chemical,* Vol. 117, No. 2, 2006., pp. 332-338.

[115]. H.-S. Kwon, K.-C. Han, K. S. Hwang et al., Development of a peptide inhibitor-based cantilever sensor assay for cyclic adenosine monophosphate-dependent protein kinase, *Analytica Chimica Acta,* Vol. 585, No. 2, 2007, pp. 344-349.

[116]. T. Y. Kwon, K. Eom, J. H. Park et al., In situ real-time monitoring of biomolecular interactions based on resonating microcantilevers immersed in a viscous fluid, *Applied Physics Letters,* Vol. 90, No. 22, 2007, pp. 223903-3.

[117]. Y. Lee, G. Lim, and W. Moon, A piezoelectric micro-cantilever bio-sensor using the mass-micro-balancing technique with self-excitation, *Microsystem Technologies,* Vol. 13, No. 5, 2007, pp. 563-567.

[118]. Q. Zhu, W. Y. Shih, and W.-H. Shih, Enhanced detection resonance frequency shift of a piezoelectric microcantilever sensor by a DC bias electric field in humidity detection, *Sensors and Actuators B: Chemical,* Vol. 138, No. 1, 2009, pp. 1-4.

Chapter 8
Nanomaterials and Chemical Sensors

Sukumar Basu and Palash Kumar Basu

8.1. Nanomaterials

Nanoscale materials are made of clusters of atoms and molecules. For example 3.5 atoms of gold or 8 atoms of hydrogen linked up in a row are nanometer long; a glucose molecule is about 1 nm size. Size of nanomaterials is intermediate between isolated atoms or molecules and bulk materials. At this scale the material shows exceptional properties. Use of quantum mechanics is required to explain the behavior of nanomaterials. Quantum effects and salient physical properties relevant at the nanoscale dimension are presented as follows [1-7].

8.1.1. Properties of Nanomaterials

The mass of nanomaterials is extremely small and so the gravitational forces are negligible. As a result the electromagnetic forces are dominant in determining the behavior of nanomaterials. Also the wavelike nature has a more pronounced effect due to extremely small mass and their position is represented by wave (probability) function.

8.1.1.1. Quantum Tunneling

According to quantum physics a particle with energy less than that required to jump the barrier has a finite probability of being found on the other side of the barrier. So it can be imagined that the particle passes into virtual tunnel through the barrier. We know that the thickness of the barrier must be comparable to the wavelength of the particle for tunneling effect and it is observed at nanometer level.

Sukumar Basu - IC Design & Fabrication Centre, Department of Electronics & Telecommunication Eng., Jadavpur University, Kolkata, India

Sergey Y. Yurish (ed.), *Modern Sensors, Transducers and Sensor Networks*
© International Frequency Sensor Association Publishing, 2012

8.1.1.2. Quantum Confinement

In case of the nanomaterials like metals or semiconductors electrons are confined in space rather than free to move in the bulk. For semiconductors the band gap becomes large as compared to the bulk ones since the electrons exist in the discrete energy level. In very small dimensions the metals may behave like semiconductors due to the quantization of the energy levels and the disappearance of the band overlap normally present in metals.

8.1.1.3. Random Molecular Motion

Molecules move due to their kinetic energy (assuming that the sample is above absolute zero) and it is called random molecular motion. This motion is present in every material and it is very small compared to the size of the object. Thus it cannot influence on how the object moves. But at the nanoscale, this motion can be of the same extent as the size of the particles and can influence on how the molecules move.

8.1.1.4. Surface and Reactivity

Nanomaterials have the distinguished properties of increased surface to volume ratio compared to the bulk materials. The atoms and molecules at the surface or at an interface are different from those that exist in the interior of a material. When the crystal size decreases it exposes more and more surface. So fraction of atoms in the grain boundary increases. It is known that the grain boundaries contain a large density of defects states like vacancies and dangling bonds that play an important role in the transport properties of electrons of materials in general and of nanomaterials in particular. In fact, the grain boundaries are nothing but metastable states and they have natural tendency to reduce their energy either by exchange or by sharing of electrons with other atoms. Therefore, the surface reactivity (or chemical reactivity) increases. Since a large fraction of atoms in nanomaterials is at the surface it influences some physical properties like melting point. For the same material, for example, the melting point is lower if it is nanosized. This happens because the surface atoms can be more easily removed than bulk atoms in the crystalline structure and so the total energy needed to overcome the intermolecular forces that hold the atoms is less in nanomaterials, resulting in the reduction of the melting point.

8.1.1.5. Mechanical Properties

Some nanomaterials such as carbon nanotubes have inherent exceptional mechanical properties. These are extremely small tubes with honeycomb structure of graphite. Carbon nanotubes are hundred times stronger but six times lighter than steel. Therefore, the nanomaterials can also have improved mechanical properties than existing materials like metal, polymer, composites etc.

8.1.2. Nanomaterials Used for Chemical Sensors

In recent times nanomaterials have become extremely popular for chemical & bio sensing, due to their interesting electrical conductivity, unique structural and catalytic properties, high loading, good stability and excellent penetrability. Carbon nanotube (CNT) (discussed later) can be used as electrode materials with useful properties for various potential applications including miniature biological devices. The electrochemical sensing using CNTs has been extensively studied and reviewed by different authors. The sensors showed higher response with lower work potential and minimum interference. Soluble carbon nanofibers have been reported to be used to modify glucose sensors that perform the electro reduction of dissolved oxygen at a low operating potential.

TiO_2, ZnO, and CuO nanotube arrays have been proved as promising functional materials for applications in chemical gas sensors and bio chemical sensors. Pt and Au nanoparticles with about 20 nm diameter and uniformly distributed in the nanotube channels can work as oxide- reaction centres to catalyze the oxidation of H_2O_2. Also the nanotubular TiO_2 can be used for the direct immobilization of glucose oxidase with a stable catalytic activity.

In recent years, researchers have explored the production of novel nano-scale metal oxides, noble metal-doped metal oxides, metal oxide-CNT nanocomposites, and metal oxide-polymer composites. Devices based on nanostructured metal oxides are cost-effective, highly sensitive due to the large surface-to-volume ratio of the Nanostructures, and show considerable selectivity to bio molecules. ZnO nanomaterial has been studied extensively in optics, optoelectronics, solar cells, sensors, and actuators owing to their semiconducting, optical and favourable electrical properties in addition to the piezoelectric and pyroelectric properties [8-12]. Like its applications for gas sensors ZnO is attractive for the fabrication of the metal-oxide based biochemical sensors, due to its good biocompatibility, chemical stability, non-toxicity, electrochemical activity and fast electron transfer rate. However, sometimes the nanostructured ZnO shows poor mechanical stability, because the ZnO nano-structure is removed from the electrode surface during functionalization. Indeed, improved stability without loss of sensitivity or selectivity is one of the big challenges for glucose monitoring with these biosensors. Wang et al [13] prepared ZnO nanoparticles and coated them onto a multiwall carbon nanotube (MWNT)-modified electrode to improve the stability. Transition metal oxides significantly enhance direct oxidation of glucose compared with other metals that attribute to the catalytic effect resulting from the multi-electron oxidation mediated by surface metal oxide layers. Transition metals such as Cu and Ni can oxidize carbohydrate easily without surface poisoning. Unlike Cu and Ni, corresponding oxides or hydroxides are relatively stable in air and solutions [14-15]. Natural abundance of copper oxide, its low production cost, good electrochemical and catalytic properties make the copper oxide to be one of the best materials for electrical, optical and photovoltaic devices, heterogeneous catalysis, magnetic storage media, gas sensing, field-emission emitters, lithium ion electrode etc. [16-18]. Recently the catalytic effect of copper oxide in relation to nonenzymetic glucose oxidation, voltammetric sensing of carbohydrates and hydrogen peroxide detection with ultra-sensitive response and good stability has been improved due to advances in nanoscience and nanotechnology [19]. The CuO nanowire can greatly promote electron transfer rate of glucose oxidation by increasing the electro catalytic active area. Moreover, the CuO modified electrodes can be used repeatedly without being contaminated by glucose oxidation by-product. Experimental data for the glucose detection are in good agreement with the results from the spectrophotometer method performed with real samples. TiO$_2$ nanomaterials are biocompatible and environmental friendly. Moreover, nanostructured TiO$_2$ provides better environment for enzyme immobilization by enlarging the surface area [20-21].

For glucose detection the attractive catalytic properties of transition metal oxides like CuO and NiO are interesting for nonenzymetic direct electro oxidation of glucose. A glucose biosensor was reported by Xie et al. [22] in which glucose oxidase was embedded inside TiO$_2$ nanotube channels where pyrrole was electropolymerized. Using pure Titania nanotube array the direct detection of hydrogen peroxide by electro catalytic reduction reaction was achieved with a detection limit down to 2.0×10^{-4} mmol/L. The glucose biosensor based on the glucose oxidase-titania/titanium electrode showed an excellent performance with a response time below 5.6 s and a detection limit of 2.0×10^{-3} mmol/L. The corresponding detection sensitivity was 45.5 µA L mmol^{-1} cm^{-2}. The high response and low detection limit of this novel biosensor could be quite suitable for potential applications.

The nano-sized Au particles show extraordinary catalytic activity as revealed by recent studies. The Au nanopillars showed high electro catalytic activity not only in the reduction of hydrogen peroxide and molecular oxygen but also in the oxidation of glucose due to its nano-sized pillar array structure. An Au nano-structured film with active Au adatom was proposed. Good sensitivity of Pt-nanotube and macro porous Pt-modified electrodes to glucose has been shown. Sensors were fabricated by Pt nanoparticles immobilized on CNTs. Polypyrrole embedded with Pt nanoclusters provides a porous, biocompatible and highly catalytic platform. Other metal nanomaterial applied in fabrication of glucose sensors is copper (Cu) that showed good selectivity and sensitivity. Also Iridium, could detect H$_2$O$_2$ released from the enzymatic reactions with a favorable signal-to-noise ratio [23-27]. The silica nanoparticles and the non-doped nanocrystalline diamond have been reported for the same purpose. There is a report on an amperometric sensor based on polypyrrole nanotube array deposited on a Pt plated nano-porous alumina template as the electrode for an efficient enzyme loading and an increased surface area for sensing. Another nanoelectrode sensor like polypyrrole nanofibers with entrapped graphene oxide (GO) was recently fabricated [28-31] and was demonstrated for good biocatalytic activity to glucose.

An Au nano wire (NW) [32] array electrode was reported to be successfully fabricated by template-assisted deposition and transfer of nanowires onto a rigid glass substrate applied for nonenzymetic glucose sensor. It was observed by electrochemical investigation that the partial oxidation of glucose on Au NW array electrode takes place at more negative potential than that on the Au film electrode. Also, the kinetics of partial oxidation of glucose on Au - NW array revealed that three different electrochemical reactions for the voltammetric detection of glucose could be used with the peak currents linear on the wide range of glucose concentrations. The amperometric detection within physiologically important range of glucose concentrations was possible with a very high sensitivity of 309.0 $AmM^{-1} cm^{-2}$.

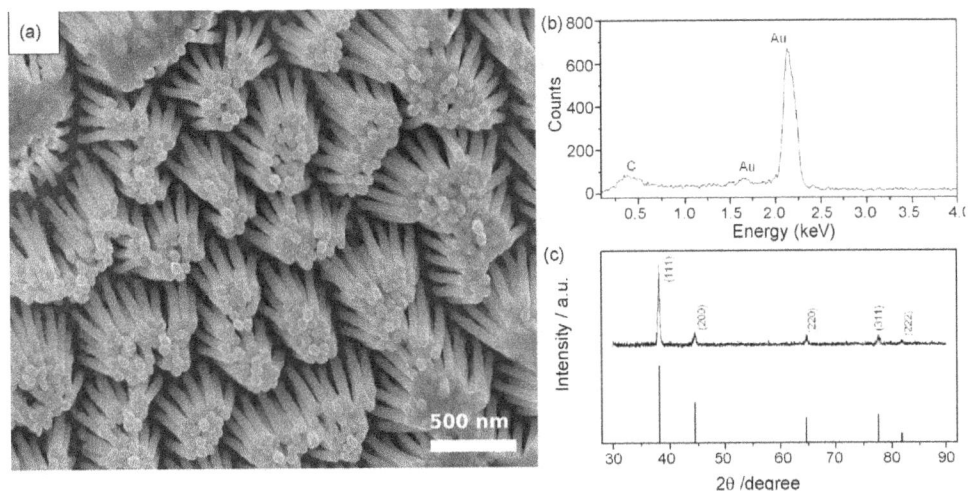

Fig. 8.1. FESEM of gold nano wire.
(Reprinted from Sensors and Actuators, B, 142, S. Cherevko, C-H. Chung ,"Gold nanowire array electrode for non-enzymatic voltammetric and amperometric glucose detection", pp 216, Copyright 2009, with permission from Elsevier).

The advantage of using a NW array electrode is that the roughness of such electrodes is smaller than the so called chronoamperometric diffusion field and therefore the faraday current of rapidly oxidizable or reducible chemicals is proportional to the apparent geometric area of the electrode and independent of its roughness. On the other hand, the faraday current for kinetically controlled electrochemical reactions is sensitive to the nanoscopic features of the electrode and, proportional to its entire area. Nanostructures such as nanowires, nanoparticles and CNTs have been used as smart building blocks for emerging electronic and sensing devices. Metallic nanowires allow higher sensitivity, higher capture efficiency and faster response time because of their large adsorption surface (large surface to volume ratio), high electrical conductivity and small diffusion time. Au nanowires promote better electron transfer between the enzymes and the electrodes [9].

Surface modified multiwall CNTs with a biocompatible polymer like polyvinyl alcohol was reported to convert the hydrophobic nanotubes surface into a hydrophilic one that facilitates efficient attachment of biomolecules. CNTs modification can improve the response of lactate electrode sensors. Pt nanoparticles were electrodeposited onto a multiwall CNTs film by a multi-potential step technique and precasted on a glassy carbon or boron-doped diamond electrode. This technique significantly improved the conductivity, stability and electro activity for detection of lactate and other biomolecules.

8.2. Chemical Sensors

Chemical sensor is a self contained probe that provides the real time information about the chemical composition of its surroundings. It has to perform two functions:

- A specific interaction with a component of the analyte sample known as Recognition or Sensing.
- A physical or a chemical property can be converted into a measurable physical signal (electrical, optical, etc.) and is known as Transduction. The signal can be used for measuring the analyte concentration through appropriate calibration procedure. As for an example, the glass electrode can recognize the hydrogen ion and measure its concentration by a selective ion exchange process using the development of a membrane electric potential in presence of hydrogen ions.

8.2.1. Advantages of Chemical Sensors

- The dedicated sensors can substitute the standard analytical procedures and can help in fast detection and analysis.
- The chemical quantities can be automatically monitored.
- The portable sensor instruments can be conveniently used for field analysis.

8.2.2. Applications

- Process analytical procedures can be efficiently adopted in industry.
- Monitoring of the chemical pollutions in the environment can be automatically performed.
- Fast and/or *in situ* monitoring of drugs and biologically important compounds such as O_2, CO_2 and glucose content in blood are possible in biomedical science.
- Automatic control of important chemical parameters, such as pH, O_2 and CO_2 content, or nutrient concentrations can be carried out in food industry and biotechnology.

Chemical sensors have a chemical or molecular target to be measured and so the biosensors can be considered as a subset of chemical sensors because of the similar transduction methods as those for chemical sensors. Chemical sensor arrays with instrumentation, like the electronic nose or electronic tongue are known to address chemically complex analyte like taste, odor, toxicity, or freshness. Therefore the definition of a chemical sensor may be ''a small device that, as the result of a chemical interaction or process between the analyte gas and the sensor device, transforms chemical or biochemical information of a quantitative or qualitative type into an analytically useful signal [33]. The wide use of chemical sensors includes applications for critical care, safety, industrial hygiene, process controls, product quality controls, human comfort controls, automotive emissions monitoring, clinical diagnostics, home safety alarms, and homeland security (Table 8.1) [34]. Both economic and social benefits of chemical sensors are possible through these applications.

Table 8.1. Application of Chemical sensors.

Application	Detected Chemicals and Gases
Automotive	O_2, H_2, CO, NOx, Hydrocarbons (HCs),
Water treatment	pH, Cl_2, CO_2, O_2, O_3, H_2S,
Food	Bacteria, biological, chemicals, fungal toxins, humidity, pH, CO_2
Agriculture	NH_3, amines, humidity, CO_2, pesticides, herbicides
Military	Agents, explosives, propellants
IAQ	CO, CH_4, humidity, CO_2, VOCs
Industrial safety	Indoor air quality, toxic gases, combustible gases, O_2
Petrochemical	HCs and conventional pollutants
Steel	O_2, H_2, CO, conventional pollutants
Medical	O_2, glucose, urea, CO_2, pH, Na_1, K_1, Ca, Cl_2, bio-molecules, H_2S,Infectious disease, ketones, anesthesia gases
Environmental	SOx, CO_2, NOx, HCs, NH_3, H_2S, pH, heavy metal ions
Utilities [gas, electric]	O_2, CO, HCs, NOx, SOx, CO_2

The chemical sensors can be classified into two categories e.g. direct and indirect sensors.

8.2.2.1. Direct Sensor

A measured electrical output by a chemical reaction or the presence of a chemical can be directly correlated to the quantity, e.g. the capacitive moisture sensor where the capacitance is directly proportional to the amount of water present between the two plates of the capacitor.

8.2.2.2. Indirect Sensor

It works on the indirect reading of the sensed stimulus e.g. optical smoke detector where a photo resistor is illuminated by a source and smoke is allowed to flow and alter the light intensity, its velocity, its phase or some other measurable property.

There are also other chemical sensors, which are more complex due to the involvement of more transduction steps. In this review paper we have classified the chemical sensors into two types, e.g. (I) Gas sensors and (II) Electrochemical sensors.

8.2.3. Gas Sensors

8.2.3.1. Metal Oxide Based Solid-state Resistive Gas Sensor

The cost of establishing and implementing ordinary monitoring systems is extremely high; use of analytical instruments are time-consuming, expensive, and can seldom be applied for real-time monitoring in the field, even though they can provide a precise analysis. Hence the need for low cost, fast responsive and highly sensitive solid-state chemical sensors to detect air pollutant has increased in this century. The wide band gap semiconducting oxides (ZnO, TiO_2, SnO_2 etc) are suitable materials for the fabrication of inexpensive gas sensitive structure because of their versatile material properties and robust nature and low cost of production [35-37].

Semiconducting oxides are the fundamentals of smart devices as both the structure and morphology of these materials can be tailored precisely and accordingly, they are referred as functional oxides. They have two structural characteristics: cations with mixed valence states and anions with deficiencies. By varying either one or both of these characteristics, the electrical, optical, magnetic, and chemical properties can be tuned, giving the possibility of fabricating smart devices. The structures of functional oxides are very diverse and varied, and there are endless new phenomena and applications. Such unique characteristics make oxides one of the most diverse classes of materials, with properties covering almost all aspects of materials science and physics, in areas like semiconductors, superconductivity, Ferro electricity, and magnetism.

Since the demonstration almost 50 years ago that the adsorption of gas on the surface of a semiconductor can bring about a significant change in the electrical resistance of the material, there has been a sustained and successful effort to make use of this change for the purposes of gas detection [38-39]. Detection of toxic and flammable (H_2, CH_4, NO_x, SO_x, CO_x, H_2S etc) gases is a subject of growing importance in both domestic and industrial environments. Metal oxides such as Ga_2O_3, SnO_2, WO_3, TiO_2, and ZnO [37, 40-44] are physically and chemically stable and are widely investigated for gas and humidity detection. Sensing performances, especially sensitivity, are controlled by three independent factors: receptor function, transducer function, and utility. Receptor function concerns the ability of the oxide surface to interact with the target gas. Chemical properties of the surface oxygen of the oxide itself are responsible for this interaction in an oxide device, but this function can be largely modified. A considerable change in the sensitivity takes place when an additive (noble metal, acidic or basic oxide) is loaded on the oxide surface. Transducer function concerns the ability to convert the signal caused by chemical interaction of the oxide surface (work function change) into electrical signal. This function can be realized by the measure of the current through a system containing an innumerable number of grains and grain boundaries, to which a double-Schottky barrier model can be applied.

It has been observed by almost all researchers working with oxide semiconductors for gas sensing that the operation of such sensors with selectivity for a particular gas is extremely difficult, specially when the change in the electrical properties are used as the sensor signal. Use of sensor arrays and artificial neural network can normally solve this problem. In fact, today's chemical sensors are much more reliable with the implementation of ANN logic to improve the selectivity [45].

8.2.3.2. Principle of Gas Sensing

It is well known that the performance of gas sensors can be improved by incorporation of noble metals on the oxide surface. SnO_2 based gas sensors in the form of thick film, porous pellets or thin films and with Pt or Pd modifications are widely applied for monitoring explosive and toxic gases in industry, urban and domestic life [46]. Such promoting effects are undoubtedly related to the catalytic activities of the noble metals for the oxidation of hydrocarbons. In case of planar type resistive gas sensors two metal contacts are taken from the metal oxide. A polycrystalline semiconductor has a structure with a large number of grains and grain boundaries. In contrast to the single crystalline materials, polycrystalline materials give rise to local potential barriers, which arise between the grains. The electrical properties of the surface of the thin film and the surface boundaries of the grains are affected by the adsorption and desorption of gaseous molecules. Oxygen ions can be found at the material boundaries. At elevated temperature O_2 is chemisorbed by gaining one further electron from the surface Due to this chemisorption the resistivity of the material increases.

$$O + e^- \longleftrightarrow O^- [150\,^\circ C - 300\,^\circ C] \qquad (8.1)$$

$$2O + e^- \longleftrightarrow O_2^- [30\,^\circ C - 150\,^\circ C] \qquad (8.2)$$

8.2.3.2.1. Reducing Gas

Reducing gases like methane & hydrogen react with chemisorbed oxygen at the material boundaries. A negative charge carrier (electron) is added to the bulk and hence the resistance decreases.

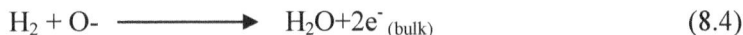

$$CH_4 + 4O^-_{(ads)} \longrightarrow CO_{2\,(air)} + 2H_2O + 2e^-_{(bulk)} \qquad (8.3)$$

$$H_2 + O\text{-} \longrightarrow H_2O + 2e^-_{(bulk)} \qquad (8.4)$$

8.2.3.2.2. Oxidizing Gas

Oxidized gases like NOx; COx etc react with chemisorbed oxygen at the material boundaries. A negative charge carrier (electron) is taken from the bulk and hence the resistance increases.

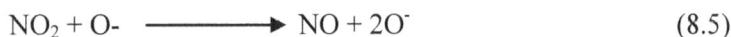

$$NO_2 + O\text{-} \longrightarrow NO + 2O^- \qquad (8.5)$$

By measuring the change in the conductivity of the semiconductor oxide thin films we can detect the reducing gases [47-48].

For p-type semiconductor the effect is just the opposite of n-type semiconductor because for p-type semiconductor the majority carrier is hole and adsorption of O⁻ on the semiconductor increases the hole conductivity and hence by reaction with reducing gases the hole conductivity decreases. In case of the oxidizing gases, the reaction with chemisorbed oxygen increases the hole conductivity further by electrons from the surface of the semiconductors.

8.2.3.2.1. Grain Size Effect

Nano crystalline feature is a single phase or multiphase of reduced size (1 nm to 100 nm) with at least of one dimension. When the crystal size is decreasing, more and more surface is exposed. So fraction of atoms at the grain boundary increases and the grain boundaries contain a large density of defects like vacancies and dangling bonds that can play an important role in the transport properties of electrons in nano materials. As the grain boundaries are metastable states they want to reduce their energy either by exchange of electrons or by sharing electrons with other atoms thereby increasing surface reactivity (or chemical reactivity). Xu et. al [36] proposed a model to explain the dependence of crystal size on the depletion layer due to adsorption of oxygen and to explain the high sensitivity of nanocrystalline metal oxide gas sensors. Later, A. Rothschild et. al [49] showed that the conductivity increases linearly with decreasing trapped charge states, and that the sensitivity to the gas-induced variations in the trapped charge density is proportional to $1/D$, where D is the average grain size. Fig. 8.2 shows a schematic of few grains of nanocrystalline metal oxide thin films and the space charge region around the surface of each grain at the intergrain contacts. The space charge region, being depleted of electrons, is more resistive than the bulk. When the sensor is exposed to reducing gases, the electrons trapped by the oxygen return to the oxide grains, leading to a decrease in the potential barrier height and thus a decrease in the resistance. The crystallites in the gas sensing elements are connected to the neighboring crystallites either by grain boundary or by necks. When the grain size (D) is greater than the depletion layer (L) width (D >> 2L) most of the volume of the crystallites is unaffected by the surface interactions with the gas phase. In this case the conductivity between two grains is due to the grain boundary contacts. In case of D> 2L, the grain size decrease in the depletion region extends deeper into the grains due to adsorption of oxygen causing a decrease in the conductivity. The conductivity depends not only on the grain boundary but also on the cross sectional area of the channels (neck control). When D<2L the depletion region extends throughout the whole grains and the crystallites are almost fully depleted of electrons. As a result the conductivity decreases through the junction and so the change of conductivity is very large in presence of reducing gases, yielding a high gas response. Fig. 8.2 demonstrates the three situations schematically. Further, nanocrystalline metal oxides can reduce the operating temperature of the gas sensors. H. Zhang and co-workers [50] reported that the surface or interfacial tension decreases with decreasing particle size because of the increase in the potential energy of the bulk atoms of the particles. Smaller particles with increased molar free energy are prone to adsorption per unit area of molecules or ions onto their surfaces in order to decrease the total free energy and to become more stable, and therefore, the smaller particles have higher adsorption coefficient for gases. Thus, the adsorption of oxidizing or reducing gases takes place relatively easily onto the nanocrystalline metal oxide surfaces.

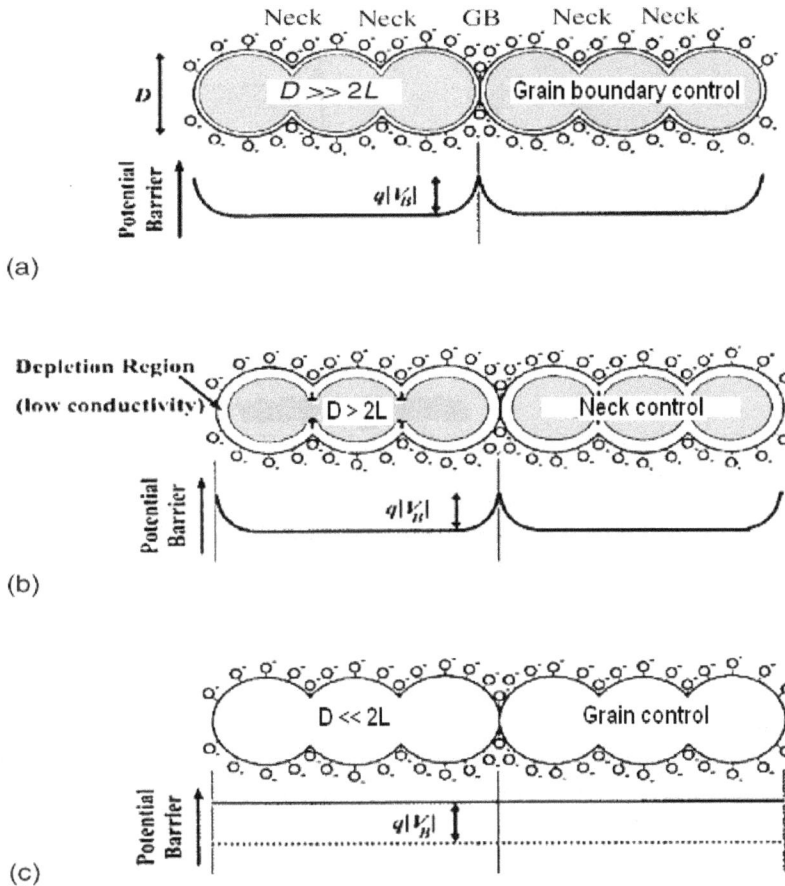

(a)

(b)

(c)

Fig. 8.2. Schematic of few grains of nanocrystalline ZnO thin films and the space charge region around the surface of each grain at the inter grain contacts.
(Reprinted with permission from "The effect of grain size on the sensitivity of nanocrystalline metal-oxide gas sensors", Journal of Applied Physics, 95 (2004) 6374 by A. Rothschild, Y. Komen ,Copyright 2004: American Institute of Physics).

8.2.3.2.2. Gas sensing behavior of Nanocrystalline Metal Oxides

Basu et al [51-52] prepared electrochemically grown ZnO-nanorod based methane and hydrogen sensors for low temperature applications. The FESEM of the grown ZnO is shown in Fig. 8.3 (a). The sample was also modified with Pd for better functional characterization of the sensors. The crystal size as determined from XRD was below 10nm. For the planar structures both contacts were taken from the Pd-Ag (26 %) electrodes on the ZnO surface and the separation between two electrodes was 3 mm. The electrical connections from the sensor device were made by using fine copper wires and silver paste. The schematic of the planar structures is shown in Fig. 8.3 (b).

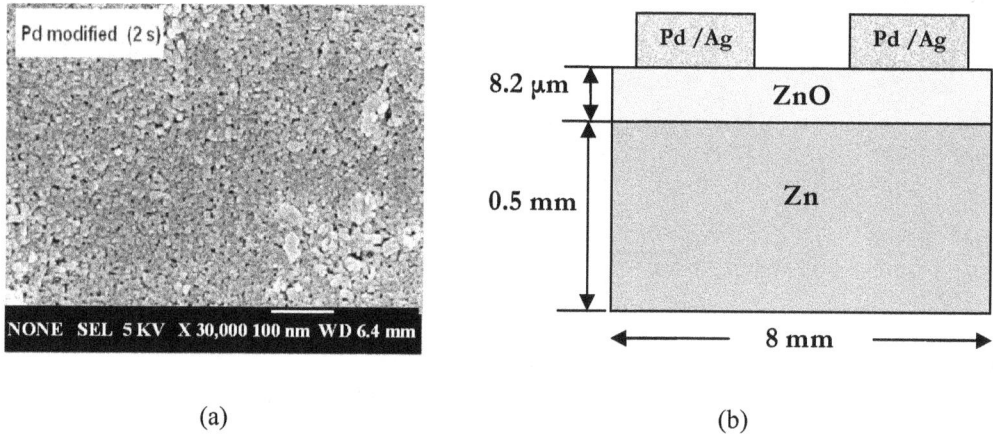

(a)   (b)

Fig. 8.3. (a) FESEM pictures of Pd modified ZnO thin films and
(b) schematic of planar configuration for gas sensing.
(Reprinted from Sensors and Actuators, B, 135, P. K. Basu, S. K. Jana, H. Sasha and S. Basu, "Low Temperature Methane Sensing by Electrochemically Grown and Surface Modified ZnO Thin films", 81, Copyright 2008: with permission from Elsevier).

The Pd loaded nanocrystalline (below 10 nm) ZnO enhances the oxygen spillover process [51-56] resulting in a large amount of chemisorbed oxygen that yields a high electrostatic potential across the Pd-Ag/ZnO Schottky interface. Subsequently methane or hydrogen reacts with this adsorbed oxygen and produces H_2O as shown in the Equations 8.6 and 8.7 respectively.

$$CH_4 + 2O_2^- \longleftrightarrow CO_2 + 2H_2O + e^- \qquad (8.6)$$

$$H_2 + 2O_2^- \longleftrightarrow H_2O + e^- \qquad (8.7)$$

The electrons on the surface of ZnO enhance the current through the electrodes. I-V characteristics of the planar sensor structures were studied with 1 % methane and 1 % hydrogen using pure nitrogen as the carrier gas over the temperature range 30 °C – 100 °C in separate experiments. The variations of the response with operating temperature and with biasing voltage have been plotted for planar sensor structures and are shown in Fig. 8.4.

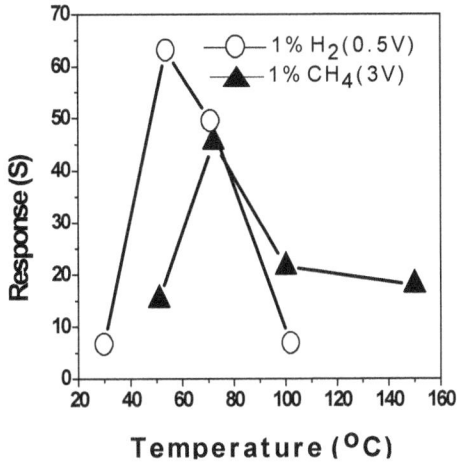 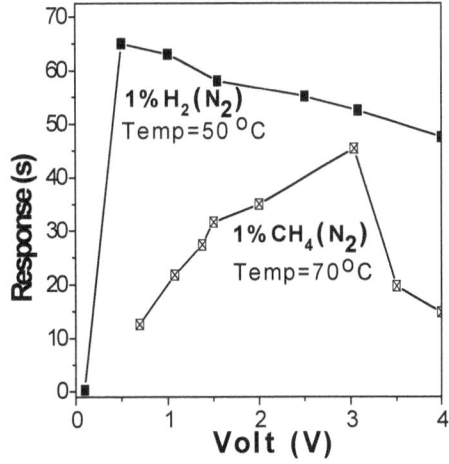

(a)   (b)

Fig. 8.4. (a) Response vs. temperature and (b) response vs. voltage curves of the planar sensor
in 1% methane and 1 % hydrogen using nitrogen as carrier gas.
(Reprinted from Sensors and Actuators, B, 135, P. K. Basu, S. K. Jana, H. Sasha and S. Basu, "Low
Temperature Methane Sensing by Electrochemically Grown and Surface Modified ZnO Thin films", 81,
Copyright 2008: with permission from Elsevier).

It is evident from the Figs. 8.4 (a) & (b) that the maximum response for methane at 3 V
bias is obtained at 70° C, the lowest temperature so far reported while hydrogen shows
maximum response at 50° C and 0.5V. For nanocrystalline structure the adsorption
activation energy is normally low and it is further reduced due to presence of dispersed
Pd acting as a catalyst over ZnO surface [57]. The transient response of the planar sensor
in different gas concentrations (0.01 %, 0.05 %, 0.1 %, 0.5 % and 1 %) and with
nitrogen and synthetic air as carrier gases respectively, in separate experiments, are
shown in Fig. 8.5.

The response time and the recovery time as derived from the transient response curves
are summarized in Table 8.2 and Table 8.3.

Table 8.2. The results of the planar sensor in different concentrations of methane
and hydrogen at 70 °C and 50 °C respectively using nitrogen as a carrier gas.

% of Gas	Response time (s)		Recovery time (s)		Response	
	Methane	Hydrogen	Methane	Hydrogen	Methane	Hydrogen
0.01	16.1	12.8	22.6	42.3	34.5	29.3
0.05	12.9	11.8	26.3	53.1	28.9	27.9
0.1	10.9	9.9	40.2	60.2	27.7	24.8
0.5	8.8	6.8	43.1	63.5	24.7	23.6
1	4.8	3.2	48.4	65.0	22.4	21.4

Table 8.3. The results of the planar sensor in different concentrations of methane and hydrogen at 70 °C and 50 °C respectively using synthetic air as a carrier gas.

% of Gas	Response time (s)		Recovery time (s)		Response	
	Methane	Hydrogen	Methane	Hydrogen	Methane	Hydrogen
0.01	19.2	22.3	9.2	24.2	34.2	46.2
0.05	17.4	16.4	14.6	32.3	32.8	44.4
0.1	16.4	8.2	18.3	40.5	29.6	34.2
0.5	15.2	5.3	21.6	48.2	27.4	29.6
1	12.6	4.2	25.7	52.0	23.4	26.8

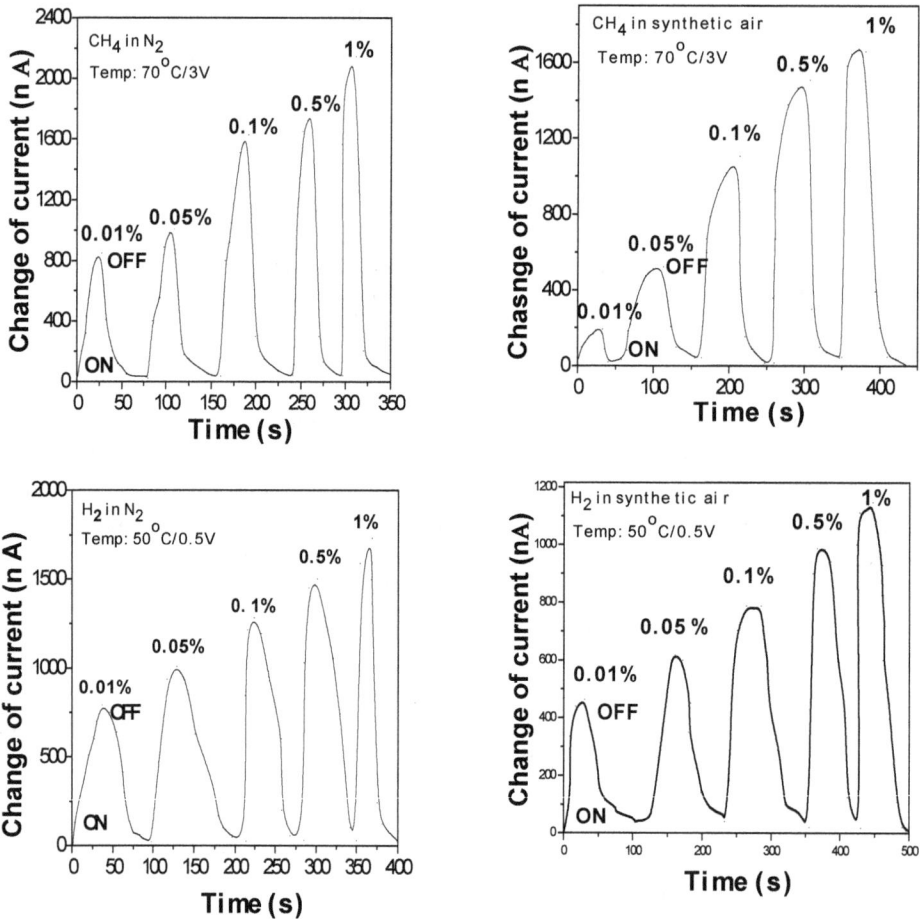

Fig. 8.5. Transient response curves of the planar structure for methane and hydrogen using nitrogen and synthetic air as carrier gases. For methane the sensor was operated at 70 °C and at 3V bias while for hydrogen it was operated at 50 °C and at 0.5 V bias.
(Reprinted from Sensors and Actuators, B, 135, P. K. Basu, S. K. Jana, H. Saha and S. Basu, "Low Temperature Methane Sensing by Electrochemically Grown and Surface Modified ZnO Thin films", 81, Copyright 2008: with permission from Elsevier)

From Fig. 8.6 it is observed that response of hydrogen gets saturated at 1000 ppm and above, most probably due to the fact that the active surface area of the sensor is already occupied. So, further increase of target gas concentration does not change the response of the sensors. This is supported by the observation of K. Christmann et al [57], which clearly states that with increasing coverage (θ) the sticking coefficient initially goes down sharply and then tends towards saturation with lower slope.

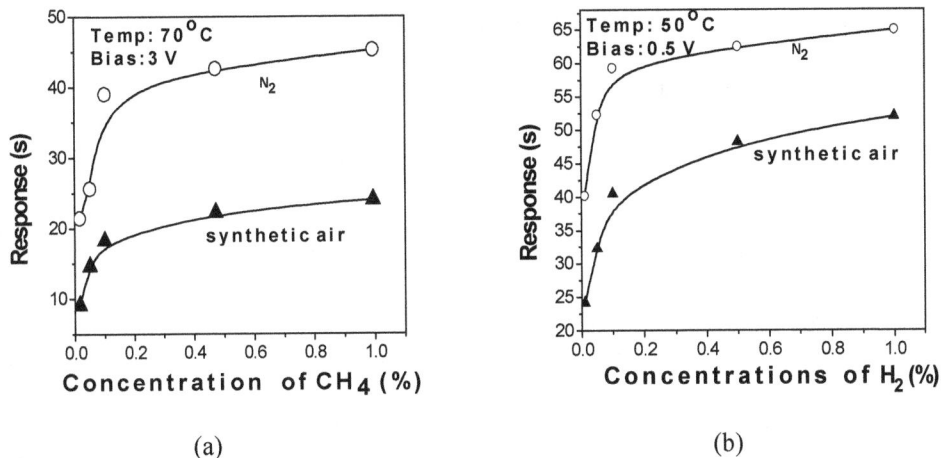

(a)   (b)

Fig. 8.6. Concentrations vs. response curves of the planar sensor for (a) methane and (b) hydrogen using nitrogen and synthetic air as carrier gases.

From Tables 8.2 and 8.3 it is found that the response time is shorter than the recovery time. When gas supply is cut off and air is allowed to flow into the sensing chamber, oxygen is adsorbed on crystalline ZnO by spillover technique as mentioned earlier and the out diffusion of adsorbed hydrogen from the junction takes place simultaneously. The chemisorbed oxygen reacts with hydrogen and produces H_2O. As a result, electrons of the chemisorbed oxygen come back to the ZnO surface and the current through the electrodes increases instead of decreasing. This process continues until all the atomic hydrogen desorbs from the junction and reacts with oxygen causing the recovery time relatively longer than the response time. It is evident from Fig. 8.5 that initially the current increases in presence of target gases and then gets saturated. After withdrawing the gas flow the current decreases with a slope less sharp than that after the gas is on. From the adsorption-desorption kinetics it can be explained that desorption always takes place at higher temperature than adsorption. Some of hydrogen atoms obtained through dissociative chemisorptions of methane or hydrogen molecules remain adsorbed on nanoporous ZnO, and as a result the current flow kinetics get sluggish causing the electrode current taking longer time to come back to its original base line value. By increasing the temperature or by supplying more oxygen the complete recovery might be possible with faster kinetics and thus faster recovery time could be achieved.

Nanotube-based sensors include metal oxides such as Co_3O_4, Fe_2O_3, SnO_2, and TiO_2 and metal tubes like Pt and Pd that have also great potential for sensing. Li et al. [58] reported that Co_3O_4 nanotubes exhibit superior gas sensing properties for H_2. It was suggested that the change of resistance is caused by the adsorption and desorption of gas molecules on the sensing surface consisting of the inside hollow structure and the porous texture of the nanotubes that provide more active sites in the three dimensional configuration. This enables access of more detecting gases. Fe_2O_3 shows high response to H_2 and C_2H_5OH at room temperature [59]. Recently, Liu and Liu [60] reported a single square shaped SnO_2 nanotube ethanol sensor provided with to two interdigitated Pt electrodes. The advantages were shown to be dramatically accelerated transport of gas/liquid in and out of the box beams, significantly increased active surface areas, increased flexibility in surface modification for chemically or biologically selective catalysis, drastically enhanced transport of ionic and electronic defects in the solid state (perpendicular to the wall thickness) due to shorter diffusion lengths, increased population of defects at the surfaces/ interfaces for fast electrode kinetics and quantum interactions at the nanoscale.

Nanowire based gas sensors are also very popular now a days [61]. Such structures have high aspect ratio (i.e. size confinement in two coordinates) & better crystallinity. They consume less power and yield higher integration densities. Due to large surface-to-volume ratio and Debye length (the distance over which a local electric field affects the distribution of free charge carriers) comparable to the size, they exhibit superior gas response at room temperature in comparison to their bulk and thin film counterparts. The applied gate potential of the nanowire configured field-effect transistors (FETs), can effectively control surface processes. Moreover, the carrier diffusion time at the surface is significantly reduced, resulting to a faster response and recovery. Some researchers have investigated devices based on zinc oxide (ZnO) nanowires configured as FETs (Fig. 8.7) to build high performance sensors [61-62].

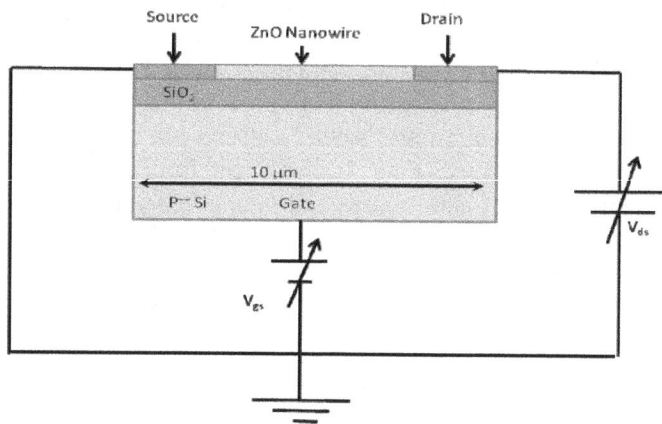

Fig. 8.7. Schematic of nanowire based FET structure.

Single crystalline ZnO nanowires (with a typical diameter of about 50 nm) were synthesized via a chemical vapor deposition method. Electrical measurements showed the ZnO nanowires to be n-type with the measured room-temperature electron concentration of the order 10^{19} cm^{-3}, and the electron mobility ranging from 20 to 100 cm^2/Vs. The charge transfer between the semiconductor and the chemical species adsorbed on the oxygen-vacancy sites is the origin of the sensing mechanism of the metal oxide and the relative change of conductance on exposure to a target gas determines the sensitivity. For example, the conductance of ZnO nanowires decreases with the introduction of NO_2 and at room temperature more than 50 % conductance change was observed on exposure to 0.6 ppm NO_2. To compare, doped ZnO thin films achieved less than 2 % conductance change when exposed to 1.5 ppm NO_2. This demonstrates a high potential of nanowire sensors with superior sensitivity.

The sensitivity can also be tuned by a transverse electric field induced by the gate of the FET configuration. Above a threshold gate voltage (the voltage at which the electrons are depleted in the channel), the sensitivity decreases with the increasing gate voltage. So the gate voltage can be used as a knob to adjust the sensitivity range. More significantly, the gate potential can be electrically tuned to desorb the gas molecules, thereby adjusting the recovery time to the desired scale.

Since the available thermal energy is usually lower than the activation energy for desorption, chemical sensing is generally not reversible at room temperature and this also causes a long recovery time. Ultraviolet (UV) illumination may be a common method for refreshing a sensor but this increases the complexity of the sensor design, and the recovery time still may be fairly long. A gate-induced refreshing mechanism may be much superior and convenient. It is known that the conductance of a nanowire based gas sensor can be electrically recovered by applying a negative gate voltage with a value much larger than the threshold voltage. This reduces the chemisorptions rate, and the hole migration to the surface (driven by the negative field) leads to a discharge of chemisorbed species thereby reverting the channel conductance and refreshing the sensor efficiently at room temperature without including any additional hardware to the devices. However, in spite of the enhanced sensitivity and faster recovery time, the nanowire sensors face the challenge of selectivity.

As already mentioned above, the enhanced response of the Q1D (Quasi One-Dimension) systems to the chemical environment is due to large surface-to-volume ratio that is related to the aspect ratio of the Q1D systems. So it is obvious that the radius of the nanowires significantly affects their chemical sensitivity. The dependence of sensitivity to O_2 on the radius of the ZnO nanowire was experimentally studied and is plotted in Fig. 8.8. In fact, this dependence can also be derived from the Drude model. Z. Fan and J. G. Liu [63] calculated the relation of conductance, G of nanowire with radius r and length l and it was expressed as:

$$G = \frac{\pi r^2}{l} n_e \mu_e,$$

where n_e and μ_e are the electron concentration and the electron mobility respectively.

$$n_e = n_o \exp\frac{2\alpha Ns}{r},$$

where n_0 is the electron concentration before the exposure of nanowires to O_2, Ns and α represent the surface density of chemisorbed O_2 molecules and the charge transfer coefficient respectively. Therefore, upon chemisorptions of the sensing gas the relation becomes

$$\frac{\Delta G}{G_0} = \frac{2}{r}\frac{\alpha Ns}{n_0}$$

The above equation apparently corroborates the plot shown in Fig. 8.8 and it clearly demonstrates that the smaller the diameter of nanowires the more effective is the chemisorptions on the nanowire surface and thus the more sensitivity for gases

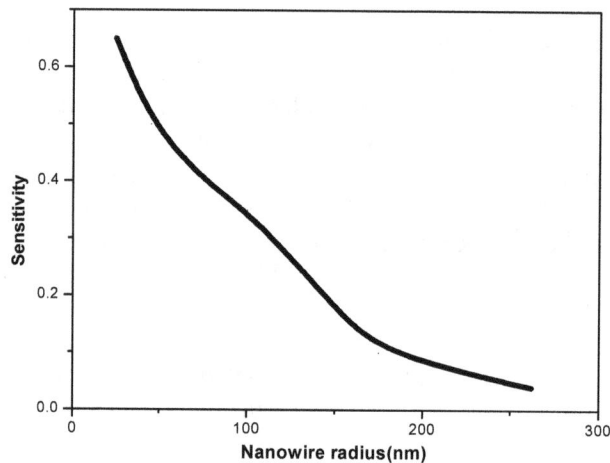

Fig. 8.8. Nanowire radius dependent sensitivity to O_2.

Metal nanowires and nanotubes have also shown great interest for sensing H_2 or hydrogen containing molecules and the hydrogen sensing by palladium nanowires has been experimentally studied in recent times [63–65]. The increase in resistivity in presence of hydrogen due to conversion of palladium to palladium hydride was reported earlier and therefore Pd alloys were used for H_2 sensing [76-69]. But recently, different sensing mechanisms were proposed depending on the properties of nanoscale materials [70–81]. The study of the hydrogen-induced percolation in discontinuous films or between nanoclusters and break junction formation in electrodeposited Pd mesowires [63, 82-83] was focused. The relatively big width of the breaks (25–35 nm) and relatively small active area of the mesowires limit hydrogen sensing below 1 %. The decrease of resistivity under hydrogen gas exposure was observed in these types of sensors and was believed to be due to percolation effect. On the other hand, both percolation and break junction formation mechanisms were assumed for a pulse-like hydrogen sensing response in Pd nanoparticle layers [84-86]. The scientists have investigated the influence of these two mechanisms on electrical resistivity of nanowires and nanotube arrays. The adsorption and diffusion of hydrogen atoms into palladium lead to an increase in resistivity due to electron scattering. On the other hand, the Pd lattice expansion because of the formation of PdHx phase induces a slight increase in the nanowire diameter and thus the increase of overall conductivity. However, when hydrogen is taken off, the inverse process takes place. In that case, the superposition of these two mechanisms creates difficulty in explaining the use of such an array as an active component in a hydrogen sensor. However, the high conductivity of continuous Pd film deposited on backside of anodic aluminium oxide (AAO) diminishes the effect of conductivity increment due to improvement of the connections between the nanowires. In fact, the sensor response of hydrogen at concentrations below 2.5 % was very small and was not very stable. The same logical arguments may be used for hydrogen sensing by nanotube array.

Pd nanotube has also a great potential to sense H_2 or hydrogen like molecules. The change of nanotube morphology on hydrogen exposure is schematically demonstrated in Fig. 8.9. Before activation the conductivity of the as deposited nanotube array is poor due to a poor contact between the nanotubes [84]. The diffusion of H_2 enlarges the diameter of the nanotubes and improves the contact. As a result the nanotube array conductivity increases.

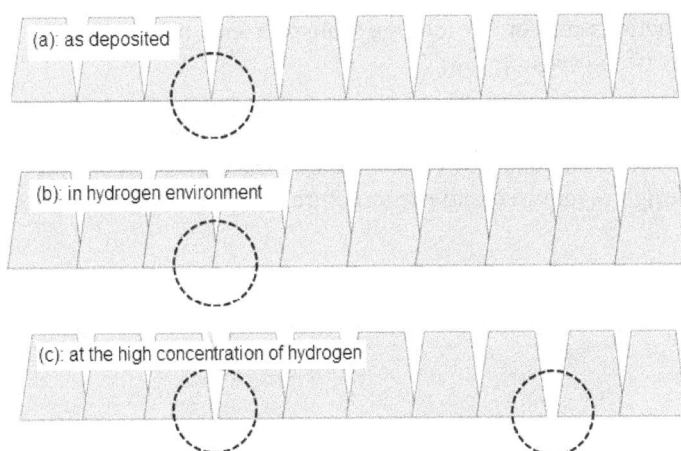

Fig. 8.9. Schematic representation of nanotube array sensor operation. (a) Continuous array of as-deposited conical Pd nanotube; (b) improvement in contact between nanotube under H_2 was exposed; (c) the array with nanocracks formed after strain induced by 1–2.5 % H_2 released. (Reprinted from Sensors and Actuators, B, 136, S. Cherevko, N. Kulyk, J. Fu, C-H. Chung "Hydrogen sensing performance of electrodeposited conidial palladium nanowire and nanotube arrays", 388, Copyright 2009: with permission from Elsevier).

This mechanism is valid until the applied hydrogen concentration is below the limit of inducing the phase transformation of Pd from alpha(Pd-α) to beta(Pd-β) form during "activation". Once the concentration is large enough for the phase transition, the reverse process generates & accumulates strain leading to the formation of the cracks as a result of tensile stress release. When H_2 is exposed to the sensor again, these nanocracks come closer and the conductivity drastically rises. The stable response to the tens of loading/unloading cycles proves again that such mechanism is valid for lower hydrogen concentrations.

8.2.4. Electrochemical Sensors

Bio molecular analyses in a chemistry laboratory are expensive and time-consuming. One of the main challenges today is the development of methods to perform the rapid '*in situ*' analyses. These methods must be sensitive, accurate, and be able to determine various substances with different properties in 'real-life' samples. Due to their high sensitivity and selectivity, portable size, rapid response and low-cost, electrochemical sensors are ideally suited for the measurement of biomolecules as analyte of interest. The applications of nanomaterials as electrochemical sensor for biochemical detection are discussed here [85-87].

8.2.4.1. Principle

"Electrochemistry implies the transfer of charge from an electrode to another phase, which can be a solid or a liquid sample. During this process chemical changes take place at the electrodes and the charge is conducted through the bulk of the sample phase. Both the electrode reactions and/or the charge transport can be modulated chemically and serve as the basis of the sensing [88]."

Electrochemical sensors may be classified on the basis of potentiometric, amperometric, or conductivity measurements. For each type of measurement with the electrochemical cell a specific design is required and the structure of electrochemical sensors is shown in Fig. 8.10.

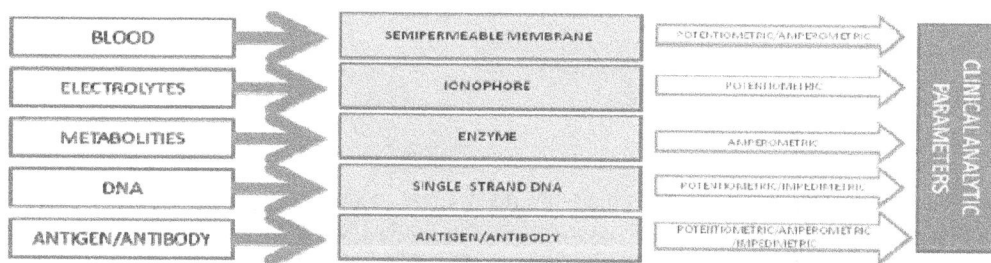

Fig. 8.10. Procedures for clinical analysis using electrochemical sensors. [85].

8.2.4.1.1. Potentiometric Sensors

In potentiometric sensors, a very small current is allowed to measure the potential difference between the reference electrode and the indicator electrode without polarizing the electrochemical cell. The reference electrode is meant for providing a constant half-cell potential while the indicator electrode exhibits a variable potential depending upon the concentration of a specific analyte in a solution. The potential change is related to the logarithm of concentration. The relation between the potential difference at the interface and the activities (concentrations) of species i in sample phases (s) and in the electrode phase (β) is given by the Nernst equation:

$$E = E_0 + \frac{RT}{Z_i F} \ln \frac{a_i^s}{a_i^\beta} .$$

where E_0 is the standard electrode potential of the sensor electrode; a_i is the activity of the ion, R is the universal gas constant; T is the absolute temperature; F is the Faraday constant and Z_i is the valency of the ion. The ion-selective electrode (ISE) is a potentiometric sensor for the measurements in the electrolyte phases. Mostly, a potentiometric sensor comprises a membrane (either a solid like glass and inorganic crystal or a plasticized polymer) with a unique composition and the ISE composition is chosen to indicate a potential that is primarily associated with the ion of interest via a selective binding process at the membrane-electrolyte interface [85-86].

Fig. 8.11 (a) shows the cell configuration of a conventional liquid junction ISE. In this type of potentiometric sensor a reversible ion or electron transport mechanism prevails at the ion-selective electrode. On the other hand, in solid contact electrode sensor illustrated in Fig. 8.11 (b) polymer membrane is deposited directly onto the solid electrode and there is no liquid electrolyte solution internally.

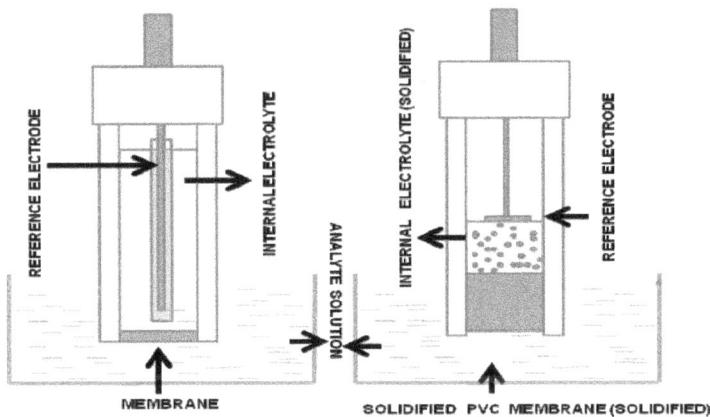

Fig. 8.11. The schematic of: (a) liquid junction ISE, and (b) solid-contact ISE (redrawn from [85]).

8.2.4.1.2. Amperometric Sensors

Amperometry is a method of electrochemical analysis in which the signal of interest is current and it is linearly dependent upon the concentration of the analyte. As the chemical species are oxidized or reduced (redox reactions) on inert metal electrodes, electrons are transferred from the analyte to the working electrode or vice versa [85-86]. The electron flow direction depends upon the properties of the analyte and it can be controlled by the potential applied to the working electrode. Two or three electrodes may be present in an amperometric cell. Usually the working electrode is a noble metal like platinum (Pt) or gold (Au) and the potential applied to the working electrode is measured and controlled by a reference electrode, usually Ag/AgCl that provides a fixed potential. Sometimes a third electrode, known as the counter or the auxiliary electrode is also used. By amperometric measurement a linear current vs. ion-concentration characteristics can be obtained using diffusion-controlled processes in the "limiting current operating mode". The measured cell current i.e. the diffusion current determines the analyte concentration quantitatively. Based on the amperometric measurements there are three "generations" of biosensors due to the different electron transfer processes. First-generation biosensors cause the electrical response because of the diffusion of normal product of the reaction to the transducer and the second-generation biosensors involve specific "mediators" between the reaction and the transducer for the improved response. In the third generation biosensors the reaction itself causes the response without the direct involvement of any product or mediator diffusion.

8.2.4.1.3. Conductometric Sensors

Conductometric sensors measure the electrolyte conductivity that varies when the cell is exposed to different environments. The sensing effect is due to the change of the number of mobile charge carriers in the electrolyte. The electrolyte shows ohmic behavior with the non-polarized electrodes in the AC supply mode operations during conductometric measurements. Since the conductivity is a linear function of the ion concentration in the electrolyte the method can be used for sensor applications. However, it is nonspecific for a given type of ion and so it functions as a non-selective sensor. The most essential conditions for using this sensor are the absence of polarization and limiting current operation mode. Thus, small amplitude alternating bias is used for the measurements with frequencies where the capacitive coupling is still not determining the impedance measurement.

8.2.4.2. Applications of Electrochemical Sensors

8.2.4.2.1. Glucose Detection

The proper absorption of glucose in the body is biologically important, and the lack of it can create diabetes with the risk for renal failure, retinal and neural complications. So for diagnosis of diabetes the detection of glucose in blood is medically important.

Principle of Electrochemical Reactions:

The electrochemical reactions proceed as follows to detect GO_x (glucose oxidase):

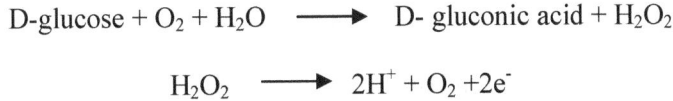

$$\text{D-glucose} + O_2 + H_2O \longrightarrow \text{D- gluconic acid} + H_2O_2$$

$$H_2O_2 \longrightarrow 2H^+ + O_2 + 2e^-$$

By adding glucose dehydrogenase (GDH) the electrochemical reactions using electron transfer mediator in the glucose sensors proceed as follows:

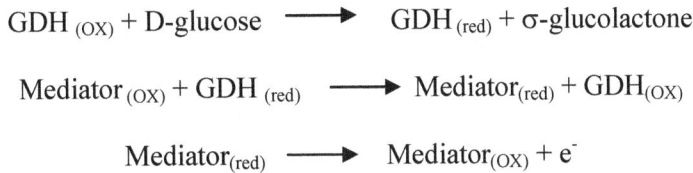

$$\text{GDH}_{(OX)} + \text{D-glucose} \longrightarrow \text{GDH}_{(red)} + \sigma\text{-glucolactone}$$

$$\text{Mediator}_{(OX)} + \text{GDH}_{(red)} \longrightarrow \text{Mediator}_{(red)} + \text{GDH}_{(OX)}$$

$$\text{Mediator}_{(red)} \longrightarrow \text{Mediator}_{(OX)} + e^-$$

Sol-gel precursor mixture of 3-glycidoxypropyltrimethoxysilane with methyltrimethoxysilane or tetraethoxysilane and ionic liquid methylimidazolium hexafluophosphate provides a unique microenvironment for the immobilization of GO_x. GO_x adsorbed in hexagonal mesoporous silica retains its bioactivity and stability [85-89]. LB (Langmuir Blodget) film was employed for GO_x immobilization because very thin film in nanoscale might produce a highly sensitive sensor with ultra fast response time. Conducting polymers have been receiving great and broad interests in clinical diagnosis of blood sugar [90-96]. There are many advantages of conducting polymers for preparing sensors like efficient transfer of electric charge and considerable flexibility of chemical structure. Researchers also indicated doping of redox enzyme within polypyrrole.

8.2.4.2.2. pH Detection

A preliminary test with surface charge field effect silicon nanowire device as a chemical sensor was performed by detection of pH level of the solution. [92, 97]. Fig. 8.12 shows the detection result of pH level by the silicon nanowire sensor of 100 nm width and 6 µm lengths.

At pH < pKa, the concentration of available proton ion (H+) is low and so the electrical conductance is low. On the other hand, when the pH > pKa, the electrical conductance becomes higher. This shows the usefulness of silicon nanowire chemical sensor based upon the concentration of surface charge around the silicon nanowire. More protonation on surface hydroxyl groups and accumulation of positive surface charge for pH < pKa, the boron doped p-type Si shows depletion of mobile charge carriers (holes) around the perimeter of the silicon nanowire. It can be stated qualitatively that less the pH value, the more protonation on the surface, and therefore the less conductance in the nanowire. If For pH above pKa, the surface hydroxyl groups follow the deprotonation process by generating negative ions on the nanowire surface. The mobile charge carrier depletion is reduced and more carriers are accumulated in the nanowire. Consequently, the electrical conductance of silicon nanowire is increased. The sensor shows an approximate linear relationship between pH level and electrical conductance of the nanowire with a sensitivity of about 79.4 nS/pH. From these experimental results, it has been verified that this silicon nanowire sensor works quite well as a chemical field effect sensor and can be used for the detection of charged protein molecules with the proper functionalization of the surface. Research is being currently pursued in this direction.

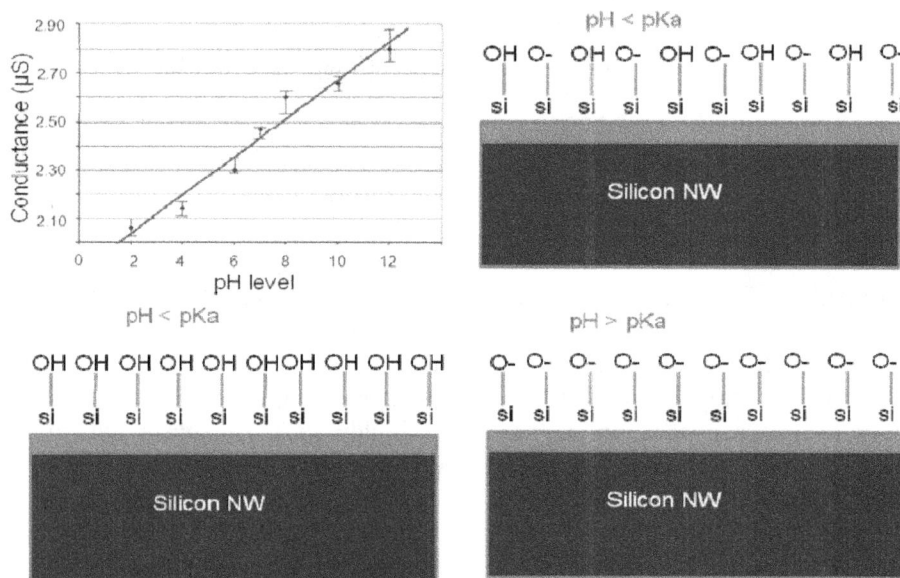

Fig. 8.12. pH level detection of solution by silicon nanowire sensor. The Change of electrical conductance with solutions of different pH level (approximately) and the proposed mechanism are displayed here. (Reprinted from Biosensors and Bioelectronics, 22, I. Park, Z. Li , X. Li, Albert P. Pisano , R. S. Williams "Towards the silicon nanowire-based sensor for intracellular biochemical detection", 2065, Copyright 2007: with permission from Elsevier).

8.2.4.2.3. pO_2 Measurement

In this sensor configuration the dissolved oxygen molecules in solution reach the electrode surface and there occurs a redox reaction. In the typical Clark-type O_2 electrode, the function of the cathode-working electrode is to reduce oxygen as given by the half-cell reaction

$$O_2 + 2H_2O + 4e^- \rightarrow 4OH^-$$

The half-cell reaction at the anode-counter electrode where the oxidation takes place and provides the return path to complete the circuit is given by [98]:

$$4Ag + 4Cl^- \rightarrow 4AgCl + 4e^-$$

The current output can be calibrated linearly with respect to the dissolved oxygen by applying a potentiostatic bias voltage of approximately 0.7 V.

8.2.4.2.4. pCO_2 Measurement

When dissolved CO_2 in solution diffuses into the internal electrolyte of the liquid junction potentiometric sensor, the following reactions occur:

$$CO_2 + H_2O \longrightarrow H_2CO_3$$

$$H_2CO_3 \longrightarrow H^+ + HCO_3^-$$

$$HCO_3^- \longrightarrow H^+ + CO_3^-$$

At the steady state chemical equilibrium, the concentration of CO_2 and H_2CO_3 are equal. The partial pressure is related to the pH of the sample because H_2CO_3 dissociates into H+ and HCO3−. When the ionic buffer, such as sodium bicarbonate (NaHCO3), is present, the pH is related to pCO_2 and the activity (concentration) of the sodium ion by

$$pH = -lg \frac{K_1 \alpha pCO_2}{\alpha_{Na^+}}$$

Here K_1 is the dissociation constant and α is the solubility coefficient for CO_2. The pH change is directly related to pCO_2 because the sodium ion activity remains almost constant.

8.3. Carbon Nano Tube Chemical Sensors

Sensors continue to make significant impact in everyday life with applications ranging from biomedical to automotive industry. Intensive research activities are continuing throughout the world for developing new materials and technologies for sensing. This is further intensified with the advancement of nanotechnology in order to develop miniaturized sensors with reduced weight, lower power consumption, and low cost materials. Of late, the discovery of carbon nanotubes (CNTs) has brought keen interest among the researchers to develop CNT-based sensors for various applications in the medical diagnosis, food industry, environmental pollutions and chemical warfare.

Small size, high strength, high electrical and thermal conductivity, and high specific surface area have enabled CNT sensors to be noted as the next-generation sensors to revolutionize the sensor industry. CNTs are hexagonal networks of carbon atoms of approximately 1 nm diameter and 1 to 100 microns length. CNT is essentially a sheet of graphite rolled-up cylindrically. There are two types of nanotubes depending upon the arrangement of the graphene cylinders, e.g. single-walled nanotubes (SWNTs) and multi-walled nanotubes (MWNTs) (Fig. 8.13). SWNTs have only one single layer of graphene cylinders; while MWNTs have many layers (approximately 50).Synthesized CNTs can be either aligned or random in nature. However, there are certain limitations of using CNTs for sensing in spite of their promising applications.

Based on the advantages of CNTs compared to other materials as mentioned above several papers have been published using CNTs as the sensing material in pressure, flow, thermal, gas, optical, mass, position, stress, strain, chemical, and biological sensors [98-99].

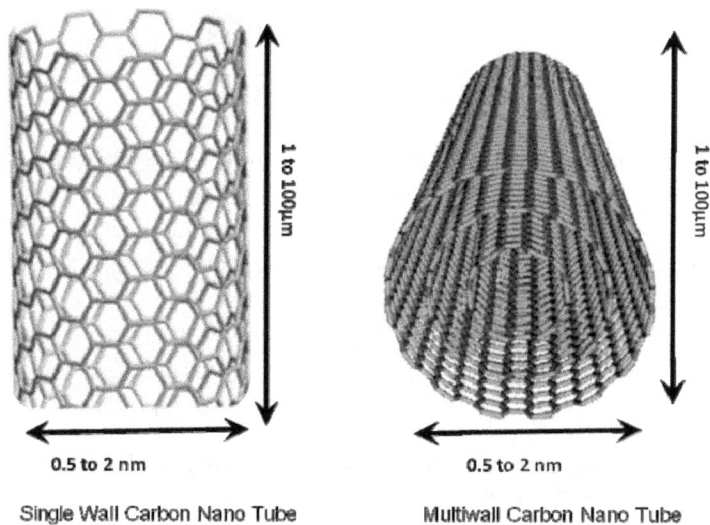

Fig. 8.13. A schematic of carbon nano tube with both single wall and multiwall configurations.

The implantable sensors may be very useful for health assessment and CNT based nanosensors are more appropriate for this purpose as they are thousands of times smaller than MEMS sensors and they consume much less power. Also, they are less sensitive to temperature variations that are necessary for reliable biomedical sensor performance. So, CNT-based nanosensors can be suitable as implantable sensors for continuous monitoring of pulse rate, temperature, blood glucose, and for diagnosing various diseases [100–102]. CNTs can also be potentially used for repairing the damaged cells or killing the tumors by chemical reactions. Implantable nanosensors can also monitor heart's activity level by regulating the heartbeats [103]. CNT based nano bio chemical sensors may also be used to detect DNA sequences in the body and DNA related to a particular disease may be detected. The use of CNT-based sensors can avoid problems of other implantable sensors with high current flowing that can cause inflammation, and can eliminate the need to draw and test blood samples. The devices can be administered through the skin and it is much less painful [104].

Sensors and bio chemical sensors are popular for widely use in food industry for quality control and for safety of food products as we know that the food contaminated by bacterial pathogens may lead to numerous diseases [105]. Efficient quality assurance is becoming increasingly important in the food industry. By the upsurge of nanotechnology and the unique properties of carbon nanotubes, CNT-based sensors show great potential for applications in the food industry including fish & meat freshness evaluation. After constant storage certain volatile components (such as, dimethyl and trimethyl amines, ethyl acetate etc) are released due to the initial bacterial putrification of fish & meat, By detecting the concentration of these chemicals, the quality control of fish & meat can be possible. Philip et al. [103] reported CNT/polymethylmethacrylate (PMMA) gas sensor for detecting different organic vapors including ethyl acetate. The use of a CNT-based gas sensor provides a non-destructive, noncontact method for food analysis, which is desirable and convenient for quality control in the food industry. CNT-based chemical sensors can also be used to detect undesired chemical residues and other environmental contaminants in drugs, food additives, herbicides, and pesticides in raw and processed foods. The promising aspect of CNT-based sensor research includes evaluation of the quality of fruits and vegetables. CNT-based gas sensors can offer improved performance for real-time monitoring of environmental pollution, combustible gas leak detection/alarms, chemical warfare agents (e.g., TNT or RDX and nerve agents such as GB or VX). CNT-based miniaturized gas sensors, in contrast to conventional solid-state gas sensors working at relatively high temperature, can be operated at room temperature. Moreover, gas sensors based on CNTs can be built in different geometrical features, not limited by the micro fabrication techniques and can offer reliable response [104-106]. The CNT sensors can ensure that the level of CO does remain within safe limit in the car park regions and in the ventilation systems coupled with an exhaust fan. By monitoring different level of CO concentration by CNT based CO sensors, the fans in ventilation system can be automatically set for starting, stopping or operating with variable speed. Also CNT-based sensors can be installed to monitor NO_2 where significant diesel traffics are prevalent. CNT-based electrochemical biosensors are suitable for wastewater monitoring [107-109]. The results of wastewater real sample tests using CNT biosensors were found to be in good agreement with the results of other genotoxicity tests .Humidity and temperature conditions impact the quantity and quality of the product directly in the agriculture and fishing industry. CNT-based humidity sensors can be effectively applied to monitor the humidity in green house agriculture. MWNT-coated quartz crystal microbalance humidity sensor has been experimentally proved to monitor relative humidity over the range, 5–97 % RH with a response and recovery time of about 60 and 70 seconds respectively. Plants cannot grow properly if the concentration of CO_2 is too high because the excess CO_2 get dissolved in water and produce carbonic acid that makes the soil more acidic. CNT -based CO_2 sensors can be used to monitor the concentration of CO_2 within the green house to maintain an optimal environment for the desirable growth of plants. An amperometric biosensor was developed by Sotiropoulou and Chaniotakis [108] using aligned MWNTs as the immobilization matrix and it was grown on platinum (Pt) substrate that served as the transduction platform for signal monitoring. The sensor arrays were treated with acid to remove the impurities like amorphous carbon that was included during the production process and with air to peel off the outer graphite layers from the nanotubes in separate experiments. The response and sensitivity of the acid treated sensor was found to be very high compared to the air-treated sensor after immobilization of the enzyme (e.g. glucose oxidase).

Kong et al [104] reported the change in electrical resistance of semiconducting SWNTs on exposure to gases like nitrogen dioxide (NO_2), ammonia (NH_3), and oxygen (O_2). The nanotube sensors exhibited the response time less than other sensors. Semiconducting SWNTs can function at room temperature with appreciably high sensitivity and this is an advantage of SWNTs-based chemical sensors.

However, Modi et al. [109] reported that the CNT gas sensors working on changes in electrical conductance have certain limitations, such as poor diffusion kinetics, inability to identify gases with low adsorption energies, and poor selectivity for gas mixtures. They also observed that the changes in moisture, temperature and gas-flow velocity affect the conductance of CNTs to large extent. They could overcome these limitations by proposing gas ionization sensors based on the electrical breakdown of a range of gases and gas mixtures at the tips of CNTs. Aluminium was used as cathode and the vertically aligned MWNT film (25–30 nm in diameter, 30 µm in length, and 50 nm separation between nanotubes) grown on SiO_2 substrate was used as anode. A glass insulator separated the electrodes. This ionization sensor showed good selectivity and sensitivity, and was not affected by the humidity, temperature, and gas-flow rates. Santhanam et al [110] used a nanocomposite of MWNTs and poly (3-methylthiophene) to develop a chemical sensor. Apart from this, nanocomposites of MWNTs and poly (3-methylthiophene) can be used to develop a chemical sensor. The sensor was sensitive to different chloromethane and changed its electrical resistance in presence of them. However, the response time of the sensor was found to be longer and was of the order 60 to 120 sec.

Of late Graphene based chemical sensors have got special attention of the researchers. Graphene is a single sheet or just a 2D structure of carbon in which the carbon atoms are SP^2 hybridized. They form a wide 2-dimensional hexagonal sheet in which each carbon is bonded to 3 other carbon atoms [111]. This sheet when rolled is called carbon Nanotube (Fig. 8.14). We can just imagine a sheet of paper rolled to form a tube like structure. These graphene sheets are arranged layer by layer in the graphite and these sheets are bonded to each other by weak van der Waals forces but these are strong enough to make graphite hard. Graphene has emerged out to be a potential semiconductor and due to its abundance we can say that this can prove out to be a boon in semiconductor electronics [112]. Imagine a sheet of material that is just one atom thick, yet super-strong, highly conductive, practically transparent and able to reveal new secrets of fundamental physics.

Graphene has few intrinsic charge carriers but remains conductive, as told by team member Andre Geim, of the University of Manchester and Nobel Laureate in Physics to *Chemistry World*. So when you add just a single electron it changes the resistivity significantly. Geim said that it should be possible to functionalize the surface of graphene to detect defined species. Graphene has the ultimate sensitivity because in principle it cannot be beaten - you cannot get more sensitive than a single molecule

[113]. Hydrogen sensors from Pd-functionalized multi-layer graphene nanoribbon networks are fabricated. The fabrication method of these networks is simple, low cost, and scalable, and their high specific surface area facilitates efficient functionalization and gas adsorption. These networks show high sensitivity to hydrogen at parts-per-million concentration levels at room temperature with a fast response and recovery time. Graphene has been attracting great interest because of its distinctive band structure and physical properties. Today, graphene is limited to small sizes because it is produced mostly by exfoliating graphite. Large-area graphene films of the order of centimeters have been grown on copper substrates by chemical vapor deposition using methane. The films are predominantly single-layer graphene, with a small percentage (less than 5%) of the area having few layers, and are continuous across copper surface steps and grain boundaries. The low solubility of carbon in copper appears to help make this growth process self-limiting. Graphene film transfer processes to arbitrary substrates have been developed, and the dual-gate field-effect transistors fabricated on silicon/silicon dioxide substrates have shown electron mobility as high as 4050 square centimeters per volt per second at room temperature. The thermally reduced graphene oxide (GO) [114, 115] shows *p*-type semiconducting behavior in ambient conditions and is responsive to low-concentration nitrogen dioxide and ammonia diluted in air at room temperature The sensitivity is attributed to the electron transfer from the reduced GO to adsorbed NO_2 & ammonia which leads to enriched hole concentration and enhanced electrical conduction in the reduced GO sheet.

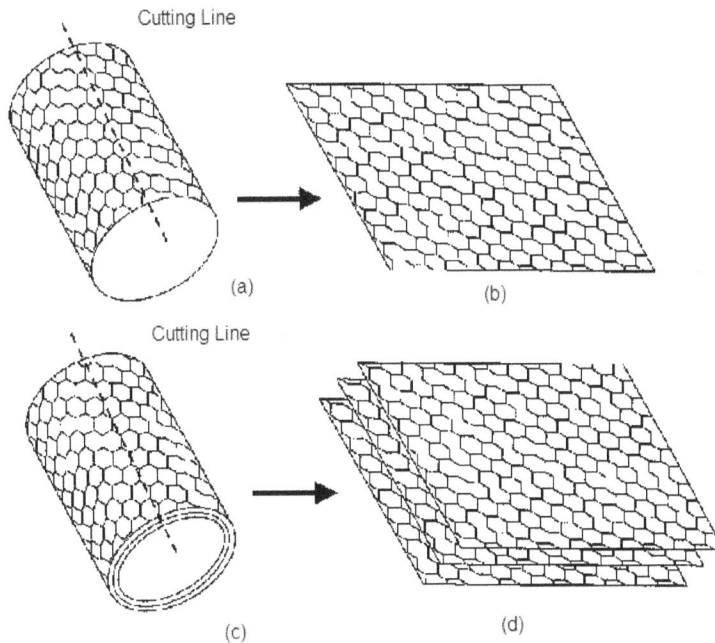

Fig. 8.14. Schematic of the (a) single wall carbon nano tube; (b) single layer grapheme; (c) multi wall carbon nano tube, and (d) multi layer graphene.

8.4. Summary & Concluding Remarks

This review has highlighted the chemical sensors e.g. gas sensors and electrochemical sensors using nanocrystalline materials. Especially, the nanotubes and nanowires of oxides, silicon and carbon have been focused. Various application possibilities of the nano-based chemical sensors have been discussed with the adequate citations of the literature. The nano chemical sensors can serve the need of different aspects of the human society like medical, food, environment and safety. The reliable automatic sensors based on nano materials and nano devices are possible to be fabricated with an economic viability. The different mechanisms of operations of the nanomaterial based chemical sensors have been discussed briefly. The merits and limitations of nanomaterials for the fabrications & operations of the chemical sensors have been discussed and the advantages of the nanomaterials for chemical sensing have been clarified with examples. In view of its recent importance and applications in the field of chemical sensors carbon nanotube (CNT) & its derivatives have been discussed in details in a separate section. The special activity on graphene based sensors has also been cursorily discussed. The principle of operations of each type of sensors with figures, mathematical relations and chemical equations, as applicable, has been presented in this chapter.

This review has revealed the importance of nanomaterials in enhancing the chemical sensing properties of different functional materials. The material property in nanoscale phase shows some exceptional chemical sensing behaviour and it is explained from basic concepts of chemistry & physics. It has been observed that the poor selectivity of chemical gas sensors cannot be substantially improved by using nanomaterials unless the artificial neural network (ANN) is applied. However, more intense research down to the molecular and atomic level can largely solve the selectivity problem of the chemical sensors and indeed, there is a global alertness on this issue.

References

[1]. NANOYOU teachers training kit-Module 1-Chapter 4, L. Filipponi and D. Sutherland, Interdisciplinary Nanoscience Centre (iNANO), *Aarhus University,* Denmark.

[2]. J. Maserjian and G. P. Petersson, Tunneling through thin MOS structures: Dependence on energy (E−κ), *Applied Physics Letters*, Vol. 25, Issue 1, 1974, pp. 50-52.

[3]. H. W. C Postma, T. Teepen, Z. Yao, M. Grifoni and C. Dekker, Carbon Nanotube Single-Electron Transistors at Room Temperature, *Science*, Vol. 293, No 5527, 2001, pp. 76-79.

[4]. G. Cao, Nanostructures & Nanomaterials: Synthesis, Properties & Applications, 1st edition, *Imperial College Press*, April 30, 2004.

[5]. P. Knauth and J. Schoonman, Nanostructured Materials: Selected Synthesis Methods, Properties and Applications, 1st edition, *Springer*, November 30, 2002.

[6]. A. S. Edelstein and R. C. Cammarata, Nanomaterials: Synthesis, Properties and Applications, *Institute of Physics Publishing*, 1998.

[7]. C. N. R Rao, A. Müller and A. K. Cheetham, The Chemistry of Nanomaterials: Synthesis, Properties and Applications, *John Wiley & Sons*, 2001.

[8]. J. Y. Park, Sun-Woo Choi, S. S. Kim, Fabrication of a Highly Sensitive Chemical Sensor Based on ZnO Nanorod Arrays, *Nanoscale Research Letters* , Vol. 5, 2010, pp. 353-359.

[9]. Xing-Jiu Huang, Yang-Kyu Choi, Chemical sensors based on nanostructured materials, *Sensors and Actuators*, B, Vol. 122, 2007, pp. 659-671.

[10]. S. Ashok Kumra, Shen-Ming Chen, Nanostructured Zincoxide particles in chemically modified electrodes for biosensor applications, *Analytical Letters*, Vol. 41, Issue 2, 2008, pp. 141-158.

[11]. M. H. Asif, S. M. Ali, O. Nur, M. Willander, C. Brännmark, P. Strålfors, U. H. Englund F. Elinder, B. Danielsson, Functionalised ZnO-nanorod-based selective electrochemical sensor for intracellular glucose, *Biosensors and Bioelectronics*, Vol. 25, Issue 10, 2010, pp. 2205-2211.

[12]. Md. M. Rahman, A. J. Saleh Ahammad, Joon-Hyung Jin, S. J. Ahn and Jae-Joon Lee, A Comprehensive Review of Glucose Biosensors Based on Nanostructured Metal-Oxides, *Sensors*, Vol. 10, Issue 5, 2010, pp. 4855-4886.

[13]. Y-T. Wang, L. Yu, Z-Q. Zhu, J. Zhang, J-H. Zhu, C-H. Fan, Improved Enzyme Immobilization for Enhanced Bioelectrocatalytic Activity of Glucose Sensor, *Sensors and Actuators, B*, Vol. 36, 2009, pp. 332-337.

[14]. X. Jiang, T. Herricks, Y. Xia, CuO Nanowires Can Be Synthesized by Heating Copper Substrates in Air, *Nano Letters*, Vol. 2, Issue 12, 2002, pp. 1333-1338.

[15]. D. Wang, C. Song, Z. Hu, X. Fu, Fabrication of Hollow Spheres and Thin Films of Nickel Hydroxide and Nickel Oxide with Hierarchical Structures, *The Journal of Physical Chemistry, B*, Vol. 109, Issue 3, 2005, pp. 1125-1129.

[16]. X. P. Gao, J. L. Bao, G. L. Pan, H. Y. Zhu, P. X. Huang, F. Wu, D. Y. Song. Preparation and Electrochemical Performance of Polycrystalline and Single Crystalline CuO Nanorods as Anode, Materials for Li Ion Battery, *The Journal of Physical Chemistry, B*, Vol. 108, Issue 51, 2004, pp. 5547-5551.

[17]. J. Zhang, J. Liu, Q. Peng, X. Wang, Y. Li, Nearly Monodisperse Cu_2O and CuO Nanospheres: Preparation and Applications for Sensitive Gas Sensors, *Chemistry of Materials*, Vol. 18, Issue 4, 2006, pp. 867-871.

[18]. A. E. Rakhshani, Y. Makdisi, X. Mathew, Deep Energy Levels and Photoelectrical Properties of Thin Cuprous Oxide Films, *Thin Solid Films*, Vol. 288, Issue 1-2, 1996, pp. 69-75.

[19]. C. B. McAuley, Y. Du, G. G. Wildgoose, R. G. Compton, The use of Copper(II) Oxide Nanorod Bundles for the Non-Enzymatic Voltammetric Sensing of Carbohydrates and Hydrogen Peroxide, *Sensors and Actuators, B*, Vol. 135, Issue 1, 2008, pp. 230-235.

[20]. Y. Kurokawa, T. Sano, H. Ohta, Y. Nakagawa, Immobilization of Enzyme onto Cellulose-Titanium Oxide Composite Fiber, *Biotechnology and Bioengineering*, Vol. 42, Issue 3, 1993, pp. 394-397.

[21]. Z. Zhang, Y. Xie, Z. Liu, F. Rong, Y. Wang and D. Fu, Covalently immobilized biosensor based on gold nanoparticles modified TiO_2 nanotube arrays, *Journal of Electroanalytical Chemistry*, Vol. 650, Issue 2, 2011, pp. 241-247.

[22]. Y. Xie, L. Zhou, H. Huang, Bioelectrocatalytic application of titania nanotube array for molecule detection, *Biosensors and Bioelectronics*, Vol. 22, Issue 12, 2007, pp. 2812-2818.

[23]. Yu Bai, Y. Sun, C. Sun, Pt–Pb nanowire array electrode for enzyme-free glucose detection, *Biosensors and Bioelectronics*, Vol. 24, Issue 4, 2008, pp. 579-585.

[24]. F. Xiao, F. Zhao, D. Mei, Z. Mo, B. Zeng, Nonenzymatic glucose sensor based on ultrasonic-electrodeposition of bimetallic PtM (M= Ru, Pd and Au) nanoparticles on

carbon nanotubes–ionic liquid composite film, *Biosensors and Bioelectronics*, Vol. 24, Issue 12, 2009, pp. 3481-3486.

[25]. D. Liu, Q. Luo and F. Zhou, Nonenzymatic glucose sensor based on gold–copper alloy nanoparticles on defect sites of carbon nanotubes by spontaneous reduction, *Synthetic Metals*, Vol. 160, Issue 15-16, 2010, pp. 1745-1748.

[26]. Q. Xu, Y. Zhao, J. Z Xu, J. J Zhu, Preparation of functionalized copper nanoparticles and fabrication of a glucose sensor, *Sensors and Actuators, B*, Vol. 114, Issue 1, 2006, pp. 379-386.

[27]. J. Shen, L. Dudik, C. C. Liu, An iridium nanoparticles dispersed carbon based thick film electrochemical biosensor and its application for a single use, disposable glucose biosensor, *Sensors and Actuators, B*, Vol. 125, Issue 1, 2007, pp. 106-113.

[28]. H. Yang, Y. Zhu, A high performance glucose biosensor enhanced via nanosized SiO_2, *Analytica Chimica Acta*, Vol. 554, Issue 1-2, 2005, pp. 92-97.

[29]. W. Zhao, J. J. Xu, Q. Q. Qiu, H. Y. Chen, Nanocrystalline diamond modified gold electrode for glucose biosensing, *Biosensors and Bioelectronics*, Vol. 22, Issue 5, 2006, pp. 649-655.

[30]. E. M. I. M Ekanayake, D. M. G. Preethichandra, K. Kaneto, Polypyrrole nanotube array sensor for enhanced adsorption of glucose oxidase in glucose biosensors, *Biosensors and Bioelectronics*, Vol. 23, Issue 1, 2007, pp. 107-113.

[31]. L. Liu, N. Q. Jia, Q. Zhou, M. M Yan, Z. Y. Jiang, Electrochemically fabricated nanoelectrode ensembles for glucose biosensors, *Materials and Science Engineering, C*, Vol. 27, Issue 1, 2007, pp. 57-60.

[32]. S. Cherevko, C-H. Chung, Gold nanowire array electrode for non-enzymatic voltammetric and amperometric glucose detection, *Sensors and Actuators, B*, Vol. 142, Issue 1, 2009, Vol. 216-223.

[33]. F. L. Dickert, P. A. Lieberzeit, O. Hayden, S. Gazda-Miarecka, K. Halikias, K. J. Mann and C. Palfinger, Chemical Sensors – from Molecules, Complex Mixtures to Cells – Supramolecular Imprinting Strategies, *Sensors*, Vol. 3, Issue 9, 2003, pp. 381-392.

[34]. Joseph R. Stetter, William R. Penrose and S. Yao, Sensors, Chemical Sensors, Electrochemical Sensors, and ECS, *Journal of The Electrochemical Society*, Vol. 150, Issue 2, 2003, pp. S11-S16.

[35]. O. Pummakarnchanaa, N. Tripathia, J. Dutta, Air pollution monitoring and GIS modeling: a new use of nanotechnology based solid state gas sensors, *Science and Technology of Advanced Materials*, Vol. 6, Issue 3-4, 2005, pp. 251-255.

[36]. J. Xu, Q. Pan, Y. Shuna, Z. Tian, Grain size control and gas sensing properties of ZnO gas sensor, *Sensors and Actuators, B*, 66, Issue 1-3, 2000, pp. 277-279.

[37]. A. Trinchi, S. Kaciulis, L. Pandolfi, M. K. Ghantasala, Y. X. Li, W. Wlodarski, S. Viticoli, E. Comini, G. Sberveglieri, Characterization of Ga_2O_3 based MRISiC hydrogen gas sensors, *Sensors and Actuators, B*, Vol. 103, Issue 1-2, 2004, pp. 129-135.

[38]. Y. L. Sandler, M. Gazith, Surface Properties of germanium, *Journal of Physical Chemistry*, Vol. 63, Issue 7, 1959, pp. 1095–1102.

[39]. S. Hoyt, J. Janata, K. Booksh, and L. Obando, Chemical Sensors For Portable, Handheld Field Instruments, *IEEE Sensors Journal*, Vol. 1, Issue 4, 200, pp. 256 -274.

[40]. J. Xu, Q. Pan, Y. Shuna, Z. Tian, Grain size control and gas sensing properties of ZnO gas sensor, *Sensors and Actuators, B*, Vol. 66, Issue 1-3, 2000, pp. 277-279.

[41]. U. Hoefer, J. Frank and M. Fleischer, High temperature Ga2O3-gas sensors and SnO_2-gas sensors: a comparison, *Sensors and Actuators, B*, Vol. 78, Issue 1-3, 2001, pp. 6-11.

[42]. X. Chen, W. Lu, W. Zhu, S. Y. Lim, S. A. Akbar, Structural and thermal analyses on phase evolution of sol–gel (Ba, Sr)TiO$_3$ thin films, *Surface and Coating Technology*, Vol. 167, Issue 2-3, 2003, pp. 203-206.

[43]. E. Comini, M. Ferroni, V. Guidi, G. Faglia, G. Martinelli, G. Sberveglieri, Nanostructured mixed oxides compounds for gas sensing applications, *Sensors and Actuators, B*, Vol. 84, Issue 1, 2002, pp. 26-32.

[44]. M. Rumyantseva, V. Kovalenko, A. Gaskov, E. Makshina, V. Yuschenko, I. Ivanovaa, A. Ponzoni, G. Faglia, E. Comini, Nanocomposites SnO$_2$/Fe$_2$O$_3$: Sensor and catalytic properties, *Sensors and Actuators, B*, Vol. 118, Issue 1-2, 2006, pp. 208-214.

[45]. T. Islam, H. Saha, Hysteresis compensation of a porous silicon relative humidity sensor using ANN technique, *Sensors and Actuators, B*, Vol. 114, Issue 1, 2006, pp. 334-343.

[46]. P. Bhattacharyya, P. K. Basu, C. Lang, H. Saha, S. Basu, Noble metal catalytic contacts to sol–gel nanocrystalline zinc oxide thin films for sensing methane, *Sensors and Actuators, B*, Vol. 129, Issue 2, 2008, pp. 551-557.

[47]. H-W. Cheong and M-J. Lee, Sensing characteristics and surface reaction mechanism of alcohol sensors based on doped SnO$_2$, *Journal of Ceramic Processing Research*, Vol. 7, Issue 1, 2006, pp. 183-191.

[48]. L. Castañeda, Effects of palladium coatings on oxygen sensors of titanium dioxide thin films, *Materials Science and Engineering*, Vol. 139, Issue 2-3, 2007, pp. 149-154.

[49]. A. Rothschild, Y. Komen, The effect of grain size on the sensitivity of nanocrystalline metal-oxide gas sensors, *Journal of Applied Physics*, Vol. 95, Issue 11, 2004, pp. 6374 -6380.

[50]. H. Zhang, R. Lee. Penn, R. J. Hamers and J. F. Bafield, Enhanced adsorption of molecules on surface of nanocrystalline particles, *The Journal of Physical Chemistry*, Vol. 103, Issue 22, 1999, pp. 4656-4662.

[51]. P. K. Basu, N. Saha, S. K. Jana, H. Saha, A. Lloyd. Spetz and S. Basu, Schottky Junction methane sensors using electrochemically grown nanocrystalline-nanoporous ZnO thin films, *Journal of Sensors*, Vol. 2009, 2009, Article ID 790476.

[52]. P. K. Basu, S. K. Jana, H. Saha and S. Basu, Low Temperature Methane Sensing by Electrochemically Grown and Surface Modified ZnO Thin films, *Sensors and Actuators, B*, Vol. 135, Issue 1, 2008, pp. 81-88.

[53]. V. R Shinde, T. P. Gujar, C. D. Lokhande, Enhanced response of porous ZnO nanobeads towards LPG: Effect of Pd sensitization, *Sensors and Actuators, B*, Vol. 123, Issue 2, 2007, pp. 701-706.

[54]. A. V. Tadeev, G. Delabouglise, M. Labeau, Influence of Pd and Pt additives on the microstructural and electrical properties of SnO$_2$-based sensors, *Materials Science and Engineering, B*, Vol. 57, Issue 1, 1998, pp. 76-83.

[55]. P. K. Basu, P. Bhattacharyya, N. Saha, H. Saha and S. Basu, Methane sensing properties of platinum catalyzed nano porous Zinc Oxide thin films derived by electrochemical anodization, *Sensor Letters*, 6, 2008, p. 219.

[56]. P. K. Basu, P. Bhattachayya, N. Saha, H. Saha and S. Basu, The superior performance of the electrochemically grown ZnO thin films as methane sensor, *Sensors and Actuators, B*, Vol. 133, Issue 2, 2008, pp. 357-367.

[57]. K. Christmann. G. Ertl, T. Pignet, Adsorption of Hydrogen on a Pt (111) surface, *Surface Science*, Vol. 54, Issue 2, 1976, pp. 365-392.

[58]. W. Y. Li, L. N. Xu, J. Chen, Co$_3$O$_4$ Nanomaterials in lithium-ion batteries and gas sensors, *Advanced Functional Materials*, Vol. 15, Issue 5, 2005, pp. 851-857.

[59]. J. Chen, L. N. Xu, W. Y. Li, X. L. Gou, Fe2O3 nanotubes in gas sensor and lithium-ion battery applications, *Advanced Materials*, Vol. 17, Issue 5, 2005, pp. 582-586.

[60]. Y. Liu, M. L. Liu, Growth of aligned square-shaped SnO2 tube arrays, *Advanced Functional Materials,* Vol. 15, Issue 1, 2005, pp. 57-62.

[61]. Jia Grace Lu, Electrically controlled nanowire-based chemical sensors, The International Society for Optical Engineering, *SPIE,* 2006.

[62]. Zhiyong Fan and Jia G. Lu, Chemical Sensing with ZnO Nanowires, in *Proceedings of the 5th IEEE Conference on Nanotechnology*, Nagoya, Japan, July 2005.

[63]. Z. Fan, D. Wang, P.-C. Chang, W.-T. Tseng, and J. G. Lu, ZnO nanowire field-effect transistor and oxygen sensing property, *Applied Physics Letters*, Vol. 85, Issue 24, 2004, pp. 5923-5925.

[64]. A. Barr, The effect of hydrogen absorption on the electrical conduction in discontinuous palladium films, *Thin Solid Films*, Vol. 41, Issue 2, 1977, pp. 217-226.

[65]. F. Favier, E. C. Walter, M. P. Zach, T. Benter, R. M. Penner, Hydrogen sensors and switches from electrodeposited palladium mesowire arrays, *Science*, Vol. 293, Issue 5538, 2001, pp. 2227-2231.

[66]. M. Wang, Y. Feng, Palladium–silver thin film for hydrogen sensing, *Sensors and Actuators, B*, Vol. 123, Issue 1, 2007, pp. 101-106.

[67]. Y.-T. Cheng, Y. Li, D. Lisi, W. M. Wang, Preparation and characterization of Pd/Ni thin films for hydrogen sensing, *Sensors and Actuators, B*, Vol. 30, Issue 1, 1996, pp. 11-16.

[68]. S. Nakano, S.-I. Yamaura, S. Uchinashi, H. Kimura, A. Inoue, Effect of hydrogen on the electrical resistance of melt-spun Mg90Pd10 amorphous alloy, *Sensors and Actuators, B*, Vol. 104, 2005, pp. 75-79.

[69]. X. M. H. Huang, M. Manolidis, S. C. Jun, J. Hone, Nanomechanical hydrogen sensing, *Applied Physics Letters*, Vol. 86, Issue 14, 2005, pp. 143103-143105.

[70]. Z. H. Chen, J. S. Jie, L. B. Luo, H. Wang, C. S. Lee, S. T. Lee, Applications of silicon nanowires functionalized with palladium nanoparticles in hydrogen sensors, *Nanotechnology*, Vol. 18, Issue 34, 2007, pp. 345502-345506.

[71]. J. Kong, M. G. Chapline, H. Dai, Functionalized carbon nanotubes for molecular hydrogen sensors, *Advanced Materials*, Vol. 13, Issue 18, 2001, pp. 1384-1386.

[72]. T. Kiefer, F. Favier, O. Vazquez-Mena, G. Villanueva, J. Brugger, A single nanotrench in a palladium micro wire for hydrogen detection, *Nanotechnology*, Vol. 19, Issue 49, 2008, pp. 125502.

[73]. K. Kim, S. M. Cho, Pd nanowire sensors for hydrogen detection, *Proceedings of IEEE on Sensors*, Vol. 2, 2004, pp. 705-707.

[74]. O. Dankert, A. Pundt, Hydrogen-induced percolation in discontinuous films, *Applied Physics Letters*, Vol. 81, Issue 9, 2002, pp. 1618-1620.

[75]. F. Rahimi, A. Irajizad, F. Razi, Characterization of porous poly-silicon impregnated with Pd as a hydrogen sensor, Journal of Physics D: *Applied Physics*, Vol. 38, Issue 1, 2005, pp. 36-40.

[76]. A. Tibuzzi, C. Di Natale, A. D'Amico, B. Margesin, S. Brida, M. Zen, G. Soncini, Polysilicon mesoscopic wires coated by Pd as high sensitivity H$_2$ sensors, *Sensors and Actuators, B*, Vol. 83, Issue 1-2, 2002, pp. 175-180.

[77]. T. Xu, M. P. Zach, Z. L. Xiao, D. Rosenmann, U. Welp, W. K. Kwok, G. W. Crabtree, Self-assembled monolayer-enhanced hydrogen sensing with ultrathin palladium films, *Applied Physics Letters*, Vol. 86, Issue 20, 2005, pp. 203104-203106.

[78]. C. L. Richard, P. V. Mangesh, A. B. Nosang, V. M. Eric, J. M. Reginald, M. P. M. Y. Yeonho Im, Investigation of a single Pd nanowire for use as a hydrogen sensor, *Small*, Vol. 2, 2006, pp. 356-358.

[79]. S. Yu, U. Welp, L. Z. Hua, A. Rydh, W. K. Kwok, H. H. Wang, Fabrication of palladium nanotubes and their application in hydrogen sensing, *Chemistry of Materials*, Vol. 17, Issue 13, 2005, pp. 3445-3450.

[80]. S. Yugang, H. H. Wang, Electrodeposition of Pd nanoparticles on single-walled carbon nanotubes for flexible hydrogen sensors, *Applied Physics Letters*, Vol. 90, Issue 21, 2007, pp. 213107-213109.

[81]. J. V. Lith, A. Lassesson, S. A. Brown, M. Schulze, J. G. Partridge, A. Ayesh, A hydrogen sensor based on tunneling between palladium clusters, *Applied Physics Letters*, Vol. 91, Issue 18, 2007, pp. 181910-181912.

[82]. E. C. Walter, F. Favier, R. M. Penner, Palladium mesowire arrays for fast hydrogen sensors and hydrogen-actuated switches, *Analytical Chemistry*, Vol. 74, Issue 7, 2002, pp. 1546-1553.

[83]. K. Manika, V. Deepak, R. M. Bodh, Pulse like hydrogen sensing response in Pd nanoparticle layers, *Applied Physics Letters*, Vol. 91, Issue 25, 2007, pp. 253121-253123.

[84]. S. Cherevko, N. Kulyk, J. Fu, C-H. Chung, Hydrogen sensing performance of electrodeposited conoidal palladium nanowire and nanotube arrays, *Sensors and Actuators, B*, Vol. 136, Issue 2, 2009, pp. 388-391.

[85]. Y. Wang, H. Xu, J. Zhang and G. Li, Electrochemical Sensors for Clinic Analysis, *Sensors*, Vol. 8, Issue 4, 2008, pp. 2043-2081.

[86]. M. Florescu, M. Barsan, R. Pauliukaite, C. M. A. Brett, Development and application of oxysilane sol-gel electrochemical glucose biosensors based on cobalt hexacyanoferrate modified carbon film electrodes, *Electro Analysis*, Vol. 19, Issue 2-3, 2007, pp. 220-226.

[87]. J. Li, J. Yu, F. Zhao, B. Zeng, Direct electrochemistry of glucose oxidase entrapped in nano gold particles-ionic liquid-N, N-dimethylformamide composite film on glassy carbon electrode and glucose sensing, *Analytica Chimica Acta* , Vol. 587, Issue 1, 2007, pp. 33-40.

[88]. Z. H. Dai, J. Ni, X. H. Huang, G. F. Lu, J. C. Bao, Direct electrochemistry of glucose oxidase immobilized on a hexagonal mesoporous silica-MCM-41 matrix, *Bioelectrochemistry*, Vol. 70, Issue 2, 2007, pp. 250-256.

[89]. H. Ohnuki, T. Saiki, A. Kusakari, H. Endo, M. Ichihara, M. Izumi, Incorporation of glucose oxidase into Langmuir-Blodgett films based on Prussian blue applied to amperometric glucose biosensor, *Langmuir*, Vol. 23, Issue 8, 2007, pp. 4675-4681.

[90]. J. Kan, S. Mu, H. Xue, H. Chen, Effects of conducting polymers on immobilized galactose oxidase, *Synthetic Metals*, Vol. 87, Issue 3, 1997, pp. 205-209.

[91]. M. Gerard, A. Chaubey, B. D. Malhotra, Application of conducting polymers to biosensors, *Biosensors and Bioelectronics*, Vol. 17, Issue 5, 2002, pp. 345-359.

[92]. I. Park, Z. Li, X. Li, Albert P. Pisano, R. S. Williams, Towards the silicon nanowire-based sensor for intracellular biochemical detection, *Biosensors and Bioelectronics*, Vol. 22, Issue 9-10, 2007, pp. 2065-2070.

[93]. N. Sinha, J. Ma, and John T. W. Yeow, Carbon Nanotube-Based Sensors, *Journal of Nanoscience and Nanotechnology*, Vol. 6, Issue 3, 2006, pp. 573-590.

[94]. A. Bhargava [Online], Available: http://www.ewh.ieee.org/r10/Bombay/news3/page4.html (1999).

[95]. M. C. Shults, R. K. Rhodes, S. J. Updik e, B. J. Gilligan, and W. N. Reining, A telemetry-instrumentation system for monitoring multiple subcutaneously implanted glucose sensors, *IEEE Transaction on Biomedical Engineering*, Vol. 41, Issue 10, 1994, pp. 937-942.

[96]. A. Bolz, V. Lang, B. Merkely, and M. Schaldach, First results of an implantable sensor for blood flow measurement, in *Proc. of the Ann. Int. Conf. IEEE Eng. Med. Biol.*, Chicago, IL, USA, 30 October-02 November, 1997, pp. 2341-2343.

[97]. R. Shandas and C. Lanning, Development and validation of implantable sensors for monitoring prosthetic heart valves: in vitro studies, *Medical and Biological. Engineering and Computing*, Vol. 41, Issue 4, 2003, pp. 416 -424.

[98]. L. C. Clark, C. Lyons, Electrode systems for continuous monitoring in cardiovascular surgery, *The New York Academy of Sciences*, 102, 1962, p. 29.

[99]. N. Gordon and U. Sagman [Online]: http://www.regenerativemedicine.ca/ (February, 20 2003), PDF, 2003.

[100]. O. P. Galaasen [Online]. Available at: http://plausible.custompublish.com/cparticle54173-5911.html, *Med. Technol.*, 2002.

[101]. D. Ivnitski, I. Abdel-Hamid, P. Atanaso, E. Wilkins, and S. Stricker, Application of electrochemical biosensors for detection of food pathogenic bacteria, *Electroanalysis*, Vol. 12, Issue 5, 2000, pp. 317-325.

[102]. N. Funazaki, A. Hemmi, S. Ito, Y. Asano, Y. Yano, N. Miura, and N. Y Amazoe, Application of semiconductor gas sensor to quality control of meat freshness in food industry, *Sensors and Actuators, B*, Vol. 25, Issue 1-3, 1995, pp. 797 -800.

[103]. B. Philip, J. K. Abraham, A. Chandrasedhar, and V. K. Varadan, Carbon nanotube/PMMA composite thin films for gas-sensing applications, *Smart Materials and Structures*, Vol. 12, Issue 6, 2003, pp. 935-939.

[104]. J. Kong, N. R. Franklin, C. Zhou, M. G. Chapline, S. Peng, K. Cho, and H. Dai, Nanotube molecular wires as chemical sensors, *Science*, Vol. 287, Issue 5453, 2000, pp. 622-625.

[105]. S. Chopra, K. McGuire, N. Gothard, and A. M. Rao, Selective gas detection using a carbon nanotube sensor, *Applied Physics Letters*, 83, Issue 11, 2003, pp. 2280-2282.

[106]. L. Valentini, C. Cantalini, I. Armentano, J. M. Kenny, L. Lozzi, and S. Santucci, Highly sensitive and selective sensors based on carbon nanotubes thin films for molecular detection, *Diamond and Related Materials*, Vol. 13, Issue 4-8, 2004, pp. 1301-1305.

[107]. G. Chiti, G. Marrazza, and M. Mascini, Electrochemical DNA biosensor for environmental monitoring, *Analytica Chimica Acta*, Vol. 427, Issue 2, 2001, pp. 155-164.

[108]. S. Sotiropoulou and N. A. Chaniotakis, Carbon nanotube array-based biosensor, *Analytical and Bioanalytical Chemistry*, Vol. 375, Issue 1, 2003, pp. 103-105.

[109]. A. Modi, N. K Oratkar, E. Lass, B. Wei, and P. M. Ajayan, Miniaturized gas ionization sensors using carbon nanotubes, *Nature*, Vol. 424, 2003, pp. 171-174.

[110]. K. S. V. Santhanam, R. Sangoi, and L. Fuller, A chemical sensor for chloromethane using a nanocomposite of multiwall carbon nanotubes with poly(3-methylthiophene), *Sensors and Actuators, B*, Vol. 106, Issue 2, 2005, pp. 766-771.

[111]. M. Inagaki, Y. A. Kim and M. Endo, Graphene: preparation and structural perfection, *J. Mater. Chem*, Vol. 21, Issue 10, 2011, pp. 3280-3294.

[112]. A. H. Castro Neto, F. Guinea, N. M. R. Peres, K. S. Novoselov, and A. K. Geim, The electronic properties of graphene, *Rev. Mod. Phys*, Vol. 81, Issue 1, 2009, pp. 109-162.

[113]. http://www.rsc.org/chemistryworld/News/2010/October/05101002.asp

[114]. Dreyer, D. R., Park, S., Bielawski, C. W. & Ruoff, R. S. The chemistry of graphene oxide, *Chem. Soc. Rev*, Vol. 39, Issue 1, 2010, pp. 228–240.

[115]. Qazi M, Vogt T, Koley G. Trace gas detection using nanostructured graphite layers, *Appl Phys Lett*, Vol. 91, Issue 23, 2007, pp. 233101-233101.

Chapter 9
Advances in Biosensors and Biochips

Sarmishtha Ghoshal, Debasis Mitra, Sudip Roy, Dwijesh Dutta Majumder

9.1. Introduction

Recent advances in nanoscience and technology have fueled a complete shift of paradigm in the physical, chemical, material, biological, healthcare, medical, and agricultural sciences, and also in mechanical, electrical, and computer engineering, as a result of attaining control of matter at the molecular or atomic level, i.e., at the single nanometer scale [1-56]. At the nanometer scale, matter is observed to exhibit numerous unique properties and phenomena, fundamentally different from their macroscopic counterparts. These properties can be harnessed to create new materials for novel applications such as biosensors or drug delivery systems, or to design new types of circuits and computers [26] using quantum dots or single-electron transistors, which are 10,000 times smaller than the current ones. Colloidal gold, iron oxide crystal, and semiconductor quantum dots having sizes 1-20 nm, have showed unique diagnostic applications in biology, medicine, and agri-biotechnology [7].

The domain of nanotechnology encompasses a very important area called "nanomedicine", which is concerned with the development of minimally invasive and targeted delivery of diagnostic, pharmaceutical and therapeutic agents to various body organs, tissues, and cells in a controlled manner [7]. Research in nanomedicine includes three important aspects: (i) nanosensors, (ii) nanofluidics and (iii) lab-on-a-chip. Nanosensors consist of nanostructured particles, or nanoparticles, or nanodevices that respond to physical, chemical, or biological stimuli. Nanofluidics is concerned with navigation, mixing, and controlled delivery of nanoliter volume of fluids through microchannels on a chip [9]. Lab-on-a-chip integrates sensors, fluidics, optics and electronics on a silicon chip to be used as a biochip for drug delivery systems or for biochemical diagnostics or DNA detection [9, 42-45, 86-88, 119, 120].

Sarmishtha Ghoshal
School of Materials Science and Engineering, Bengal Engineering and Science University, Howrah, India

During the early eighties, powerful microscopes were invented, which could scan the surface of the specimen using physical probes to produce images of the surface at the nanometer region. Several simultaneous interactions can also be imaged with these microscopes. Some microscopes used electron beams for studying surface topography, compositions and also other properties of a sample. These powerful microscopes like Atomic Force Microscope (AFM), Scanning Electron Microscope (SEM), Tunneling Electron Microscope (TEM), etc., helped not only to understand and study materials at the nanoscale region, but also to pick them up and move them around to form basic nanostructures, allowing some materials to be built molecule by molecule.

For the past several years, various materials have been developed whose dimension lies in the nanoscale region, such as, inorganic nanocrystals or quantum dots (QD), nanoparticles, nanocomposites, different nanostructured materials and many other for sensor applications. In the nano region, materials nomenclatures, i.e. wire, dots, quantum well etc., are defined according to the size of the crystals and the way the atoms are arranged in it. Nanostructured materials show interesting optical, electronic and catalytic properties.

Among the nanostructured materials, porous silicon (PS) shows very amazing features like biocompatibility, biodegradable, electroluminescence (EL), and photoluminescence (PL) at room temperature. It also shows non-toxic behavior when applied to human body. These unique properties of PS make it particularly suitable for biosensor application and also as a drug delivery material for in vivo applications. Sailor and his group reported a review on PS where it has been used as a drug delivery material [84]. Nanostructured porous silicon (PS) is also used to create optical biosensors, DNA detection sensors and photodetectors. A review on the scope of PS in nanotechnology has been reported by C. A. Betty [81]. As PS is fabricated from Si wafers, its production cost is considerably low and can be easily integrated with electronic equipments to produce a link between CMOS technology and photonic devices to create smart sensors and biochips.

From the inorganic nanocrystal category, the luminescent semiconductor nanocrystals or quantum dots (QDs) appears to be very promising material in the biosensor industry. Quantum dots generated a huge interest in the biosensing industry due to their excellent fluorescent properties which may help in eliminating the problems faced during the use of conventional organic or protein based fluorophores.

Luminescent porous silicon nanoparticles were also synthesized and studied as a replacement of fluorescent dyes and for in vivo applications as a drug delivery system.

In this review, we will discuss how QDs, PS, and Si-nanoparticles, can be used as biosensors and transducers or as drug delivery materials in biomedical studies. Recent advances in microarrays and microfluidic biochips have also been reviewed.

9.2. Nanobiosensors

A biosensor is a device that *detects*, *transmits* and *records* the information regarding an analyte that combines a biological component with a physiochemical detection system. A nanosensor is a biosensor which acts on the nano-scale region. There are different types of nanobiosensors – optical biosensors, electrical biosensors, electrochemical biosensors, nanowire biosensors, nanotube based biosensors, viral nano biosensors and nanoshell biosensors.

Fig. 9.1 shows a basic biosensor assembly, which includes a bioreceptor, i.e., a biological recognition element, a transducer and a processor [118]. The biological recognition elements used are living biological systems like, cells, tissues, or whole organism and biological molecular species such us antibody, enzyme, protein etc. The transducer essentially acts like a translator which recognizes the biological or chemical event from the biological component and transforms it into another signal for interpretation by the processor, which then converts it into a measurable output.

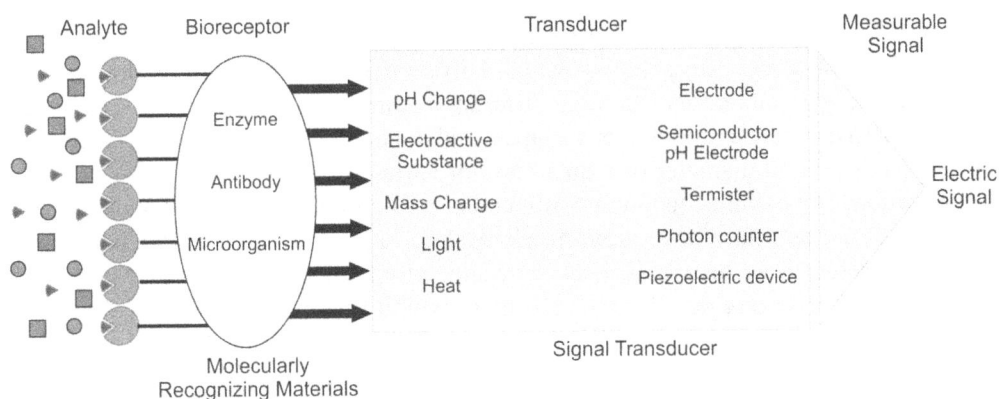

Fig. 9.1. Components of a biosensor.

The transducers may have different forms depending on the type of parameters being measured. These are (a) Amperometric transducers [76]; (b) Potentiometric transducers [77]; (c) Piezoelectric transducers [78]; (d) Thermal transducers [79]; (e) Optical transducers [80]. Transducers act as an interface, which measures the physical changes occurred at the bioreceptor [102], and transforms that energy into a readable output.

Nanobiosensors have wide applications in the field of biology and environment. Among the biological applications, there are (a) DNA sensors: genetic monitoring of diseases, (b) Immunosensors: HIV, hepatitis, other viral disease, drug testing, environmental monitoring, (c) Cell-based sensors: functional sensors, drug testing, (d) Point-of-care sensors: blood, urine, electrolytes, gases, steroids, drugs, hormones, proteins and others, (e) Bacteria (*E-coli*, streptococcus, other) sensors: food industry, medicine, environmental and others, (f) Enzyme sensors: diabetics, drug testing, and others are important. The environmental applications are (a) Detection of environmental pollution and toxicity, (b) Agricultural monitoring, (c) Ground water screening, (d) Ocean monitoring.

Nanosensors can be used to measure biotargets in a living cell without affecting its viability in a major way [44]. The effect of nanotechnology is enormous on the sensor industry because most chemical, biological and even physical sensors depend on the interactions occurring at the nanoscale level. The improvements required in sensor designing is low cost, small size, lower weight, greater sensitivity and better specificity all of which may be achieved through nanotechnology.

9.3. Types of Biosensors

9.3.1. Quantum dot Biosensors

Quantum dots (QD) appears to be the most promising functional and reproducible nanostructures available in the nanotechnology research circle. Quantum dots are very small and the smallest objects that can be synthesized on the nanoscale. Its structure is like a small dot which suggests its name. Different kinds of quantum dots can be synthesized in the laboratory through different techniques and the shapes normally appear to be pyramids, cylinders, lens shapes, and spheres. It confines electrons in three dimensions. The total diameter of a quantum dot varies between 3-60 *nm* depending on its application. QD exhibits quantum confinement properties in all three dimensions, i.e., the electrons are not allowed to move freely around in any direction. As its behavior is similar to atoms, it is also called an 'artificial atom'. This has a lot of important consequences for researchers. First of all, they exhibit quantized energy levels like an atom. For a particular incident radiation, for instance, a quantum dot will only emit certain specific spectra of light. The quantum theory also predicts that with the decreasing diameters of quantum dots, there will be a corresponding increase in energy of emitted light. From the solution of Schrödinger equation for an electron confined in an 1-dimensional box of length L, the energy difference between two successive levels E_n and E_{n+1} can be given as, $\Delta E = (2n+1) \{h^2/8mL^2\}$. This equation shows that if the length of the box decreases the energy difference between the levels increases and for $L \to \infty$ (ΔE will be 0, i.e., the electrons are delocalized and there is no quantization). This particular emission property of QDs has huge applications in diagnostics. Quantum dots are already in use as markers that are inserted into patients' body. These markers can be seen under medical scanners helping detection of biological processes as they occur.

Quantum dots can be fabricated with either top-down or bottom-up techniques. Top down techniques are very effective for generating a uniform distribution of diameters. This is important if it is desirable to create a large array of dots that will emit the same wavelength of light. The top down approaches like lithography are diffraction limited and cannot create dense networks of quantum dots. This approach inherently implies material damage and many quantum dots produced with these techniques have defects that reduce their effectiveness.

The commonly used methods for producing quantum dots are bottom-up approaches. This can be done either with chemical vapor deposition or molecular beam epitaxy on a highly mismatched substrate. By layering a desired material that does not fit properly with the lattice of the substrate, high strain occurs at the interface and that layer will start nucleating into small quantum dots. Bottom-up approaches are acceptable ways to create quantum dots in dense arrays that will self-assemble in an orderly manner. However, the uniformity of their size distribution is not as precise as that produced through top-down approach mainly because it's impossible to control their formation as strictly.

From early eighties, quantum dots are being deployed in nano-scale computing applications, where light is used to process information. However, this technology is now being used in medicine. The QD crystals are one ten-millionth of an inch in size and can be dissolved in water, which when illuminated, act as molecule-sized LEDs, and can be used as probes to track antibodies, viruses, proteins, or DNA within the human body. Biomolecules labeled with luminescent colloidal semiconductor quantum dots (QDs) have various applications to fluoro-immunoassays and biological imaging. Because of their small size, quantum dots can be used to visualize, measure, and track individual molecular events using fluorescence techniques, as they have the ability to visualize and track dynamic molecular processes over long time scales, which is a unique property. A review on QD biosensors was reported earlier by Sapsford, et al. [93].

It has been found that single QD incorporated nanosensors for DNA detection can reduce significantly or even eliminate the complication of background fluorescence encountered by conventional molecular fluorescence resonance energy transfer (FRET) technique [35]. Zhang, et al. [35] reported the extraordinary performance characteristics of a QD-FRET nanosensor for DNA detection with ultrahigh sensitivity, discrimination capacity and great simplicity. Fig. 9.2 shows how FRET induced QD DNA nanosensor works. The induced fluorescence emission between Cy5 acceptors and QD donors is shown in Fig. 9.2 (a). The experimental setup and the formation of nanosensor assembly in the presence of targets are shown in Figs 9.2 (b) & 9.2 (c) respectively [35]. Tran, et al. [38] described in their paper how CdSe-ZnS QDs can be designed and used as nanosensors for detection both in water soluble and in solid phase conditions using Förster energy transfer method.

Quantum dot technology presents a promising tool in neuroscience research [104]. Several researchers are trying hard to produce new quantum-dot-based tools for applications to neurobiology. Triller, et al. [98] used antibody functionalized quantum dots to study diffusion of glycine receptors in cultures of primary spinal cord neurons. Vu, et al. [99] tagged nerve growth factor to quantum dots and used them to promote neuronal-like differentiation in cultured pheochromocytoma 12 (PC12) cells. This method could be used to visualize and track functional responses in neurons. A technique of producing biocompatible water-soluble quantum dot micelles that retain the optical properties of individual quantum dots, was developed by Brinker, et al. [100]. Ting, et al. [101] developed a modified quantum dot labeling approach that presented the relatively large size of antibody–quantum-dot conjugates and the instability of some quantum-dot–ligand interactions. The problems with semiconductor QDs are its toxic effects, which prevent it from being used in-vivo applications.

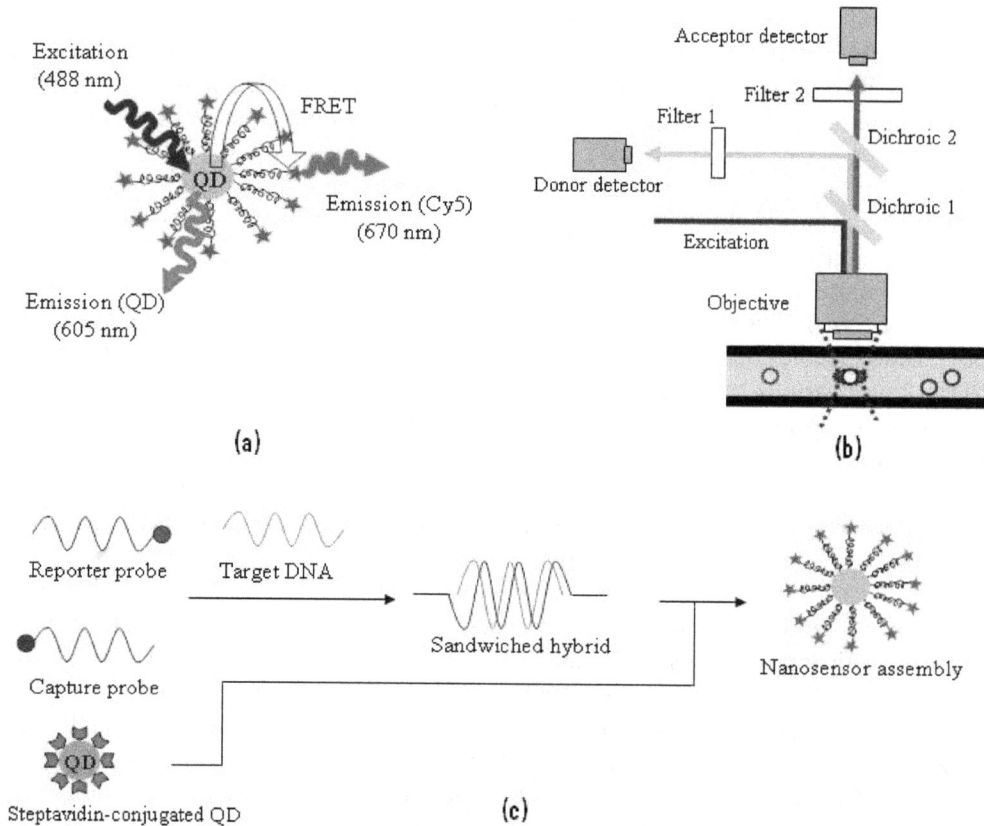

Fig. 9.2. Single QD-based DNA nanosensor and its experimental setup.

9.3.2. Porous Silicon Biosensors

Porous silicon (PS) and Si-nanocrystal have amazing properties that are particularly suitable for applications [57, 59] in the biosensor industry. Both PS and Si-nanocrystals have potential applications to optical biosensors, DNA detection sensors, or photodetectors [32, 46-51]. Sensors based on PS offer enhanced sensitivity, reduced power demands and low cost. A review article explaining various applications of PS as a transducer material has been reported recently by Andrew Jane, et al. [95]. The interesting features about PS are its high surface area and reactive surface chemistry. Si-nanocrystals can also be obtained from PS [54-55] in aqueous form.

Porous Silicon is an electrochemically derived nanostructured material consisting of nanometer-sized silicon regions surrounded by empty space, and can be prepared as quantum wires or quantum dots. The quantum confinement of Si atoms in PS leads to interesting optical, chemical, and electronic properties. The visible room temperature photoluminescence (PL) and the electroluminescence properties of PS, along with the simplicity of its fabrication process, make it extremely convenient and useful material for several opto-electronic and sensor applications. The wavelength of the photoluminescent light can be changed by simply increasing or decreasing the porosity of the material. For example, a highly porous sample (70-80 % porosity) will emit green/blue light while a less porous sample (40 %) will emit red light. The most acceptable theory about this photoluminescence (PL) property is the quantum confinement effect where by confining the matter in the nanoscale dimension, the interaction between matter and light can be limited in nano dimension (as described in Section 9.3.1).

PS can be divided into three main categories based on their pore size: (i) for microporous porous silicon the pore width is less than 2 *nm*, (ii) for mesoporous the pore width is in between 2 *nm* to 50 *nm*, (iii) for macroporous the pore width is greater than 50 *nm*. With appropriate modification of the electrochemical process, PS can also be fabricated to behave as 1-D photonic crystals [58]. The intensity and wavelength of the reflected light is determined by the nanostructure, and these optical properties can be deployed in sensing of chemical and biological agents like viruses and bacteria [34]. Because of their non-invasive and non-radioactive nature, they promise versatile applications to medical diagnostics, pathogen detection, gene identification, and DNA sequencing [12-13, 39].

The standard procedure for fabrication of nanostructured porous silicon is the electrochemical etching method in hydrofluoric (HF) acid solution. The etching resulted in a system of disordered pores with nanocrystals remaining in the inter-pore region. The pores propagate primarily in the <100> direction of the crystal. Almost all properties of PS, such as porosity, porous layer thickness, pore size and shape, and microstructure morphology, strongly depend on the fabrication conditions. In the case of anodization, these conditions include HF concentration, chemical composition of electrolyte, current density (and potential), wafer type and resistivity, crystallographic orientation, temperature, time, electrolyte stirring, illumination intensity, and wavelength, etc. Thus, a complete control of the fabrication is complicated and all possible parameters should be taken into account. Some of these parameters also depend on each other.

The average diameter of the pores can be tuned from a few nanometers to several micrometers. Tuning the pore diameters and chemically modifying the surface allow developers to control the size and type of molecules adsorbed [60, 111]. The large surface area enables bio-organic molecules to adhere to the surface of the PS [81, 85]. Aqueous HF is suitable for the etching process because the silicon surface is hydrophobic. The porous layer can be made more structurally uniform if an ethanoic solution is used – this increases the wettability of the silicon and allows more surface penetration by the acid. Fig. 9.3 shows the top-view and cross-sectional views of *p*-type boron-doped PS with different resistivity wafers captured by a scanning electron microscope (SEM) [106, 107, 123].

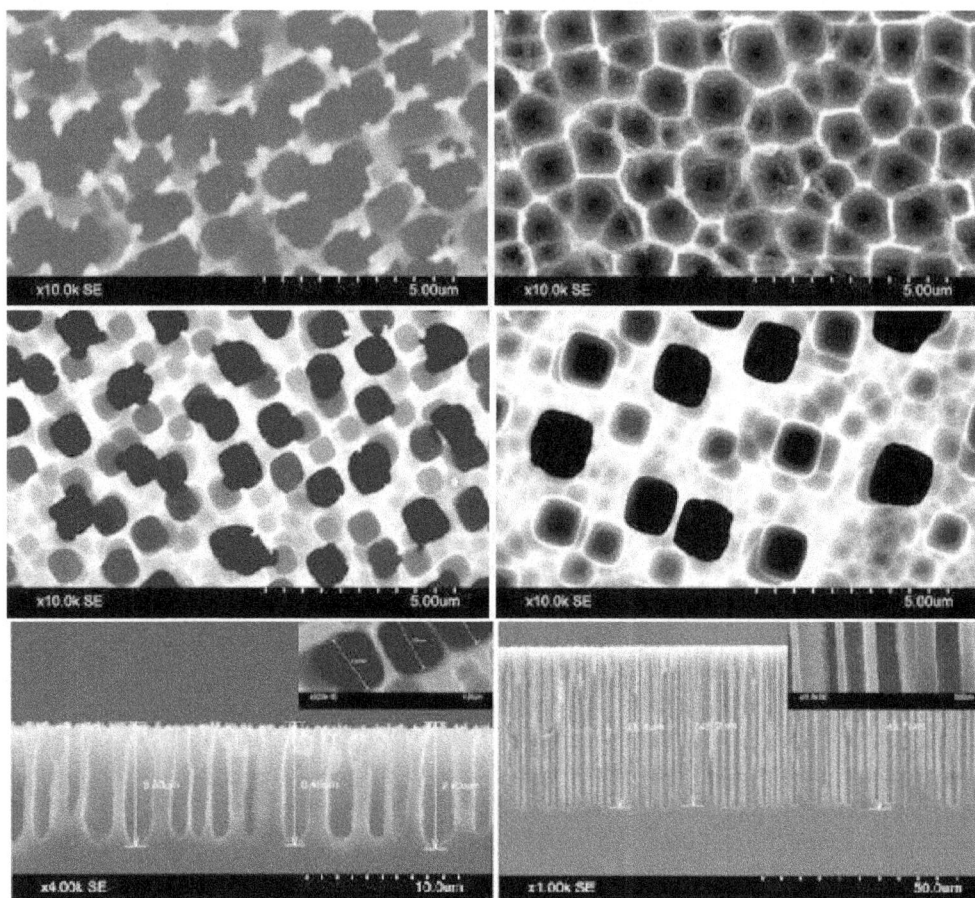

Fig. 9.3. Top-view and cross-sectional views of p-type PS for different resistivity wafers.

Although the electrochemical anodization is commonly used in the fabrication of PS, several other fabrication methods have been introduced. Stain etching method is one of them. The stain films are produced by immersion of Si substrate in HF solutions without any electrical bias [89-91]. This method is even simpler than the previously presented anodization. However, the control of the porosity, layer thickness and pore size of PS is quite limited. In addition, in the stain etching of Si microparticles, it is quite difficult to control the porosification of particles. Incomplete porosification of Si particles might cause problems in drug delivery applications. Porous silicon fabricated by stain etching method shows low photoluminescence efficiency than the electrochemically etched one.

Nanoporous silicon consists of a complicated network of silicon threads of 2-5 nm thickness with an internal surface area-to-volume ratio of around 500 m^2/m^3. Thus PS can absorb large amounts of foreign molecules onto its surface eventually changing the effective refractive index of the semiconductor porous material. Because of the quantum confinement effect, strong luminescence at room temperature is observed from the tiny pores. This photoluminescence (PL) intensity changes when PS is exposed to various chemicals and biological samples and the final photoluminescence efficiency depends on the dipole moment of the molecules attached to the pores. Similarly the effective dielectric constant and the conductivity of PS layer changes when the pores are filled with some other molecule. This property helps in developing electrical and optical PS biosensors adsorbing foreign materials on its surface.

It has been shown that porous silicon can be used as a base material for passive or active optical devices like Fabry-Perot interferometers, Bragg filters and optical microcavities [82], because of its lower effective refractive index than that of bulk silicon. It can also be used as an antireflection coating for silicon solar cells. A wide range of refractive index varying from 1.25 to 3, allows this material for many optical application. The PS Fabry-Perot film with two planer and parallel interfaces can produce high contrast optical fringes. Shift in the fringe occur when an analyte binds to the surfaces in the pores, providing a sensitive transduction modality [114]. Multilayer devices, like Bragg filters, can be prepared by periodically varying the current density during the etching process. Such multiplayer structures act as 1-D photonic crystals with reflectivity maxima that depend on the refractive index gradient and the periodicity of the superlattice. Porous silicon 1-D photonic crystals used as a label free optical sensor for detection of bacteria has been reported earlier [103].

The freshly etched PS surface is hydrogen terminated and hydrophobic in nature. Impurities like, carbon and fluorine are also found attached to the surface. The PS surface in unmodified form is unstable for sensor application and also very much fragile. For biosensor applications the PS surface needs to be stabilized. Different surface treatments have been reported to achieve a stable and hydrophilic surface [60, 96-97]. Mild oxidation removes the Si-H bonds which stabilizes and protects the surface. For biological molecules to be attached to the PS surface silanization or hydrosilanization treatment is to be done [94, 96-97].

The enormous medical application of silicon was recognized very recently. Researchers investigate PS material as a transducer in sensing systems [67, 68] because of its physical and structural properties. High sensitivity results have been obtained using PS by monitoring changes in optical properties, such as photoluminescence [69-70] and ellipsometry [71]. The special features of PS material which led to its applications in the sensor industry are large surface area within a small volume, controllable pore sizes, convenient surface chemistry and compatibility with conventional silicon microfabrication technologies [72]. Scientists used these properties to develop PS sensors to detect toxic gases, volatile organic compounds, explosives, DNA and proteins. Porous Silicon is a well known material for sensing layers in different gas and humidity sensors. It shows great effectiveness when combined with titanium, ceramics, composites, polymers and other materials, which are mainly used for biological implants.

The PS optical biosensors normally measure the change in the average refractive index of the device when a bioconjugation event takes place [67], because the immobilization of the probe and the target biological sample changes the effective refractive index of the PS surface, thus modifying the interference pattern on the output. In the case of label free optical biosensors, the biological probe is attached with a signaling material, which automatically transduces the hybridization effect into an optical signal. The label free optical detection of single strand of DNA (ssDNA) and its complementary (cDNA) conjugation is carried out on the PS chip by comparing the signals taken after the surface modification, then after probe immobilization on the chip surface and finally after its hybridization with the cDNA. In each step of the chip preparation, the optical path length changes which is recorded in the reflectivity spectrum [68-69]. Vicky Vamvakaki, et al. [113] developed PS DNA sensors, which can be used for label-free detection of oligonucleotides in DNA microarrays and microfabricated PS field effect sensors. Francia et al. [109] reported photoluminescence measurements for label-free optical porous silicon DNA sensors.

Singh, et al. [112] in their work, showed how PS films with good mechanical and optical properties can be effectively used for the biofunctionalization purpose for its possible application in immunosensors.

Measurement techniques of molecular binding interactions have been patented by Rauh-Adelmann and his coworkers [117], where ligands are immobilized within pores of a PS interaction region produced in a Si subtrate, after which analytes suspended in a fluid are flowed over the PS region. A large surface area with easily modified chemistry makes porous silicon an effective transducer for optical and electrical biosensing. Porous Silicon optical biosensors sensitivity and performance depend strongly on their nanomorphology and calculated as a function of the pore size [66].

Due to its biocompatibility and biodegradability properties, PS can be injected inside the body which over time releases it without any harm. Current research targets on how to find out the possible applications of PS as a biodegradable material in the field of medicine, for slow release of drugs or essential trace elements for in vivo applications [84]. PS can be used to treat everything from broken bones to cancer. Label –free optical biosensor using PS for detection of immunoglobulin G (IgG) in serum and whole blood sample were also reported [105]. Salonel and his co-workers [116] studied the effect of size reduction of PS particles from micro to nanosize, which affects in vitro cytotoxicity and biochemical mechanism of toxicity when these PS particles are applied to human cells. According to their findings, this cytotoxicity depends on the particle size and also on the surface chemistry of the PS particles.

Porous silicon has potential of several nanomedical applications, particularly, as a biomaterial in cancer detection because of its property of reflectivity and its resistance to stomach acid. Reflectivity of PS increases in the presence of cancer related chemicals in the blood, which indicate possible growth of tumors in the body. A silicon capsule containing the required drug can thus be directly administered orally to reach colon through the stomach without biodegradation therein [110].

An array of micro test tubes and micro beakers on p-type Silicon substrate can be fabricated by electrochemical etching method, which can be coated with superparamagnetic iron oxide nano (SPION) particles. As both silicon and SPION are biocompatible, these SPION coated macroporous silicon chips offer viable platforms for biosensor applications [123]. These PS devices will have numerous applications to magnetic resonance imaging, for targeted delivery of drugs or genes, and in hyperthermia. Further research is in progress for manipulating the superparamagnetic behavior of iron oxide nano particles by embedding them into porous silicon matrix [124-127].

9.3.3. Silicon Nanoparticle Sensors

To prepare Si nanoparticles, first PS is obtained by electrochemical etching of single-crystal silicon wafers in ethanolic HF solution. This PS layer was then lifted off and ultrasonicated to get silicon nanocrystals. A silicon oxide layer then grows on these nano crystals. These crystals, in aqueous solution, generate visible luminescence at room temperature due to quantum confinement effect.

In the case of medical or biological imaging, dyes are used as markers, which are not photostable. The dyes can break down under photoexcitation or visible light or at higher temperatures. The amazing property of visible room temperature luminescence of PS created an interest among the scientists for synthesizing and characterizing silicon nanoparticles. In addition to its luminescence property, PS is biocompatible and stable against photobleaching. These properties are ideal for replacing fluorescent dyes with silicon nanoparticles. Silicon nanoparticles can even replace highly toxic cadmium quantum dots for in vivo applications [92]. For biomedical applications, it is essential that they have high stability, a substantial photoluminescence quantum yield in the visible region, and solubility in aqueous media. Nanomaterials that can circulate inside the body, have great advantage for disease diagnosis and treatment. These nanomaterials ought to be harmlessly eliminated from the body shortly after they carry out their diagnostic or therapeutic functions.

Nanoparticle-based sensors and drug delivery systems have considerable potential for various types of medical treatment. The important technological advantages of nanoparticles used as drug carriers are high stability, high carrier capacity, feasibility of incorporation of both hydrophilic and hydrophobic substances, and feasibility of variable routes of administration, including oral application and inhalation. Nanoparticles can also be designed to allow controlled (sustained) drug release [92] from the matrix. These properties of nanoparticles enable improvement of drug bioavailability and reduction of the dosing frequency, and may resolve the problem of nonadherence to prescribed therapy.

Despite efforts to improve their targeting efficiency, significant quantities of systematically administered nanomaterials are cleared by the mononuclear phagocytic system before finding their targets, increasing the likelihood of unintended acute or chronic toxicity. However, there has been little effort to engineer for self-destruction of errant nanoparticles into non-toxic, systematically eliminated products. M. J. Sailor and his group [92] showed that luminescent porous silicon nanoparticles (LPSiNPs) producing near infrared luminescence can be used as drug payload for in vivo monitoring. The most interesting property manifested by these particles is, when tested on mouse, they self-destruct and are cleared from the body within a short period of time without producing any toxic effect. Their work presents a new type of multifunctional nanostructure for in vivo applications with low toxicity. Other uses of silicon nanoporous particles as effective carriers for in vivo simultaneous application for different nanotherapeutics have been reported recently [116].

It is a great challenge for the researchers to produce monodispersed Si nanoparticles with distinct visible luminescence and proper surface functionalization for detection of biological samples, targeted drug delivery and for bio-imaging. Various chemical and physical methods are reported in the literature for synthesis of Si nanoparticles along with their physical properties, and how their surface can be functionalized for both in-vitro and in-vivo medical applications [128]. Highly stable aqueous suspensions of Si QDs are produced with phospholipid micelles for use as luminescent marker for detection of pancreas cancer [129]. Silicon QDs as fluorescent tag for the detection of Vero cells by illuminating with UV radiation have also been reported [130, 131]. It is also a difficult task to functionalize the surface of Si nanopartices in order to ensure better solubility and increased photoluminescence compared to those of organic dyes for use as biological labels for cell imaging [132]. By one-step melt synthesis procedure, water soluable Si naoparticles can be produced, which provide intense blue and green emission with a short lifetime, suitable for biological imaging and detection of cells [133]. Notable advantages of porous structure of silicon for drug loading and release and its favorable optical properties applicable for diagnostic purpose have been reported [134, 135]. Clorgyline drug loaded silicon nanoparticles are fabricated for controlled drug delivery, which would help treating neurodegenerative diseases like Parkinson's, Alzheimer's, and Central nervous disease (CNS) disorder [136]. These particles offer an enormous potential in the field of drug delivery because they can carry huge amount of low solubility drugs keeping their activity intact inside the pores. The hyperpolarized Si nanoparticles are considered as a sensitive imaging agent for MRI contrast enhancement as it increases the proton relaxation time [137,138].

9.4. Biochip or Lab-on-a-Chip

Apart from the above-mentioned porous silicon based sensors, there exist another wide category of healthcare devices known as biochips. Some of them also use a PS platform. Dinh, et al. [42] presented an overview of the various types of biosensors and biochips that were developed till 2000 for biological and medical applications. Although there are various classification schemes for different biosensors in recent literature, they classified biosensors and biochips either by their bioreceptor or their transducer type. Junquan, et al. published another review paper on advances and applications of biochip technologies [139]. Ravindra, et al. reported a survey on biosensors providing an overview of the fundamentals, applications, as well as the market trends of biosensors [140]. They also provided a partial list of international manufacturers of biosensors.

9.4.1. Classification of Biochips

Biochips can be broadly classified into two categories (see Fig. 9.4): (a) lab-on-a-chip (LoC) or *in-vitro* (also known as non-implantable or external) biochip, and (b) implantable or *in-vivo* biochip, which is implanted within living organisms. Furthermore, LoCs are of two kinds: (i) microarray-based chips and (ii) microfluidic-based chips.

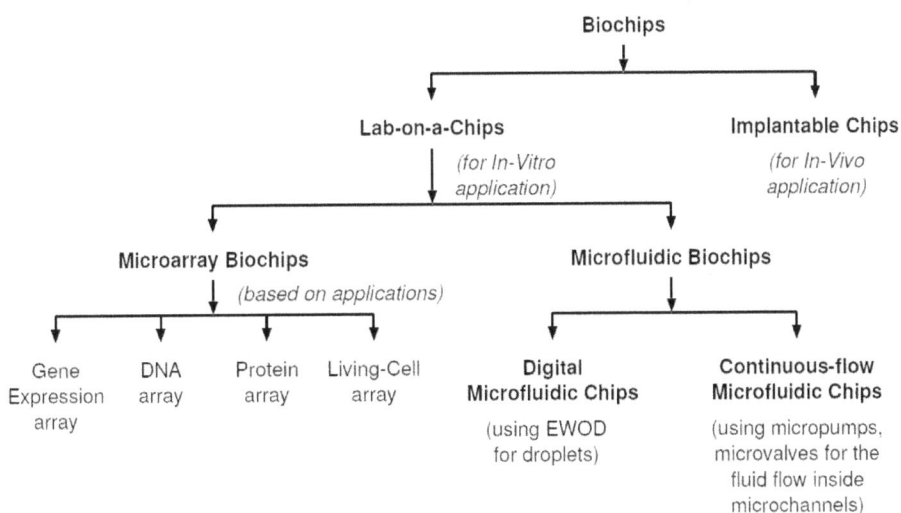

Fig. 9.4. Classification of biochips.

9.4.1.1. Lab-on-a-chip

A "laboratory-on-a-chip" (known as "lab-on-a-chip" or LoC) is a miniaturized analytical device that integrates one or several laboratory functions (biochemical operations) on a single chip (or miniaturized lab) of only millimeters to a few square centimeters in size [141]. LoCs deal with the handling of extremely small fluid volumes down to less than pico liters. LoC devices are the subset of micro-electro-mechanical systems (MEMS) devices and often indicated by "Micro Total Analysis Systems" (μTAS) as well. The significance of such miniaturized devices lies in their potentiality of automating laboratory procedures, which highly reduces the cost and time of biomedical tests and laboratory work. In [142] over 300 research articles on this rapidly growing area of μTAS (or LoCs) or miniaturized analysis systems are reviewed with the theory of miniaturization and technology.

LoCs are used for instant home tests of illnesses, food contaminants and toxic gases. LoC design is a multidisciplinary approach of adapting classical biochemical assays to a miniaturized platform by exploiting advances in microelectronic and microfluidic technologies. This emerging technology combines electronics with biology, fluidics and various sensorics to open new areas of research and field applications, e.g., point-of-care diagnosis, on-chip DNA analysis and drug screening. LoCs are capable of analyzing biochemical liquid samples, like solutions of metabolites, macromolecules, proteins, nucleic acids, or cells and viruses. Examples of applications of LoCs are: real-time PCR (Polymerase Chain Reaction), protein crystallization assay, immunoassay (to detect bacteria, viruses and cancers based on antigen-antibody reactions), etc. In [143], the authors review the published articles on CMOS-based capacitive sensors for LoC applications and present the recent significant advances for capacitive detection LoCs. Another survey [144] summarizes numerous fabrication methods and procedures for producing LoC devices and also envisages future evolution.

A. Microarray Biochip

A microarray is a solid support (e.g., glass slide or nylon membrane) on which DNA spots of known sequence is deposited in a regular grid like array. A generic microarray consists of multiple features (spots) of DNA which are used to determine the levels of mRNA expression in a collection of cells. As the loaded DNA spots are not transported during its operation, this type of biochips is static in nature. A schematic diagram of a DNA microarray chip and its analysis from the microscopic scanned image has been shown in Fig. 9.5.

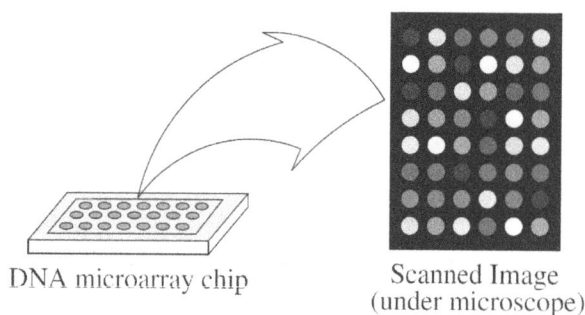

DNA microarray chip   Scanned Image
                             (under microscope)

Fig. 9.5. A microarray based chip.

Microarray technology has also been extended to proteomic research, specifically for protein identification and quantification (e.g., miniaturized and parallelized immunoassays), protein-interaction analysis (e.g., protein-protein interactions, enzyme-substrate assays, protein-DNA interactions), as well as cell analysis in gene activities, protein expression and cell surface molecules. A wide set of applications of microarrays include: gene expression arrays [145], DNA sequence arrays [146], protein arrays [147], and living-cell arrays [148].

In microarray type of chips, the term "gene expression" is used to describe the transcription of the information contained within the DNA, into messenger RNA (*m*RNA) molecules that are then translated into the proteins that perform most of the critical functions of the cells. In our body, all genes are not "expressed" in the same cell, though almost all cells contain the same gene. Many genes represent unique features to a particular type of cell. For example, liver cells express genes for enzymes that detoxify poisons, while pancreas cells express genes for making insulin. Scientists are working in these areas to identify which genes are expressed by each type of cells.

In a microarray, *m*RNA molecules bind specifically to a complementary DNA, to hybridize and to form a double helix structure. By using an array containing many DNA samples, scientists can determine, in a single experiment, the expression levels of hundreds or thousands of genes within a cell by measuring the amount of *m*RNA bound to each site on the microarray. The amount of *m*RNA bound to the spots on the microarray is precisely measured by a microprocessor attached to it, generating a profile of gene expression in the cell.

In a microarray, nucleic acid sensing is done by immobilizing single stranded oligonucleotide (5 to 50 nucleotides long) probe onto transducer surface forming a recognition layer that binds its complementary (target) DNA sequence to form a hybrid for the purpose of expression profiling, monitoring expression levels for thousands of genes simultaneously or for comparative genomic hybridization. The hybridization reaction means coupling of any four different nucleotides, adenine (A), thymine (T), guanine (G), and cytosine (C) with its complementary one e.g. the complementary sequence of G-T-C-C-T-A is C-A-G-G-A-T. Fig. 9.6 [122] shows how a hybridization reaction takes place. This process of hybridization helps in identifying diseases, where fluorescently labeled nucleic acid molecules are used as mobile probes to identify the complementary molecular sequences that are able to base-pair with one another.

DNA microarrays, which are also called gene chips, are used to capture the genome sequence information to analyze the structure and function of tens of thousands of genes at a time. The first DNA chip is commercialized by Affymetrix (GeneChipR Technology) in the late 1980s [146]. Barbulovic-Nad, et al. provided a review on the current state of microarray fabrication [149]. There are two different techniques used for spot formation – "contact printing" and "non-contact printing".

B. Microfluidic Biochip

The microfluidic-based biochip is widely used for on-chip implementation of several biochemical laboratory assays, for sample preparation, dilution and mixing [119, 120]. These chips use only nanoliter volumes of fluids and thus offer the advantages of low sample and reagent consumption, high throughput and sensitivity, and minimal intervention. The fluidic operations can be performed on-chip either in a continuous fashion (continuous-flow microfluidic chips), or in a discrete fashion (digital microfluidic biochips). Their applications include clinical diagnostics, enzymatic analysis, e.g., glucose and lactate assays, DNA analysis, immunoassays, and environmental toxicity monitoring. The third type of biochips are those, which can be implanted inside the human body or administered orally, for drug release or for controlling/monitoring some biological functioning, *in vivo*. All these biochips need several types of sophisticated optical and electronic sensors as interface.

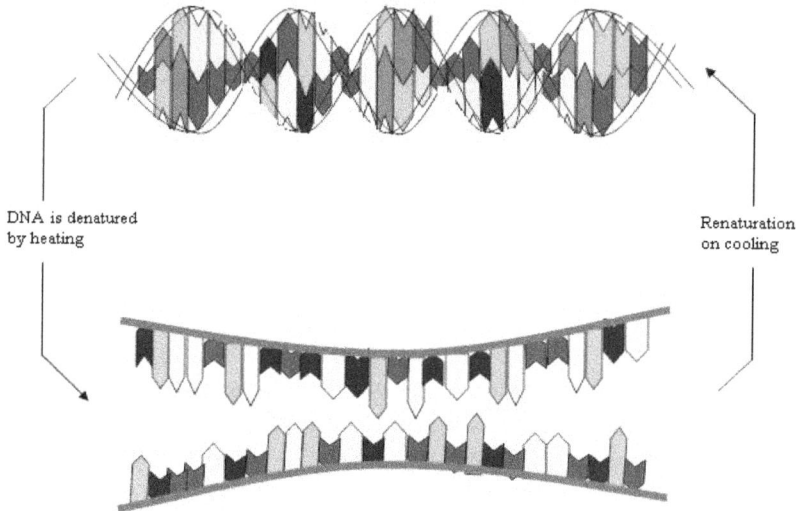

DNA is denatured by heating

Renaturation on cooling

Fig. 9.6. Schematic diagram of hybridization reaction.

A microfluidic chip deals with the behavior, precise control and manipulation of fluids that are geometrically constrained to a small, typically sub-millimeter scale. In contrast to microarrays, transport phenomena (e.g., electrical, fluidic, optical, acoustical, thermal, etc.) of the fluid sample in microfluidic chips are of primary interest and pivotal to system performance. Because of its ability to transport fluid, microfluidic biochips are dynamic in nature, compared to a microarray biochip, which is a static one.

Compared to the traditional biochemical laboratory bench-top procedures, microfluidic biochips offer the advantages of low sample and reagent consumption, less likelihood of error due to minimal human intervention, high-throughput and high-sensitivity [150]. Microfluidics is a multi-disciplinary field intersecting engineering, physics, chemistry, microtechnology and biotechnology, with practical applications to the design of systems in which small volumes of fluids are manipulated to perform some biochemical assay on-chip. This field has emerged in the beginning of 1980s and is used in the development of inkjet printheads, DNA chips, LoC technology, micropropulsion, and micro-thermal technologies. Some key application areas of microfluidics are enzymatic analysis (e.g., glucose and lactate assay), DNA analysis (e.g., polymerase chain reaction and high throughput sequencing) and proteomics (large-scale study of proteins, particularly their structures and functions). This field offers exciting possibilities for high-throughput DNA sequencing analysis, protein crystallization, drug discovery, immunoassay, and environmental toxicity monitoring and the detection of airborne chemicals and detection of explosive such as TNT. Another emerging application of microfluidic biochips is clinical diagnostics, i.e., the immediate point-of-care diagnosis of diseases.

Squires and Quake present an elaborate review of the physics of small volumes (nanoliters) of fluids and important parameters need to develop microfluidic systems [151]. In [152], Auroux, et al. summarize a number of standard operations (namely sample preparation, sample injection, sample manipulation, reaction, separation, and detection) as well as some biological applications of μTASs (namely cell culture, polymerase chain reaction, DNA separation, DNA sequencing, and clinical diagnostics) using over 350 articles. Eijkel, et al. describe the recent state of research in this field and discuss possible directions of development, starting from the background of nanofluidics [153]. Again, MEMS processes are applied to new materials resulting into new approaches for fabrication of microchannels in microfluidic devices. Another review paper [154] explores these developments emerged from the increasing interaction between the MEMS and microfluidics worlds and showed how MEMS technologies can play an important role in the future development of microfluidics.

In general, there are two kinds of microfluidic biochips: (1) continuous-flow microfluidics (CMF), and (2) digital microfluidics (DMF). In the consecutive subsections, we will discuss them in detail. However, several mechanisms are experimentally demonstrated for the transportation of fluids (flow or droplets) in a microfluidic biochip (see Fig. 9.7). Depending on these mechanisms of fluid transportation, microfluidics has three broad variations such as fluid-flow inside microchannels (i.e., CMF) (Fig. 9.8 a), droplets on electrode array (i.e., DMF) (Fig. 9.8 b), droplets in microchannels (Fig. 9.8 c) and microfluidic bubble in microchannels (Fig. 9.8 d). Fluid movements inside the microchannels are performed by pressure-driven forces using micropumps and microvalves. Moreover, the droplet movements on the array of electrodes are performed by periodically applying voltages on the electrodes (a phenomenon similar to the electrowetting-on dielectric or EWOD) (Fig. 9.8 b).

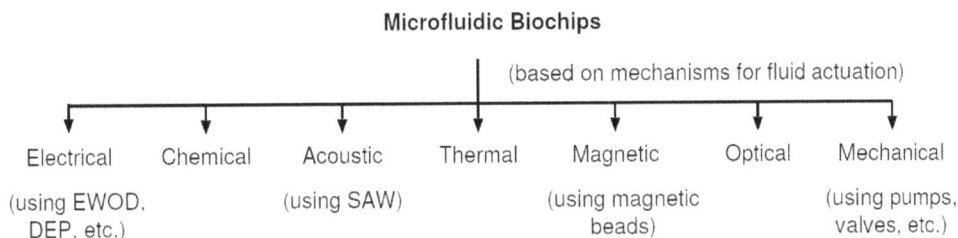

Microfluidic Biochips

(based on mechanisms for fluid actuation)

Electrical	Chemical	Acoustic	Thermal	Magnetic	Optical	Mechanical
(using EWOD, DEP, etc.)		(using SAW)		(using magnetic beads)		(using pumps, valves, etc.)

Fig. 9.7. Several mechanisms for fluid movement in microfluidic biochips.

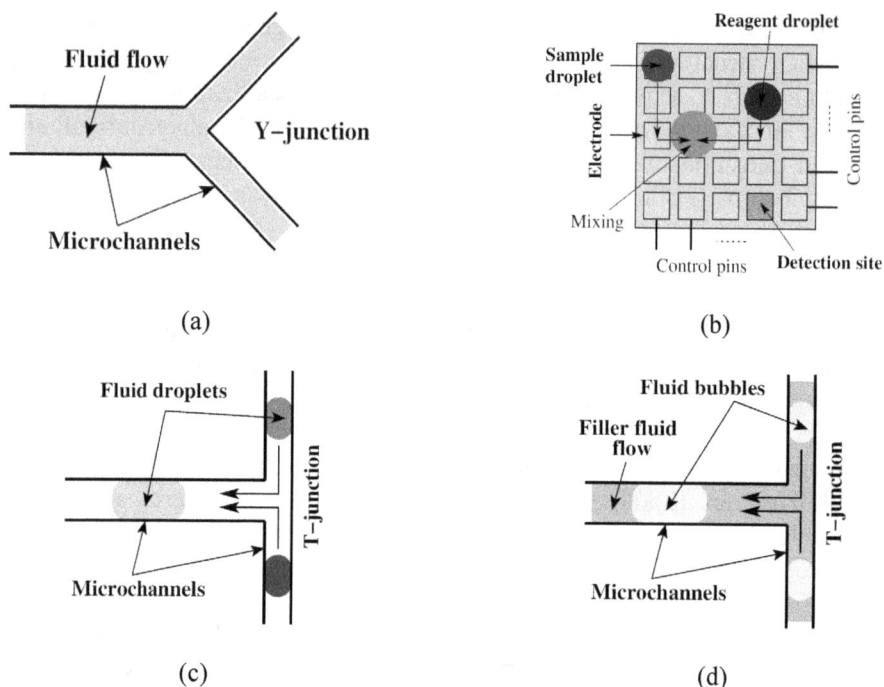

Fig. 9.8. (a) Continuous-flow microfluidics, (b) droplet-based microfluidics, (c) droplets in microchannels, (d) microfluidic bubbles in microchannels.

B.1. Continuous-flow Microfluidics (CMF): The first generation LoCs were developed as continuous-flow microfluidic devices, which consist of several micrometer scale components including fixed microchannels, microvalves, micropumps, microactuators, microsensors, etched on it. In CMF biochips, fluid-flow is controlled either using micropumps and microvalves or using electrokinetics [155]. Till date most commercially available biochips rely on either continuous fluid-flow in etched microchannels or microarrays [154].

Fluidic Operations: The most fundamental fluidic operations in CMF are mixing of two (or more) different fluid-flows and splitting of a fluid-flow into two (or more). Several kinds of microchannel junctions have been designed and experimentally verified for microfluidic flow mixing and splitting/separation. Some of them are T-junction [156], Y-junction [157], cross-junction [158], Y-junction, F-shape mixing units [159], etc. Figs. 9.8 and 9.9 schematically depict the structures of these microchannel junctions.

Here, we provide a brief survey of recently published articles on the fundamental microfluidic operation, i.e., mixing. Ottino and Wiggins briefly reviewed the main issues associated with mixing at the microscale and provided some basic mathematics of mixing in channel-based microfluidic devices [160]. Locascio [161] presented a short review on mixing of fluids in CMF biochips and presents how passive diffusive mixing in CMF has evolved from a simple to a highly efficient mixing process with improved performance and dramatically decreased mixing times. Campbell and Grzybowski discussed various microfluidic mixers using microfabricated or MEMS devices with different fluidic actuation techniques and device geometry [162]. Park, et al. [163] described the design and characterization of a new five-inlet port mixer, which uses hydrodynamic focusing technique for uniform mixing of fluids in microchannels. A new computational approach to the modeling and design of efficient microfluidic mixers was demonstrated by Howell, et al. [164]. This mixer can provide rapid mixing than previous designs. Suh and Kang [165] defined the terminology necessary to understand the fundamental concept of mixing and introduce quantities for evaluating the mixing performance, such as mixing index and residence time. They reviewed various designs of mixers developed for use in microfluidic devices and classify the designs in terms of the driving forces, including mechanical, electrical and magnetic forces, used to control fluid-flow upon mixing. The most suitable design of mixing is selected depending on the specific application in mind. However, droplet mixing is ranked Number one judged from the aspects regarding the mixing performance as well as the cost of fabrication. Lee, et al. reported in an extensive review on microfluidic mixers that mixing schemes can be categorized as either "active", where an external energy force is applied to perturb the sample species, or "passive", where the contact area and contact time of the species samples are increased through specially-designed microchannel configurations [166]. They describe some of the more significant proposals for active and passive microfluidic mixers.

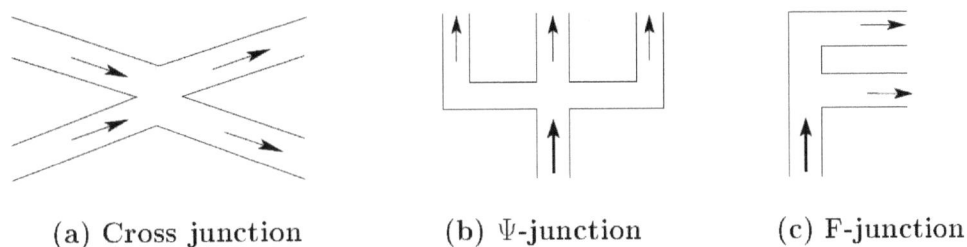

(a) Cross junction   (b) Ψ-junction   (c) F-junction

Fig. 9.9. Microchannels used in CMF biochip (arrows show the directions of fluid-flow).

B.2. *Digital Microfluidics (DMF):* An alternative category of microfluidic biochips relies on the principle similar to electrowetting-on-dielectric (EWOD) and known as "droplet"-based digital microfluidic (DMF) chips. In contrast to CMF, DMF works much the same way as traditional bench-top protocols, only with much smaller volumes and much higher automation. The droplets are the digitized form of a liquid and the most common carriers of bio-chemical agents in a LoC design. These are minute amounts (nano-, pico- or femtoliter i.e., *nl*, *pl* or *fl*) of biochemical samples drawn from individual sample reservoirs. This droplet-based LoC technology is referred to as digital microfluidics. However, just like fluid droplets "digital" microfluidic systems can also be developed using fluid bubbles or particles or cells.

A schematic diagram of a DMF biochip along with its computer control is shown in Fig. 9.8 b, which provides the top view of the DMF biochip. The cross-sectional view of a basic detection cell in such a DMF biochip is elaborated in Fig. 9.10.

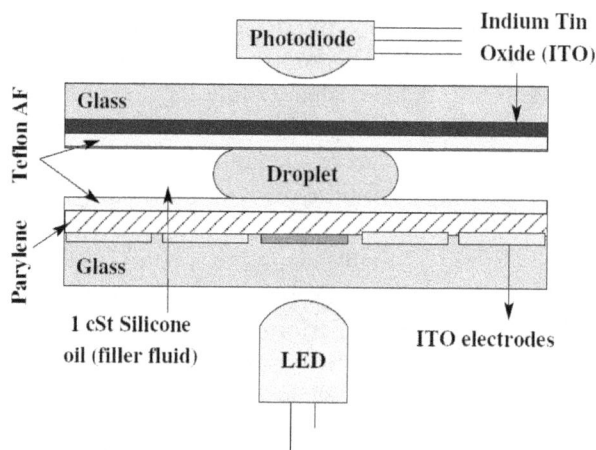

Fig. 9.10. Cross-sectional view of a cell at detection site.

There are a number of review papers on DMF and its applications. Fair discusses the basics of electrowetting-based DMF in terms of its suitability for true LoC applications in a review paper [120]. He presents a brief overview of the basics of actuating droplets by electrowetting on dielectric (EWOD). Also, the types of applications that are possible using DMF architecture and the fluidic operations are explored to understand the opportunities and limitations. The, et al. [167] provided a comprehensive review of DMF focusing on the various droplet operations, as well as the numerous applications of the system. Abdelgawad and Wheeler presented a summary of the state-of-the-art in DMF, describing device formats and fabrication, the physics of droplet actuation, and a sampling of the myriad applications to which this technology is being applied [168]. They broadly classified all the applications as biological and nonbiological. X. C. i Solvas and A. deMello reported recent advances in the field of DMF highlighting the unique features of such platforms for high-throughput experimentation and describing functional components that afford complex analytical processing [169]. They reported some of the applications of DMF chips in synthesis, high-throughput screening, cell biology and in synthetic and systems biology.

Since discrete droplets of nanoliter volumes can be manipulated in a "digital" manner on a 2-D electrode array, this technology is referred to as "digital microfluidics". By using discrete unit-volume droplets, a microfluidic function can be reduced to a set of repeated basic operations, i.e., moving one unit of fluid over one unit of distance. This "digitization" method facilitates the use of a hierarchical and cell-based approach for DMF biochip design. Therefore, DMF offers a flexible and scalable system architecture as well as high fault-tolerance capability. Since each droplet can be controlled independently, these systems also have dynamic reconfigurability, whereby groups of unit cells in a microfluidic array can be reconfigured to change their functionality during the concurrent execution of a set of bioassays. Droplet motion in such devices is typically controlled by a system clock, which is similar in operation to a digital microprocessor.

DMF mechanisms for fluidic movements: There are several actuation methods developed for DMF devices for fluid droplet movements. Some of them are electrowetting-on-dielectric (EWOD), dielectrophoresis (DEP), etc. Biochip designers use the electrohydrodynamic forces to accomplish the droplet operations by applying EWOD and DEP principles

Electrowetting-on-dielectric (EWOD) is based on changing the wettability of liquids on a dielectric solid surface by varying the electric potential. The contact angle of a fluid droplet changes as the voltage is applied to the dielectric surface on which the droplet is sitting. The contact angle of a droplet placed on a dielectric layer coated electrode can be changed by EWOD when a voltage is applied across a dielectric layer as described by the Lippmann-Young equation. This principle can handle fundamental droplet operations in DMF.

EWOD works without any mechanical part and the only moving mass in such DMF chip is the fluid droplet itself. This is technologically much easier to implement, the manufacturing process requires only one step to pattern a metallic layer, whereas other type of microfluidic devices (with micropumps or microvalves) require a number of lithographic steps and complex etching procedures. Hence, EWOD is the most common droplet actuation method used in DMF and a number of published articles describe the study on this mechanism and its use in DMF chips such as [170-174]. Lienemann, et al. [175] presented a modeling-and-simulation methodology for EWOD effects that can help a designer for dimensioning and layout of the biochips. J. H. Song et al. [176] provide a detail study of scaling model for EWOD microfluidic actuators and the physics behind the EWOD method under scaling of the device implementing the mechanism.

Dielectrophoresis (DEP) is another technique for droplet actuation and it is basically dielectric polarization due to non-homogeneous electric field. In this phenomenon a force is exerted on a dielectric (neutral) particle (or fluid droplet) when it is subjected to a non-uniform (DC or AC) electric field. This method does not require the particle to be charged and it can be used for separation of cells (or fluid droplets) [177, 178]. There are a number of published articles providing details about this mechanism and its use in DMF chips such as [179-184].

Techniques based on surface acoustic wave (SAW) or magnetic field can be used for droplet transportation [185,186,187,188]. Optoelectrowetting is a recently reported technique in which optical illumination causes the droplet to translate towards the illuminated region and thus the illuminated area acts as a *virtual* electrode. This technique allows large-scale, parallel manipulation of arbitrarily-sized droplets. Using this technique, Pei, et al. successfully demonstrated a new light-actuated DMF device that can perform droplet translation, splitting, merging, and other fluidic operations on-chip [189].

9.4.1.2. Implantable Biochips

Hu, et al. reported that biochips can also be used as implantable physiological probes in patients [190]. Implantable or *in-vivo* biochips are implanted inside the human or animal body (living organisms) for continuous monitoring of some biological parameters to diagnose abnormalities and to start treatment immediately by releasing drugs in a controlled fashion. An example in this direction is a human microchip implant for the use of unique identification [191]. In [192] the authors described an implantable diagnostic device that senses the local *in-vivo* environment during biopsy. This serves as a monitoring device for soluble cancer biomarkers *in-vivo*. For designing such biochips, low-energy and biodegradability issues are to be addressed as they are required to be implanted inside the animal body (living organisms).

9.4.2. Integration of Sensors

One of the major challenges in nanobiosensor industry is to design efficient sensors and integrating them on a biochip. The information captured by the sensors must be converted into a readable form [32]. Several optical, electrical, chemical, and biological data from the nanosensors are to be transformed into signals for processing, analysis, and for deciding actions. A review article on protein based lab-on-a-chip sensors reported by Borini, et al. [65], describes different approaches for fabrication of biochips with PS and their future perspectives. Lingang, et al. [73] in their paper, reported a new technique for fabrication of a biochip on porous silicon and its application for detection of small molecule–protein interactions with desorption/ionization on PS (DIOS) [74]. Other applications of PS for building DNA sensors have also been reported earlier [108]. Specially designed gold-based electrochemical sensor chips and detectors are frequently used for various biochemical and molecular biology applications [193, 194]. Several semiconductor and/or polymer based micro/nano mechanical cantilever type of sensors are also shown to very useful for a wide class of cardiac diagnostics and related biochemical applications [195-200]. A recent review on L-o-C and microfluidic based biosensor technologies for the detection of cardiac biomarkers appears in [198]. Needless to say, integration of sensing technology and controlling probes with the fluidic, mechanical and the electronic world remains the main challenge while designing a multi-function biochip.

9.5. Optical Detection Techniques

For functioning of biochips, one of the most important criteria to be fulfilled is to attach a powerful transduction or signal processing unit to the system, which can directly and accurately detect the biological event and convert it in to a human readable output. The biological events like, antibody/ DNA binding, oxidation/reduction, etc., need to be transduced into a format understandable by a computer (voltage, light, intensity, mass, etc.), for analyzing and processing the signal to produce the final output. Several optical techniques are commonly used for detecting and quantifying biomolecules [62-63] as one does not require electrical contacts with the system for capturing data with optical devices. Instead, one can use fluorescent tags either with the probe or with the analyte to detect any change in the system. The major advantages of the optical transduction methods are that the devices are small, lightweight and portable due to the integrability of all optical components.

Several optical transduction methods such as FRET (fluorescence resonance energy transfer), SERS (surface enhanced Raman spectroscopy), and fluorescence spectroscopy are used for detection of biological samples. Spectroscopic techniques are used, for detecting biological samples or events occurring in it because cells or tissues can absorb or emit light, thereby producing a signal or spectrum, which is a characteristic of that particular event. From this fingerprint spectrum, one can directly identify or quantify the sample or the event.

Fluorescence Resonance Energy Transfer (FRET) is a nonradiative energy transfer process from excited state donor molecule to an acceptor molecule, when appreciable overlap exists between the emission spectrum of the donor and the absorption spectrum of the acceptor. This radiation-less transfer of energy, when the excited state fluorophore and the second chromophore lie within a range of approximately 10 *nm*, provides vivid structural information about the donor-acceptor pair. This is a quantum mechanical process that does not require a collision and does not involve production of heat. When energy transfer occurs, the acceptor molecule quenches the donor molecule fluorescence, and if the acceptor is itself a fluorochrome, increased or sensitized fluorescence emission is observed [35, 36, 38]. Fig. 9.11 shows the underlying principle of FRET. The information obtained by this method is unique because the surrounding solvent shell of a fluorophore does not affect the FRET measurements.

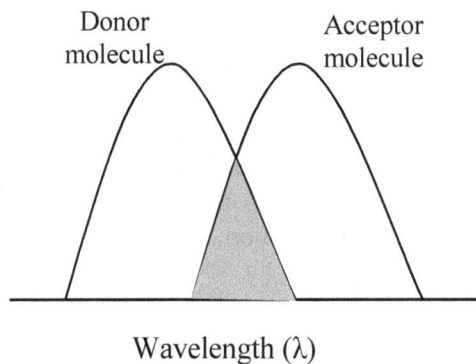

Fig. 9.11. Fluorescence resonance energy transfer.

The *Surface Enhanced Raman Spectroscopy* or *Surface Enhanced Raman Scattering (SERS)* is a surface sensitive technique that results in the enhancement of Raman scattering by molecules adsorbed on rough metal surfaces. The vibrational modes of the adsorbates on the roughened surface are sometimes observed to have about one million times the intensity that would be predicted by comparison with their Raman spectra in the gaseous phase [39].

Fluorescent measurement techniques are commonly used for the detection of biomolecules. In fluorescence spectroscopy, fixed or living cells or single stranded DNA probes are often labeled with fluorescent tags or fluorophores, each specific to a particular intercellular component, which absorps light at one wavelength (excitation), followed by a subsequent emission of secondary fluorescence at a longer wavelength. The excitation and emission wavelengths are usually separated from each other by tens to hundreds of nanometers.

Cellular components are labeled with specific fluorophores to identify their localization within fixed and living parameters. In microarray systems, the target molecules are labeled, which is a process of covalently binding a molecule or particle to the target DNA strand, for generating transducer signal. This approach takes care of the issue related to safety and disposal associated with radioactive markers and allows the researchers to study several experimental parameters simultaneously with multiplex samples. In the case of multiple probes, different dyes are attached to different probes which can be simultaneously detected at different wavelengths using optical filters. After hybridization, the fluorescent signals from a DNA chip are studied using specific instruments. However, this method of labeling with fluorophores is not possible everywhere [64] because the optical labels are costly and unreliable and also the optical scanners are expensive and the procedure of extracting information from the data is complicated [70]. Researchers are trying to work it out with label free techniques or reagent-less optical biosensors where the target sample can be detected in a heterogeneous solution without adding anything but the sample [96].

9.6. Conclusion

In this review article, attempts have been made to highlight how nanomaterials or nanostructured materials like PS, Si nanoparticles, and QDs are being used as highly efficient biosensors for several biomedical applications. Inorganic nanocrystals or QDs have been used in nano scale computing for the past several years. These crystals dissolve in water and fluoresce when exposed to light radiation, and are usable as markers or trackers in biological recognition events. However, these QDs have toxic effects and therefore, are unsuitable for in vivo applications. In contrast, PS and Si nanoparticles are non-toxic and are easily removed from the body without leaving any trace of harmful effect. They are very effective as a biosensor/transducer material and because of their luminescence properties and strong biocompatibility, they have wide applications in drug delivery systems and in the design of biomedical implants. We have also reviewed the recent advances in biochip research including the wide categories of microfluidic-based LoCs. Other research endeavors aiming at producing low-cost sensor material are currently being explored.

Acknowledgement

Dr. S. Ghoshal acknowledges Department of Science and Technology (DST), India for financial support under WOS-A scheme.

References

[1]. M. T. Bohr, Nanotechnology goals and challenges for electronic applications, *IEEE Transactions on Nanotechnology*, Vol. 1, Issue 3, 2002, pp. 56-62.

[2]. M. A. Reed, J. M. Tour, Computing with molecules, *Sci. Amer,* Vol. 282, Issue 6, 2000, pp. 86-93.

[3]. J. C. Ellenbogen, J. C. Love, Architectures for molecular electronic computers: Logic structures and an adder designed from molecular electronic diodes, *Proc. IEEE,* Vol. 88, Issue 3, 2000, pp. 386- 426.

[4]. H. Bachtold, D. Nakanishi, Logic circuits with carbon nanotube transistors, *Science,* Vol. 294, Issue 11, 2001, pp. 1317- 1320.

[5]. D. Huang, L. Cui, L. Kim, Logic gates and computation from assembled nanowire building blocks, *Science,* Vol. 294, Issue 11, 2001, pp. 1313-1317.

[6]. T. V. Dekker, Room-temperature transistor based on a single carbon nanotube, *Nature,* Vol. 393, Issue 5, 1998, pp. 49-51.

[7]. D. Dutta Majumder, R. Banerjee, Ch. Ulrichs, I. Mewis, A. Samanta, A. Das, S. K. Mukhopadhayay, S. Adhikary, A. Goswami, Nano-Fabricated Materials in Cancer treatment and Agri-biotech Applications: Buckyballs in Quantum Holy Grails, *IETE Journal of Research*, Special issue on Nanoscience, 2006, pp. 339-355.

[8]. C. N. R. Rao, A. Govindaraj, S. R. Vivekchand, Inorganic nanomaterials: Recent developments and future directions, *Annual Reports of the Royal Society of Chemistry (London)*, 2006.

[9]. A. C. R. Grayson, I. S. Choi, B. M. Tyler, P. P. Wang, H. Brem, M. J. Cima, R. Langer, Multi-pulse drug delivery from a restorable polymeric microchip device, *Nature Materials*, Vol. 2, 2003, pp. 767-772.

[10]. S. Tyagi, F. R. Kramer, Molecular beacons: probe that fluoresces upon hybridization, *Nature Biotechnology*, Vol. 14, 1996, pp. 303-308.

[11]. T. A. Taton, G. Lu, C. A. Mirkin, Two-colour labeling of oligonucleotides array via size-selective scattering of nano-particle probes, *Journal of American. Chemical. Society.,* Vol. 123, 2001, pp. 5164-5165.

[12]. F. A. Jaffer, R. Weisslender, Seeing within-molecular imaging of cardiovascular system, *Circulation Research*, Vol. 94, 2004, pp. 433-445.

[13]. J. M. Perez, L. Josephson, R. Weissleder, Use of magnetic nanoparticles as nanosensors to probe for molecular interactions, *ChembioChem*, Vol. 5, 2004, pp. 261-264.

[14]. A. Saleh, M. Schroeter, C. Jinkmanns, U. Modder, S. Jander, In vivo MRI of brain inflammation in human ischaemic stroke, *Brain,* Vol. 127, 2004, pp. 1670-1677.

[15]. A. Watson, X. Wu, M. Bruchez, Lighting up cells with quantum dots, *Biotechniques*, Vol. 34, 2003, pp. 296-303.

[16]. W. C. Chan, D. J. Maxwell, X. Gao, R. E. Bailey, M. Han, S. Nie, Luminescent quantum dots for multiplexed biological detection and imaging, *Current Opinions in Biotechnology*, Vol. 13, 2002, pp. 40-46.

[17]. J. K. Jaiswal, H. Mattoussi, J. M. Mauro, and S. M. Simon, Long-term multiple colour imaging of live cells using quantum dot bioconjugate, *Nature Biotechnology*, Vol. 21, 2003, pp. 47-51.

[18]. D. R. Larson, W. R. Zipfel, R. M. Williams, S. W. Clark, M. P. Bruchez, F. W. Wise, W. W. Webb, Water-soluble quantum dots for multi photon fluorescence imaging in vivo, *Science*, Vol. 300, 2003, pp. 1434-1436.

[19]. B. Dubertert, P. Skourides, D. J. Norris, V. Noireaux, A. H. Brivanlou, A. Libchaber, In vivo imaging of quantum dots encapsulated in phospholipid micelles, *Science*, Vol. 298, 2002, pp. 1759-1762.

[20]. B. Ballou, B. C. Langerholm, L. A. Ernst, M. P. Bruchez, A. S. Waggoner, Noninvasive imaging of quantum dots in mice, *Bioconjugate Chemistry*, Vol. 15, 2004, pp. 79-86.

[21]. X. Gao, Y. Cui, R. M. Levenson, L. W. K. Chung, S. Nie, In vivo cancer targeting and imaging with quantum dots, *Nature Biotechnology*, Vol. 22, 2004, pp. 969-976.

[22]. M. E. Akerman, W. C. W. Chan, P. Laakkonen, S. N. Bhatia, E. Ruoslahti, Nanocrystal targeting in vivo, *Proceedings of National Academy of Sciences, USA*, Vol. 99, 2002, pp. 12617-12621.

[23]. E. B. Voura, J. K. Jaiswal, H. Mattoussi, S. M. Simon, Tracking metastatic tumor cell extravasation with quantum dot nanocrystals and fluorescence emission-scanning microscopy, *Nature Medicine*, Vol. 10, 2004, pp. 993-998.

[24]. M. Dahan, S. Levi, C. Luccardini, P. Rostaing, B. Riveau, A. Triller, Diffusion dynamics of glycine receptors revealed by single-quantum dot tracking, *Science*, Vol. 302, 2003, pp. 442-445.

[25]. X. Wu, H. Liu, J. Liu, K. N. Haley, J. A. Treadway, J. P. Larson, N. Ge, F. Peale, M. P. Bruchez, Immunofluorescent labeling of cancer marker Her2 and other cellular targets with semiconductor quantum dots, *Nature Biotechnology*, Vol. 21, 2003, pp. 41- 46.

[26]. R. S. Williams, Computing at the nanoscale, *Trends in Nano Technology (TNT)*, September 4-8, 2006.

[27]. D. S. Sutherland, Surface Nanostructures to Control Properties and Cells, *Trends in Nano Technology (TNT)*, September 4-8, 2006.

[28]. B. Bourlon, J. Wong, C. Miko, L. Forro, M. Bockrath, Carbon nanotube based flow sensor, *Trends in Nanotechnology (TNT)*, September 4-8, 2006.

[29]. L. Samuelson, Semiconductor Nanowires for Emerging Nanoelectronics Applications, *Trends in Nano Technology (TNT)*, September 4-8, 2006.

[30]. P. Majewski, Functionalized Nanoparticles for Cancer Diagnosis and Treatment, *Workshop on Nanotechnology (Nanofair)*, November 2006.

[31]. G. Y. Jung, et al., Circuit Fabrication at 1.7 nm Half-Pitch by Nanoimprint Lithography, *Nanoletters*, Vol. 6, 2006, pp. 351-354.

[32]. D. L. Carrillo, Is the sensor industry ready for the nanotechnology boom? – Part II, *Fiber Optic Technology* (www.fpnmag.com), 2003.

[33]. H. T. M. Pham, C. R. de Boer P. M. Sarro, Non-Catalyst and Low Temperature Growth of Vertically Aligned Carbon Nanotubes for Nanosensor Arrays, in *Proceedings of the 13th international Conference on Solid-State Sensors, Actuators and Microsystems*, Seoul, Korea, June 5-9, 2005, pp. 97- 100.

[34]. M. J. Sailor, Biological Nanosensors, *AFOSR Annual Program Review Nanoelectronics, Negative Index Materials and Superconducting Electronics,* July 26-28, 2005.

[35]. C. Y. Zhang, H. C. Yeh, M. T. Kuroki, T. H. Wang, Single-quantum-dot-based DNA nanosensors, *Nature Materials*, Vol. 4, 2005, pp. 826-831.

[36]. B. Dubertret, Quantum dots – DNA detectives, *Nature Materials*, Vol. 4, 2005, pp. 797-798.

[37]. I. L. Medintz et al., Self-assembled nanoscale biosensors based on quantum dot FRET donors, *Nature Materials*, Vol. 2, 2003, pp. 630–638.

[38]. P. T. Tran, et al., Use of luminescent CdSe-ZnS nanocrystal bioconjugates in quantum dot-based nanosensors, *Phys. Stat. Sol. (b),* Vol. 229, 2002, pp. 427-432.

[39]. T. Vo-Dinh, Biosensors, Nanosensors, and Biochips: Frontiers in Environmental and Medical Diagnosis, in *Proceedings of the 1st International Symposium on Micro & Nano Technology,* March 2004, Honolulu, Hawaii, USA.

[40]. I. L. Medintz, H. T. Uyeda, E. R. Goldman, H. Mattoussi. Quantum dot bioconjugates for imaging, labelling and sensing, *Nature Materials,* Vol. 4, 2005, pp. 435–446.

[41]. T. Vo-Dinh, D. L. Stokes, G. D. Griffin, M. Volkan, U. J. Kim, M. I. Simon, Surface-Enhanced Raman Scattering (SERS) Method and Instrumentation for Genomics and Biomedical Analysis, *J. Raman Spectrosc.,* Vol. 30, 1999, pp. 785-793.

[42]. T. Vo-Dinh, B. M. Cullum, Biosensors and Biochips, Advances in Biological and Medical Diagnostics, *Fresenius Journal of Analytical Chemistry*, Vol. 366, 2000, pp. 540-551.

[43]. T. Vo-Dinh, J. P. Alarie, B. M. Cullum, G. D. Griffin, Antibody-based Nanoprobe for Measurements in a Single Cell, *Nature Biotechnology*, Vol. 18, 2000, pp. 764-767.

[44]. T. Vo-Dinh, M. Askari, Microarrays and Biochips: Applications and Potential in Genomics and Proteomics, *Journal of Current Geonomics*, Vol. 2, 2001, pp. 399-415.

[45]. T. Vo-Dinh, B. M. Cullum, D. L. Stokes, Nanosensors and biochips: Frontiers in biomolecular diagnostics, *Sensors and Actuators*, B, 74, 2001, pp. 2-11.

[46]. R. R. K. Reddy, I. Basu, E. Bhattacharya, A. Chadha. Estimation of triglycerides by a porous silicon based potentiometric biosensor, *Current Applied Physics*, Vol. 3, 2003, pp. 155.

[47]. R. R. K. Reddy, A. Chadha, E. Bhattacharya, Porous silicon based potentiometric triglyceride biosensor, *Biosensors and Bioelectronics*, Vol. 16, 2001, pp. 313-317.

[48]. P. Kurup, H. Sun, J. Chen, An electronic nose for nanosensors, *Proceedings of the 8th International Conference on Nanostructured Materials (Nano 2006)*, IISc., Bangalore, Aug. 20-25, 2006.

[49]. J. R. Link, M. Sailor, Smart dust: Self-assembling, self-orienting photonic crystals of porous silicon, *Proceedings of the National Academy of Sciences (PNAS), USA*, Vol. 100, Issue. 9, 2003, pp. 10607-10610.

[50]. T.-P. Nguyen, P. Le Rendu, K. W. Cheah, Optical properties of porous silicon/poly(p phenylene vinylene) devices, *Physica E: Low-dimensional Systems and Nanostructures*, Vol. 17, April 2003, pp. 664-665.

[51]. Y. Y. Li, F. Cunin, J. R. Link, T. Gao, R. E. Betts, S. H. Reiver, V. Chin, S. N. Bhatia, M. J. Sailor, Polymer replicas of photonic porous silicon for sensing and drug delivery applications, *Science*, Vol. 299, 2003, pp. 2045-2047.

[52]. S. Ghosh, A. K. Sood, N. Kumar, Carbon nanotube flow sensors, *Science*, Vol. 14, 2003, pp. 1042-1044.

[53]. J. W. Aylott, Optical nanosensors- an enabling technology for intracellular measurements, *Analyst*, Vol. 128, 2003, pp. 309-312.

[54]. A. Dmytruk, A. Kasuya, S. Mamykin, Y. S. Park, V. Ovechko, A. Schur, A. Watanabe, N. Ohuchi, Metal nanoparticles reductively grown in silicon nanoparticle solution and in porous silicon, in *Proceedings of the 1st International Workshop on Semiconductor Nanocrystals, SEMINANO*, September 12-15, 2005, Budapest, Hungary.

[55]. E. Froner, R. Adamo, Z. Gaburro, B. Margesin, L. Pavesi, A. Rigo, M. Scarpa, Luminescence of porous silicon derived nanocrystals dispersed in water: dependence on initial porous silicon oxidation, *Journal of Nanoparticle Research*, Published online, May 4, 2006.

[56]. D. D. Majumder et al., Nano-materials: Science of bottom-up and top-down, *IETE Technical Review*, Vol. 24, Issue 1, 2007, pp. 9-25.

[57]. O. Bisi et al., Porous silicon: a quantum sponge structure for silicon based optoelectronics, *Surface Science Report*, Vol. 38, 2000, pp. 1-126.

[58]. S. M. Weiss, P. M. Fauchet, Porous silicon 1-D photonic crystals for optical signal modulation, *IEEE J. Selected Topics in Quantum Electronics*, Vol. 12, Issue 6, 2006.

[59]. M. J. Sailor, L. T. Canham, Properties of porous silicon, *IEE INSPEC*, 1997, pp. 364.

[60]. F. P. Mathew, E. C. Alocilja, Fabrication of porous silicon–based biosensor, *IEEE*, 2003, pp. 293-298.

[61]. F. S. Ligher, Optical Biosensors, C. A. Rowe Tailt, *Elsevier*, Amsterdam, The Netherlands, 2004.

[62]. R. P. Hangland, The Handbook: A guide to fluorescence probes and labeling technologies, *Invitrogen Corporation*, San Diego, 2005.

[63]. J. R. Lakowicz, Principles of fluorescence spectroscopy, *Kluwer Academic/Plenum Publishers*, New York, 1999.

[64]. S. S. Salitermkan, Fundamentals of BioMEMS and medical microdevices, *Wiley-Interscience, SPIE PRESS Bellingham*, Washington USA, 2006.

[65]. S. Borini, et al., Advanced nanotechnological approaches for designing protein-based Lab-on-a-Chip sensors on porous silicon wafer, *Recent Patents on DNA Gene Sequences*, Vol. 1, 2007, pp. 1-7.

[66]. H. Ouyang et al., Quantitative analysis of the sensitivity of porous silicon optical biosensors, *Appl. Phys. Lett.*, Vol. 88, 2006, pp. 163108 - 63108-3.

[67]. V. S. Y. Lin, et al., A porous silicon based optical interferometric biosensor, *Science*, Vol. 278, 1997, pp. 840-843.

[68]. L. De. Stefano, et al., Time resolved sensing of chemical species in porous silicon optical microcavity, *Sensors and Actuators B*, Vol. 100, 2004, pp. 168-172.

[69]. V. Mulloni, L. Pavesi, Porous Silicon microcavities as optical chemical sensors, *Appl. Phys. Lett.*, Vol. 76, 2000, pp. 901-903.

[70]. T. M. Benson, et al., Progress towards achieving integrated circuit functionality using porous silicon optoelectronic components, *Mat. Sci. Eng. B*, Vol. 69-70, 1999, pp. 92-95.

[71]. C. Wongmanered et al., Determination of pore size distribution and surface area of this porous silicon layer by spectroscopic ellipsometry, *Applied Surface Science*, Vol. 172, 2001, pp. 117-125.

[72]. L. Canham, Properties of porous silicon, *EMIS Datareviews, ed. B. Wiess*, Vol. 18, 1997.

[73]. Hu Ligang, et al., Preparation of a biochip on a porous silicon and application for label-free detection of small molecule-protein interactions, *Rapid Commun. Mass Spectrum.*, Vol. 21, 2007, pp. 1277-1281.

[74]. J. Wei, et al., Desorption-ionization mass spectroscopy on porous silicon, *Nature*, Vol. 399, 1999, pp. 243-246.

[75]. E. Marshall, Getting the noise out of Gene Arrays, *Science*, Vol. 36, Issue 10, pp. 630- 631.

[76]. K. C. Ho, et al., Amperometric detection of morphine at a Prussian blue-modified indium tin oxide electrode, *Biosens. Bioelectron*, Vol. 20, 2004, pp. 3-8.

[77]. Ben-Dov, et al., Piezoelectric immunosensors for urine specimens of chlamydia trachomatis employing quartz-crystal-microbalance microgravimetric analyses, *Anal. Chem.*, Vol. 69, 1997, pp. 3506-3512.

[78]. M. Murata, et al., Piezoelectric sensors for endocrine-disrupting chemicals using receptors-co-factor integration, *Anal. Sciences*, Vol. 19, 2003, pp. 1355-1357.

[79]. B. Xie, Mini/micro thermal biosensors and other related devices for biochemical/clinical monitoring, *TRAC*, Vol. 19, 2000, pp. 340-349.

[80]. M. Mehrvar, Fibre-Optic biosensor-trends and advances, *Anal. Sciences*, Vol. 16, 2000, pp. 677-692.

[81]. C. A. Betty, Porous silicon: A resourceful material for nanotechnology, *Recent Patents on Nanotechnology*, Vol. 2, Issue 2, 2008, pp. 128-136.

[82]. S. Chan, et al., Nanoscale silicon microcavities for biosensing, *Material Science and Engineering: C*, Vol. 15, Issue 1-2, 2001, pp. 277-282.

[83]. R. Prabakaran, et al., The effects of ZnO coating on the photoluminescence properties of porous silicon for the advanced optoelectronic devices, *J. Nano-crystalline Solids*, Vol. 354, No. 19-25, 2008, pp. 2181-2185.

[84]. E. J. Anglin, et al., Porous silicon in drug delivery devices and materials, *Advanced Drug Delivery Reviews*, Vol. 60, No. 11, 2008, pp. 1266-1277.

[85]. J. Gole, E. S. Lewis, Nanostructure and morphology modified porous silicon sensors, *SPIE Proc. Series*, Vol. 5732, 2005, pp. 573-583.

[86]. V. Srinivasan, et al., An integrated digital microfluidic lab-on-a-chip for clinical diagnostics on human physiological fluids, *Lab Chip*, Vol. 4, No. 4, August 2004, pp. 310-315.

[87]. V. Srinivasan, et al., A digital microfluidic biosensor for multianalyte detection, *Proceedings of the IEEE Micro Electro Mechanical System (MEMS) Conference*, Kyoto, Japan, 2003, pp. 327-330.

[88]. F. Su, K. Chakraborty, R. B. Fair, Clinical diagnostics on human whole blood, plasma, serum, urine, saliva, sweat, and tears on a digital microfluodic platform, *Proceedings of µTAS*, Squaw, Vally, CA, 2003, pp. 1287-1290.

[89]. R. W. Fauthauer, et al., Visible luminescence from silicon wafers subjected to stain etches, *Appl Phys Lett.*, Vol. 60, 1992, pp. 995–997.

[90]. S. Shih, et al., Photoluminescence and formation mechanism of chemically etched silicon, *Appl Phys Lett.*, Vol. 60, 1992, pp. 1863–1865.

[91]. M. T. Kelly, et al., High efficient chemical etchant for the formation of luminescent porous silicon, *Appl Phys Lett.*, Vol. 64, 1994, pp. 1693–1695.

[92]. Ji-Ho Park, et al., Biodegradable luminescent porous silicon nanoparticles for in vivo applications, *Nature Materials*, Published on line, February 29, 2009, pp. 331-336.

[93]. K. E. Sapsford, et al., Biosensing with luminescent semiconductor quantum dots, *Sensors*, Vol. 6, 2006, pp. 925-953.

[94]. S. K. Bhatia, et al., Use of thiol-terminal silanes and heterobyfunctional crosslinkers for immobilization of antibodies on silica surface, *Anal Biochem*, Vol. 178, 1989, pp. 408-413.

[95]. A. Jane, et al., Porous silicon biosensors on the advance, *Trends in Biotech.*, Vol. 27, No. 4, 2009, pp. 230-239.

[96]. L. D. Stefano, et al., DNA optical detection based on porous silicon technology: from biosensors to biochips, *Sensors*, Vol. 7, 2007, pp. 214-221.

[97]. D. Zhang, E. C. Alocilja, Characterization of nanoporous silicon-based DNA biosensors for the detection of salmonella enteritidis, *IEEE Sensors Journal*, Vol. 8, No. 6, 2008, pp. 775-780.

[98]. M. Dahan, et al., Diffusion dynamics of glycine receptors revealed by single-quantum dot tracking, *Science*, Vol. 302, 2003, pp. 442–445.

[99]. T. Q. Vu, et al., Peptide-conjugated quantum dots activate neuronal receptors and initiate downstream signaling of neurite growth, *Nano Lett*, Vol. 5, 2005, pp. 603–607.

[100]. H. Fan, et al., Surfactant-assisted synthesis of water-soluble and biocompatible semiconductor quantum dot micelles, *Nano Lett*, Vol. 5, 2005, pp. 645–648.

[101]. M. Howarth, et al., Targeting quantum dots to surface proteins in living cells with biotin Ligase, *Proceedings of Natl Acad Sci USA*, Vol. 102, 2005, pp. 7583–7588.

[102]. Vikas, et al., Biosensors: Future Analytical Tools, *Sensors & Transducers,* Vol. 76, Issue 2, 2007, pp. 937-944.

[103]. C. Wu, et al., Label-free optical detection of bacteria on a 1D photonic crystal of porous silicon, *Proc. SPIE*, Vol. 7167, 71670Z, 2009.

[104]. S. Pathak, et al., Quantum Dot Applications to Neuroscience: New Tools for Probing Neurons and Glia, *The Journal of Neuroscience*, 26, 7, February 15, 2006, pp. 1893-1895.

[105]. L. M. Bonanno, L. A. De Louise, Whole blood optical biosensor, *Biosensors and Bioelectronics*, Vol. 23, Issue 3, 2007, pp. 444-448.

[106]. J. Salonel, V. P. Lehto, Fabrication and chemical surface modification of mesoporous silicon for biomedical applications, *Chem. Eng. J.,* Vol. 137, 2008, pp. 162-172.

[107]. V. Lehmann et al., On the morphology and the electrochemical formation mechanism of mesoporous silicon, Mater. Sci. Eng., B B69, No. 70, 2000, pp. 11-22.

[108]. F. Bessueille et al., Assessment of porous silicon substrate for well-characterized Sensitive DNA chip implement, *Biosensors and Bioelectronics*, Vol. 21, December 2005, pp. 908-916.

[109]. G. D. Francia et al., Towards a label-free optical porous silicon DNA sensor, *Biosensors and Bioelectronics*, Vol. 21, 2005, pp. 661–665.

[110]. C. Dickson, Porous Silicon, (www.met.kth.se/mattechnol/FUMA2002/PorousSilicon.doc)

[111]. J. M. Buriak, Organometallic Chemistry on Silicon and Germanium Surfaces, *Chem. Rev.,* Vol. 102, Issue 5, 2002, pp. 1272-1308.

[113]. Singh, et al., Nanostructured porous silicon as functionalized material for biosensor applications, *J Mater Sci Mater Med.*, Online publication, July 3, 2008.

[114]. V. Vamvakaki, et al., DNA stabilization and hybridization detection on porous silicon surface by EIS and total internal reflection FT-IR spectroscopy, *Electroanalysis*, Vol. 20, Issue 17, 2008, pp. 1845-1850.

[115]. M. Ghadiri, Porous semiconductor–based optical interferometric sensor, *US Patent* 2004/0152135 A1, Aug. 5, 2004.

[116]. E. Tasciotti, C. Chiappini, J. Martinez, X. Liu, M. Ferrari, Multistage Mesoporous Silicon Particles for Biomedical Applications, in *Proceedings of the 11th International Conference on Advance Materials*, Rio de Janeiro, Brazil, Sep. 20-25, 2009.

[117]. H. A. Santos, J. Salonel, J. Riikonen, T. Heikkila, L. Peltonen, V-P Lehto, J. Hirvonen, Size-dependent in vitro toxicity of porous silicon micro- and nanoparticles, in *Proceedings of the Nanotech Conference & Expo*, May 3-7, 2009, Houston, TX, USA.

[118]. C. Rauh-Adelmann, et al., US 2006063178, 2006.

[119]. http://www.dddmag.com/images/0409/HTS1_lrg.jpg

[120]. R. B. Fair, Digital microfludics: Is a true lab-on-a-chip possible?, *Microfluidics and Nanofluidics*, Vol. 3, March 2007, pp. 245-281.

[121]. K. Chakrabarty, F. Su, Digital Microfluidic Biochips: Synthesis, Testing, and Reconfiguration Techniques, *CRC Press*, 2007.

[122]. http://transcriptome.ern.fr/sgbd/presentation/principle.php

[123]. S. Ghoshal, et al., Superparamagnetic iron oxide nanoparticle attachment on array of micro-test tubes and micro-beakers formed on p-type silicon substrate for biosensor applications, *Nanoscale Research Letters*, 2011, 6:540.

[124]. P. Granitzera, et al., Porous silicon/Fe_3O_4-nanoparticle composite and its magnetic behaviour, *ECS Transactions*, Vol. 16, Issue 3, 2008, pp. 91-99.

[125]. P. Granitzer, K. Rumpf, Porous Silicon—A versatile host material, *Materials*, Vol. 3, 2010, pp. 943-998.

[126]. J. M. Kinsella, et al., Enhanced magnetic resonance contrast of Fe_3O_4 nanoparticles trapped in a porous silicon nanoparticle host, *Adv. Mater.* Vol. 23, 2011, pp. H248–H253.

[127]. P. Granitzer, K. Rumpf, Magnetic nanoparticles embedded in a silicon matrix, *Materials*, Vol.4, 2011, pp. 908-928.

[128]. N. O'Farrell, et al., Silicon nanoparticles: applications in cell biology and medicine, *International Journal of Nanomedicine*, Vol. 1, Issue 4, 2006, pp. 451–472.

[129]. F. Erogbogbo, et al., Biocompatible Luminescent Silicon Quantum Dots for Imaging of Cancer Cells, ACS Nano., Vol. 2, Issue 5, 2008 May, pp. 873-878.

[130]. R.D. Tilley, et al., Micro-emulsion synthesis of monodisperse surface stabilized silicon nanocrystals, *Chem Commun (Camb)*. Vol. 14, Issue 14, 2005 Apr, pp. 1833-1835.

[131].J. H. Warner, et al., Water-Soluble Photoluminescent Silicon Quantum Dots, *Angewandte Chemie International Edition*, Vol. 44, Issue 29, 2005, pp. 4550-4554.

[132]. Z. F. Li, E. Ruckerstein, Water-Soluble Poly (acrylic acid) Grafted Luminescent Silicon Nanoparticles and Their Use as Fluorescent Biological Staining Labels, *Nano Letts.*, Vol. 4, No. 8, 2004, pp. 1463- 1467.

[133]. B. A. Manhat, et al., One-Step Melt Synthesis of Water-Soluble, Photoluminescent, Surface-Oxidized Silicon Nanoparticles for Cellular Imaging Applications, *Chem. Mater.*, dx.doi.org/10.1021/cm200270d.

[134]. D. Sathis Kumar, et al., Nanostructured Porous Silicon – A Novel Biomaterial for Drug Delivery, International *Journal of Pharmacy and Pharmaceutical Sciences*, Vol. 1, Issue 2, Oct-Dec. 2009, pp. 8-16.

[135]. J. Salonen, et al., Mesoporous silicon in drug delivery applications, *J. Pharm. Sci*, Vol. 97, 2008, pp. 632–653.

[136]. M. Pradeepa, et al., Fabrication of Porous Silicon Nanoparticles to Attach Clorgyline for Drug Delivery, *2011 International Conference on Bioscience, Biochemistry and Bioinformatics IPCBEE*, Vol. 5, 2011, pp. 327-330, © (2011) IACSIT Press, Singapore.

[137]. D. Reeves, A Study of Silicon Nanoparticles with Applications to Medical Resonance Imaging, *NNIN REU Research Accomplishments,* 2008, pp. 90-91.

[138]. D. Smith, Functionalization of Silicon Nanoparticles for Hyperpolarized Magnetic Resonance Imaging, *National Nanotechnology Infrastructure Network*, 2007 REU Research Accomplishments, pp. 78-79.

[139]. J. Xu, et al., Research and applications of biochip technologies, *Chinese Science Bulletin*, Vol. 47, 2000, pp. 101–108.

[140]. N. Ravindra, et al., Advances in the manufacturing, types, and applications of biosensors, *Journal of the Minerals, Metals and Materials Society (JOM),* Vol. 59, 2007, pp. 37–43.

[141]. K. E. Herold and A. Rasooly, *Lab-on-a-Chip Technology* (Vol. 1): Fabrication and Microfluidics, Caister Academic Press (2009), Vol. I.

[142]. D. R. Reyes, et al., Micro total analysis systems. 1. Introduction, theory, and technology, *Analytical Chemistry*, Vol. 74, 2002, pp. 2623–2636.

[143]. E. Ghafar-Zadeh, et al., CMOS based capacitive sensor laboratory-on-chip: a multidisciplinary approach, *Analog Integrated Circuits and Signal Processing*, Vol. 59, 2009, pp. 1–12.

[144]. A. T. Giannitsis, Microfabrication of biomedical lab-on-chip devices. A review, *Estonian Journal of Engineering,* Vol. 17, 2011, pp. 109–139.

[145]. R. Shyamsundar, et al., A DNA microarray survey of gene expression in normal human tissues. *Genome Biol.*, Vol. 6, 2005, pp. R22.

[146]. GeneChip, McDevitt Research Laboratory, http://www.gene-chips.com/

[147]. D. A. Hall, et al., Protein Microarray Technology. *Mech Ageing Dev.*, Vol. 128, 2007, pp. 161–167.

[148]. M. L. Yarmush and K. R. King, Living-Cell Microarrays, *Annual Review of Biomedical Engineering,* Vol. 11, 2009, pp. 235–257.

[149]. I. Barbulovic-Nad, et al., Bio-Microarray Fabrication Techniques-A Review, *Crit. Rev. Biotechnol.* Vol. 26, 2006, pp. 237–259.

[150]. K. Chakrabarty and T. Xu, Digital Microfluidic Biochips: Design and Optimization. *CRC Press* , 2010.

[151]. T. M. Squires and S. R. Quake, Microfluidics: Fluid physics at the nanoliter scale, *Reviews of Modern Physics,* Vol. 77, 2005, pp. 977–1026.

[152]. P.-A. Auroux, et al., Micro total analysis systems. 2. Analytical standard operations and applications, *Analytical Chemistry*, Vol. 74, 2002, pp. 2637–2652.

[153]. J. C. T. Eijkel and A. van den Berg, Nanofluidics: what is it and what can we expect from it? *Microfluidics and Nanofluidics*, Vol. 1, 2005, pp. 249–267.

[154]. E. Verpoorte and N. F. D. Rooij, Microfluidics meets MEMS, *Proc. of the IEEE*, Vol. 91, No. 6, 2003, pp. 930–953.

[155]. Y. Wang, et al., Composable Behavioral Models and Schematic-Based Simulation of Electrokinetic Lab-on-a-Chip Systems, *IEEE Trans. Comput.-Aided Design Integr. Circuits Syst.* Vol. 25, 2006, pp. 258–273.

[156]. J. H. Xu, et al., Correlations of droplet formation in T-junction microfluidic devices: from squeezing to dripping, *Microfluidics and Nanofluidics*, Vol. 5, 2008, pp. 711–717.

[157]. M. L. J. Steegmans, et al., Characterization of Emulsification at Flat Microchannel Y Junctions, *Langmuir*, Vol. 25, 2009, pp. 3396–3401.

[158]. H. Liu and Y. Zhang, Droplet formation in microfluidic cross-junctions, *Phys. Fluids*, Vol. 23, 2011, pp. 082,101.1–082,101.12.

[159]. J. M. Li, et al., Multi-layer PMMA Microfluidic Chips with Channel Networks for Liquid Sample Operation, *Elsevier Journal of Materials Processing Technology*, Vol. 209, 2009, pp. 5487–5493.

[160]. J. M. Ottino and S. Wiggins, Introduction: mixing in microfluidics, *Phil. Trans. R. Soc. Lond.*, Vol. 362, 2004, pp. 923–935.

[161]. L. E. Locascio, Microfluidic mixing, *Anal Bioanal Chem.,* Vol. 379, 2004, pp. 325–327.

[162]. C. J. Campbell and B. A. Grzybowski, Microfluidic mixers: from microfabricated to self-assembling devices, *Phil. Trans.*, Vol. 362, 2004, pp. 1069–1086.

[163]. H. Y. Park, et al., Achieving Uniform Mixing in a Microfluidic Device: Hydrodynamic Focusing Prior to Mixing, *Analytical Chemistry*, Vol. 78, 2006, pp. 4465–4473.

[164]. P. B. Howell, et al., A combinatorial approach to microfluidic mixing. *Journal of Micromechanics and Microengineering,* Vol. 18, 2008, pp. 115019.

[165]. Y. K. Suh and S. Kang, A Review on Mixing in Microfluidics, *Micromachines*, Vol. 1, 2010, pp. 82–111.

[166]. C.-Y. Lee, et al., Microfluidic Mixing: A Review, *International Journal of Molecular Sciences*, Vol. 12, 2011, pp. 3263–3287.

[167]. S. Y. The, et al., Digital Microfluidics. *Lab-On-a-Chip*, Vol. 8, 2008, pp. 198–220.

[168]. M. Abdelgawad and A. R. Wheeler, The Digital Revolution: A New Paradigm for Microfluidics, *Advanced Materials*, Vol. 21, 2009, pp. 920–925.

[169]. X. C. i Solvas and A. deMello, Droplet microfluidics: recent developments and future applications, *Chem. Commun.,* Vol. 47, 2011, pp. 1936–1942.

[170]. M. G. Pollack, et al., Electrowetting-based Actuation of Liquid Droplets for Microfluidic Applications, *Applied Physics Letters*, Vol. 77, 2000, pp. 1725–1726.

[171]. J. Lee, et al., Electrowetting and electrowetting-on-dielectric for microscale liquid handling, *Sensors and Actuators A: Physical*, Vol. 95, 2002, pp. 259–268.

[172]. M. G. Pollack, et al., Electrowetting-based actuation of droplets for integrated microfluidics, *Lab-on-a-Chip*, Vol. 2, 2002, pp. 96–101.

[173]. H. Moon, et al., Low voltage electrowetting-on-dielectric, *Journal of Applied Physics*, Vol. 92, 2002, pp. 4080–4087.

[174]. J. Wu, et al., Droplets actuating chip based on electrowetting-on-dielectric, *Frontiers of Electrical and Electronic Engineering in China*, Vol. 2, 2007, pp. 345–349.

[175]. J. Lienemann, et al., Modeling, Simulation, and Optimization of Electrowetting, *IEEE Trans. Comput.-Aided Design Integr. Circuits Syst.*, Vol. 25, 2006, pp. 234–247.

[176]. J. H. Song, et al., A Scaling Model for Electrowetting-on-Dielectric Microfluidic Actuators, *Microfluidics and Nanofluidics*, Vol. 7, 2008, pp. 75–89.

[177]. H. A. Pohl and J. S. Crane, Dielectrophoresis of Cells, *Biophysical Journal*, Vol. 11, 1971, pp. 711–727.

[178]. H. Morgan, et al., Separation of submicron bioparticles by dielectrophoresis, *Biophysical Journal*, Vol. 77, 1999, pp. 516–525.

[179]. H. A. Pohl, The Motion and Precipitation of Suspensoids in Divergent Electric Fields, *Journal of Applied Physics*, Vol. 22, 1951, pp. 869–871.

[180]. H. A. Pohl, Some effects of nonuniform fields on dielectrics, *Journal of Applied Physics*, Vol. 29, 1958, pp. 1182–1188.

[181]. T. B. Jones, et al., Dielectrophoretic liquid actuation and nanodroplet formation, *Journal of Applied Physics,* Vol. 89, 2001, pp. 1441–1448.

[182]. A. Castellanos, et al., Electrohydrodynamics and dielectrophoresis in microsystems: scaling laws, *Journal of Physics D: Applied Physics*, Vol. 36, 2003, pp. 2584–2597.

[183]. P. R. C. Gascoyne, et al., Dielectrophoresis-based programmable fluidic processors, *Lab-On-a-Chip*, Vol. 4, 2004, pp. 2584–2597.

[184]. R. Ahmed and T. B. Jones, Optimized liquid DEP droplet dispensing, *Journal of Micromechanics and Microengineering*, Vol. 17, 2007, pp. 1052–1058.

[185]. H. Li, et al., Driving Cell Seeding Using Surface Acoustic Wave Fluid Actuation, *Proc. of the Australasian Fluid Mechanics Conference,* 2007, pp. 625–629.

[186]. Z. J. Jiao, et al., Scattering and attenuation of surface acoustic waves in droplet actuation, *Journal of Physics A: Mathematical and Theoretical*, 2008, Vol. 41, pp. 1–9.

[187]. A. A. Garcia, et al., Magnetic movement of biological fluid droplets, *Journal of Magnetism and Magnetic Materials*, Vol. 311, 2007, pp. 238–243.

[188]. J. Schneider, et al., Automated Digital Magnetofluidics, *Journal of Physics: Conference Series,* Vol. 127, 2008, pp. 012006.

[189]. S. N. Pei, et al., Light-actuated digital microfluidics for large-scale, parallel manipulation of arbitrarily sized droplets, *Proc. of the IEEE MEMS*, 2010, pp. 252–255.

[190]. W. Hu and L. Tang, In Vivo Diagnosis: A Future Direction for Biochip Technology, *Journal on Biochips and Tissue chips*, Vol. 47, 2011, pp. 101–108.

[191]. VeriChip, Positive ID Corporation, http://www.positiveidcorp.com/.

[192]. K. Daniel, et al., Implantable diagnostic device for cancer monitoring, *Biosens Bioelectron.*, Vol. 24, 2009, pp. 3252–3257.

[193]. S. Sato, K. Fujita, M. Kanazawa, K. Mukumoto, K. Ohtsuka, M. Waki, and S. Takenaka, Electrochemical assay for deoxyribonuclease-I activity, *Anal. Biochem*, Vol. 15, 381, 2, 2008, pp. 233-239.

[194]. M. Kanazawa, S. Sato, K. Ohtsuka, and S. Takenaka, Ferrocenylnaphthalene diimide-based electrochemical ribonuclease assay, *Anal. Sci,* Vol. 23, 12, Dec. 2007, pp. 1415-1419.

[195]. M. Joshi, N. Kale, S. Mukherji, R. Lal, and V. Ramgopal Rao, Affinity cantilever sensors for cardiac diagnostics, *Indian Journal of Pure and Applied Physics,* Vol. 45, 04, April 2007, pp. 287- 293.

[196]. V. Seena, N. Kale, S. Nag, M. Joshi, S. Mukherji, and V. Ramgopal Rao, Developing a polymeric microcantilever platform technology for biosensing, *International Journal of Micro and Nano Systems,* Vol. 1, 1, 2009, pp. 65-70.

[197]. M. Joshi, N. Kale, R. Lal, S. Mukherji, and V. Ramgopal Rao, Development of a biochip for cardiac diagnostics, in Bionanotechnology: Global Prospects, *Crc Press*, 2009, pp. 141-160.

[198]. M.-I. Mohammed and M. P. Y. Desmulliez, Lab-on-a-chip based immunosensor principles and technologies for the detection of cardiac biomarkers: a review, *Lab Chip*, Vol. 11, 2011, pp. 569-595.

[199]. S. G. Surya, S. Nag, N. M. Duragkar, D. Agarwal, G. Chatterjee, S. Gandhi, S. Patil, D. K. Sharma, and V. Ramgopal Rao, Low-power instrumentation system for nano-electromechanical-sensors for environmental and healthcare applications, *Journal of Low Power Electronics* (JOLPE), to appear, 2012.

[200]. V. Seena, A. Nigam, P. Pant, S. Mukherji, and V. Ramgopal Rao, Organic cantiFET: a nanomechanical polymer cantilever sensor with integrated OFET, *Journal of Microelectromechanical Systems,* Vol. 21, 2, April 2012, pp. 294 – 301.

Chapter 10
Distributed Information Extraction from Large-scale Wireless Sensor Networks

Elena Gaura, James Brusey, John Halloran, Tessa Daniel

10.1. Introduction

The increase in availability and affordability of wireless technology has led to a proliferation of large wireless sensor networks (WSNs) with increasing numbers of nodes deployed to resolve complex "informational" problems. Typically, however, nodes in these networks have limited resources including energy (given the limited battery power), memory, computing power and communication bandwidth. The coupling of high complexity global tasks and the reduced resource nodes make WSNs a rich research domain. The use of WSNs has been explored since the 1990's, in a number of application domains with deployments ranging from scientific research to battlefield surveillance for the military. Some early examples of successful deployments include habitat monitoring applications [1, 2]; agricultural monitoring [3]; healthcare [4]; disaster management and detection [5] and military applications [6]. Moreover, some newer applications involve the integration of multiple WSN systems spanning a variety of disciplines. The High Performance Wireless Research and Education Network [7] and ROADNet [8] projects, both in California, are two examples of early days multidisciplinary applications.

Although specific application requirements may differ from one WSN system to another, essential to all WSN systems is the ability to acquire data and convert this to usable information in order to fulfill the application needs. Many of the WSN information extraction approaches currently in operation are based on applicative query mechanisms where requests are initiated via queries written in an appropriate declarative language, posed to the network, and data or information generated as a result. Data processing in such applications usually follows one of two approaches: centralized or distributed. For some systems, resources are not as severely constrained; hence, minimizing energy usage is not a major concern. In such unconstrained systems, a centralized approach to processing is often used where sensed values are pushed to a

Elena Gaura
Cogent Computing Applied Research Centre, Coventry University, UK

power-rich location for processing, which may involve cleaning and querying the data as part of more in-depth analysis.

In a large number of systems, however, constraints like computing and battery power dictate that applications be developed with energy-efficiency in mind. For many of these applications human intervention can be both time-consuming and expensive, for example, in terms of changing batteries or adding new nodes. Therefore, a key design objective is to extend the lifetime of the network as much as possible. It is well understood that power usage costs in WSNs are dominated by communication as opposed to computation [9-11]. Therefore, techniques that promote a decrease in communications while using in-network data reduction have been identified as key to creating more energy-efficient WSN applications. Distributed processing is proposed as a method that promotes processing of data on nodes in the network [10, 12]. This in-network processing takes advantage of nodes' processing power and may include techniques like data aggregation, fusion or elimination of redundant values through filtering [13, 14]. The net result is more energy-efficient applications since data transmission is reduced in exchange for in-network computation. While this is apparently of obvious benefit towards long-life systems design, in-network computation is still in its infancy.

Existing information extraction procedures (whose survey is at the core of the chapter) can be categorized into three main approaches: agent-based, query-based and macroprogramming. Of the three, query-based systems are the most popular mainly because they provide a usable, high level interface to the sensor network while abstracting away some of the low level details like the network topology and radio communication. They are very useful and provide a solution in cases where data needs to be retrieved from the entire network. With this ease of use, however, come a number of limitations.

The first limitation is in terms of what queries can be posed to the network. In general, the query-based applicative systems in use allow the issuing of restricted queries, ranging from those targeting raw sensor readings on nodes in the network to those requiring the computation of simple aggregates like average, maximum, and minimum over some attribute of interest, for the entire network. Second, the query languages used cannot easily express spatio-temporal characteristics, which are an important aspect of the data generated in WSNs. Third, it is quite difficult if not impossible to construct information requests that represent higher level behaviour, involve just a "subset" of the network (whether physical or logical), or require more complex in-network interactions between "subsets" in order to generate information. Furthermore, as distributed computation is not the main focus of SQL-based query languages (which support most WSN query-based approaches) implementing arbitrary aggregation, for example, is quite difficult [15].

In contrast, macroprogramming has been proposed as an approach to information extraction that provides a more general-purpose approach to distributed computation compared to traditional query-based approaches. As applied to WSNs, macroprogramming approaches focus on programming the network as a whole rather than programming the individual devices that form the network. Many macroprogramming systems provide the ability to create programs that represent higher-level behaviour, a level of abstraction beyond that of the more popular query-based approaches. Global behaviour can be specified, programmed and then translated to node level code. Ideally, the programmer is not concerned with low level details like network topology, radio communication or energy capacity.

Of interest with some macroprogramming systems are the application-defined, in-network abstractions (some based on local node interactions) that are used in data processing. One example is the Regiment system [16] in which a programming abstraction called a region is used. A region is described as a collection of spatially distributed signals with an example being the set of sensor readings from nodes in a geographic area. Regions as opposed to individual nodes are programmed (for example, an rfold operator is used to aggregate the values in a region into a single signal which can then be communicated to the user or used in further computation).

Macroprogramming, however, still presents a number of challenges. First, creating a powerful macroprogram requires a learning curve for the programmer. Expressing high-level requirements are not necessarily as intuitive to the user as perhaps SQL-based approaches are. Second, although proposed as an approach that eliminates the need for node-level programming, many of the current macroprogramming systems provide node-specific abstractions, undermining the rapid development and productivity advantages macroprogramming is meant to provide. Finally, because of the wide semantic gap that exists between the high level program and the node level code, compiler construction is quite challenging. The code generated as a result of compilation has to cater for not just computation but node-level communication as well.

A third approach to information extraction looks at the extracting of information in a network-aware manner and tailors the mechanism to the type of information needed and the configuration of the network it needs to be extracted from. These models use agents to perform tasks, make decisions and collaborate to achieve more complicated tasks. An agent is simply a piece of software that performs a task without the need of user invocation to function. Many agent-based models are multi-agent systems in which multiple agents sometimes coded to function in different ways, collaborate, coordinate and organize to perform complex tasks and generate information. These models therefore introduce expressiveness and flexibility in terms of what functions they can facilitate in the network as well as in facilitating the ability for agents to make decisions while in the network. Agents can act autonomously, multiple agents can run on a node at the same time, and multiple applications can co-exist in the network. Mobile agents can move, clone themselves and act to deal with unexpected changes in the environment [17, 18].

Although attractive, in theory, the agent-based approach presents a number of challenges particularly in terms of the difficulty they present for non-expert users to design and program. First, given the complexity of distributed systems, it is a challenge to model accurately what interactions will be taking place between components in the network when designing agents. Second, when implementing the agents, particularly mobile agents, the code must be able to run in unpredictable, resource-constrained environments. Forecasting accurately exactly what conditions will exist in the network is difficult and therefore difficult to program for. Third, there is difficulty in testing and debugging, given the distributed nature of the system. In mobile agent systems, tracking and understanding execution in the face of agents moving from node to node makes debugging and testing even more of a challenge. Given these difficulties, practical implementations of agent-based systems, particularly multi-agent and mobile agent systems, are not as widespread as query-based systems are. Each of the three approaches above is suitable for a number of applications. They each have strengths but pose challenges as well. The agent-based approach provides a high degree of expressiveness and flexibility as do the macroprogramming approaches, but they both are more difficult to implement into deployable WSN systems. The query-based systems have, as their key feature, ease of use but are limited in the types of queries that can be posed to the network. Macroprogramming tries to address this by providing powerful constructs for capturing higher level behaviour through global programs but introduce a steep learning curve to the user.

The authors here argue that it is desirable to take a hybrid approach that retains the simplicity and ease of use of traditional query-based approaches while allowing the inclusion of useful logical abstractions provided by macroprogramming approaches to facilitate construction and resolution of more powerful queries. Such an approach would need to incorporate some of the principles of agent-based systems, such as collaboration and decision making in the network, in assisting in query resolution. In this context it can be hypothesized that end-users of WSN applications could be provided with the ability to construct and pose higher-level information requests instead of simple queries requiring the collection of raw sensed values or the calculation of simple aggregates over the entire network. Complex query type constructs that allow the expression of phenomenological spatio-temporal characteristics would provide the means for exploiting the networked sensing concept at large scale. Moreover, providing the end-user with a system that produces responses to these higher level requests for information within the network instead of as a result of post-collection analysis of all data would most certainly demolish one of the largest road-blocks of the WSN technology in its route to adoption: ensuring user acceptability.

The chapter is structured as follows: Section 10.2 below concentrates on Agent-based approaches to information extraction; Section 10.3 surveys query-based systems and macroprogramming approaches and Section 10.4 looks at the open research issues the community faces with respect to information extraction, raises research question relating to the usefulness of in-network processing from an informational viewpoint and concludes the survey. In both Sections 10.2 and 10.3, example developments are identified, highlighting those that have been used in practical deployments. Wherever possible the architecture and mode of information processing (whether centralized or distributed) is identified. The key issues for discussion lie with:

- Identifying which of the methods proposed in the literature have been evaluated at implementation level. (This analysis is needed as a large proportion of the research effort is at theoretical level and not readily suitable for practical deployments.)
- Identifying which of the methods proposed in the literature make use of or support in-network information extraction. (This is important, as approaches that utilize in-network processing such as aggregation and filtering have been shown to be more energy efficient and therefore more desirable.)

10.2. Agent Based Approaches to Information Extraction

10.2.1. Agent-based Approaches and Architectures

The agent-based approach to information extraction tasks a network by injecting into it a program with some type of processing or decision making function. In this approach there is no attempt to devise ways of transmitting data to "collection" or "aggregation" points for processing, rather, the application is sent to the data instead [19]. The agent is able to collect local data and perform any necessary data aggregation. Autonomous agents can make decisions without user input. In the case of an autonomous agent, which also happens to be mobile, that decision may involve whether the agent should move to a particular node and, if yes, which node it should migrate to. Autonomous mobile agents moving from one node to another can be designed to be cognizant of issues that may exist in the network [20]. For instance, they can make decisions, perhaps to clone themselves, if necessary, to avoid faulty regions or nodes in the network. If an agent detects that a node has a problem and is unreachable it can adapt its route so that it avoids the faulty node [21]. The mobile agent therefore presents added flexibility in terms of decision-making and reliability in dealing with faults in the network.

[22] report three architecture models for agent-based wireless sensor networks. The classification is based on the number and type of agents being used and the mode of operation of the agent-based system.

In the first model, all nodes respond to a single agent usually housed at the base station or central processor. The advantage is that there is a single control point with the benefit being the ability to access the entire network and an increase in efficiency since transmissions from this central point could be better synchronized and collision reduced as a result. Collisions during transmission require a retransmission, which is quite costly. Therefore, synchronization brings with it direct energy savings. There are, however, a number of limitations to this model.

The first limitation is lack of scalability. In this model, the agent queries each node for a sensed value, which forms the input to the agent's deliberation (the agent follows a perceive-deliberate-act cycle). As the number of nodes increases, however, the time taken to deliberate also increases which can lead to time delays. Also, with larger networks it is more likely that multihop communication is needed. This can lead to even more time delays during data gathering as well as increased power consumption.

A second problem is observed in cases where there is a need for more than one agent to access nodes in the network simultaneously, for example, to execute different application tasks. Here an increase in the time delay occurs because one agent would have to wait for another to complete its task before accessing that node or set of nodes. This leads to delays in agents getting the data needed for processing and acting if needed. This model can be considered to utilize a centralized processing approach as all data is transmitted to a central point (albeit an agent) for further processing. In-network processing is absent.

In the second model, each node in the network hosts an agent that will determine how that node will behave. Control of the network is therefore distributed and agents can coordinate and collaborate to achieve goals both at a local and global level. A main advantage in using this model is its scalability. The ability to perform local computation without needing multihop messages to a central controller results in a reduction in energy usage. A second advantage is the ability of the network to perform tasks concurrently given the presence of multiple agents. However, in practice, it is difficult to create agent-based systems that solve global tasks using only local information. The result, therefore, is an increase in resource and energy usage for the necessary inter-agent communications, a cost avoided by the centralized model presented above. This multi-agent approach clearly incorporates in-network processing techniques for information generation. The "environmental nervous system", [23] a multi-agent system, is based on this model. It is a small (three node) autonomic wireless sensor network aimed at environmental sensing for intruder detection. This system is described in greater detail in Section 10.2.1.2.

The third model uses mobile agents. Here an agent is able to migrate to a node or nodes, collecting, aggregating and fusing data before it sends a message back to the central processor. A primary advantage here is a reduction in the cost of computation since only one node is active at a time. A second advantage is that communication cost is low since migration of agents only involves one transmit and receive event. The agent in this model is responsible for both collecting data and the itinerary related to the task in operation. The main disadvantage, however, is that if the node to which an agent has migrated fails then all data gathered up to that point is lost. There are, however, techniques that could be implemented to combat this problem, for instance leaving copies of the data on previously visited nodes so that the agent could restore its state if needed. A second disadvantage is the lack of concurrency this approach brings. In large networks, in particular, this could be a problem if the agent takes too long to collect all the data required to perform the given task. The mobile agent-based deployment reported by [17] is based on this model. This network was aimed at ground vehicle classification using acoustic sensors and is described in greater detail in Section 10.2.1.1.

Each of the three models described exhibits both advantages and disadvantages. According to [22] the strengths of the three methods were simplicity in the case of the centralized approach, robustness in the case of the multi-agent approach, and efficiency in the case of the mobile agent approach. The weaknesses were identified as scalability, complexity and latency. Two hybrid approaches were further proposed and described by [22] that combined the features of the three in ways that attempted to minimize the impact of the disadvantages described.

Much of the reported research in the area of agent-based information extraction mechanisms for WSNs describes variations of the agent approach outlined above. In the following, some examples of mobile and non-mobile agent system architectures are presented.

10.2.1.1. Mobile Agent-based Distributed Sensor Networks (MADSN)

The Distributed Sensor Network (DSN) model was first proposed by [24]. In this model, referred to as fusion architecture, all sensor data is sent to a central location where it is fused [20]. A number of variations followed [25-27], each making some improvement on the DSN architecture but holding to a common client/server paradigm or a centralized approach to information processing.

In a WSN with a large number of deployed sensors, each with its own piece of data, it is impractical to have all that information flowing to a central store. First, it has been identified that communication dominates power usage in sensor networks and that the cost of transmission of data from individual nodes in a network to a base station or sink far exceeds the cost of in-network processing of data before transmission [28, 29, 11]. With resource constraints like limited battery power on the nodes in the network, there is a need to reduce the cost of communication if at all possible. The energy availability problem is further exacerbated by the fact that often not all data that is transmitted is critical to ensuring quality information. There is often a large quantity of redundant or erroneous data. Other problems include network bandwidth limitations, unreliable connectivity and noise in the network. Given these constraints, DSN and other client/server type architectures cannot meet all the challenges of WSNs. Qi, Iyengar and Chakrabarty proposed an improved DSN architecture that used mobile agents (MADSN) [30, 17]. Whereas in DSN, sensor nodes collect data and transmit to the base station node or sink (the central processor), with MADSN the computation code is transmitted to the node where the data resides. The need to transmit masses of data is removed and replaced by transmission of only a small piece of code representing the agent. Additional benefits are that the agents can be programmed to perform fusion tasks to consolidate data increasing the extensibility of the system and the mode of information collection is more stable since it is not affected by fluctuations in connectivity. If a network connection fails, the agent can wait till it is re-established before returning its results.

A comparison between MADSN and DSN showed up to 90 % savings in data transfer for MADSN over DSN due to avoiding transfer of raw data [30]. It is not clear however if this was despite the overhead introduced by having agents created and dispatched. In another study, a model that incorporated collaborative signal and information processing (CSIP) techniques was applied and an analysis of performance between MADSN and

DSN conducted to measure energy consumption and execution time. Again, in that study MADSN outperformed DSN [31].

[32] studied the use of mobile agents for data fusion in a WSN running MADSN and focused on creating an optimum design of the itinerary component of the agent to improve performance and minimize resource usage. [33] showed that MADSN could be successfully applied to the real-world problem of vehicle classification in an unattended ground sensor system. The study showed, however, that classification accuracy was low (about 23 %) when only one sensor was used but improved to about 80 % when a multisensor array was used. (Target classification accuracy with a single sensor was high (about 75 %) only when the target was in close proximity to the sensor.)

10.2.1.2. Autonomic Wireless Sensor Networks (AWSN)

[23] introduced the concept of the autonomic wireless sensor network. They proposed that mobile agents could be deployed in the network to facilitate autonomic computing. Autonomic computing implies the ability to self-manage, self-maintain as well as interact with a user of the network at a policy rather than hardware level. The purpose of interaction would be to allow interoperability with legacy systems if needed. Mobile agents in that context would support cooperation and negotiation in the system via relevant protocols and a suitable agent communication language. The mobile agents would introduce flexibility in terms of the ability to modify agent code through agent migration or agent adaptation. The idea here is that an agent could evolve as system policies evolved. In modeling autonomic behaviour, an AWSN also uses the self-knowledge, self-protection and self-interest characteristics inherent in mobile agents of multi agent systems.

Autonomic behaviour in Marsh's system includes the ability to handle sensors that have been rendered inoperable due to some type of damage, and the system can put into effect strategies to deal with that. For instance, the authors suggest that the system could make an estimate of data lost due to system damage and effect nearby sensors to increase their sampling rate to compensate. It could also mean a reconfiguration of the routing topology to minimize or eliminate message loss. The AWSN concept is closely related to the multi agent system framework and incorporates some of the same ideas including that of entities migrating from node to node while making "intelligent" decisions.

[34] and [35], building on this idea of an autonomic wireless sensor network, proposed an agent-based approach to implementing intelligent power management. In their work they promote the use of agents in intelligently activating or deactivating nodes for power conservation based on interpolation. For example, the following criteria are used to decide whether a node should be put to sleep: a node is considered redundant if the remaining nodes can interpolate the temperature at its location to a desired degree of accuracy. Redundancy in this context means that the node is not needed since the other nodes or the network can operate effectively without it.

In Marsh's scenario, a coordinator agent is appointed and aggregates all calculations performed by other agents in deciding whether deactivation should occur or not. The coordinator agent sends out requests to other agents to calculate the components of the interpolation function used and send that result back to the coordinator. The coordinator then calculates the actual interpolated temperature and then requests the temperature from the appropriate agent. If the two values fall within a given error range the node is put into sleep mode.

It must be noted, however, that the issue of power management is influenced by a number of other network characteristics such as coverage, latency and longevity. Any power management scheme would have to balance the tradeoffs in making decisions on how power can be conserved. In the above example, the inherent autonomic characteristics of the network itself are used for efficiently managing power usage. Although a promising approach, this concept has not yet been implemented in agent-based WSN deployments.

10.2.1.3. Mobile Agent-based Wireless Sensor Networks (MAWSN)

[19] identified that the operation of MADSN was really based on three main assumptions. First, that the network had a cluster-based architecture, second, that each source node (nodes with data) was one hop away from the clusterhead, and third, that the redundant data collected could all be fused into one packet of fixed size [19], [36]. Those restrictions in a real-world context seemed to impose too strict limitations on the applications that could make use of the architecture and excluded many systems that did not boast all of these features. Hence, [19] proposed MAWSN as an alternative architecture to address networks where the features needed for the MADSN approach to work were not evident. Previous work done by [36] showed that in multi-hop environments without clusters, mobile agents could be employed to eliminate data redundancy by employing context-aware local processing techniques, use data aggregation to eliminate spatial redundancy with sensors that were of close proximity to each other, and reduce communication overhead by combining tasks. They were able to reduce information redundancy at three levels: the node level, the task level and the combined task level.

At the node level mobile agents were assigned processing code specifically targeting the requirements of the specific application. The result was that the mobile agent only needed local processing of that data requested by the application thereby reducing the amount of data transmitted as only relevant data was extracted from nodes for transmission. At the task level, the mobile agent aggregated sensed data from individual nodes when visited. Finally, at the combined task level, the mobile agent used the packet unification technique to unify shorter data packets to one longer packet again reducing the number of transmissions. MAWSN built on the encouraging results in [36] and experimentally showed improvements in energy consumption for data packets transmission over the more traditional client-server based WSN.

Concurrently with the above work, [37] proposed a framework aimed at implementing a wide variety of sensor network applications. Their aim was to create an architecture that allows multiple applications to share a network and permits scalable deployments that can evolve over time. In their design autonomous agents are deployed in the network with a particular agent having the ability to be on one or more nodes. Their framework is built atop TinyOS and its associated virtual machine Mate [38] and agents run on Crossbow motes. The agents were shown to use resources only on those motes they visited so resources in the network were used more efficiently. Agents were able to read sensor values, start devices and sent radio packets if required.

In a data collection experiment aiming to find the maximum value in a field of sensor readings, an agent was injected into the network via broadcast. After arriving at a node, it re-broadcasts itself if the reading at the current node was greater than that at the previous node. Agents stopped propagating once the region with the maximum sensor reading was reached. This was done by incrementing a count value when an agent arrived in the area. Once that value went over a pre-defined threshold the last agent to enter the area could issue a notification message to the processor.

Analysis of the experiment showed that as the size of the network increased, a smaller proportion of the network needed to be active. This meant that in large networks agents stayed on nodes only long enough to perform their task and quickly release resources for other task execution. The conclusion of the experiment was that the agent model presented was successful in showing that it could scale up without having to occupy all nodes on the network in performing tasks; however some areas for future work were identified. The first is security to ensure that unauthorized agents are not able to mount energy-depletion and denial-of-service attacks among others. The second involves giving agents the capability to query nodes for information such as processing power and energy. The third area involved giving agents the ability to cache their code on a node to reduce the number of agent transmissions.

In a similar approach [39] describe the agent oriented programming paradigm for development of intelligent sensor networks. They implemented a test application using the Java Agent Development Framework (JADE) [40] aiming to develop an intelligent ground sensor network. The application consisted of an Unattended Ground Sensor Network (UGSN) used to monitor moving targets with agents analyzing the data being acquired by nodes. In the experiment, information from the sensors were correlated and fused by reasoning agents using a fusing algorithm (the majority voting ordinary technique) suited for distributed information fusion. A special agent called the Target Agent (TA) is created on the node where the first sensor trigger occurs. The TA then reasons with its neighbours and after correlating their sensor data, decides what node it should migrate to next. The objective of migration is to get closer to the target being tracked. After each move the TA performs data fusion using the new data available on its host node. Through the TA, the network as a whole, decides when external applications need to be updated and what information should be sent to those applications.

The experiment was successful in its sole aim of showing that agents were suitable for distributed information fusion. Distributed information fusion is the idea, quite similar to aggregation, that data generated by nodes in a network can be efficiently combined in-network instead of transmitting all raw data to a sink for processing. The main advantages of this approach were identified as: reduction in data redundancy since we have agents deciding whether data is important enough to be used; reduction in power consumption in having all data sent directly back to a central location; conservation of communication bandwidth by limiting the exchange of data between nodes as much as possible.

The advantages of agent-based distributed information fusion parallel those exhibited by other systems, particularly query-based systems that incorporate in-network aggregation.

10.2.1.4. Multi-Agent Systems

Although the idea of using agents in wireless sensor networks is widely supported by the literature [27, 31, 41, 32, 36], the actual use of multi-agent systems in WSNs is not as widespread as one might expect [42]. There are of course the big problems of deployment, testing and debugging of such systems on what is essentially a distributed application with minimal interfaces to support user interaction. There have been real deployments using agents [20, 37, 43, 36] but multi-agent system deployments are still rare. [42] propose a methodology for multi-agent deployment. They start with a base station implementation for the agent that leads to a distributed implementation where agents are mapped to nodes, using either a one-to-one, many-to-one or one-to-many mapping. This mapping is used to model the interaction expected between agents and the base station, as well as iron out communication and interaction rules between agents. These first two stages are carried out on a device with more processing power like a laptop or desktop. In the third and final stage, the statements and rules governing agent behaviour are translated to the language of the WSN system hardware that will be hosting the agent. The result of the last stage is a topology in which the load of executing an algorithm is distributed among the agents, which can allow faster response times for complex algorithms. In addition, it can also result in a reduction in the number of transmissions in the network since an agent on a node or on a neighbouring node can now process packets that previously had to be routed to the base station for processing.

In the examples described above, agents are pieces of code injected into the system, gathering data, making decisions and migrating between nodes in some cases. Some research, however, has investigated the concept of having nodes themselves act as intelligent agents. [44] use intelligent agent theory to model sensor behaviour, viewing the WSN as a network of distributed autonomous nodes with the capability of making decisions. They also propose using artificial intelligence strategies on nodes to enable decision making, selecting a rule-based expert system that makes use of locally collected information on a node to make step-wise decisions. The concept, however, has not yet been implemented and deployed.

10.2.2. Agent-based Middleware

In general terms, middleware sits between the operating system and the application with its main function being to provide support for development, deployment, execution and maintenance of sensing applications [45]. There are a number of categories of middleware based on the objectives of the approach and the way in which the wireless sensor network is viewed. With some, the main goal is to provide a dynamic reprogramming capability while others seek to provide a platform-independent model on which sensor network programs can be written and executed. Other approaches try to provide a cross-layer management approach and provide functionality for manipulating other services like routing, etc. Information extraction, which is ultimately the goal of the sensor network applications built using these models, is supported by the services provided by the particular middleware approach. Approaches include the use of virtual machines which in many cases are application-specific but reduce the amount of code that has to be transmitted and therefore reduces the cost of communication, as well as modular programming approaches which allow easier and more efficient reprogramming of sensor nodes [46]. Some agent-based middleware will be described below.

10.2.2.1. Mate

Virtual machines (VMs) provide one middleware solution to the problems posed by wireless sensor networks. In WSNs the greatest drain on energy resources is the cost of communication, so if the traffic needed to reprogram motes could be reduced this could lead to a longer network lifetime as well as the ability to reprogram more frequently. Mate is a middleware approach that abstracts high-level operations into virtual machine bytecodes (see Fig. 10.1). As a result a VM program can be very short, bytes long instead of kilobytes. Mate, therefore, is a virtual machine designed for sensor networks [38] and forms the basis for many agent-based information-gathering applications.

Mate is a bytecode interpreter, running on TinyOS and provides a high-level interface for creating small (< 100 bytes) but complex programs. Mate is focused on energy conservation in transmission of code and breaks code into small sections called capsules (referred to as agents by some), each 24 bytes long. Programmers write applications and the system injects them into the network using certain algorithms that minimize energy and resource usage. The small size of the modules makes it easier to distribute into the network. Mate specifically targets the constraints of limited bandwidth and power by allowing adjustments or reprogramming of capsules before they are issued to the network.

Mate is a general architecture that allows a user to create a wide variety of virtual machines. A user builds a Mate VM in three steps [45]: a) the user selects a language that defines a set of VM bytecodes representing the basic functionality; b) the user selects the execution events. Each event has its own execution content which runs when an event is triggered; c) the user selects the primitives to be used. These are operations that provide functionality beyond that of the selected language.

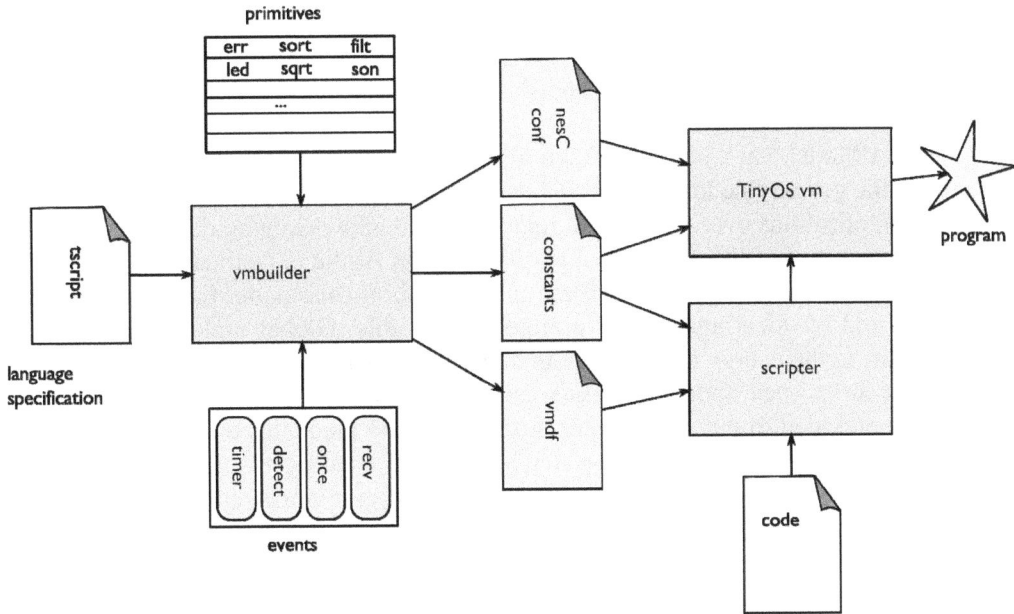

Fig. 10.1. Process of generating a program using Mate.
Revised image used with permission of P. Levis [47].

A set of files is generated after these steps have been executed. Some will build the VM that will run on the motes, others will build a scripter program with information on the language and primitives selected. The user can now construct programs in the scripter that will be compiled down to the VM-specific binary code. These bytecodes are then issued to the network and processed by a VM there, which will execute it.

To reprogram motes, the user needs only add one mote with the new capsule to the network. When a mote hears a new capsule it immediately starts forwarding it until every mote in the network has been updated.

The advantages of Mate, therefore, are the simplicity of programming for the sensor network as well as the ease with which the network can be reprogrammed. Mate's authors have identified future work as being the creation of propagation algorithms for larger programs, security and scripting languages that can be compiled to a Mate VM.

10.2.2.2. Agilla

Agilla is a mobile agent middleware based on Mate that allows quick deployment of adaptive applications [18], [48]. Instead of creating capsules that are flooded throughout the network, Agilla allows the user to create mobile agents that can be injected into the network. Agents are dynamic and intelligent, and can coordinate and collaborate locally or through remote invocation. These agents can clone themselves and move from node to node as they perform tasks. Agilla affords more flexibility because it allows applications to decide how agents in the system should migrate and spread in the network while maintaining their code and state.

In the Agilla Model, each node can have a maximum of four agents and addressing is done using the geographic location of nodes instead of IDs. Agilla can therefore be used in running applications over geographic regions and to allow geographic routing for any multi-hop interactions [18]. Inter-agent coordination in Agilla is facilitated using a tuple space and an acquaintance list, which are maintained on each node. Each node's tuple space is shared by local agents and remotely accessible. Global tuple spaces are not supported due to the energy and bandwidth constraints, rather separate local tuple spaces are maintained by each node and agents can access remote tuple spaces using special instructions provided in the middleware. Each node also maintains an acquaintance list containing the location of all one-hop neighbours and local agents can access it via a number of provided functions. Fig. 10.2 shows the programming model for the Agilla system.

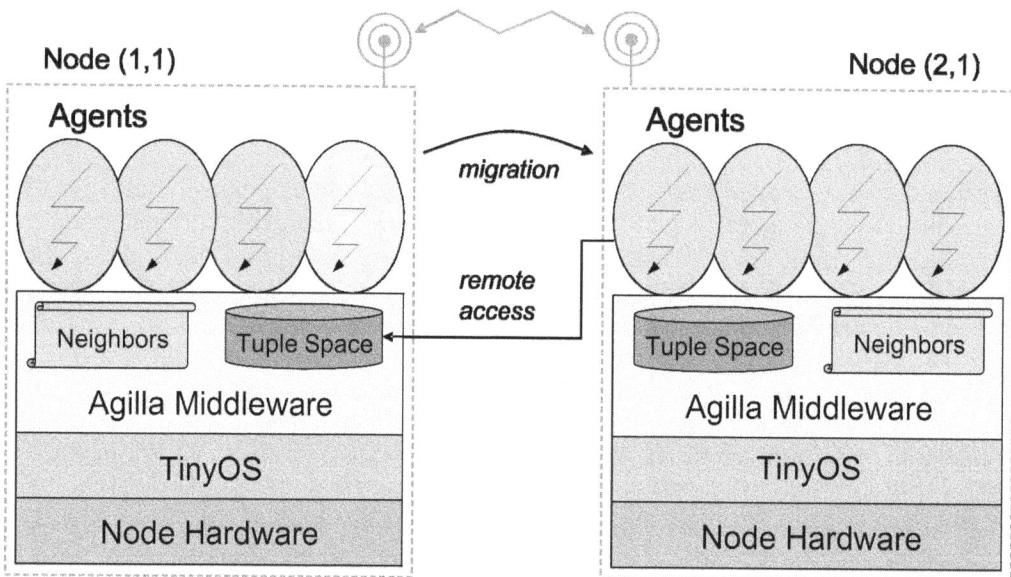

Fig. 10.2. The Agilla programming model. Original image used
with permission from C-L. Fok [48].

Fok [48] used Agilla to successfully deploy fire and tracker agents in a fire tracking application and demonstrated the reliability and efficiency of the exercise in this case study [18]. The scenario is as follows: a fire starts in a region of the network and as it spreads tracker agents crowd round it repeatedly cloning themselves until a perimeter is formed. Once that perimeter is formed a fire fighter is informed. The fire fighter injects a guidance agent who is responsible for guiding the fire fighter along a safe path to the fire. The study focused on the tracker agent, leaving the guidance agent and the development of a safe-route algorithm to future work.

Two types of fire modeling agents were used, static fire agents and dynamic fire agents. Static agents simply insert a fire tuple into the local tuple space and then repeatedly blink a red LED (visual indicator of network state). These agents are used to create fires of different shapes. The dynamic agent models a spreading fire. It inserts a fire tuple upon arrival at a node, blinks the red LED a number of times, clones itself onto a random non-burning neighbour and repeats the blinking. The cloning and blinking process is repeated until every node is on fire. The rate of spread can be controlled by the number of times a node blinks between cloning operations.

A fire-tracking agent is responsible for discovering a fire and forming a boundary around it. The tracker agent inserts a tracker tuple when it arrives at a node. Other tracker agents will use this tuple to check whether a neighbouring tracker agent is still present on the node. If the node a tracker agent is on catches on fire the tracker agent dies. A reaction is registered when this occurs, the tracker agent will turn off all LEDs, remove its tuple and stop functioning.

The life cycle of a tracker agent is as follows: it repeatedly checks whether any of its neighbours are on fire. If there are none, it moves to a random neighbour and repeats the check. If, however, a neighbour is on fire it enters a tracking mode and lights up its green LED and repeats the following. It determines the locations of all neighbours that are on fire and for each non-burning neighbour within a certain distance of the fire clones itself to those nodes. This process is repeated until the fire dies. Periodically checking for neighbours close to the fire allows the agent to adjust the perimeter. Once a fire was started, a tracker agent was injected next to it. This allowed the investigators to focus solely on the efficiency of perimeter formation and not on fire discovery as well.

The experiment was successful in a number of ways. First, it demonstrated the use of Agilla in deploying complex applications. Second, it showed that multiple applications could share the network at the same time (fire-simulation and tracker application). Third, it showed that mobile agents can be used to successfully program WSNs. Experiments on a 26-node MICA2 network showed that tracker agents each 101 bytes long were able to form a perimeter around a static as well as a dynamic fire. It was also evident that the efficiency of perimeter formation, in the case of static fires, depended to a great extent on the amount of agent parallelism in the system. The authors have identified areas of future work aimed at allowing new instructions to be added to the network after deployment and also allowing the instruction set to be customized depending on the agent being deployed. [18] describes the experiment and its results in greater detail.

10.2.2.3. Impala

Impala can be considered another mobile agent-based middleware approach. Its middleware architecture allows a user to create modular, adaptable and maintainable applications for WSNs. The adaptation facilities allow changes to be made at runtime in an effort to improve energy-efficiency, reliability and performance of running applications. The idea is that given the need for long-term management of a sensor application, a middleware layer that can update and adapt applications dynamically, switch to new protocols easily at runtime is a big advantage. Impala, therefore, is a middleware layer that acts as an operating system, resource manager and event filter upon which applications can be installed and executed [49]. It was specifically designed for a wildlife-monitoring project, ZebraNet, also detailed in [49].

The system architecture is depicted in Fig. 10.3. The architecture's upper layer contains all application protocols and programs for ZebraNet. In the lower layer are three middleware agents: Application Updater (AU), Application Adapter (AA) and Event Filter (EF). The AU receives and transmits software updates to the node via the wireless transceiver, the AA adapts the protocols to different conditions at runtime in a bid to improve energy-efficiency and performance and the EF collects and sends events to the system for processing.

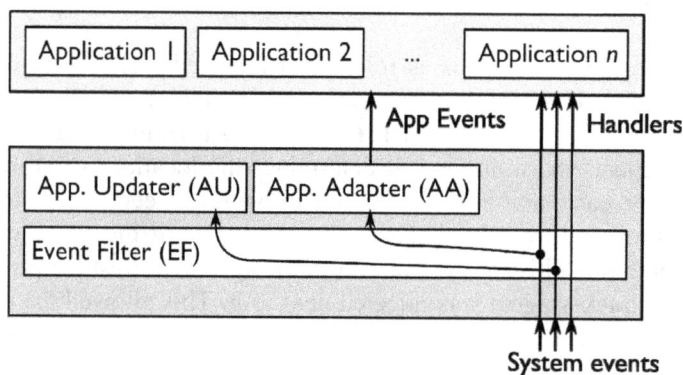

Fig. 10.3. Architecture of the Impala system [49].

A program module is compiled into binary code before it is injected into the network. Linking is performed by the updater on every node and when all modules in an update have been received the module is linked to the main program. Modules are linked independently. Agents work autonomously making the network more fault-tolerant and better able to self-organize. It is not suitable, however, for resource-limited systems.

Impala uses an event-based programming model where the AA, AU and applications are programmed as a set of event handlers called by the EF when appropriate events occur. The EF controls various operations and starts a number of processing chains including timer, send, done data and device events. Based on different scenarios such as improving energy efficiency or performance, the AA will adapt a particular application. The AU is then responsible for effecting software updates while being cognizant of such issues as resource or bandwidth constraints and node mobility. If, for example, the radio transceiver on a node was to fail, the EF would call on an appropriate handler in the AA. The AA would determine the impact of that failure and decide whether an updated or new application that takes into account these issues should be issued to the network. The AU would then be responsible for carrying out the software update itself.

10.2.2.4. SensorWare

Although not strictly considered mobile agent architecture by some, SensorWare [50] is based on a scriptable lightweight run-time environment suited for nodes with energy and memory constraints. It is formally called an active sensor framework (ASF) where mobile scripts less than 180 Kbytes are used to make the network programmable and open to users.

The difficulty in designing an ASF has been identified as determining how to accurately define the abstraction of the run-time environment. Any abstraction would have to result in compact code, resource sharing, multiple users being able to access the system and portability to different platforms. With these in mind they designed a node abstraction that allows multiple users access to the modules on a node while still being able to create new modules.

The authors focused on defining a framework that allowed the description and deployment of distributed algorithms for wireless ad-hoc sensor networks. SensorWare's language model can effectively express these algorithms while simultaneously shielding the user from low level details while allowing sharing of node resources among several concurrently running applications or many users. The language model focuses on the properties of efficient algorithms for sensor networks while application development for a real network is underway. Fig. 10.4 shows the architecture of the SensorWare system.

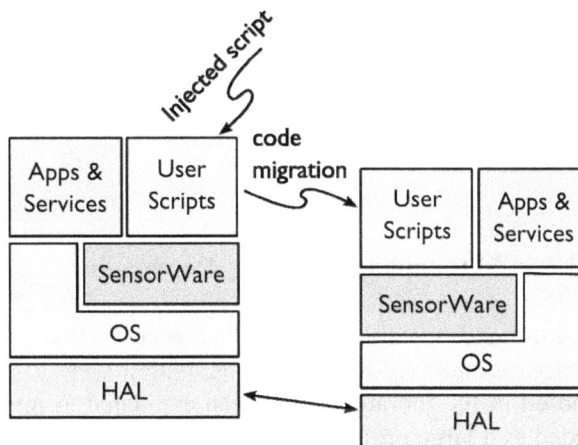

Fig. 10.4. Architecture of sensor nodes running SensorWare [50].

277

SensorWare is also quite effective at dynamically deploying the distributed algorithms represented in the language. Given that the nodes are memory-constrained and cannot store every application in local memory the method of dynamic deployment has to be effective. Rather than each node being programmed by the user the SensorWare approach gets the nodes themselves to program their neighbours. The user injects the program into the network and the program is autonomously distributed to the nodes that should receive it. SensorWare is described in detail in [51].

10.2.2.5. TinyLIME

Unlike many of the examples described above that use a single controlling processor or sink node for data collection and processing, TinyLIME instead promotes a distributed approach. Multiple mobile clients or agents are distributed with the ability to receive data only from the sensors that they are directly connected to. The clients can share this locally collected data through their wireless interconnections [52]. TinyLIME is built on the original Linda in a Mobile Environment (LIME) architecture [53] and deployed on Crossbow motes. All base stations run an instance of moteAgent, a LIME agent that among other functions maintains data freshness and historical information on recent sensed values. TinyLIME also supports data aggregation both over values sensed by multiple sensors and those sensed by a single sensor.

10.2.3. Remarks

With respect to agent-based approaches to information extraction, it is forthcoming from the survey that many of the described techniques and architectures supporting them have not yet been sufficiently developed and evaluated other than theoretically and are, hence, far from being deployable (or even successfully implementable) on resource constrained nodes such as those found most commonly in WSN applications.

It is, however, encouraging to see the wealth of research and proposals related to agent, multi-agent and mobile agent systems, leading, no doubt, in the future, to energy efficient, hardware independent, robust methods of relaying information drawn from large scale WSNs, to the user.

10.3. Query-based and Macroprogramming Approaches

Having examined agent-based approaches in the last section, this section surveys both query-based and macroprogramming-based systems, with a view to first identify which of the methods proposed in the literature have been evaluated at implementation level. (This analysis is needed as a large proportion of the research effort is at theoretical level and not readily suitable for practical deployments.) Second, looking out to common constraints shared by WSN systems (such as limited energy and communication resources for example), it is aimed here to identify which of the methods proposed in the literature make use of or support in-network information extraction. (This is important, as approaches that utilize in-network processing like aggregation and filtering have been shown to be more energy efficient and therefore more desirable).

10.3.1. Query-based Information Extraction

Many researchers have taken the view that the sensor network can be considered a database from which information has to be requested and retrieved. The reasons for this are as follows: first, the network is a collection of data albeit that data is dispersed across multiple nodes. Second, data needs to be retrieved from the network, and like a conventional database one needs to be able to query the network for what is desired and get a response. Third, the nodes in the network are of interest primarily in as far as the data they generate, this can also be termed a data-centric view of the WSN.

Many of the sensor network databases use a query language similar to the Structured Query Language (SQL) for constructing queries.

SQL is the most popular language used for creating, modifying and retrieving information from relational databases. It consists of a number of clauses including SELECT - FROM - WHERE and GROUPBY clauses that allow selection, join, projection and aggregation respectively. SQL-type queries constructed for WSNs are quite similar to those of traditional SQL both syntactically and semantically. The FROM clause, for example, can be used to either specify sensors or data stored in tables. The GROUPBY clause allows a SELECT statement to collect data across multiple records and group the results by one or more attribute values. In the WSN querying context, for example, records can be grouped by the node's id or by locations, etc. if these are attributes retrieved in the query.

In traditional database systems, data is accessed via an application or directly via a front end. Such applications insulate the user from the inner workings of the database system while allowing easy and meaningful queries to be applied to it in retrieving the necessary information. In a similar way, the sensor network can also be interfaced using a suitable application. The application would act as a query processor-type interface to the network taking into account the power and computation resource limitations on the devices in the network. According to [54] the challenge in query processing is not in processing data as quickly as possible but rather in figuring out a way to effectively respond to queries while transmitting as little data as possible. Several researchers have noted the benefits of this query-processor interface approach.

[55] describe a view of the network as a database with two methods for processing queries issued to the network: warehousing and distributed approaches. With the warehousing approach the processing of a query is separated from the interaction with the network itself. Essentially, the data is extracted and stored at a centralized location for processing. Queries in this case are pre-defined and usually ask for aggregate historical data, for example, "for each rainfall sensor, display the average level of

rainfall for 1999" [57]. This model really mirrors a client/server approach to data collection as is found in a traditional distributed network [56].

There are two main disadvantages to this approach, however. First, it is not well suited to requests for continuous data, either because it is not possible to retrieve the required data from the node or because data is not retrieved frequently enough within the network to be able to answer a query [57]. Second is the high-energy consumption that occurs when transmitting large quantities of data from the node to the centralized database and even more so in the case of a continuous stream of data, making the process very energy-inefficient [57, 55].

Alternatively, [57, 55] propose a view of the network as a set of distributed databases where the composition of the query in effect determines what data is retrieved. Such queries have the advantages of efficiency because only what is required is actually retrieved, as well as flexibility since various queries can be formulated depending on what information is needed. This approach also takes advantage of the processing power on the node itself and where possible processes queries or parts of queries on them.

[58] note that data generation and routing in the sensor network can be seen as similar to the concepts of data storage and query processing in traditional databases and so also view the sensor network as a database. They examine the challenges in implementing the network as such and identify the need for robustness of data access given the possibility of node failure and noise that can affect readings. The idea is that by creating a standard querying interface that allows less restrictive query semantics, approximate results can be obtained which can help in producing an energy-efficient implementation of the "sensornet database application". [58] are also of the view that query optimization is closely linked to routing and they propose an adaptive approach that takes into account the volatile nature of data and communication in the network.

Several major query-based systems are described in detail below.

10.3.1.1. COUGAR

The COUGAR approach has been proposed in [57, 55, 58] and more fully described in [59]. It builds on the Cornell PREDATOR object-relational database system [60] and models each sensor in the network as an abstract data type while signal-processing functions are modelled as abstract functions returning sensor data. The network is viewed as a system where each node is a "mini database" holding part of the data contained by the whole. COUGAR allows the issuing of declarative queries for information requests and uses a query optimizer to plan an in-network processing strategy. Queries in Cougar are SQL-based and are posed to a virtual table called sensors with one row per node per instant in time and one column for every attribute. (The same model is used in TinyDB and is described in Section 10.3.1.5.).

A query proxy layer is placed on each node and interacts with both the routing and application layers in the system. A query optimizer is placed on a gateway node and distributes the query processing plans after it receives the queries from the user. COUGAR uses catalogue information and query specification to create the query plan that describes not only the flow of data between sensors but also how information is to be computed on individual sensors. Once the plan is complete it is sent out to all relevant nodes. During execution, data is returned to the gateway node. In addition, a query proxy can inject queries into the network from arbitrary or specified nodes as it performs high-level services.

Fig. 10.5 shows how the system executes a simple aggregate query. COUGAR is especially useful for executing continuous queries.

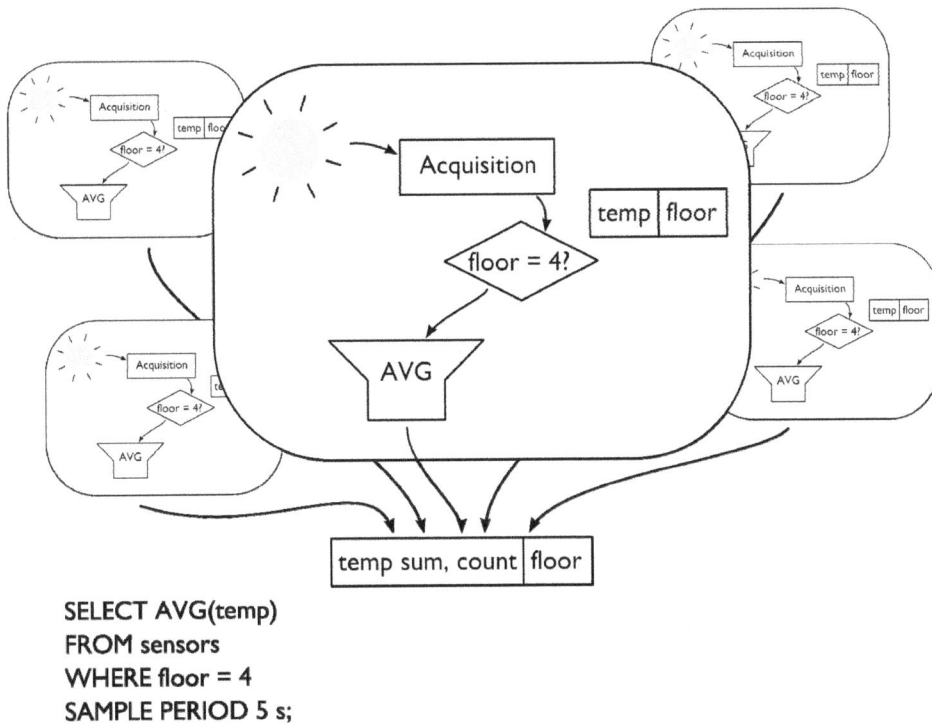

SELECT AVG(temp)
FROM sensors
WHERE floor = 4
SAMPLE PERIOD 5 s;

Fig. 10.5. A sensor network running the Cougar system executes a simple aggregate query [61].

A big advantage of COUGAR is that it shields the user from needing to have any knowledge of the underlying network, or how data is generated and processed. A second advantage is that it increases efficiency in terms of energy consumption since in-network processing is allowed thereby reducing the amount of data that has to be sent back to the gateway node [62]. There are other architectures that promote this distributed approach to query processing, for example, IrisNet, described in [63], SINA and TinyDB. Both TinyDB and Cougar follow a very similar query processing model and address the resolution of both simple and aggregate queries. Neither, however, facilitates the processing of other types of complex queries such as spatio-temporal or queries that target only a subset of the network.

10.3.1.2. Sensor Information Networking Architecture (SINA)

Like COUGAR, the SINA architecture described in [64] promotes a distributed database query interface that emphasizes in-network processing and reduction in power consumption in the network. It views the network as a collection of datasheets (a spreadsheet) with each datasheet containing the attributes relevant to the node it describes. A cell in this scheme represents an attribute and the collection of datasheets form what is called an associative spreadsheet or spreadsheet database representing the network. Each cell is therefore referred to using an attribute-based naming method.

SINA also makes use of clustering of low-level information in the network to increase efficiency. Clusters are formed from aggregations of sensors based on their power levels and proximity. Recursive aggregation can also be used to produce a "hierarchy of clusters". A cluster head within a cluster can then be elected to perform key functions like information filtering, fusion as well as aggregation [64]. Fig. 10.6 shows a model of a network with the SINA middleware.

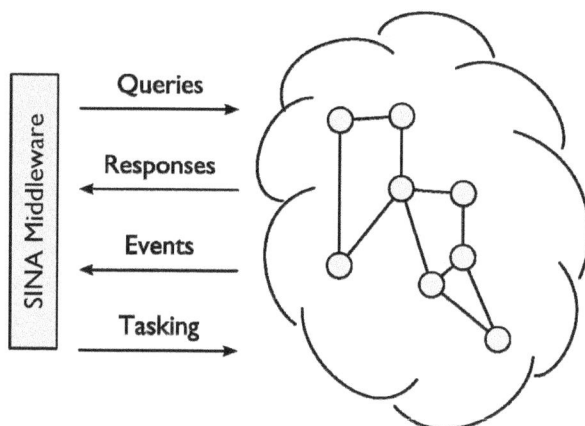

Fig. 10.6. Model of a sensor network and SINA middleware [64].

SINA uses the sensor programming language Sensor Query and Tasking Language (SQTL), which serves as the programming interface between applications and the middleware. SQTL is a procedural scripting language built from Structured Query Language (SQL) and messages written in the language can be interpreted and executed by any node in the network, although messages can target specific nodes if required. SQTL also supports the generation of information as a result of the occurrence of, or change in, some phenomena (events). Three types of events are supported by SQTL. The first is an event generated when a node receives a message, the second is an event triggered periodically by a timer, and the third is an event triggered by the expiration of a timer.

SINA modules run on each node in the network and allow querying as well as monitoring of events. Nodes are aggregated to form clusters and elected cluster heads perform filtering, fusion and aggregation tasks. The user issues a query and the SINA architecture selects the most suitable methods for information gathering and distribution based on the type of query issued as well as the current status of the network. A node receiving the query will interpret it and request information from neighbouring nodes in evaluating it.

A number of information gathering mechanisms are employed to help reduce resource consumption and increase response quality, as follows:

1. Sampling Operation - for some applications a query may need to target the entire network in eliciting information of interest. Responses from all nodes in a network may result in response implosion [65], however, the effect can be reduced if nodes are able to make decisions on whether they should respond or not. SINA implements this "decision-making" capability by assigning each node a response probability that determines whether they respond or not.

2. Self-Orchestrated Operation - in networks with a small number of nodes it is sometimes necessary that all nodes respond in order to improve the accuracy of a result. SINA uses what is called self-orchestration in an effort to reduce the problem of response implosion mentioned before. Here each node suspends its response transmission for a particular period of time to help reduce the likelihood of collisions occurring as could happen if nodes transmitted simultaneously.

3. Diffused Computation Operation - this method is used to implement aggregation functionality in the network. Each node is first assumed to have knowledge of its immediate neighbours and aggregation logic already programmed into the SQTL scripts is used to perform aggregation before routing the result to a designated node.

SINA is therefore particularly suited for aggregate queries for replicated data in a cluster-based network configuration but does not address other complex queries like spatial or temporal queries which may also be suited for processing in a cluster-based configuration.

10.3.1.3. Data Service Middleware (DSWare)

Like SINA and Cougar, DSWare [66] is a data-centric middleware approach to providing information retrieval services within a WSN. DSWare, however, is aimed at handling real-time events and integrates a number of real-time data services while providing a database-like abstraction. In this way it can be considered another database-like approach adapted to sensor networks. It provides services for reliable data-centric storage and implementing alternative methods for improving real-time execution performance, as well as reliable data aggregating and decreased communication that is aimed at conserving the limited resources available within the network. Its architecture separates the routing layer from the DSWare and network layers as DSWare provides components for improving power-awareness and real-time awareness of routing protocols.

There are six services/components available in this middleware approach (Fig. 10.7):

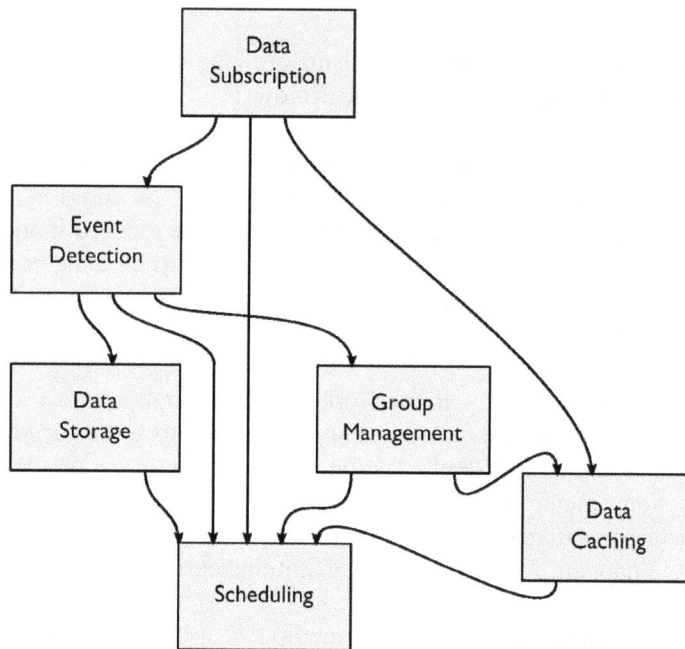

Fig. 10.7. The DSWare Framework [66].

1. Data Storage - this component allows storage of location-linked data so that future queries for similar data would not mean having the query re-issued or flooded through the entire network. In DSWare data lookup is implemented using a hashing function that maps data to physical storage nodes via a unique identifier while robustness is facilitated by using replicating needed data in several physical nodes which can then be mapped to a single logical node. Queries targeting that data can be directed to any one of these nodes in an effort to avoid collision and sustained overload on a single node.

2. Data Caching - this service provides copies of data that is requested often and spreads it through the network using the routing path. This serves to accelerate execution of the query while reducing communication aimed at accessing data for query processing.

3. Group Management - this component facilitates cooperation among nodes to accomplish more complex tasks. In some cases, for example, requested information requires multiple sensors combining their values to calculate a result. The group facility therefore can create a group when a query is issued. Nodes receiving some information on this group can decide whether they match the particular criteria described for that group. The group then manages the values of relevant sensors in accomplishing the given task after which the group is dissolved. In some cases, the query itself may expire before the group can be dissolved.

4. Event Detection - DSWare is based on event detection, an event being an activity that can be monitored or detected in the environment and is of interest to the application. An event is pre-registered depending on the application and can be classified as either atomic events (events that can be determined based on the observation made by sensor) or compound events (events that cannot be determined directly but rather need to be inferred from other atomic "subevents"). Like COUGAR, DSWare uses statements in a SQL-similar language for registering and cancelling events. After parsing the statement defining the event, DSWare generates the query execution plan and calls on the methods required to register, execute and cancel the event.

5. Data Subscription - this service is used to aid in more efficient data dissemination. If multiple nodes request the same data, the subscription service puts copies at intermediate nodes that can then be accessed by the requestor nodes. This helps in reducing communication and helps conserve resources within the network.

6. Scheduling - this component provides a real-time scheduling mechanism as the default method with the option of adding on an energy-aware mechanism if the first has already been successfully implemented. All components of DSWare are scheduled using this component.

DSWare, because of its emphasis on providing event detection services and the mechanisms it provides for data caching, group management and data subscription, is particularly suited to aggregate queries as well as queries for replicated data. These aggregate queries, however, are simple aggregates over an attribute and do not include more complex aggregate queries where data from multiple clusters, for example, could be combined to resolve such queries.

10.3.1.4. Framework in Java for Operators on Remote Data Streams (Fjords)

The Fjords architecture presented in [67] shows how processing multiple queries can be managed over multiple sensors while allowing more efficient resource usage and keeping query throughput high. The focus in the research is on creating the underlying architecture that will support the processing of multiple queries over sensor data streams.

The architecture comprises operators configured for streaming data and Fjord operators called sensor-proxies whose function is to serve as a mediator between the query processing plan and sensors. Data is allowed to flow into the Fjord from the sensor and then pushed into the query operator. Query operators are not active pulling in data but rather wait till data is sent to them from the sensors. The fjord, in addition to combining streaming and fixed data also processes multiple queries and combines them into a single plan.

The Fjords architecture is therefore suited to continuous queries and addresses a number of issues important to query processing. It does this in two ways. First, by using proxies, non-blocking operators and query plans which allow streaming data to be pushed through operators that pull from data sources, it allows merging of the both stream data and local data. Second, by letting proxies serve as mediators between query plans and sensors it allows query processing while at the same time taking into account power, communication and processor restrictions in the network.

The work here is similar to the COUGAR project in that they both focus on processing of streams of sensor data, however, COUGAR although it does incorporate in-network processing in the form of data aggregation, it does not focus specifically on energy efficiency and the resource constraints on sensor devices but rather on modeling the streams of data using abstract types. Fjords, however, look more closely into improving efficiency for processing of data streams. Although a viable approach for continuous queries it does not address complex queries or even aggregate type queries to any great extent and is more concerned with managing simple query processing over multiple sensor data streams.

10.3.1.5. TinyDB

All the approaches to query processing described above have one thing in common: they view query processing in the network as a modified version of that in traditional databases. In essence, each node in the network is seen as a generator of named data against which queries can be issued. [68] describe a distributed database approach that attempts to improve energy consumption by applying varying techniques for aggregation and optimization within the network. This aggregation service is called Tiny Aggregation (TAG) and is designed specifically for TinyOS [69] motes. TinyDB has been one of the more widely used and popular query processing systems and considerable work has gone into extending its capabilities both in terms of the types of queries that can be issued as well as improving its operation by integrating different routing protocols. TinyDiffusion [70] is one such example. It has been widely used in a number of real-life deployments as well as inspired the research into other query processing systems and hence will be discussed in full in this section.

TinyDB uses acquisitional query processing (ACQP) as described in [71] for in-network query processing. In terms of practical applications ACQP focuses on the location and cost of accessing data as a way of reducing power consumption. There is no assumption made that data exists at a particular location rather queries are only issued where it is known that data exists.

TinyDB can therefore be considered an ACQP engine, a distributed query processor that runs on all nodes in the network. TinyDB was designed to be deployed on the Berkeley Mica mote platform, which runs the TinyOS operating system (Mica motes are arguably one of the most popular of WSN hardware platforms.)

Query Syntax: Queries in TinyDB are quite similar to SQL with the FROM clause being used to either specify sensors or data stored in tables (called materialization points).

Sensor tuples are stored in a table (sensors) with one row per node per instant in time and one column representing each attribute. An attribute could be light, temperature, sound, etc. Records in this table are stored for a short period usually and only "acquired" when needed to resolve a query. An example of a query could be:

> SELECT nodeid , light
>
> FROM sensors
>
> SAMPLE PERIOD 1s FOR 10s

This query states that each node should return its own id and light reading from the sensors table once per second for ten seconds. The results are either returned to the user or logged. The sample period is used to indicate when data collection should begin. The sensors table is an unbounded, continuous data stream of values and can only be sorted or used in joins via a specified window.

Time Synchronization: TinyDB adopts the Timing-Sync protocol presented by [72] which attempts to provide network-wide time synchronization in the sensor network. The algorithm proceeds in two stages: the Level Discovery Phase establishes a hierarchical topology for the network and occurs at the point of network deployment. In the second phase, the Synchronization Phase, a two-way message is used to synchronize two nodes. The time synchronization process begins with the root node first sending a time-sync packet after which receiving nodes initiate the message exchange with the root node as described above in the synchronization phase. Once the nodes receive acknowledgment from the root node they adjust their clocks to that of the root node. This process is carried out until all nodes are synchronized with the root node.

Aggregation Queries: TinyDB supports aggregation of data, which in turn allows reduction of data prior to transmission. The main difference between such queries in TinyDB and SQL is that the output to a query is a stream of values for the former while it is an aggregate value for TinyDB. The aggregate operators allowed are the same as that allowed in a normal relational database system, for example AVG, MAX, MIN which compute the average, maximum value and minimum values respectively for the attribute they are applied over.

Temporal Aggregates: The TinyDB system acknowledges that aggregates over values within a common sample interval are not the only type of aggregate values that a user may want. In some cases a user will need to perform some type of temporal calculation. For example, users of a habitat monitoring system may want to monitor the temperature as a frost moves through a geographical area. This can be done by measuring the maximum temperature over a period of time and reporting that temperature at set intervals. In TinyDB this can be implemented as a sliding window query, for example:

SELECT WINAVG(TEMP, 60s, 5s)

FROM SENSORS

SAMPLE PERIOD 1s

The above query would report the average temperature over the last 60 seconds every 5 seconds with a sampling rate of once per second.

Event-Based Queries: TinyDB also supports event-based queries which are generated either by another query or by some part of the operating system, the language provides an EVENT clause for that purpose and the code that generates the event is compiled into the node [73].

Continuous Queries: Continuous or lifetime-based queries are also supported using the LIFETIME clause as well as nested queries and offline delivery queries which allow data to be logged for non-real time delivery. Materialization points are used to implement offline logging.

Query Optimization: Queries in TinyDB are first parsed at the base station and then optimized to select the ordering of joins, selections and sampling. The optimizer determines the lowest power consumption that takes into account not only the cost of data acquisition but also processing and radio communication. Data acquisition, however, is the main source as it includes actions like sampling of sensors and transmission of query results. Query optimization, therefore, is focused on reducing the number and cost of data acquisition. Once a query has been optimized it is issued into the network in a binary format, where it is instantiated and executed.

Query Dissemination and Routing: TinyDB uses semantic routing trees (SRT) to help determine whether a query is forwarded or not.

Query Processing: after a query has been optimized and disseminated the query processor executes it. Execution follows a simple sequence of sleep, sampling, processing and delivery. Nodes sleep for as much time in an epoch as possible to conserve power and wake up to sample sensors and deliver results. The nodes are synchronized so parent nodes can ensure that child nodes are awake to process messages. Results are sampled and filtered based on the plan provided by the optimizer and then routed to the aggregation and join operators further up the query plan.

In summary, the TinyDB approach does allow quite a good deal of flexibility in the types of queries that can be processed and shows increased efficiency in terms of power consumption using a number of query optimization techniques. Some further work needs to be done on optimization of multiple queries, as these are not addressed as well as implementing more sophisticated data delivery prioritization techniques. The authors stress, however, that with TinyDB using the ACQP method it is essential that data delivery, query language and optimization techniques take into account the underlying hardware capability and semantics in order to increase the likelihood of successful deployments.

Add-ons aimed at simplifying the deployment and development of applications for wireless sensor networks, are also being developed. The Tiny Application Sensor Kit (TASK) is built on top of TinyDB for that purpose [74]. This kit contains a relational database used for storage of sensor readings, a server to act as a proxy for the network on the Internet as well as a front-end that facilitates data selection and recording. Other architectures with a similar purpose also exist, for example jWebDust described in [75].

Although useful, like the other query-based approaches in use TinyDB has its drawbacks. The queries that can be issued are limited to those that target "low level" sensed values on nodes in the network or those requiring the computation of simple aggregates like average, maximum, and minimum over some attribute. It does not allow the creation of nested queries or even nested, aggregate queries. In addition, the query language used cannot easily express spatio-temporal characteristics which are an important aspect of the data generated in WSNs. Therefore, there is still room for improvement and expansion in a query processing system that follows the same model but able to issue and process more powerful queries.

10.3.1.6. Active Query Forwarding (ACQUIRE)

Common to all of the query-based approaches described above is the clear distinction between the dissemination phase when the query is sent out and the response stage when results are sent back to the querying node. ACQUIRE described in [76], does not distinguish between these two stages but rather issues what is called an active query.

An active query, in addition to the simple queries for an attribute or attributes, also includes nested queries. Each subquery can be a query of interest in a different variable or value. The active query is propagated through the network node to node. At any point in time, an active node (the node carrying the active query) uses data from a set of neighbouring nodes within a look-ahead of x number of hops and tries to resolve the query or part of the query if it can. The number of hops used can vary, but has been analyzed experimentally to determine what the most effective value is. The experiments in effect measured the average number of nodes from which new information is obtained during query forwarding based on different values for x and set that as the effective look-ahead. Fig. 10.8 gives an illustration of the ACQUIRE mechanism at work.

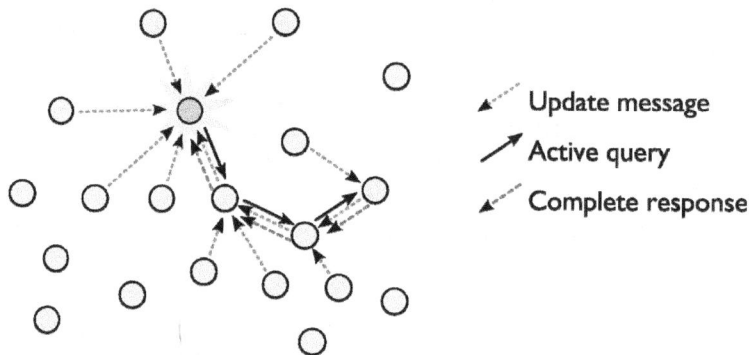

Fig. 10.8. Illustration of the ACQUIRE mechanism given a one hop look ahead. At each stage of the active query propagation the node carrying the query uses information gained to partially resolve the query [76].

Analysis of ACQUIRE has been restricted to networks exhibiting a regular grid topology. Given these restrictions, however, ACQUIRE worked better than flooding-based querying (FBQ) mechanisms like Directed Diffusion as well as had a 60-75 % savings in energy consumption when compared to Expanding Ring Search (ERS) [76]. ACQUIRE is most suited to complex one-shot queries for replicated data and needs to be analyzed on more realistic networks where the topology is irregular or dynamic. The reported evaluation of ACQUIRE is further limited as it uses mathematical modeling to analyze the performance of the mechanism in terms of energy costs. It does not address implementation. Further, the complex queries although including nested queries do not include queries where spatial or temporal characteristics may be of interest.

10.3.2. Macroprogramming Approaches

In the applicative query approaches described above, the user must have some knowledge of what type of queries or operations are to be issued as well as what raw data needs to be targeted in order to retrieve the information. Although information is retrieved, in many cases, the user still has to make sense of what it means and whether it is relevant or not. For example, a user could issue a query in TinyDB requesting the average temperature over the network for a particular period of time. Once the results are received, however, the user would then have to look at this "information" in deciding whether it indicates that a fire has started. If it does, the user then has to decide what to do next, for instance, issue additional queries. The ability to create a "program" where events are anticipated and response (further queries) issued is not allowed in the system.

Macroprogramming approaches to information extraction aim to treat the wireless sensor network as a whole by directly specifying global behaviour instead of programming individual nodes. It attempts to address the issues raised above in the form of a global program that is defined in a high level language that can capture operations going on in the network at a global level. A number of systems have been developed which incorporate global abstractions and in some cases couple them with node-level abstractions in creating useful macroprograms. In this Section some of the more popular macroprogramming systems will be examined.

10.3.2.1. Node-level Abstractions

The underlying principle of macroprogramming is the creation of a global program that can capture network level operations. These macroprograms ultimately have to target node level data in order to function globally. A number of node-level abstractions have been proposed in the literature for modeling node level data and are used by macroprograms to target the data within the network. The data is not accessed on a low level, node-by-node basis but through these abstractions that provide access via logical "collections" of nodes. The logical node-level abstractions proposed are usually based on one or more attributes within the WSN. This can be a dynamic attribute like a sensed value (temperature, for example) or a static attribute like geographic location (in the case of stationary nodes). Following is a description of some of the more popular node-level abstractions put forward in the literature.

Hood: Whitehouse et al. [77] describe a neighbourhood abstraction called a hood (which is defined by a set of criteria for selecting neighbours) and a set of variables to be shared within the hood. Hoods are mainly defined geographically with nodes being included in hoods based on the number of hops from a node. For example, a hood could be a one-hop or two-hop neighbourhood, over which temperature readings are shared or a one-hop neighbourhood over light and temperature readings. A node can therefore define multiple neighbourhoods over different variables.

Whitehouse et al. use a broadcast/filter mechanism to allow data sharing and neighbourhood discovery. Shared attributes are broadcast and receiving nodes cache any attributes of interest from valuable neighbours. A neighbour's value is based on how the neighbourhood has been defined. For example, a node may define a routing neighbourhood that caches location information of valuable routing nodes [77]. Once a neighbourhood has been defined and attributes broadcast, interested observers add the broadcasting node to its neighbour list and cache the attribute or attributes of interest. Fig. 10.9 shows a model for an application with two neighbourhoods and two shared attributes.

Abstract Regions and Region Streams: Welsh and Mainland [78] propose abstract regions, spatial operators used to hold local communications in regions which can be defined in terms of radio connectivity or geographic location. A region in that sense would serve as an identifier for neighbouring nodes who can then share data, identify each other, and reduce data among them. Nodes could then be manipulated via their common interface. An abstract region could therefore be defined as a relationship between a node and its neighbours and the main aim of this work is to move away from the need to construct low-level mechanisms for routing and data collection and instead focus on a higher level interface that is flexible enough to allow a user to implement queries. Such a system would not, for instance, have to consider routing, data dissemination or state management in order to function effectively, however it would still allow a programmer to create applications that can adjust to changing network conditions as well as adjust energy consumption to improve accuracy and latency.

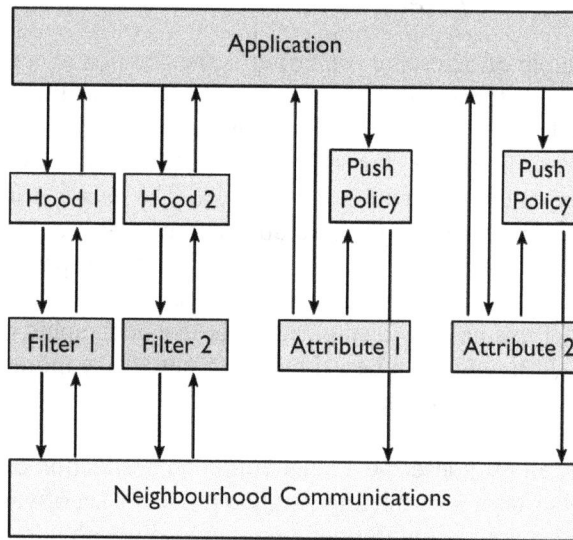

Fig. 10.9. Component model for an application with two neighbourhoods and two shared attributes [77].

10.3.2.2. Semantic Streams

Semantic streams, a system proposed by [79] allows a user to issue queries over semantic values directly without identifying what data should be targeted or what operations should be used on that data. It allows the user to interact with the sensor network using declarative statements, for example, "I want the ratio of cars to trucks in the parking garage". There is no need to construct code to actually check the data for the existence of cars or trucks rather components referred to as inference units are used to facilitate interpretation of sensor data. This is in direct contrast to other approaches that issue queries over raw sensor data like TinyDB and Cougar, and is of interest as it is similar in principle to the work proposed on complex queries in this research.

The Programming Model: The semantic streams framework uses the semantic services programming model, which contains two main elements, inference units and event streams. An event stream is simply a flow of asynchronous events, each of which represents some real-world entity (for example, in the car park query given earlier, a car detection has properties such as the location at which it was detected, its speed and direction). An inference unit is a process that operates on an event stream. It infers semantic information about its environment and either adds it as a property to an existing event stream or generates a new one. Fig. 10.10 shows the semantic streams model within its service-based architecture.

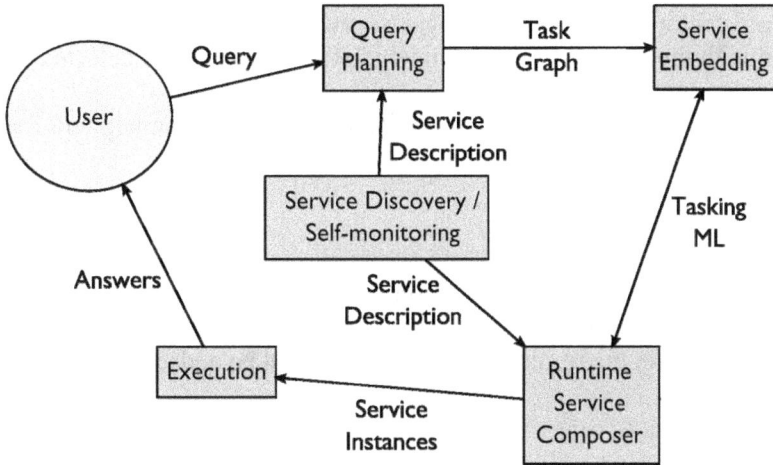

Fig. 10.10. The semantic streams query processing model [79].

As a stream moves from a sensor through inference units its events acquire new semantic properties. A more detailed example, a vehicle detector service, follows.

The Query Language: Each inference unit is specified using a first-order logic description of the semantic information it needs in its input stream and that it adds to its output streams as well as any relationships between the two. A mark-up language (the Semantic Streams mark-up and query language) is used to construct a logical description of that semantic information. A number of predicates are used to describe sensors and services: the sensor predicate defines the type and location of each sensor; the service, needs and creates predicates describe a service, the semantic information that it needs and creates. In query processing these are treated as rules and their pre- and post-conditions; the stream, isa and property predicates describe an event stream and the type and property of its events. An example of a sensor declaration is as follows:

sensor(magnetometer, [[60,0,0], [70,10,10]])

This defines a magnetometer that covers a three-dimensional cube defined by the pair of 3-D coordinates.

An example of a service would be:

service(magVehicleDetectionService,

needs(sensor(magnetometer, R)),

creates(stream(X), isa(X, vehicle),

property(X, T, time), property(X, R, region)))

This vehicle detector service uses a magnetometer to detect vehicles and generates an event stream with the time and location in which the vehicles were detected.

Once a set of sensors and services has been declared the user can begin to issue queries. An example of a simple query would be:

stream(X), isa(X, vehicle)

This query would result in true if a set of services could be composed to generate events X that are known to be vehicles. All known possible service compositions would be generated. To constrain the result more predicates could be added. For instance,

stream(X), isa(X, car)

property(X, [[10,0,0], [30,20,20]], region)

This would restrict the result to cars found in the region described by the pair of 3-D coordinates given.

Query Processing: An inference engine is used to determine which sensors and services will provide the required semantic information. The inference engine employs the backward-chaining algorithm [80] using the pre-conditions and post-conditions of the services. An attempt is made to match each element of the query to the post-condition of the service; if this is successful the pre-conditions are added to the query. Once all pre-conditions have been matched to actual sensor declarations (without pre-conditions) the process terminates. In other words, the process terminates once physical sensors have met all inference units' requirements. A description of a real-world application of the semantic streams framework is described fully in [79].

In the semantic model described, all services and their descriptions are maintained in a central repository or query server. Resource usage is optimized because the inference engine reuses services when possible and the query language used allows great flexibility in how the query processor executes queries. The language also allows the user to specify constraints. When a user issues a query as an event stream the query-planning engine generates a task graph, which is then assigned to a set of nodes in the network. The service runtime on each node then accepts the graphs and instantiates services as required, resolves any conflicts between tasks and available resources and then executes the query.

One major limitation of the system is the semantic mark-up language used. Although suitable for simple examples it cannot capture more complex applications. Another drawback of the system is that the query processor cannot reason at runtime and so two applications that have to access the same device, for instance, would not be able to run simultaneously. It does not address sensor network issues that are not semantic transformations. For example, routing data between nodes in a sensor network would not change the semantics of the data being routed. Semantic Streams cannot differentiate between different routing algorithms and so cannot take advantage of their features if needed.

10.3.2.3. The Regiment Macroprogramming System

Regiment, proposed by [81] is a functional programming language for sensor networks. Its data model is based on the concept of region streams, where sensor nodes are represented as streams of data that can be grouped into regions for the purpose of in-network aggregation or event detection. Collections of nodes, therefore, can be represented both spatially and temporally. A region stream is used to describe a collection of nodes with a logical, topological or geographical relationship. For example, a programmer may be interested in nodes that fall within certain 3D coordinates. The corresponding region stream would represent all sensor values for those nodes.

Operations allowed on region streams include fold, which aggregates values across nodes in the region to an anchor node, and map which applies a function over all values in a region stream. A fold requires inter-node communication while a map does not. Regiment, as is the case with other functional programming languages, allows functions to take functions as arguments.

The Regiment compiler converts the high level program into node-level code that targets an abstract machine model called the token machine. This is an intermediate language that links token handlers to a token name which represents a task to be executed by a sensor node once it receives a token of that name. A token may be generated locally or through a radio message. The token machine language described by [82] elaborates on these concepts and defines the Token Machine Language, which is based on an abstract machine model called Distributed Token Machines (DTMs). Typed messages with a small payload (a fixed size buffer) are referred to as token messages and used by DTMs for communication. Each token has an associated token handler, which is executed when a token message is received by a node. DTMs, in addition, allow for concurrency, state management and communication. The aims of TML are to provide abstractions for key requirements in the sensor network including data dissemination and communication and provide a framework within which complex algorithms, such as those that allow in-network aggregation, can be implemented.

[16] describes the Regiment Macroprogramming System, which extends the work done on the Regiment language and token machine interface. It uses "signals" which represent attribute values for nodes, for example, the temperature read by a sensor, and

groups signals into regions that are then used as a programming abstraction upon which different operations can be performed (aggregation for instance). Region membership is not static and may vary over time, due to changes in sensed values, node or communication failure or new nodes being added to the network. Key here is that the system abstracts away details of storage, communication and data acquisition from the user and allows the compiler to map global operations on regions and signals to local network structures [16].

The Regiment system has been evaluated in a number of simulated chemical plume detection macroprograms with a network size of 250 nodes over an area of 5000 x 5000 m. The results demonstrated the flexibility of the system in maintaining good communication performance [16].

10.3.2.4. Kairos

In contrast to some programming approaches that focus on providing high-level abstraction for local node behaviour in a distributed computation, the Kairos macroprogramming system [83] focuses on abstractions for specifying global behaviour. This distributed computation encompasses the entire network but uses a centralized approach. The abstraction considers the sensor network as a collection of nodes that can be tasked at the same time from one program. The programming model is based on shared-memory-based parallel programming models over message passing architectures and three programming abstractions are made available in the system. Fig. 10.11 shows the architecture of the Kairos macroprogramming system.

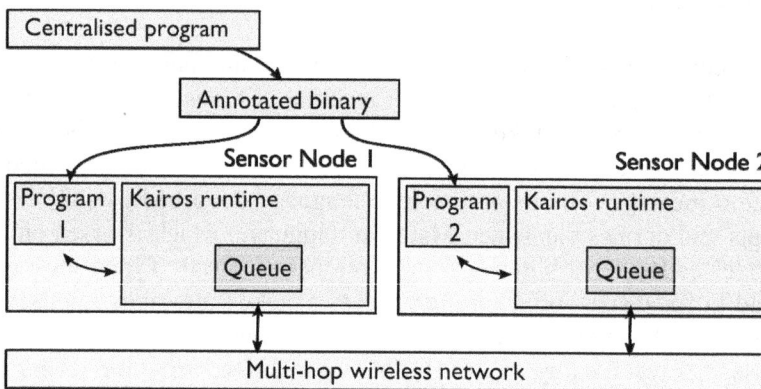

Fig. 10.11. The Kairos programming architecture [83].

Programming Abstractions: Kairos provides three abstractions for manipulation by the programmer. The first is the node abstraction where programmers are allowed to manipulate individual nodes or lists of nodes. Nodes have integer-based identifiers and are represented by a node data type which makes available operators like equality, ordering and type testing. There are also functions made available for manipulating lists of nodes.

A second abstraction is the set or list of all one-hop neighbours of a specified node, which is made available using a get_neighbours() function. This abstraction is quite similar to the region and hood abstractions and it reflects the natural construct of radio neighbourhood of the nodes in the network.

The third abstraction is remote data access, which allows a programmer to read from variables at a named node using a variable node construct. In Kairos only a node may write to its variable although multiple nodes may have read access.

Programming Mechanism: A distinguishing characteristic of the Kairos system is that a programmer writes a single centralized program representing the distributed computation. First, the program is pre-processed to produce annotated source code that is then compiled into a binary. During this stage, the Kairos pre-processor identifies and translates remote data references to calls to the Kairos runtime. The binary can then be distributed to all nodes in the network using a code distribution facility. Although the initial program is a global level one, after compilation a node-level version is created that details what an individual node's functions are, when and what data remote and local it manipulates.

When a copy of the node-level program is instantiated and executed on a node, the runtime exports and manages variables that are owned by a particular node but are referenced by remote nodes.

Kairos has been used to implement programs for a variety of problems from routing tree construction to vehicle tracking [83].

10.3.2.5. Knowledge-Representation for Sentient Computing

An interesting approach is proposed by [84] called a scalable knowledge representation and abstract reasoning system for Sentient Computing (See Fig. 10.12). Sentient Computing promotes the view that an application can be made more aware by perceiving the environment and reacting to changes in it. A gap exists, however, between the level of abstraction in the knowledge of the sentient world such a system requires in order to function and the low-level data produced by sensors in the environment. [84] propose the creation of a deductive component that would interact with the low level data, perform some type of reasoning function in order to deduce a higher-level abstract knowledge. The application layer can then use this knowledge. The deductive knowledge base layer therefore is really a domain specific language component positioned as a type of middleware layer between the application and hardware layers.

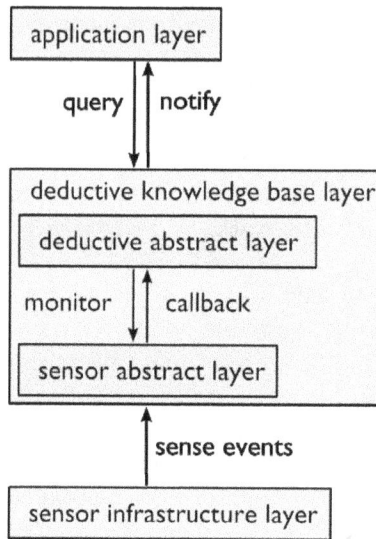

Fig. 10.12. System architecture for Sentient Computing [84].

Knowledge Representation: An event in this context is defined as an occurrence of interest taking place instantaneously at a specific time. The Sentient environment refers to the physical environment. The current logical state of the Sentient environment includes all known facts about the sentient environment from an initial event to a terminal event. These events can include any events of interest to the Sentient Application layer. The Sensor Abstract Layer (SAL) keeps a low-level view on the current logical state as sensors in the environment update it through sensed events. The Deductive Abstract Layer (DAL) maintains a high-level or abstract knowledge of the current logical state through its interaction with the SAL.

The two layers use a monitor-callback communication scheme to interact. The application layer issues a monitor call that causes the SAL to filter through to the DAL those low-level changes that affect the abstract knowledge in the DAL. The DAL, therefore, does not have to monitor all the data and is updated at a much lower rate than the SAL. This is depicted in Fig. 10.13.

Both DAL and SAL are described using first order logic and the application layer retrieves the stored knowledge about the Sentient environment using queries, which are quite similar to SQL SELECT statements. Experiments conducted with a prototype implementation showed that the two-layered architecture where the low-level and high-level interactions are separated (SAL and DAL) was more efficient than a single-layered one for a number of reasons [84]. First, queries executed in the DAL were of a simpler form with fewer conditions so used less computational resources. Second, the knowledge update rate triggered by assert or retract commands in the DAL is lower than that of the SAL making DAL more computationally efficient. Experiments with more complex queries have not been carried out.

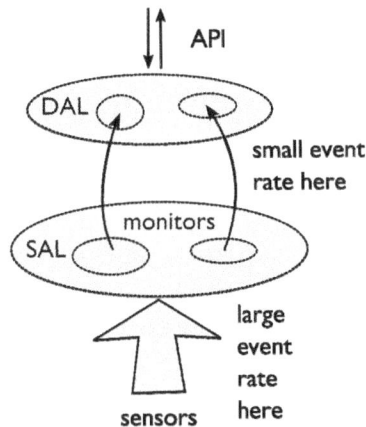

Fig. 10.13. Interaction between DAL and SAL, derived from [84].

10.3.3. Remarks

The three approaches to information extraction surveyed over Sections 10.2 and 10.3 show systems and techniques with varying degrees of ease of use, flexibility, informational complexity and expressiveness. While the querying approaches are easier for a user to manipulate they are restricted by the lack of expressiveness of the query languages they provide for requesting information. This in turn restricts the type and level of information that can be retrieved. Although it is clear that simple queries for attributes and simple aggregates are possible and extensively demonstrated in the deployments to date, informational query-based systems that go beyond that (for example spatially and temporally aware queries) are practically non-existent. Query-based systems are also limited as they do not allow users the flexibility to target smaller areas of interest within the network but in effect force the user to task the entire network in order to get a response, an issue in terms of energy efficiency. An advantage however is that these systems have been extensively used, the languages used familiar to users of sensor network systems and are relatively easier to design and implement when compared to other information extraction mechanisms.

In contrast, the agent-based approaches are more expressive and flexible in terms of what functions can be performed and the ability they incorporate to make decisions while in the network. Agents can act autonomously, multiple agents can run on a node at the same time, and multiple applications can co-exist in the network (this can be a problem with many query-based and macroprogramming systems). Mobile agents can move, clone themselves and can act to deal with unexpected changes in the environment. Agent-based systems, however, are hindered by the difficulty they present for non-expert users to design and program. It is also apparent that despite the many agent-based techniques and models proposed in the literature, actual deployments have been minimal and restricted to very small networks (less than 10 nodes) with the agents being, at most, just pieces of mobile code with minimal autonomy or decision-making capability.

The Macroprogramming approach appears to be one solution to the problem of the limitations of query-based systems while at the same time promoting the expressiveness inherent in agent-based systems. Some researchers consider macroprogramming an extension to query-based applicative approaches and attempt to adhere to the ease of usability mantra of query-based systems, while at the same time providing users the ability to access higher level information. The reality, however, is that the systems are not nearly as intuitive as many of the SQL-based querying systems and require a learning curve for the programmer. In addition, the approach seeks to eliminate the need for low-level programming on the node but in reality has to provide node-specific abstractions, which undermine rapid program development. Finally, the high abstraction level although advantageous on the one hand in terms of shielding the user from the underlying workings of the network (radio communication and network topology, for example) presents a challenge when constructing compilers that need to cater not just for computation but node level communication as well. Macroprogramming therefore presents similar challenges to the agent-based approaches.

Overall, the approaches to information extraction examined highlight that trade-offs must be made between ease of use and expressiveness coupled with flexibility (for instance in terms of decision-making capabilities within the network) in selecting a suitable information extraction system. Clearly, the query-based and macroprogramming approaches are useful in a number of applications but raise a number of challenges as outlined above. There is a definite gap between the ease of use provided by the more popular query-based mechanisms and the expressiveness of the global and node level abstractions provided by macroprograms as well as the flexibility of agent-based systems. For purposes of any applied research in WSNs, the idea of usability and utility are paramount. Future research directions and open issues are presented in Section 10.4. To that end, the authors here put the development of a declarative query-based system forward as being a viable option. Given the limitations already described above, a number of features have to be incorporated into such a query system. In order to make the querying approach more attractive, the development of a query language that allows a user to more richly express high level information requests which can target the full breadth of collectable information presented by sensors in the network is promoted. Complex queries could be suitable constructs that can allow the expression of spatio-temporal characteristics and requests requiring more involved in-network interactions to generate information. Incorporating within this query processing system useful elements that will aid the processing of these complex queries, such as abstractions already employed in some macroprogramming approaches to facilitate resolution of this higher-level information requests would be necessary. Such a hybrid approach, it is anticipated, will retain the simplicity and ease of use of the traditional query-based approaches while simultaneously allowing the inclusion of useful logical abstractions provided by macroprogramming approaches to facilitate dissemination, processing and resolution of more powerful queries than those available in query processing systems today.

10.4. Towards a Hybrid Approach

10.4.1. Introduction

In this Section, the case is made for a hybrid approach that retains the simplicity and ease of use of the more traditional query-based approaches while allowing the inclusion of useful logical abstractions provided by macroprogramming approaches to facilitate construction and resolution of more powerful queries. Such an approach would need to incorporate some of the principles of agent-based systems, such as collaboration and decision making in the network, in assisting in query resolution. In this context it can be hypothesized that end-users of WSN applications could be provided with the ability to construct and pose higher level information requests instead of simple queries requiring the collection of raw sensed values or the calculation of simple aggregates over the entire network [85].

Complex query type constructs that allow the expression of phenomenological spatio-temporal characteristics would provide the means for exploiting the networked sensing concept at large scale. Moreover, providing the end-user with a system that produces responses to these higher level requests for information within the network instead of as a result of post-collection analysis of all data would most certainly demolish one of the largest road-blocks of the WSN technology in its route to adoption: ensuring user acceptability. In the quest for testing the above hypothesis, this Section looks at some of the open research issues the community faces with respect to information extraction, raises research question relating to the usefulness of in-network processing from an informational viewpoint and concludes the chapter.

Section 10.4 is structured as follows: the first subsection identifies a family of applications which would benefit from a WSN-integrated higher level information extraction and delivery mechanism and surveys the state of the art of deployments in the identified application area. Next, the requirements for an integrated high-level information extraction mechanism are described and reviews of prior art in complex querying and in-network processing are provided. The following subsection identifies WSN topologies, architectures and protocols suitable for advanced informational systems; this is followed by the description of a new, hybrid approach, termed a Distributed Complex Query Processor.

10.4.2. Information Extraction in Monitoring Applications

Military applications were the motivation for much of the initial research into WSNs with the Defense Advanced Research Projects Agency (DARPA) providing funding for a variety of projects from early 1990s to the present day. Today, however, the use of WSNs has expanded to encompass a large number of application domains. A particular group of WSN applications, denoted as monitoring applications, make up a large proportion of the WSN systems being deployed today. Though most monitoring systems are generally characterized by the need for attributes of interest to be observed, sensed and then relayed to a user, most WSN applications of this type show very little genericity in design. Rather, the WSN systems deployed are application-specific, with designs intimately linked to the particular application requirements. This section will identify some of the common types of monitoring application and describe examples of each, highlighting the information gathering model used in each example. An application acting as a motivator for higher-level information extraction procedures is also described.

10.4.2.1. Habitat and Environmental Monitoring

Habitat and environmental monitoring form a particular set of monitoring applications that have grown over recent years. This is in part because of the great benefit they claim to provide to science and education as well as the fact that funding for improving science education is a priority. There are a number of WSN deployments in this category.

On the environmental side, the ARGO project [86] uses a sensor network to monitor the salinity and temperature of the upper ocean. The aim of the project is to generate a quantitative model of the changing state of the upper ocean while monitoring ocean climate patterns over the short to longer term. The data gathered is expected to act as input to ocean and ocean-atmosphere forecast models for data assimilation and model testing. Nodes are dropped from aircraft and cycle between the surface and a depth of about 2000m every ten days collecting data. When nodes rise to the surface, data collected by nodes are individually transmitted to a satellite.

The Great Duck Island project [85] is an example of a habitat monitoring application where a WSN was deployed to monitor the breeding behaviour of small birds called petrels. Scientists are interested in the usage patterns of nests as well as the environmental changes in and outside of nests during the breeding season. Measurements are taken of humidity, temperature, pressure and light level. Nodes are clustered with each cluster connected to a base station via a long-range antenna. Sampling is done every minute and readings are sent directly to the base station, which connects to a database via a satellite link.

Another environmental monitoring application is the GLACSWEB project [87], which uses a sensor network to monitor sub-glacier environments in Briksdalsbreen, Norway. Holes are drilled at different depths in the ice and sensor nodes carrying pressure, temperature and tilt sensors are deployed in them. The nodes communicate with a base station on top of the glacier that measures supra-glacial displacements. The collected data is transmitted via GSM with no information processing occurring within the network itself.

For all of the applications above, raw sensed values are sent via base stations or servers to a user and processing occurs at that point.

10.4.2.2. Agricultural Monitoring

Increasingly, WSNs are being used to monitor conditions that affect plant growth. These networks are used in precision agriculture, which aims to improve crop yields, reduce pollution and monitor the general health of crops. Variables monitored include temperature, soil moisture, humidity and light and are usually referred to as microclimate variables. The Climate Genie system [88] is one such commercial example and is used to monitor vineyards. Nodes are spread over the vineyard in a wireless mesh configuration and are equipped with sensors for measuring temperature, moisture and light. The data gathered is summarized (aggregates that reflect grape quality and vitality) within the network and sent via Wi-Fi, cellular or satellite to servers for viewing anywhere using a web browser. This system therefore incorporates some level of in-network processing although limited to the calculation of aggregates that are used in decision making by the observer.

Another system was developed to monitor in real-time field conditions including leaf moisture, soil temperature, soil moisture and CO_2 [89]. Field monitoring servers (FMSs) similar to web servers were deployed in rice paddy fields, collected data automatically and transmitted the data for permanent storage in publicly accessible databases. Real-time data was made accessible via a web browser. Like the habitat and environmental monitoring applications described in Section 10.4.2.1, raw sensed values are sent to the databases for storage and analysis occurs from that point on.

10.4.2.3. Structural Health Monitoring

Structural health monitoring (SHM) refers not only to the state of health of a given structure whether a building or bridge, for example, but also the detection of changes that may affect the structure's health in the future. There are two types of SHM systems: systems that monitor disaster response after some catastrophe has occurred, for instance, an earthquake and dedicated to continuous health monitoring which may check for signs of stress, monitor vibrations, wind, etc.

Disaster response systems are still a young area of WSN research. [90] describe a centralized WSN for structural-response data acquisition called Wisden. Wisden collects structural response data from a multi-hop network and relays and stores it in a base station. [91] also proposed a wireless monitoring system aimed at detecting damage to civil structures after a disaster (such as an earthquake) has occurred. In other research, technology developed by the CodeBlue project [92] is being used in the AID-N project [93] at Johns Hopkins Applied Physics Laboratory in developing systems aimed at disaster response.

For continuous, structural health monitoring applications the focus is on high sample accuracy with minimal distortion, high frequency sampling, time synchronization of readings and efficient data collection as opposed to energy efficiency through reduced power consumption. [94] describes a SHM system deployed on the Golden Gate Bridge.

64 nodes were deployed over a 4200 ft long length and measurements taken of ambient structural vibrations. All collected data were relayed to a base station for analysis.

10.4.2.4. A Motivating Scenario

A WSN application frequently put forward is that of forest fire detection and monitoring [95]. This application is attractive because forestry is a major industry in many parts of the world, and forest fires are a major cause of loss of wood (in the USA the annual average loss to fire is 17,000 km^2). Early warning of a fire event and the manner in which it spreads are hence necessary.

Theoretically, any practical forest fire detection system is likely to exceed the scale of present WSNs by a considerable margin, with an expectation that hundreds of thousands of nodes would be needed for detailed monitoring and precise fire detection and localization. Considering a network of this scale, it is clear that real-time data searches using conventional centralized query mechanisms are not an option. For each fire detection cycle, hundreds of thousands of readings must be returned to the sink and processed. Following detection of the fire, new queries must be generated and directed to the nodes in the area, which would be, ideally, geographically mapped. Finally, those nodes need to return the infrared data required for the map. A more efficient proposition would be for the initial event to be detected within the network, with the subsequent queries for the map data being generated locally, without returning data to the host.

A number of deployments have tackled some of the very issues described above but for practical reasons, the deployments have taken very different approaches to those proposed in the scenario described above. The FireWxNet system [96] a wireless sensor system aimed at monitoring weather conditions in woodland fire environments is one such example. The system monitored a variety of weather conditions that influence fire behaviour with an aim to using them to predict fire behaviour. The application, as is the case with most monitoring applications, was driven by a list of requirements acquired through consultation with both fire fighters and fire researchers. The implemented WSN system was deployed in the Bitterroot National Forest in Idaho (USA) and consisted of 3 sensor networks totalling 13 nodes, 5 wireless access points, 2 web cameras and 5 long-range links.

The deployment was distinguished by the rugged environment within which the WSN had to function as well as the sparseness of the deployment itself. The authors note that sparse coverage was a deliberate choice aimed at strategically placing nodes to cover as much meaningful terrain with as few nodes as possible. Rather than scale, the focus was on creating an extremely robust design, which included not just robust equipment but robust routing protocols as well. Once the system was launched, a number of challenges were encountered particularly given the harsh deployment environment but the system overall was a success. During its operation, over 80,000 measurements were streamed in real-time, with operations applied post-collection to transform that data into usable information.

This real-life deployment presents a marked contrast to the motivating scenario and approach described above: the problems are well characterized given the input of domain experts; there is quite detailed knowledge of the application domain and what the corresponding WSN application requirements are. However, on-the-ground challenges usually mean compromises have to be made and as a consequence real-life monitoring systems so far, rarely aim to deploy such large quantities of nodes as put forward in the motivating scenario here.

The FireWxNet example highlights the limitation of the vast majority of monitoring applications today. They are limited in that they are conceptualized, designed and implemented as primarily data collection systems. In-network processing is absent and the systems can be considered automated to the extent that there is little user interaction. In essence, events are defined, queries may be deployed for sensed readings and all data is simply relayed to a centralized location for further analysis. Very few systems look beyond this simple sense-and-send model to incorporate some sort of analysis,

in-network, prior to communicating results. Granted, with some systems it is difficult to define beforehand what events or processes may be of interest and these applications by their nature have to be more exploratory at least at the pilot deployment stage, with feedback influencing subsequent iterations. However, with some applications, as in the FireWxNet example, the problems are better characterized and therefore more amenable to some more sophisticated analysis or processing within the network. This could make the application not only more useful but more efficient as well.

Hence, the view put forward here is that in many of monitoring applications (the fire monitoring scenario being only one example) it is possible to define high-level information requirements that could incorporate in-network processing techniques in resolving the requests. Besides the efficiency benefits in terms of reduced transmissions, this approach could transform a data collection system to an information generating system, extending the application scope while still fulfilling the basic application requirements.

As an example, an informational fire monitoring system would allow the processing of high-level queries aimed at tracking the fire. The queries would need to allow the expression of spatial and temporal characteristics and the querying system would need to support a level of autonomy in terms of some in-network decision-making by the nodes given the impracticality of streaming all data back to the sink. Such a system would perhaps allow a user to zoom in on particular problem spots and query for even more detailed information on-the-fly. Complex queries are proposed as extremely suitable for use in the scenario above as well as other monitoring application scenarios and a degree of autonomy is introduced through the implementation of in-network logical abstractions for query processing and resolution.

10.4.3. Requirements for a Higher Level Information Extraction System

With a view of the above, a higher-level information extraction system should to be able to:
- Enable the user to construct and disseminate complex queries;
- Allow a user to program the sensor network in a similar way to current applicative approaches;
- Enable a means for in-network distributed query processing that allows information to be generated within the network;
- Be implementable on constrained resource nodes.

10.4.3.1. Catering for Complex Queries

Towards the requirements above, some research has addressed specifically the need for complex queries and for the ability to process these queries within the network. [113] for example, define a complex query as a query consisting of one or more subqueries that are combined by conjunctions or disjunctions in an arbitrary manner. In their work on the Active Query Forwarding mechanism (ACQUIRE), they promote the use of these "nested queries" and describe a mechanism that seeks to resolve the query in-network, generating information as a response. The work, however, does not address the implementation of the mechanism and instead presents a mathematical model that is used to analyze the performance of the approach in terms of energy cost.

[98] also identify the need for nested queries (which they too call complex queries) and highlight the problems with evaluating such queries especially in cases where aggregation dependencies exist between the nested queries. They put forward the idea of the query itself supporting abstractions that can then be used in query resolution. In their example the abstractions are geographical regions. Their research resulted in a qualitative study of the requirements of such a system and did not extend to implementation.

Beyond the work reported above, a complex query, in the authors' view, is a query which:
- Consists of one or more subqueries (nested queries) and/or
- Contains multiple operations such as aggregates and/or
- Contains spatial and/or temporal elements.

The query-based systems currently in use neither provide the facility to construct these complex queries nor give users the ability to process them. Such queries, for instance, queries with dependencies (nested queries) would require more complex in-network interactions than those supported by current query-based systems. One example of a complex query, which forms the object of the research here, would be:

"What is the average temperature in those areas in the network where the humidity is greater than 95 and the air pressure is between 900 and 1000 mbar".

This query exhibits a number of complex elements. First, the query language would have to be able to accommodate the expression of the spatial elements described as "areas" in the example above. Second, the query is a dependent or nested query and can only be answered after some reasoning within and between the defined spatial entities. In effect, parts of the query depend on a previous question being answered and only become relevant if a particular answer is obtained.

With the primary goal of creating a system that exhibits simplicity and usability, using a declarative language already familiar to users of existing applicative query mechanisms as well as traditional database systems is considered a worthwhile approach. Identifying and investigating existing SQL-based query languages towards creating a language capable of expressing the complex query requirements described above is essential.

10.4.3.2. Catering for In-Network Complex Query Processing

As will be shown in Section 10.4.4.5, a number of in-network processing techniques have already been proposed in the literature and used in existing information extraction systems. These include techniques like aggregation, fusion and filtering which have been shown to improve energy efficiency in WSN systems and have been incorporated extensively in both query-based and agent-based systems. In addition, a powerful feature of many of the macroprogramming systems has been the creation and use of node or network level logical abstractions to facilitate in-network information processing. The literature has shown that so far logical abstractions have not been considered for application in query processing systems. The authors believe that these logical constructs can provide a more powerful means for processing the complex queries identified as being of interest to monitoring applications. The usefulness of abstractions that can be constructed logically within the network and used in conjunction with the in-network processing techniques mentioned above is examined in Sections 10.4.5.5 and 10.4.5.6. The abstraction proposed in Section 10.4.5 can be based on two types of attributes: Static attributes that do not vary over time (such as, the type of reading a node provides) and dynamic attributes which do vary over time (such as, the current sensor reading). The key idea here is that the abstractions would be a component of the query itself and constructed prior to the dependent query being posed. These abstractions will drive the manner in which the query is both disseminated and processed within the network.

Consider an example where a user is interested in monitoring the soil acidity and relative humidity in "hot patches" of the vineyard. The query posed would first need to define what the "regions of interest" are. In this case, regions would be logically constructed over areas where the temperature level registers above a given threshold. Once these regions have been defined, the body of the query, aimed at retrieving soil acidity and humidity readings, will be disseminated to the relevant nodes via the region construct and not to any node within broadcast range, for instance. The core concept is that the regions are queried rather than individual nodes. Another key feature is that query-dependent logical abstractions will be used to facilitate query dissemination and processing in the network. The following sections review and identify suitable WSN topologies and adequate architectures and protocols that might enable the implementation of such an information extraction system.

10.4.4. WSN Topologies, Routing Protocols and Architectures

10.4.4.1. WSN Topologies

The most basic WSN topology is the centralized, sink-based topology, sometimes referred to as a flat or single-tier architecture where nodes in the network are homogeneous. That is, nodes are identical in terms of hardware complexity, battery power [99, 100], and bandwidth management.

Data collected by the nodes are directed toward the sink or base station (usually the only "node" more powerful both computationally and in terms of energy capabilities) using single or multi-hop communication [102]. Fig. 10.14 shows a diagram of a typical flat WSN topology. This configuration brings a number of challenges particularly if there are a large number of nodes. These include management of energy consumption, energy optimization, routing, information gathering and general management of the sensor nodes themselves [99], [100]. The University of California at Berkeley's Redwood forest deployment [103] is one example of a WSN system exhibiting a flat architecture. The 33-node deployment used TASK [111], a self-contained sensor network system based on the TinyDB query processor [104] to monitor the microclimate (temperature, relative humidity and solar radiation) of a redwood tree over a 44-day period.

While centralized, sink-based topologies are relatively simple to support and implement, they do not fulfill a topical requirement of WSN designs: scalability. Scalability has been described as one of the key design requirements for both conventional communication networks and wireless sensor networks. As the number of nodes increases with a flat topology, however, the sink node may become overloaded leading to increased latency as well as severe energy usage [104]. As a result of the apparent problems posed by flat networks, hierarchical heterogeneous architectures were proposed. The simplest example of this type of network consists of two layers: the first contains groups of homogeneous nodes, called clusters, connected to a dedicated micro-server or cluster head [102]. Cluster heads are sparsely distributed and serve as aggregators of data and managers of nodes within their individual clusters. They also serve as communicators both to a gateway or sink node and with other cluster heads in accomplishing application goals. Fig. 10.15 shows an example of a hierarchical architecture.

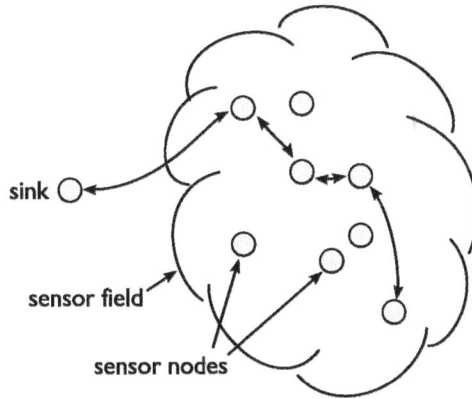

Fig. 10.14. A typical, flat WSN architecture.

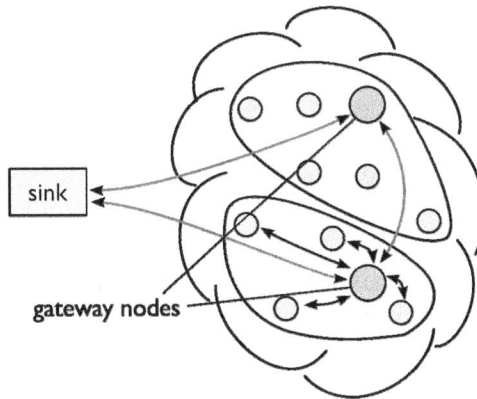

Fig. 10.15. A hierarchical, heterogeneous WSN architecture where sensor nodes are arranged in clusters with a cluster head or gateway node managing the cluster and performing all communication external to the cluster.

The introduction of cluster heads with special functionality (usually supported by enhanced hardware) gives the hierarchical structure some advantages over the flat topology particularly with respect to energy consumption.

Note: In this respect, multihop communication has been identified as a more favourable strategy over single hop, given that energy consumption is directly proportional to the square of the distance [106]. Hence, there is potentially more energy conserved with multiple short hops from node to sink, for example, as opposed to a direct long hop from a node to the sink [107]. A problem occurs, however, with the nodes closest to the sink since they are under heavier traffic and more likely to have their energy drained more quickly. High-energy consumption can also be a problem if clusters where multihop communication is used are large.

Cluster heads provide additional functions like caching and forwarding of data to the required destination as well as performing data aggregation and fusion in order to decrease the number of transmissions to the sink or gateway. This is of particular importance in data gathering networks. It has been shown that for some applications, using a hierarchical structure can bring about significant improvement in performance in terms of reliability, longevity and flexibility of the network [102]. The Great Duck Island habitat monitoring application [85] is one example of a deployment that used a hierarchical architecture.

Building on the basic two-tier hierarchical model, varying multi-tier architectures have been put forward to capitalize on the apparent advantages. Examples are those proposed in [108] and [109]. Finally, the SENMA architecture [110] proposes a novel two-tier architecture where the upper layer consists of mobile access points that are used for data acquisition from homogeneous sensors within the sensor layer. Sensors communicate with mobile agents who periodically move within radio communication range eliminating the need for multihop communication. This architecture, because of the reduction in multihop communication, showed a significant improvement in energy efficiency.

10.4.4.2. Routing Protocols for WSNs

Given the unique characteristics of WSNs the underlying routing protocol used is an important consideration when looking at information extraction. The efficiency of the information extraction mechanism mainly depends on the routing method used. In some cases the routing mechanism itself includes techniques for query dissemination and data aggregation as in Cougar [97] and ACQUIRE [113]. Routing is therefore an important aspect of information extraction in WSNs.

Many routing protocols are currently being used in a variety of wired and wireless networks [112]. Although protocols can be classified in different ways, for example, based on the network structure or perhaps on the protocol operation [113] many follow the address-centric (AC) model where routes are found and followed between pairs of addressable nodes. In mobile ad-hoc networks, AC protocols are used widely (MANETS) [114]. Here each source independently sends data along the shortest path to a sink. This path is usually based on the initial route a query took to get to that source node. Although MANETs are similar to WSNs in that they both involve multi-hop communication, their routing requirements are quite different for a number of reasons.

First, communication in a WSN comes from multiple data sources to the sink instead of just between a pair of nodes. Second, redundancy in sensed data is common in WSN since multiple sensors may be sensing the same phenomena; a fact that data-centric protocols take into account in devising optimal routes. This is not a major consideration in MANETs. Finally, in WSN the major constraint is energy making it essential that data communication rates be made as efficient as possible, much more so than in MANETs.

Given these reasons among many others, the traditional end-to-end protocols used for MANETS are not appropriate for WSNs. Some alternative protocols propose a different model, the data-centric model, where sources send data to a sink, but the content of the data is examined and some processing (whether aggregation or fusion) is executed on that data en-route to the sink. The result in using such protocols is that a better transmission/information ratio is achieved. In the next section some data-centric routing protocols will be examined in more detail.

Although protocol development is usually seen as being beyond the scope of information extraction research, it is important to have an awareness of what protocols are used within the query processing systems in use and their impact on the information extraction mechanism. Data routing, for example, is an important consideration when looking at information extraction as it affects the overall efficiency of the mechanism. For purposes of building advanced informational systems fitting the requirements in Section 10.4.3, identifying a suitable cluster-based routing protocol that facilitates the implementation and testing of the logical abstractions is key.

The choice of architecture for a WSN system appears to depend strongly on the application requirements. Pure data collection applications tend to work with hierarchical topologies with cluster heads aggregating and relaying data as it is generated. In such systems information-processing requirements are simply sense data collection or at most calculation of aggregates. This is not always the case, however, as some WSN systems like the Redwood Forest deployment [103] which for the most part exhibit a flat topology are also used for data collection. In such systems energy conservation is either not a major concern or techniques for optimizing query processing are incorporated to minimize energy consumption.

Again, cluster-based topologies (single or multihop) appear to be more amenable to logical region-based query processing approaches. Region leaders can be considered somewhat similar to cluster heads in terms of functionality therefore enabling a mapping of the logical construct to the network construct. It is expected that regions, in this type of architecture will be easier to implement and test experimentally.

A review of candidate query processing architectures is given below.

10.4.4.3. Query Processing Architectures

Database-Type Architectures: Database-type query processing systems are some of the most popular put forward in the literature. [115] propose a generic database-based query processing architecture for sensor networks (see Fig. 10.16) and describe the components they feel are required at each layer to make the architecture practical.

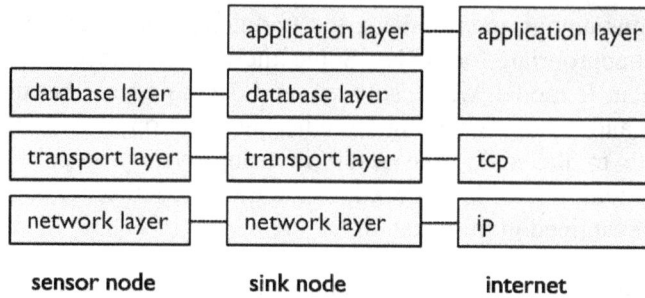

Fig. 10.16. A database-based query processing architecture based on [115].

Application Layer: at the base station this layer provides an interface for posing of queries to the sensor network. These queries can be represented either as an SQL-type message or even a SOAP-like web service. The application layer in the base station is designed to communicate with the TCP/IP stack and so an Internet-based host can easily access query results.

Database Layer: on the base station, the database layer serves the application layer by receiving queries from it and returning results to it. The database layer on the base station contains a number of components including: a Parser – creates the query tree based on the query received from the application layer; Catalog Manager – maintains the relational schema, location and distribution of sensors and monitors nodes' status, for instance, node power levels and node connectivity; a Query Optimizer - generates the optimized query execution plan.

Due to resource limitations, the database layer's function at node level is much more limited. It receives the query tree either from the base station or another node and may need to further optimize the tree based on local catalog information. The database layer returns results in the form of relations.

Transport Layer: this layer provides for efficient end-to-end communication. At the base station it should be able to communicate with TCP entities on Internet hosts although alternative methods may be needed for nodes since nodes are address-less.

Network Layer: this layer should provide energy-efficient and data-centric routing algorithms. The routing decisions made should be based not only on the current network conditions but also the given query execution plan.

10.4.4.4. Example Systems

Query processing systems like TinyDB [104] and Cougar [97] are just some of the database-type systems put forward in the literature. They all follow to varying extents the architecture put forward by [115]. The query processing architecture is similar for both, consisting of server-side software running at a base station and responsible for parsing and delivering queries into the network, as well as collecting the results as a stream out of the network.

The architecture of the TinyDB system, both server-side and node-side, is illustrated in Fig. 10.17. Architecturally the query layer sits between the application and network layers in the case of the server side component (base station) and sits on top of the network layer at node level. As in Fig. 10.16, the application layer allows the construction and posing of queries with the database layer, here referred to as the query layer, below parsing the query, constructing query execution plans and delivering the results to the network. The query layer at the server side also collects the results for presentation to the application layer.

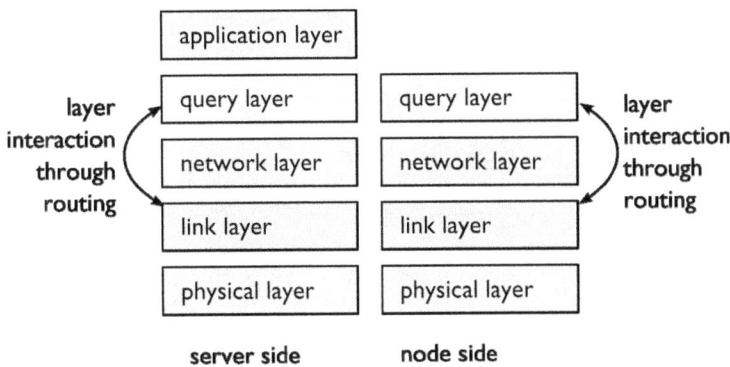

Fig. 10.17. A high level view of the query processing architecture used in Cougar and TinyDB.

The query layer houses a number of components some of which have already been described in the architecture put forward by [115] and are illustrated in Fig. 10.18.

The query layer is the backbone of the query processing system and provides a number of critical functions:

- The processing of the query itself. The query layer uses sophisticated techniques like catalogue management, query optimization to abstract the user from the physical details of the network in processing queries.
- Performing in-network processing to reduce communication given the need to preserve resources like energy and bandwidth. To do this the query layer generates different query plans with different trade offs for requirements such as accuracy, energy consumption and latency.
- Interacts with the routing layer to facilitate in-network processing. With traditional routing, the network layer on a node will automatically forward packets to the next hop towards the destination with the upper layers, completely unaware that data packets are moving through the node. In implementing in-network aggregation, for example, the query layer needs to be able to communicate with the network layer when it wants to intercept packets destined for the sink or leader node. In the

processing system described above, filters are used to first access the packets then modify and delete if necessary before passing on to the next hop onto the destination.

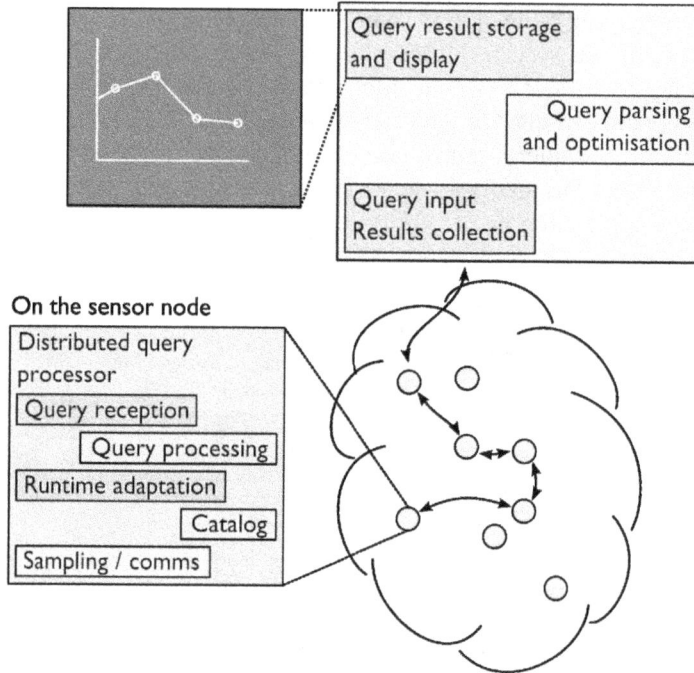

Fig. 10.18. A detailed view of the query processing architecture of TinyDB and Cougar [61].

Although they can probably be implemented on both hierarchical and flat network topologies, these database type approaches have so far been implemented on flat networks.

Middleware Architectures: Another approach to query processing in WSNs is that of query-based middleware systems. Two of the more popular are DSWare and SINA described in Sections 10.2 and 10.3 of the chapter. Although inspired by database systems in that they use SQL-based languages for construction of queries, these systems are considered middleware approaches as they also provide services that go beyond database query processing systems. Architecturally, these middleware approaches are similar to the database approaches. In both cases, the middleware layer sits between the application and network layers and in addition to providing facilities for query processing, supplies the services necessary for ensuring adequate WSN functionality. In the case of DSWare, the application layer allows the construction of SQL-based event type queries that are then registered, parsed, and used to generate execution plans. The underlying DSWare middleware layer is then responsible for registering and executing these events. The DSWare layer also interacts with the routing layer for improving network functioning, for example, improving power awareness.

SINA also comprises a middleware layer that sits above the network layer and provides an application-programming interface to the middleware layer via SQTL scripts. The middleware layer runs on all nodes in the network and like the database approach it selects the most suitable distribution of the query-based on the current network status and the type of query being executed. In addition to query processing functions, SINA also provides services for accessing sensor hardware and communication and supporting changing network topology (for example, sink mobility).

Both DSWare and SINA promote processing of aggregation-based queries and are suited to cluster-based topologies.

10.4.4.5. In-Network Processing Techniques for WSNs

Experiments have shown that the major part of power consumption costs is due to communication rather than computation [92-94, 107] describe in-network processing as the pushing of operations, particularly selections and aggregation of data, into the network to reduce communication. [97] describe it as the moving of computation from outside to inside the network in an effort to reduce energy consumption and improve network lifetime. The main goal in effect is to reduce communication in exchange for some form of computation within the network. In-network processing techniques are therefore accepted as being essential to improving energy efficiency in gathering information within the WSN and critical to any system that has improved network lifetime and energy efficiency as goals. The two most common in-network processing techniques used for reducing communication in WSN systems are packet merging and partial aggregation [99]. In addition, a number of other widely used techniques exist. These are discussed below.

Filtering: Filters are constructs used within the network (on nodes) to assist in processing [97]. A filter registers what kind of data it handles through matching and is triggered each time that type of data enters the node. Once it has been activated, the filter can manipulate the data by, for example, determining whether it is forwarded on or even generating a new message. Filtering is sometimes used in conjunction with data aggregation. [97] proposed using a filter to detect concurrent detections of four-legged animals from different sensors. It could then record what the desired interval was and ensure that only one response per interval, suppressing responses from other sensors. [116] take advantage of event properties of monitoring queries, and carry out data filtering during in-network processing.

Packet Merging: Packet merging, described by [99] combines a number of smaller records into a few larger ones in a bid to reduce the number of packets needing to be transmitted and in turn reducing the cost of communication as the cost of transmitting several smaller packets is greater than transmitting just one large packet. The Cougar query processing system is an example of a system that uses packet merging for reducing communication in processing some types of aggregate queries.

Partial Aggregation: Partial aggregation refers to the computation of intermediate results as the node receives readings. Communication is reduced as this combined partial result is transmitted on instead of all individual records. Aggregation can be achieved in various ways based on the type of correlation that exists in the WSN. This may be spatial correlation (due to physical proximity of nodes and therefore similarity in the readings), temporal correlation (if there is little variation in the sensed attribute based on the sampling frequency) or correlation in the data itself (due to overlap in sensor coverage) [117].

Tree-based Aggregation: In tree-based aggregation one node is designated the root node. A broadcast message is sent out with data on the ID of the node sending the message and its depth in the tree. In the case of the root the depth would be zero. When nodes receive this message they assign themselves a level by adding one to the value in the message received and assign the ID in the message as their parent. Broadcasting of the message continues until all nodes within range have received it and assigned themselves a level and parent. TinyDB is one query processing system that uses a tree-based aggregation service called TinyAGgregation (TAG) [118, 99].

There are a number of advantages to the TAG approach. First, a reduction in the number of communications needed to calculate aggregates when compared to aggregation that occurs at one centralized location. Second, as data moves up the tree back to the base station nodes usually are required to transmit a maximum of one message. This is in direct contrast with centralized aggregation methods where the number of transmissions increases dramatically as data moves towards the root node. This can of course lead to a decrease in the lifetime of the network as battery power is quickly drained. Third, it allows aggregation even in networks where connectivity is intermittent or disconnections occur since disconnected nodes can reconnect by listening to other nodes' partial state records as they flow up the tree. This is possible since the partial state record includes information on the query that was issued.

TAG, therefore, presents a simple interface, flexible naming and generic operators for constructing aggregate queries and is not application specific as with other approaches using aggregation like Directed Diffusion [119] and Greedy Aggregation [120]. Further, it separates the logic of aggregation from routing details so that the focus is squarely on the application and leaves routing decisions to the system. This is in contrast to Directed Diffusion, which puts aggregation mechanisms within the routing layer. The TAG approach allows a stream of aggregate values that change as sensor readings and the underlying network layout change and does this in an energy-efficient and bandwidth-efficient way.

TAG, however, does have limitations, as it does not allow joins and cannot respond to events that occur within the network. The authors have indicated that future work will focus on developing an efficient way of aggregating the results of those event-based queries across nodes before transmitting that information to interested nodes.

Cluster-based Aggregation: In some cases nodes are in close physical proximity to each other and queries may need to retrieve information based on this spatial correlation. For example, a query may want to determine the average humidity over a particular area. Typically, clusters are formed consisting of nodes in close proximity based on some metric, for example, signal strength. Clusterheads are elected and act as the data aggregator and router for cluster members. SINA and Cougar, are examples of query processing systems that incorporate cluster-based aggregation in processing aggregate type queries.

Discussion: Given the importance of considering in-network processing when developing efficient WSN applications, this section examined some of the more popular techniques used. Data aggregation is most widely used, with the type of aggregation dependent on the network topology and to some extent the type of queries being issued. Systems that incorporate in-network processing and are therefore information-generating systems (as opposed to simple data gathering) have tended to be built on cluster-based topologies. In heterogeneous configurations, cluster heads can be configured to be more computationally powerful and are conducive to supporting data aggregation. There have, however, been, information-generation systems based on flat topologies that allow tree-based aggregation. Where scalability is an issue, and where in-network processing is computationally intensive, reported systems have had the tendency to lean towards a cluster-based configuration, this being more energy efficient and practical in creating a system with an extended lifetime.

Clustered networks have the advantage of conveniently allowing aggregation at the clusterhead and are most effective in networks that are static and where the cluster structure stays unchanged for a considerable period. With dynamic clusters, however, problems can occur with energy expended in continuously updating nodes in order to keep the clusters consistent with the underlying network topology. Tree-based aggregation is simple and useful but can lead to problems. Given a node failure in the network, for example, a packet lost at a given level of the tree can lead to all data being aggregated up that node's sub-tree being lost as well. This can lead to incomplete data making its way up the tree particularly if nodes only have one route to the sink and connectivity gaps occur.

Having set the requirements, and outlined the topologies, architectures, and protocols in this section, the next section describes a new, hybrid, approach that combines query-based and macroprogramming concepts and evaluates it in light of the requirements previously discussed.

10.4.5. A Distributed Complex Query Processor

The database-type architecture forms the basis for the complex query processing architecture put forward by the authors here. The two-part architecture consists of server-side software, which is accessed by the user and used for constructing and posing declarative-type queries, and node-side software, which is in effect the distributed complex query processor (DCQP). A key component of the architecture will be the Region and Query Management Layer that will be responsible for not only query processing but will have additional tasks related to the creation, maintenance and update of regions within the network.

A number of factors influenced this selection. First, a key feature of the distributed complex query processor (DCQP) proposed is ease-of-use and familiarity for users. An SQL-based, database-like approach is therefore preferred. Although attractive, a middleware approach was not selected as the aim with the DCQP is to create a system that is separated as far as possible from network level configuration issues. The idea is that the system will sit on top of a functioning network. Second, the need for in-network processing for resolution of complex queries dictates that a query layer that allows in-network processing of queries is essential. The approaches above have shown that distributing the query layer to all nodes and providing functions for query processing and optimization are effective. The DCQP query layer therefore, will allow the formulation of query execution plans, dynamic optimization of these plans based on the attributes of interest and node state as maintained by the regions. Third, the DCQP architecture will incorporate a query layer that is decoupled as far as possible from the underlying layers. The idea here is to make the system as portable as possible by not strictly linking it to any particular routing protocol, for example, but instead identifying features that would make particular protocols more suitable than others.

A number of candidate architectures were investigated as a starting point for a distributed query processing system that incorporates best practices for in-network processing and resolution of complex queries while at the same time allowing the incorporation of novel techniques and strategies for the types of queries proposed. The next subsections describe in more detail the design choices and assumptions made and give reasons for these selections.

10.4.5.1. The Network Model

Key to the research work is the creation of logical regions within the network. These regions constitute a critical component in making a decision on where a query should be disseminated, how the query should be processed within and between regions and where the complex query will be ultimately resolved. The regions will have one "leader node" each which acts as a manager, data aggregator and communicator to a gateway as well as other region leaders when needed. This role is remarkably similar to that of clusterheads in network-level clusters [102]. Additionally, like some clusters, regions should be and are likely to be, in the approach proposed here, dynamic. Consequently, it makes sense that these logical region leaders map to physical clusterheads. These requirements, therefore, directly influence the choice of network model for the proposed work.

First, the network model proposed has a hierarchical topology. Sensors are randomly deployed and the transmission range is identical for all devices. For simplicity, an ideal

MAC layer is assumed and node death considerations are not taken into account at this stage. Also assumed is that a route has been established between nodes in the network and the gateway node. (Gateway node here is defined as the node from which a query is disseminated and through which results are acquired. Gateways are not fixed; any node could potentially become a gateway at some point in time.) To facilitate region-based routing, a cluster-based routing protocol that allows dynamic cluster formation and supports inter-node communication and communication with the external world, is selected. Two possible choices identified so far are the Hybrid Energy-Efficient Distributed Clustering protocol (HEED) presented by [121], which focuses on scalable data aggregation and increased network lifetime; and a dynamic clustering algorithm (DCRR) presented by [122] in which cluster heads are dynamically selected in the region where an event occurs according to their residual energy.

10.4.5.2. A Distributed Complex Query Processing Architecture

The architecture proposed for the DCQP consists of server-side software which is accessed by the user and used for constructing, posing and parsing of declarative-type queries and node-side software which is in effect the distributed complex query processor (DCQP) implementing query processing functions in-network.

Server-side Software: This component will host an application layer that will allow the construction of queries along with an underlying Region Query Management Layer that will be responsible for parsing of these queries and dissemination to the network. Additionally, the server-side software will allow the receipt of query results for both presentation purposes and for persistent storage. This architecture is equivalent to that of TinyDB and Cougar as shown in Fig. 10.17.

A key output of the work described here will be the creation of a system that will allow the user to either specify new logical regions as a component of a query being constructed or make use of regions already in existence within the network. This will involve the extension of a SQL-based query language to incorporate the region construct and the development of an associated parser for the language. The region abstraction will be a part of the query language used, and is essential to query dissemination as well as the creation of query execution plans.

Node-side Software: This component will also host a region query management layer which will from an architectural standpoint sit above the network layer. The node-side component is really a distributed query processor hosted by all nodes in the network.

The region query management layer will deal with query reception, dissemination, processing and delivery of results. In addition, this layer will need to manage the logical regions, which are implemented as part of the complex query. This functionality will be implemented as a region management component of the query management layer itself.

Region management will include the creation, maintenance and updating of region information on each node within the network. This may involve, for example, a region leader upon receipt of a query, informing its members of a new attribute of interest; or perhaps the termination of a previously posed query along with its associated regions; or the handing-off of a region leader's duties to a backup leader when the node's energy level falls below a certain threshold. Once a query is received, execution plans are created dynamically based on the required region formation or existing regions that need to be accessed. Region management is therefore an integral part of query plan formulation and query optimization. The query plan is then executed. Once results have been acquired these are communicated to the server-side component.

As in the Cougar system, the query management layer may have to interact with the routing mechanism in the network layer in order to enable in-network processing, which is a key feature of the region-based query processing approach proposed. This interaction, however, is anticipated to be minimal since a cluster-based routing protocol that maps the logical regions to physical clusters is proposed as a way of minimizing intrusion into the network layer by the query management layer.

Fig. 10.19 shows diagrammatically, the proposed complex query processing system architecture.

In the following section, the DCQP approach is evaluated via simulation.

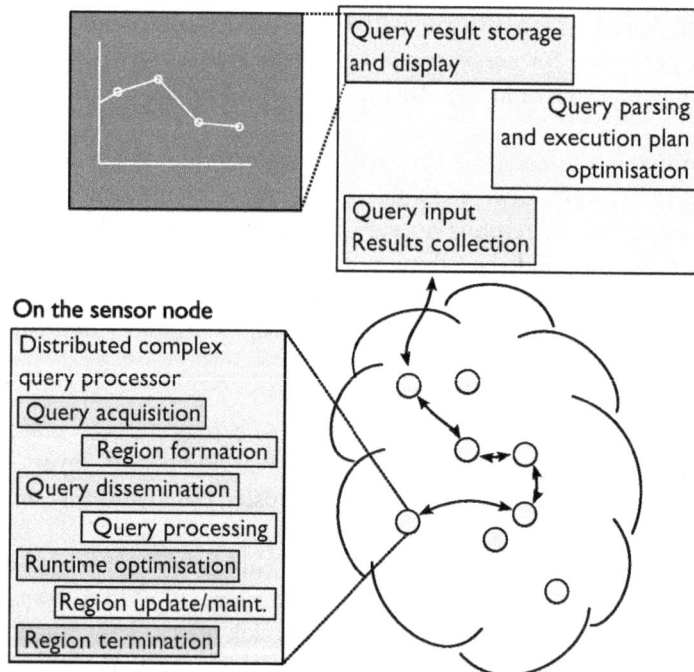

Fig. 10.19. A detail view of the functions provided by the complex query processing system.

10.4.5.3. Acoustic Monitoring Using Region-based Querying

Habitat monitoring is already a rich area of research in the area of WSNs particularly because of the benefits to science and education. Within this group are applications based on acoustic sensing, which are concerned with event detection and classification as well as monitoring and localization. Bioacoustics research is a specific example and acoustic sensing in that context can be used to help scientists acquire acoustic data which can then be used to distinguish between animals, species and census counts.

The processing required for such applications usually involves complex signal processing operations and therefore present a number of challenges and requirements that are not evident in traditional monitoring systems. The challenges include the heavy computation needed due to very high data rates and the need for development of specific algorithms to facilitate on-line processing of very large amounts of data.

One such acoustic monitoring application, VoxNet focused on creating a system that allowed the detection of marmot alarm calls. The work was informed by the requirements of scientists to detect these marmots in the field and then localize their positions. These calls, therefore, were used to help determine the location of the animal at the time the call was detected, relative to known burrow locations. Although for some systems simple recording and offline analysis of the data fulfils application requirements, in the case of this bioacoustic monitoring system it was important that the system produce timely results. The acoustic event detection and localization application consisting of eight nodes was deployed over an area of about 9800 sq meters (2.4 acres). A gateway node was then positioned about 200 m away from the nearest node. A map of the deployment is displayed in Fig. 10.20.

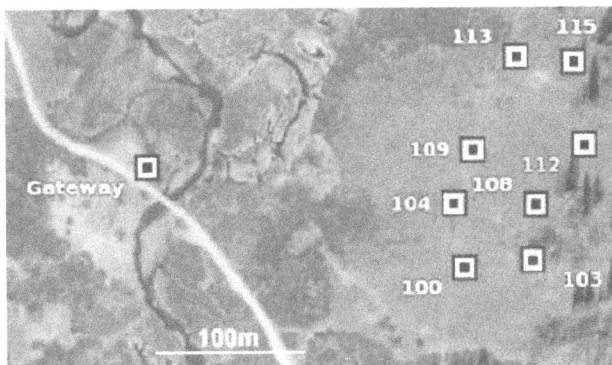

Fig. 10.20. Map of VoxNet deployment area.

Localization of Acoustic Signals – the VoxNet approach: Once a node detects an acoustic signal, it is timestamped and an attempt made to determine the location of the sound. The localization algorithm used, the Approximated Maximum Likelihood algorithm (AML) [123] functions as follows: a stream of audio is processed through a "fast path" detector to identify possible alarm calls; the algorithm then estimates bearing to the caller from multiple points followed by fusion of those estimates to produce an estimate of the sound location.

The algorithm was expressed as a WaveScript program, which is a logical dataflow graph of stream operators, connected by streams. Once the program runs on these input streams, results are streamed back to data storage components, in this case the gateway or sink node. This execution of the WaveScript program constitutes the extent of in-network processing with the result being an estimate of the direction of arrival (DOA) of the acoustic signal. Nodes that detect acoustic events then send timestamped, DOA values to the sink. In some cases, depending on time availability, AML may not be executed on the node to produce DOA estimates and instead streams of raw detections are sent instead. The size of a DOA packet is 800 bytes as compared to 32 kB for a raw detection.

In the application, a minimum of 3 acoustic signals or DOA estimates are needed to localize a sound. At the sink, these DOA estimates along with the timestamps are used to determine, first, whether the signals are indeed from the same acoustic event and if they are, they can be combined to determine the location at which the sound occurred.

For purposes of this work, metrics from the VoxNet deployment representing the number of data packets transmitted in localizing one event, the number of hops over which these were sent as well as the total distance travelled were used. A comparison was then made between the real-life deployment results and the results of the simulations incorporating in-network, region-based processing.

A Region-based Approach to Acoustic Signal Localization: The authors maintain that an approach that incorporated in-network processing using the concept of dynamic regions would be just as effective, if not more, than sending all raw data or partially processed data, that is, DOA estimates, back to the gateway or sink for analysis. In this set of experiments, given the limitations of the simulator used, estimation of query processing times and generation of random events were not possible. Instead, metrics not dependent on time, like packet size, hop count and number of packets transmitted were used to compare the efficiency and feasibility of the approach as compared to that taken in the VoxNet application in the field.

Simulation Scenario and Setup: For all simulations, a number of controls were maintained:
- The deployment of nodes in the simulator mirrored the actual real-life deployment topology. Nine nodes including the sink were positioned in the simulation window at the same relative locations as they were positioned in the actual deployment.
- All nodes were initialized with an equal and consistent amount of energy.
- Radio broadcast range was set at a 200 unit radius (a unit corresponding to a meter). This was the broadcast range of nodes in the VoxNet deployment.
- Each node's location was unique within the two-dimensional plane.

- A list of acoustic events, the ID of nodes that should detect these events and the intervals at which these events are to occur was defined prior to the start of each simulation. Lists contained between 3 and 5 events.

10.4.5.4. Region Based Query Resolution

A continuously running event detection task (referencing the event list) is implemented on all nodes. At each time interval (a system clock tick), each node checks the event list. If an event for that node exists and the current time matches the scheduled event then an acoustic event is triggered.

The detecting node logs the time of the event and checks if it has any related inquiries in its "inquiry cache". An inquiry indicates that another node has previously detected a possibly correlated event and has requested and is possible awaiting a response. If a node with an inquiry message in its inquiry cache detects a related event (that is, it is within the valid time slot) it sends a response to the node that sent the inquiry. If the node detecting the event does not have any inquiries, it sets itself as a region leader and sends an inquiry message to its one hop neighbours. This inquiry message contains the ID of the node that has detected the event and the event's timestamp. A node receiving this message registers the inquiry if it has not had an event that matches that inquiry. A matching or related event is one that falls within a time slot that was determined experimentally using the simulator. If the node detecting the event does have an inquiry, it checks to see if its event has occurred within the valid time slot (a time range is used to indicate whether two or more signals are correlated and therefore considered as relating to the same acoustic event; this is referred to as the valid time slot) and a "DATA" message is transmitted to the node who sent the inquiry message (the region leader). Any event occurring out of the valid time slot will not be sent to the region leader in response to its inquiry message.

The region leader, upon receipt of a DATA message, records that message and if it has already received the minimum (3) required or more sends the result to the sink via a RESULT message. In the simulation, if the region leader has already received 3 messages it still waits for a period of time before sending a result on. The reason is that the greater the number of messages, the better the accuracy of the measurement (this is also the model used in the VoxNet system). The time a region leader waits was again determined through use of the simulator although in the real life deployment this value was determined experimentally and is referred to as a fuzz factor.

The simulation was allowed to run until either a result was sent to the sink or if the time for detection of a particular event had elapsed. For example, an insufficient number of detections (less than 3) were detected within the valid time slot. The results of this simulation in comparison to the in-the-field results are analyzed in Sections 10.4.5.5. and 10.4.5.6.

10.4.5.5. Efficiency of Region-based Querying

The results obtained indicate that regions can be used to support in-network query processing. In analyzing how much more efficient, if at all, the region-based approach is as compared to the in-network aggregation and centralized approaches, energy efficiency is evaluated using one parameter, that is, the number of messages generated in returning a query response to the sink.

The results of running a number of 50-node simulations indicate that on average region-based query processing resulted in an almost 86 % decrease in the number of transmissions over the centralized approach and a 68 % decrease on average over the approach using in-network aggregation.

For a 100-node simulation the results were equally impressive with over 91 % reduction over the centralized approach and an almost 65 % decrease when compared to the simulation using in-network aggregation. For a 150-node simulation the results for the region-based approach were more marked when compared to the centralized approach with over 95 % reduction in the number of messages while there was an over 45% decrease when compared to the simulation using in-network aggregation. The results are summarised in Fig. 10.21.

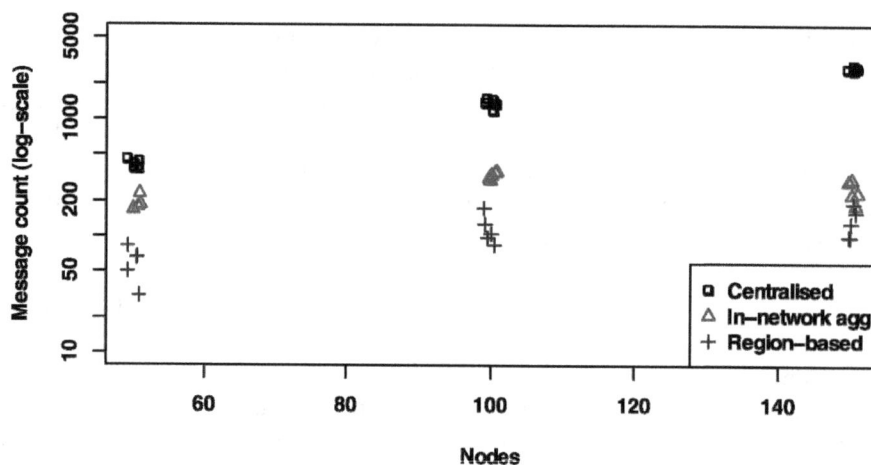

Fig. 10.21. Comparison of number of messages required for query resolution
in 50, 100 and 150-node networks.

The results showed that the efficiency of the region-based querying approach increased as the size of the network increased when compared to the centralized approach, going from 86 % to 91 % to 95 % for a 50, 100 and 150 node network respectively. In comparison to the in-network aggregation approach, however, this was not the case. Although communication was less, the relative percentages showed an overall decrease, going from 68 % to 65 % to 45 % for a 50, 100 and 150 node network respectively. It would have been interesting to run these algorithms on even larger networks in determining whether this trend would continue, however, the limitation of the simulator used made this impossible at this time. Based on these results, however, although feasible, the scalability of the region-based approach is an issue to be considered and examined more closely in future work.

10.4.5.6. Effectiveness of Region-based Querying

Again, the results obtained indicate that regions can be used to support in-network query processing and are a viable approach in the context of a real-life deployment scenario. In analyzing the effectiveness and feasibility of the region-based processing approach compared to that used in the VoxNet application a number of measures were taken. First, the number of data packets was measured and along with the packet size was used to calculate the total data transmission required for query resolution. This was done in multiple runs for cluster/region sizes of 3 and 4 nodes in the case of the VoxNet and region-based approaches respectively, and an average taken. In addition, the average total data transported was calculated for the VoxNet system in scenarios where all raw data was sent to the sink and also in cases where in-network processing for DOA estimation was carried out.

The results, displayed in Fig. 10.22, clearly indicate that in terms of data transmission savings the region-based approach exhibits an advantage over the query processing approach used in the VoxNet system. This was evident even when some in-network processing was carried out. For a 3 and 4-node region there was approximately a 28 % and 38 %, reduction respectively in the amount of data transmitted over the VoxNet approach with in-network processing. The decrease is even more marked with figures of 61 % and 67 % for the VoxNet processing approach with no in-network analysis.

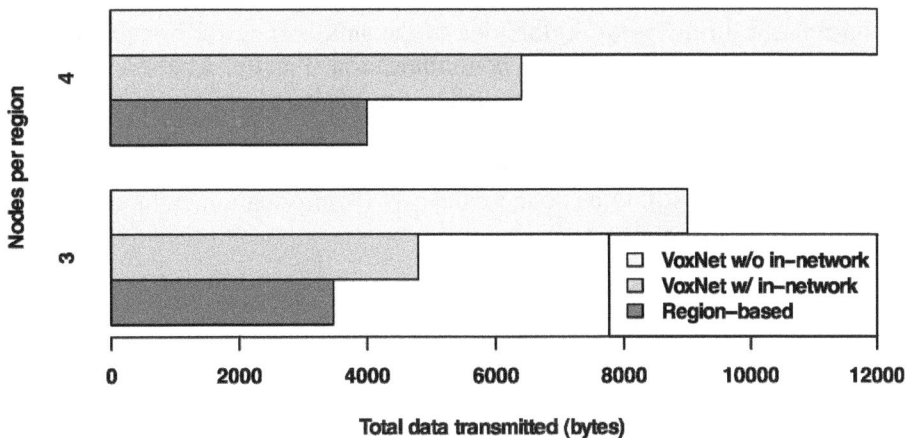

Fig. 10.22. Total amount of data transmitted in query resolution.

The simulation again confirmed the feasibility of the approach although a number of issues need to be investigated in future work. One, as identified before, is the scalability of the approach. In this simulation the network was quite small (9 nodes) and the regions created as a consequence also limited in size. An interesting exercise for the future would be to investigate the approach using larger regions and also to test the region query-processing algorithm in the field.

10.5. Conclusions

An extensive literature review has shown the absence of complex query mechanisms in the query processing systems in use today. The review did highlight, however, applications in which these queries could be used along with in-network processing to both extend and improve WSNs' deployment value.

Both the usefulness of complex queries and the feasibility of using the proposed logical abstractions (called regions) to facilitate query processing in WSN systems have been positively assessed. The preliminary experiments have produced promising results but also highlighted a number of areas that present the community with open research questions: What are the means for efficient query-based region setup ? How can dynamic regions be implemented which can change while query processing is occurring? (This becomes extremely important in cases where regions are established over an attribute that is changing with time.) What is the network size/scale at which the cost of region set-up is justified? (Combining it with cluster formation to reduce energy consumption is perhaps one approach.) For simplicity, the work here considered a scenario where a node could only be a member of a maximum of one region. Ideally, nodes should be able to be members of multiple regions. A node's single region membership can be an issue in facilitating the processing of multiple queries simultaneously or processing complex queries containing subqueries with additional attributes of interest. Moreover, here, once a response was sent to the sink the querying process terminated. In future work, the idea of the sink issuing further queries needs to be investigated as well as inter-region communication. Finally, what are the hardware support requirements to enable a system such as the one here be reliably deployed ?

This is clearly a fruitful domain and although much has been achieved, it is the authors' belief that there is much still to be done to make WSN information extraction acceptable and accessible to a wider audience.

References

[1]. A. Mainwaring, J. Polastre, R. Szewczyk, D. Culler, J. Anderson, Wireless sensor networks for habitat monitoring, in *Proceedings of the ACM International Workshop on Wireless Sensor Networks and Applications*, Atlanta, GA, United States, 2002, pp. 88–97.
[2]. R. Guy, B. Greenstein, J. Hicks, R. Kapur, N. Ramanathan, T. Schoellhammer, T. Stathopoulos, K. Weeks, K. Chang, L. Girod, D. Estrin, Experiences with the extensible sensing system, *Tech. Rep.* 61, CENS, 2006.

[3]. K Langendoen, A. Baggio, O. Visser, Murphy loves potatoes experiences from a pilot sensor network deployment in precision agriculture, in *Proceedings of 20th International Parallel and Distributed Processing Symposium, IPDPS' 2006.*

[4]. Thaddeus R. F. Fulford-Jones, Gu-Yeon Wei, and Matt Welsh, A portable, low-power, wireless two-lead EKG system, in *Proceedings of Annual International Conference of the IEEE Engineering in Medicine and Biology*, San Francisco, CA, United States, Vol. 26, III, 2004, pp. 2141–2144.

[5]. Mauricio Castillo-Effen, Daniel H. Quintela, Ramiro Jordan, Wayne Westhoff, and Wilfrido Moreno, Wireless sensor networks for flash-flood alerting, *In Proceedings of the IEEE International Caracas Conference on Devices, Circuits and Systems, ICCDCS'04*, Dominican Republic, 2004, pp. 142–146.

[6]. K. Pister, The 29 palms experiment - tracking vehicles with a UAV-delivered sensor network, 2007, http://robotics.eecs.berkeley.edu/~pister/29Palms0103

[7]. *HPWREN*, High performance wireless research and education network, 2007, http://en.wikipedia.org/wiki/High_Performance_Wireless_Research_and_Education_Net work

[8]. *ROADNet*, Real-time observatories, applications and data management network, 2007, http://roadnet.ucsd.edu

[9]. W. R. Heinzelman, A. Chandrakasan, H. Balakrishnan, Energy-efficient communication protocol for wireless microsensor networks, in *Proceedings of the 33rd Annual Hawaii International Conference on System Sciences,* Vol. 2, Maui, HI, USA, 2000, p. 10.

[10]. S. Madden, R. Szewczyk, M. J. Franklin, D. Culler, Supporting aggregate queries over ad-hoc wireless sensor networks, in *Proceedings of the 4th IEEE Workshop on Mobile Computing Systems and Applications*, Callicoon, NY, USA, 2002, pp. 49–58.

[11]. G. J. Pottie, W. J. Kaiser, Wireless integrated network sensors, *Communications of the ACM,* 43, 5, 2000, pp. 51–8.

[12]. P. Bonnet, J. Gehrke, P. Seshadri, Querying the physical world, *IEEE Personal Communications,* 7, 5, 2000, pp. 10–15.

[13]. Y. Yao, J. Gehrke, Query processing for sensor networks, *In Proceedings of Conference on Innovative Data Systems Research (CIDR),* 2003.

[14]. J. Heidemann, F. Silva, C. Intanagonwiwat, R. Govindan, D. Estrin, D. Ganesan, Building efficient wireless sensor networks with low-level naming, *Operating Systems Review (ACM)*, Vol. 35, Banff, Alta., Canada, 2002, pp. 146–159.

[15]. R. Newton, Compiling Functional Reactive Macroprograms for Sensor Networks, Masters Thesis, *Massachusetts Institute of Technology*, 2005.

[16]. R. Newton, G. Morrisett, M. Welsh, The regiment macroprogramming system, in *Proceedings of the 6th International Conference on Information Processing in Sensor Networks,* 2007, Cambridge, Massachusetts, USA, pp. 489-498.

[17]. H. Qi, Xiaoling Wang, and K. Chakrabarty, Multisensor data fusion in distributed sensor networks using mobile agents, in *Proceedings of the 5th International Conference on Information Fusion*, Annapolis, MD., 2001.

[18]. C. L. Fok, G. C. Roman, and C. Lu, Mobile agent middleware for sensor networks: an application case study, in *Proceedings of Fourth International Symposium on Information Processing in Sensor Networks* (IEEE Cat. No. 05EX1086), Los Angeles, CA, USA, 2005, pp. 382–387.

[19]. M. Chen, T. Kwon, Y. Yuan, and V. C. M. Leung, Mobile agent based wireless sensor networks, *Journal of Computers*, 1, 1, 2006, pp. 14–21.

[20]. Q. Wu, N. S. V. Rao, J. Barhen, S. S. Iyenger, V. K. Vaishnavi, H. Qi, and K. Chakrabarty, On computing mobile agent routes for data fusion in distributed sensor networks, Knowledge and Data Engineering, *IEEE Transactions on*, 16, 6, 2004, pp. 740–753.

[21]. D. Massaguert, C. Fok, Nalini Venkatasubramanian, G. Roman, C. Lu, Exploring sensor networks using mobile agents, in *Proceedings of the International Conference on Autonomous Agents,* Hakodate, Japan, 2006, pp. 323–325.

[22]. R. Tynan, G. M. P. O'Hare, D. Marsh, and D. O'Kane, Multi-agent system architectures for wireless sensor networks, in *Proceedings of 5th International Conference on Computational Science, ICCS 2005,* (Lecture Notes in Computer Science Vol. 3516), Atlanta, GA, USA. Springer-Verlag, 2005, pp. 687–94.

[23]. D. Marsh, R. Tynan, D. O'Kane, and G. M. P. O'Hare, Autonomic wireless sensor networks, *Engineering Applications of Artificial Intelligence,* 17, 7, 2004, pp. 741–748.

[24]. Robert Wesson, Frederick Hayes-Roth, John W. Burge, Cathleen Stasz, and Carl A. Sunshine, Network structures for distributed situation assessment, *IEEE Transactions on Systems, Man and Cybernetics, SMC-11,* 1, 1981, pp. 5–23.

[25]. D. N. Jayasimha, S. Sitharama Iyengar, and R. L. Kashyap, Information integration and synchronization in distributed sensor networks, *IEEE Transactions on Systems, Man and Cybernetics,* 21, 5, 1991, pp. 1032–1043.

[26]. L. Prasad, S. S. Iyengar, R. L. Kashyap, and R. N. Madan, Functional characterization of sensor integration in distributed sensor networks, in *Proceedings of the Fifth International Parallel Processing Symposium* (Cat. No. 91TH0363-2), Anaheim, CA, USA, 1991, pp. 186–93.

[27]. H. Qi, S. Iyengar, and K. Chakrabarty, Multiresolution data integration using mobile agents in distributed sensor networks, Systems, Man and Cybernetics, Part C: Applications and Reviews, *IEEE Transactions on,* 31, 3, 2001, pp. 383–391.

[28]. Yao Yuxia, Tang Xueyan, and Lim Ee-Peng, In-network processing of nearest neighbor queries for wireless sensor networks, Database Systems for Advanced Applications, in *Proceedings of 11th International Conference, DASFAA'2006.* (Lecture Notes in Computer Science Vol. 3882), Singapore, 2006, pp. 35–49.

[29]. K. Sohrabi, J. Gao, V. Ailawadhi, and G. J. Pottie, Protocols for self-organization of a wireless sensor network, *IEEE Personal Communications,* 7, 5, 2000, pp. 16–27.

[30]. H. Qi, S. S. Iyengar, and K. Chakrabarty, Distributed multi-resolution data integration using mobile agents, in *Proceedings of IEEE Aerospace Conference-2001* (Cat. No. 01TH8542), Big Sky, MT, USA, 2001, pp. 3–1133.

[31]. Hairong Qi, Yingyue Xu, and Xiaoling Wang, Mobile-agent-based collaborative signal and information processing in sensor networks, in *Proceedings of the IEEE,* 91, 8, 2003, pp. 1172–1183.

[32]. Hairong Qi and Wang Feiyi, Optimal itinerary analysis for mobile agents in ad hoc wireless sensor networks, in *Proceedings 13th International Conference on Wireless Communications,* Calgary, Alta., Canada, 2001, pp. 147–53, TRLabs/Univ. Calgary.

[33]. H. Qi, Xiaoling Wang, S. Sitharama Iyengar, K. Chakrabarty, High performance sensor integration in distributed sensor networks using mobile agents, *International Journal of High Performance Computing Applications,* 16, 3, 2002, pp. 325–335.

[34]. G. M. P. O'Hare, D. Marsh, A. Ruzzelli, and R. Tynan, Agents for wireless sensor network power management, in *Proceedings of International Workshop on Wireless and Sensor Networks (WSNET-05),* Oslo, Norway, 2005.

[35]. R. Tynan, D. Marsh, D. O'Kane, G. M. P. O'Hare, Intelligent agents for wireless sensor networks, in *Proceedings of the International Conference on Autonomous Agents,* Utrecht, Netherlands, 2005, pp. 1283–1284.

[36]. C. Min, K. Taekyoung, C. Yanghee, Data dissemination based on mobile agent in wireless sensor networks, in *Proceedings of Conference on Local Computer Networks, LCN,* Sydney, Australia, 2005, pp. 527–528.

[37]. L. Szumel, J. LeBrun, J. D. Owens, Towards a mobile agent framework for sensor networks, in *Proceedings of 2ⁿᵈ IEEE Workshop on Embedded Networked Sensors, EmNetS-II,* Sydney, Australia, 2005, pp. 79–87.

[38]. Philip Levis and David Culler, Mate: A tiny virtual machine for sensor networks, in *Proceedings of International Conference on Architectural Support for Programming Languages and Operating Systems - ASPLOS,* San Jose, CA, United States, 2002, pp. 85–95.

[39]. B. Karlsson, Oscar Bäckström, Wlodek Kulesza, and Leif Axelsson, Intelligent sensor networks: An agent-oriented approach, in *Proceedings of REALWSN,* Stockholm, Sweden, 2005.

[40]. Jade, http://jade.tilab.com, 2006.

[41]. V. K. Vaishnavi, Q. Wu, N. S. V. Rao, H. Qi, K. Chakrabarty, S. S. Iyenger, and J. Barhen, On computing mobile agent routes for data fusion in distributed sensor networks, *IEEE Transactions on Knowledge and Data Engineering,* 16, 6, 2004, pp. 740–53.

[42]. Richard Tynan, Antonio Ruzzelli, and G. M. P. O'Hare, A methodology for the deployment of multi-agent systems on wireless sensor networks, in *Proceedings of 17ᵗʰ International Conference on Software Engineering and Knowledge Engineering, (SEKE '07),* Taiwan, China, IJSEKE press, 2005.

[43]. Kenneth N. Ross, Ronald D. Chaney, George V. Cybenko, Daniel J. Burroughs, and Alan S. Willsky, Mobile agents in adaptive hierarchical bayesian networks for global awareness, in *Proceedings of the IEEE International Conference on Systems,* Man and Cybernetics, San Diego, CA, USA, 1998, pp. 2207–2212.

[44]. M. Mertsock, D. Stawski, Wireless sensor nodes as intelligent agents: Using expert systems with directed diffusion, 2005, www.mertsock.com/downloads/wsn-ia-dd.pdf

[45]. K. Romer, O. Kasten, F. Mattern, Middleware challenges for wireless sensor networks, *SIGMOBILE Mob. Comput. Commun. Rev.,* 6, 4, 2002, pp. 59–61.

[46]. Ryo Sugihara and Rajesh K. Gupta, Programming models for sensor networks: A survey, *ACM Transactions on Sensor Networks,* 4, 2, 2008.

[47]. Mate, Mate: Building application-specific sensor network language runtimes, 2003, http://www.cs.berkeley.edu/~pal/research/mate.html

[48]. Chien-Liang Fok, Gruia-Catalin Roman, and Chenyang Lu, Rapid development and flexible deployment of adaptive wireless sensor network applications, in *Proceedings of International Conference on Distributed Computing Systems,* Columbus, OH, United States, 2005, pp. 653–662.

[49]. Liu Ting and M. Martonosi, Impala: a middleware system for managing autonomic parallel sensor systems, *SIGPLAN Notices,* 38, 10, 2003, pp. 107–118.

[50]. A. Boulis, C. Han, R. Shea, M. B. Srivastava, Design and implementation of a framework for efficient and programmable sensor networks, in *Proceedings of the 1ˢᵗ International Conference on Mobile Systems,* Applications and Services, San Francisco, California, 2003, pp. 187-200.

[51]. A. Boulis, C. Han, R. Shea, M. B. Srivastava, Sensorware: Programming sensor networks beyond code update and querying, *Pervasive and Mobile Computing,* 3, 4, 2007, pp. 386–412.

[52]. C. Curino, M. Giani, M. Giorgetta, A. Giusti, A. Murphy, G. Pietro Picco, Mobile data collection in sensor networks: The tinylime middleware, *Pervasive and Mobile Computing,* 1, 4, 2005, pp. 446–469.

[53]. A. L. Murphy, G. P. Picco, G. C. Roman, Lime: a coordination model and middleware supporting mobility of hosts and agents, *ACM Transactions on Software Engineering and Methodology,* 15, 3, 2006, pp. 279–328.

[54]. A. Silberstein, Push and pull in sensor network query processing, in *Proceedings of Southeast Workshop on Data and Information Management (SWDIM '06)*, Raleigh, North Carolina, 2006, http://www.cs.duke.edu/~adam/

[55]. P. Bonnet, J. Gehrke, and P. Seshadri, Towards sensor database systems, Mobile Data Management, in *Proceedings of the 2nd International Conference, MDM 2001*, (Lecture Notes in *Computer Science* Vol., 1987), Hong Kong, China, Springer-Verlag, 2001, pp. 3–14.

[56]. R. Schollmeier, A definition of peer-to-peer networking for the classification of peer-to-peer architectures and applications, in *Proceedings of the 1st International Conference on Peer-to-Peer Computing*, Linkoping, Sweden, 2002, pp. 101–102.

[57]. P. Bonnet, P. Seshadri, Device database systems, in *Proceedings of the International Conference on Data Engineering*, San Diego, California, 2000, p. 194.

[58]. R. Govindan, J. M. Hellerstein, W. Hong, S. Madden, M. Franklin, and S. Shenker, The sensor network as a database, Technical Report 0-771, Computer Science Dept., *University of Southern California.*, 2000.

[59]. Y. Yao, J. Gehrke, The cougar approach to in-network query processing in sensor networks, *SIGMOD Rec.*, 31, 3, 2002, pp. 9–18.

[60]. P. Seshadri, Predator: A resource for database research, *SIGMOD Rec*, 27, 1, 1998, pp. 16–20.

[61]. J. Gehrke, S. Madden, Query processing in sensor networks, *IEEE Pervasive Computing*, 3, 1, 2004, pp. 46–55.

[62]. A. Demers, J. Gehrke, R. Rajmohan, N. Trigoni, Y. Yong, The cougar project: a work-in-progress report, *SIGMOD Record*, 32, 4, 2003, pp. 53–59.

[63]. S. Nath, Y. Ke, P. B. Gibbons, B. Karp, and S. Seshan, Irisnet: An architecture for enabling sensor-enriched internet service, Technical IRP-TR-03-04, *Intel Research*, June 2003.

[64]. Shen Chien-Chung, C. Srisathapornphat, and C. Jaikaeo, Sensor information networking architecture and applications, *IEEE Personal Communications*, 8, 4, 2001, pp. 52–59.

[65]. L. B. Ruiz, J. M. Nogueira, and A. A. F. Loureiro, Manna: a management architecture for wireless sensor networks, *IEEE Communications Magazine*, 41, 2, 2003, pp. 116–125.

[66]. L. Shuoqi, S. H. Son, J. A. Stankovic, Event detection services using data service middleware in distributed sensor networks, in *Proceedings of Information Processing in Sensor Networks. 2nd International Workshop, IPSN 2003*, Lecture Notes in Computer Science, Palo Alto, USA, Vol. 2634, 2003, pp. 502–517.

[67]. S. Madden, M. J. Franklin, Fjording the stream: an architecture for queries over streaming sensor data, in *Proceedings of 18th International Conference on Data Engineering*, San Jose, CA, USA, 2002, pp. 555–566.

[68]. S. R. Madden, M. J. Franklin, J. M. Hellerstein, W. Hong, Tinydb: an acquisitional query processing system for sensor networks, *ACM Transactions on Database Systems*, 30, 1, 2005, pp. 122–73.

[69]. J. Hill, R. Szewczyk, A. Woo, S. Hollar, D. Culler, and K. Pister, System architecture directions for networked sensors, in *Proceedings of International Conference on Architectural Support for Programming Languages and Operating Systems - ASPLOS*, Cambridge, MA, 2000, pp. 93–104.

[70]. SCADDS, Tinydiffusion, Technical report ISI-TR-638, *USC/Information Sciences Institute,* 2007.

[71]. S. Madden, M. J. Franklin, J. M. Hellerstein, W. Hong, The design of an acquisitional query processor for sensor networks, in *Proceedings of the ACM SIGMOD International Conference on Management of Data*, San Diego, CA, United States, 2003, pp. 491–502.

[72]. S. Ganeriwal, R. Kumar, M. B. Srivastava, Timing-sync protocol for sensor networks, in *Proceedings of the 1st International Conference on Embedded Networked Sensor Systems SenSys'03,* Los Angeles, CA, United States, 2003, pp. 138–149.

[73]. J. M. Hellerstein, H. Wei, S. Madden, K. Stanek, Beyond average: toward sophisticated sensing with queries, in *Proceedings of Information Processing in Sensor Networks. Second International Workshop, IPSN'2003.* (Lecture Notes in Computer Science, Vol. 2634), Palo Alto, CA, USA, 2003, pp. 63–79.

[74]. P. Buonadonna, D. Gay, J. M. Hellerstein, W. Hong, and S. Madden, Task: Sensor network in a box, in *Proceedings of the 2nd European Workshop on Wireless Sensor Networks, EWSN 2005,* Istanbul, Turkey, 2005, pp. 133–144.

[75]. I. Chatzigiannakis, G. Mylonas, S. Nikoletseas, Jwebdust: A java-based generic application environment for wireless sensor networks, *Lecture Notes in Computer Science,* Vol. 3560, Marina del Rey, CA, United States, 2005, pp. 376–386.

[76]. N. Sadagopan, B. Krishnamachari, A. Helmy, Active query forwarding in sensor networks, *Ad Hoc Networks,* 3, 1, 2005, pp. 91–113.

[77]. K. Whitehouse, C. Sharp, Eric Brewer, D. Culler, Hood: A neighborhood abstraction for sensor networks, in *Proceedings of the 2nd International Conference on Mobile Systems, Applications and Services MobiSys'2004*, Boston, MA, United States, 2004, pp. 99–110.

[78]. M. Welsh, G. Mainland, Programming sensor networks using abstract regions, in *Proceedings of 1st Symposium on Networked Systems Design and Implementation (NSDI '04),* San Francisco, CA, USA, 2004, pp. 29–42.

[79]. K. Whitehouse, F. Zhao, Jie Liu, Semantic streams: A framework for composable semantic interpretation of sensor data, *Lecture Notes in Computer Science*, Vol. 3868, Zurich, Switzerland, 2006, pp. 5–20.

[80]. R. Poli, W. B. Langdon, Backward-chaining evolutionary algorithms, *Artificial Intelligence,* 170, 11, 2006, pp. 953–982.

[81]. R. Newton, M. Welsh, Region streams: functional macroprogramming for sensor networks, in *Proceedings of the 1st International Workshop on Data Management for Sensor Networks,* Toronto, Canada 2004, pp. 78-87.

[82]. R. Newton, D. Arvind, M. Welsh, Building up to macroprogramming: An intermediate language for sensor networks, in *Proceedings of 4th International Symposium on Information Processing in Sensor Networks, IPSN 2005,* Los Angeles, CA, United States, 2005, pp. 37–44.

[83]. R. Gummadi, O. Gnawali, R. Govindan, Macro-programming wireless sensor networks using Kairos. Distributed Computing in Sensor Systems, in *Proceedings of 1st IEEE International Conference, DCOSS 2005.* Lecture Notes in Computer Science, Marina del Rey, Vol. 3560, CA, USA, 2005, pp. 126–140.

[84]. E. Katsiri and A. Mycroft, Knowledge representation and scalable abstract reasoning for sentient computing using first-order logic, in *Proceedings of 1st Workshop on Challenges and Novel Applications for Automated Reasoning,* Miami, FL, 2002, pp. 73–82.

[85]. T. Daniel, R. M. Newman, E. Gaura, and Mount S, Complex query processing in wireless sensor networks, in *Proc. 2nd ACM International Workshop on Performance Monitoring, Measurement, and Evaluation of Heterogeneous Wireless and Wired Networks, PM2HW2N'07,* 2007, pp. 53–60.

[86]. Argo. Argo: part of the integrated global observation strategy, 2007, www-argo.ucsd.edu

[87]. K. Martinez, R. Ong, J. Hart, Glacsweb: A sensor network for hostile environments, in *Proceedings of 1st Annual IEEE Communications Society Conference on Sensor and Ad Hoc Communications and Networks,* Santa Clara, CA, United States, 2004, pp. 81–87.

[88]. Grape Networks, Grape networks' climate genie, 2007, http://findarticles.com/p/articles/mi_m0EIN/ is_2007_Nov_12/ai_n21094407

[89]. M. Hirafuji, T. Fukatsu, T. Haoming, T. Watanabe, S. Ninomiya, A wireless sensor network with field-monitoring servers and metbroker in paddy fields, in *Proceedings of World Rice Research Conference,* Tokyo and Tsukuba, Japan, 2004.

[90]. Ning Xu, Sumit Rangwala, Krishna Kant Chintalapudi, Deepak Ganesan, Alan Broad, Ramesh Govindan, and Deborah Estrin, A wireless sensor network for structural monitoring, in *Proceedings of the 2nd International Conference on Embedded Networked Sensor Systems,* Baltimore, MD, United States, 2004, pp. 13–24.

[91]. Jerome Peter Lynch, Kincho H. Law, Anne S. Kiremidjian, Thomas Kenny, and Ed Carryer, A wireless modular monitoring system for civil structures, in *Proceedings of the International Society for Optical Engineering, SPIE'02,* Los Angeles, CA, United States, 2002, pp. 1–6.

[92]. Codeblue: Wireless sensor networks for medical care, 2007, http://www.eecs.harvard.edu/~mdw/proj/codeblue

[93]. AID-N. Aid-n Project 2007, http://www.aidn.org.au

[94]. Kim Sukun, Pakzad Shamim, Culler David, Demmel James, Fenves Gregory, Glaser Steven, and Turon Martin, Health monitoring of civil infrastructures using wireless sensor networks, in *Proceedings of the 6th International Symposium of Information Processing in Sensor Networks, IPSN,* April 2007, pp. 254-263.

[95]. B. Son, Yong-Sork Her, and Jung-Gyu Kim, A design and implementation of forest-fires surveillance system based on wireless sensor networks for South Korea mountains, *International Journal of Computer Science and Network Security,* 6, 9B, 2006, pp. 124–130.

[96]. Carl Hartung, Richard Han, Carl Seielstad, and Saxon Holbrook, Firewxnet: A multi-tiered portable wireless system for monitoring weather conditions in wildland fire environments, in *Proc. of the of 4th International Conference on Mobile Systems, Applications and Services (MobiSys 2006),* Uppsala, Sweden, 2006, pp. 28–41.

[97]. Narayanan Sadagopan, Bhaskar Krishnamachari, and Ahmed Helmy, Active query forwarding in sensor networks, *Ad Hoc Networks,* 3, 1, 2005, pp. 91–113.

[98]. Atish Datta Chowdhury and Shivashankar Balu, Consensus: A system study of monitoring applications for wireless sensor networks, in *Proceedings of the Conference on Local Computer Networks,* LCN, Tampa, FL, United States, IEEE Computer Society, Los Alamitos, CA United States, 2004, pp. 587–588.

[99]. I. Carreras, I. Chlamtac, H. Woesner, and H. Zhang, Nomadic sensor networks, in *Proceedings of the 2nd European Workshop on Wireless Sensor Networks, (EWSN' 2005),* Istanbul, Turkey, 2005, pp. 166–175.

[100]. H. Karl and A. Willig, A short survey of wireless sensor networks, Technical TKN-03-018, Telecommunication Network Group, *Technische Universitat,* 2003.

[101]. I. F. Akyildiz, W. Su, Y. Sankarasubramaniam, and E. Cayirci, Wireless sensor networks: a survey, *Computer Networks,* 38, 4, 2002, pp. 393–422.

[102]. Ali Iranli, Morteza Maleki, and Massoud Pedram. Energy efficient strategies for deployment of a two-level wireless sensor network, in *Proceedings of the International Symposium on Low Power Electronics and Design,* San Diego, CA, United States, 2005, pp. 233–238.

[103]. Gilman Tolle, Joseph Polastre, Robert Szewczyk, David Culler, Neil Turner, Kevin Tu, Stephen Burgess, Todd Dawson, Phil Buonadonna, David Gay, Wei Hong, A Macroscope in the Redwoods, in *Proceedings of the 3rd international conference on Embedded networked sensor systems, (SenSys '05),* New York, NY, USA, 2005, pp. 51-63.

[104]. K. Akkaya and M. Younis, A survey on routing protocols for wireless sensor networks. *Ad Hoc Networks,* 3, 3, 2005, pp. 325–349.

[105]. M. Younis, M. Youssef, and K. Arisha, Energy-aware routing in cluster-based sensor networks, in *Proceedings of the 10th IEEE International Symposium on Modeling, Analysis, and Simulation of Computer and Telecommunications Systems*, 2002, p. 129.

[106]. Nauman Israr and Irfan Awan, Multihop clustering algorithm for load balancing in wireless sensor networks, *International Journal of Simulation: Systems, Science and Technology*, 8, 3, 2007, pp. 13–25.

[107]. C. Chen, J. Ma, Ke Yu, Designing energy-efficient wireless sensor networks with mobile sinks, 2006, http://www.sensorplanet.org/wsw2006/1_Chen_WSW06_final.pdf

[108]. P. Gupta and P. R. Kumar, The capacity of wireless networks, *IEEE Transactions on Information Theory*, 46, 2, 2000, pp. 388–404.

[109]. Rahul C. Shah, Sumit Roy, Sushant Jain, and Waylon Brunette, Data mules: Modeling and analysis of a three-tier architecture for sparse sensor networks, *Ad Hoc Networks*, 1, 2-3, 2003, pp. 215–233.

[110]. Gokhan Mergen, Qing Zhao, and Lang Tong, Sensor networks with mobile access: Energy and capacity considerations, *IEEE Transactions on Communications*, 54, 11, 2006, pp. 2033–2044.

[111]. Yong Yao and Johannes Gehrke, The cougar approach to in-network query processing in sensor networks, *SIGMOD Rec.,* 31, 3, 2002, pp. 9–18.

[112]. S. Ramanathan and M. Steenstrup, A survey of routing techniques for mobile communications networks, *Journal of Special Topics in Mobile Networks and Applications (MONET)*, 1, 2, 1996, pp. 89–104.

[113]. J. N. Al-Karaki and A. E. Kamal, Routing techniques in wireless sensor networks: a survey, *IEEE Wireless Communications*, 11, 6, 2004, pp. 6–28.

[114]. B. Krishnamachari, D. Estrin, and S. Wicker, Modeling data-centric routing in wireless sensor networks, in *Proceedings of IEEE InfoCom*, 2002.

[115]. R. Eskicioglu, S. Ahmed, S. Hussain, A query processing architecture for sensor networks, in *Proceedings of WICON Workshop on Information Fusion and Dissemination in Wireless Sensor Networks (SensorFusion)*, Budapest, Hungary, 2005.

[116]. Yang Xiaoyan, Lim Hock Beng, Tamer M. Özsu, and Tan Kian Lee, In-network execution of monitoring queries in sensor networks, in *Proceedings of SIGMOD Conference*, 2007, pp. 521-532.

[117]. K. Ramamohanarao, L. Kulik, S. Selvadurai, B. Scholz, U. Roehm, E. Tanin, A. Viglas, A. Zomaya, and C. Leckie, A survey on data processing issues in wireless sensor networks for enterprise information infrastructure, May 2006, http://www.eii.edu.au/files/EIIAnnualRpt2006.pdf

[118]. S. Madden, M. J. Franklin, J. M. Hellerstein, and Hong Wei, Tag: a tiny aggregation service for ad hoc sensor networks, in *Proceedings of the Fifth Symposium on Operating Systems Design and Implementation (OSDI'02)*, Boston, MA, USA, 2002, pp. 131–46.

[119]. C. Intanagonwiwat, R. Govindan, and D. Estrin, Directed diffusion: a scalable and robust communication paradigm for sensor networks, MobiCom 2000, in *Proceedings of the 6th Annual International Conference on Mobile Computing and Networking*, Boston, MA, USA, 2000, pp. 56–67.

[120]. Chalermek Intanagonwiwat, Deborah Estrin, Ramesh Govindan, and John Heidemann, Impact of network density on data aggregation in wireless sensor networks, in *Proceedings of International Conference on Distributed Computing Systems*, Vienna, Austria, 2002, pp. 457–458.

[121]. O. Younis and S. Fahmy, Heed: A hybrid, energy-efficient, distributed clustering approach for ad hoc sensor networks, *IEEE Transactions on Mobile Computing*, 3, 4, 2004, pp. 366–379.

[122]. Guo Bin, Li Zhe, and Meng Yan, A dynamic-clustering reactive routing algorithm for wireless sensor networks, in *Proceedings of 1st International Conference on*

Communications and Networking in China, ChinaCom '06, Beijing, China, 2007, pp. 4149783.

[123]. A. M. Ali, Yao Kung, T. C. Collier, C. E. Taylor, D. T. Blumstein, and L. Girod, An empirical study of collaborative acoustic source localization, in *Proceedings of the 6th International Symposium on Information Processing in Sensor Networks*, Cambridge, MA, USA, 2007, pp. 41–50.

Chapter 11
Software Modeling Techniques for Wireless Sensor Networks

John Khalil Jacoub, Ramiro Liscano, Jeremy S. Bradbury

11.1. Introduction

A Wireless Sensor Network (WSN) consists of small wireless units called motes, which are attached to specific type of sensors. The sensors measure an environment phenomenon (e.g., humidity, temperature, or soil moisture) and the measured value is expressed as an analog signal generated by the sensors. According to Akyildiz et al. [1], WSN applications can be classified into 2 types; those that have mobile sensors and those that have static sensors. This classification is too broad because within the static WSN category we one can also identify 3 distinct style of static WSNs; Collector WSNs, aggregation WSNs, and actuation WSNs.

Collector WSNs consist of a number of sensors that gather data and this data is collected at one or more nodes that typically are a gateway to the Internet. These WSNs typically take advantage of tree style of routing. Aggregation WSNs perform some data aggregation and filtering within the network nodes and publish this filtered information for other collector nodes to acquire. Actuator WSNs not only sense the phenomenon, but also react in response to the sensed data. These networks are typically referred to as Wireless Sensor Actuator Networks (WSANs) or at times Wireless Sensor Actor Networks (WSANs) and typically have more stringent real time constraints on the delivery of the data.

WSN systems can be complex and many different challenges can arise during the design of a WSN such as: (i) the distribution of nodes and sensors in a physical environment that may result in lost and delayed data; (ii) the inclusion of real-time behavior within a distributed WSAN; (iii) memory management within the sensor nodes (the small size of the nodes leads to physical limitations that restrict the available memory and therefore memory management is often required;) (iv) operational reliability (WSNs often consist of self-powered nodes in environmentally challenging domains imposing strong

John Khalil Jacoub
University of Ontario Institute of Technology, Oshawa, Ontario, Canada

reliability requirements;) and (v) improving the network performance, i. e. reduce network delays, packet loss, while increasing throughput (network performance improvements may involve the use of concurrency and event driven communication, which can add additional complexity to the system.)

The design of WSN systems usually occurs at the implementation level and does not involve design at higher levels of abstraction. This leads to a decrease in code portability and to platform-specific implementations [2]. A WSN system produced using this approach is prone to both design and implementation errors and very challenging code debugging (user interfaces to sensor nodes are very limited so even simple text output is challenging). If errors are not detected during the implementation and verification stages of development then they may appear once the system is deployed and is operational. The nodes of an operational WSN application are generally difficult to access once they are deployed in their working locations.

The challenges of developing WSN systems can be mitigated by leveraging higher system level-design and analysis. The use of modeling languages and techniques can drive the design through different abstraction layers and analysis tools can help refine the model. In this chapter we survey 9 modeling techniques for WSNs. For each technique we examine how a WSN is modeled at the node and system-level. We also describe a WSN system called Sensor Infrastructure for Viticulture introduce (SensIV) that we use as a case study to assess how each modeling approach would be used to develop the SensIV system. SensIV is considered a collector WSN that was designed and built by our research team to monitor the temperature in a vineyard field. The deployed nodes are static there is no direct action taken in response to the sensed data.

The rest of the chapter is organized into 8 sections. Section 11.2 describes the SensIV case study. Section 11.3 gives an overview of the modeling techniques reviewed in this survey. Section 11.4 discusses the modeling of WSN elements including sensors, nodes and hardware. Section 11.5 presents WSN modeling at the system level. Section 11.6 discusses the supporting tools and the importance of each tool for WSN design. Section 11.7 presents related work (i.e. other sensor modeling surveys.) Finally section 11.8 presents the conclusion and future work.

11.2. Case Study

In this section we present SensIV [3] project as a case study for the modeling techniques surveyed. Using a case study helps in understanding how the modeling approaches presented in this chapter could be applied to help develop and analyze the software for a system that we are fairly familiar with.

11.2.1. System Overview

The aim of the SensIV project is to monitor the temperature of a vineyard field. For this survey understanding the vineyard application is not crucial and therefore SensIV can be considered as simply a collector WSN consisting of a set of sensor nodes that collect temperature data across a field and transmit this data to a collection point using a wireless sensor network. Each sensor node supports 4 temperature sensors that measure the gradient in the vertical direction at one spot in the field. The motes communicate with each other wirelessly until the data reaches the gateway where the temperature values are stored in a database for further analysis.

11.2.2. Sensor Physical Layer

Each node consists of the following hardware:

- Four temperature sensors: The sensors used are thermal sensors of type LM135 [4] The sensor accuracy is ± 1 °C and can measure a temperature range in between -55 °C to 150 °C.
- An MDA300CA acquisition board: The MDA300CA data acquisition board from Crossbow has expansion connectors for 7 single-ended and 4 differential ADC channels.
- A Wireless Iris mode: The Iris wireless module uses the Atmel processor with 128 Kb program memory and 8 Kb RAM. In terms of radio communication, it uses 2.4 GHz (IEEE802.15.4) with a range of 500 meters. This type of mote also supports a low-power mode of operation for the micro-processor, radio, and logger.
- A solar cell is used to recharge the 2 AA alkaline batteries that power the sensor node. The system was designed for continuous operation with sunlight conditions encountered in Southern Ontario.

11.2.3. Routing Protocol

The protocol used for routing a data packet to the collector is the Collector Tree Protocol (CTP) [5] CTP is an address free protocol that maintains a tree routing topology among the sensor network to the collector. The version of CTP used is the one available in the TinyOS deployment that is fairly reliable. The parent of a node is chosen based on the link estimation value that is continuously updated as data is being sent through the network.

11.2.4. Sensor Software

The sensor software leverages the TinyOS 2.1 [6] operating system which provides many useful services for sensor networks some of these which are: a task scheduler, boot sequence, timers, memory storage, serial and radio communication, radio management, and multi-hop routing protocols.

There are 2 distinct sensor software packages. One package is for the collector and another package is for the sensor nodes in the field. For the study case we will focus primarily on the software for that resides on the sensor nodes on the field. We also have designed a software component on sensor node that captures "control messages" sent from the collector that are transmitted using a dissemination protocol that is also provided by the TinyOS community. For the purpose of the case study we focus only on the software modules that collect data and transmit this data to the collector and do not present any logic or models that relate to the capture of the "control messages".

In TinyOS there is a task execution process coordinated by a time scheduler. If there is no task in the task queue, the timer scheduler will put the processor into sleep mode. The processor is woken up by either a message arriving signal from the dissemination protocol or the timer firing event used to collect data (the system was designed for a 5 minute data acquisition period.)

11.2.5. Location of the Sensor Nodes

The sensor nodes were deployed in a field at the University of Ontario Institute of Technology in Oshawa, Canada. We have captured the coordinates of the nodes and these are shown in a Google map image in Fig. 11.1. The approximate average distance between the sensor nodes is about 32 m. The terrain slopes downwards from nodes 4, 9, 2, 8, 7, and 3 towards 5, 1, 6, and 10. The collector node (marked as Base) is located about 4 m of the ground in a 2nd floor office of a building. From a system modeling perspective the location of the nodes is important for the performance analysis of the sensor network system.

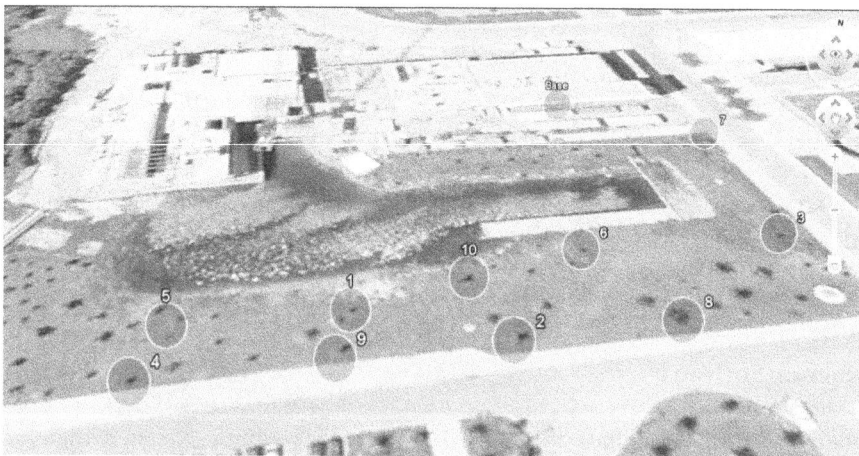

Fig. 11.1. Nodes Physical Location.

11.3. Overview of the Software Modeling Techniques for Sensor Networks

In this section, we provide an overview of each of the software modeling techniques for sensor networks included in our survey. Some of those modeling techniques have been used in the application development process, while others take WSNs as a case study for their modeling approach. Some use standard software modeling notations, such as UML, while others use their own custom notation. We also attempt for each approach to describe how the SensIV system would be modeled using that technique.

The techniques use different basic elements, such as channels, processes, modules, and components, to express a WSN as a model. A channel is used to represent the communication between two elements of a WSN. For example, channels can represent the characteristics of sensor-node communication, node-node communication, and node-gateway communication. Processes, modules, and components are used to represent the sensors and nodes of a WSN. We will describe details of the modeling elements later when we discuss the individual modeling techniques.

The modeling techniques surveyed also vary in terms of the scope of modeling. For example, some techniques are intended to model a single node or communication between a pair of nodes while others are intended to model the entire WSN.

11.3.1. HL-SDL

HL-SDL [7] is a modeling language that uses the Specification and Description Language (SDL) [8], which is normally used to model and simulate communication protocols. SDL has been adapted by Dietterle, et al., to model TinyOS components using SDL processes (i.e., extended finite state machines). The system is modeled as a collection of channels and processes. The model can be used to generate nesC source code, which is commonly used in WSNs based on the TinyOS environment. While generating nesC code, each process (which is the smallest unit of the model) represents a component in TinyOS In their work, Dietterle et al. used manual optimization to enhance the generate code

Their approach proposes that a TinyOS component be considered as the minimal entity to encapsulate the behavior of an SDL process by implementing the process state machine. Communication between SDL processes is via asynchronous signals that could be mapped to either a TinyOS command or event handler. They also propose that the process state machine logic can be captured by using one TinyOS task that services the incoming commands and events, performs some state changes in the component, and calls other components in the system. The model can be used to generate nesC source code, which is commonly used in WSNs based on the TinyOS environment. In their work, Dietterle et al. used manual optimization to enhance the generated code.

As related to the SensIV case study we designed the SensIV software using one component named WhiteRabbitC that services 21 events and launches 12 tasks as shown in the class diagram in Fig. 11.2 (Tasks are shown with the symbol *T* in front of them and the event listeners are marked with the letter *E*.)

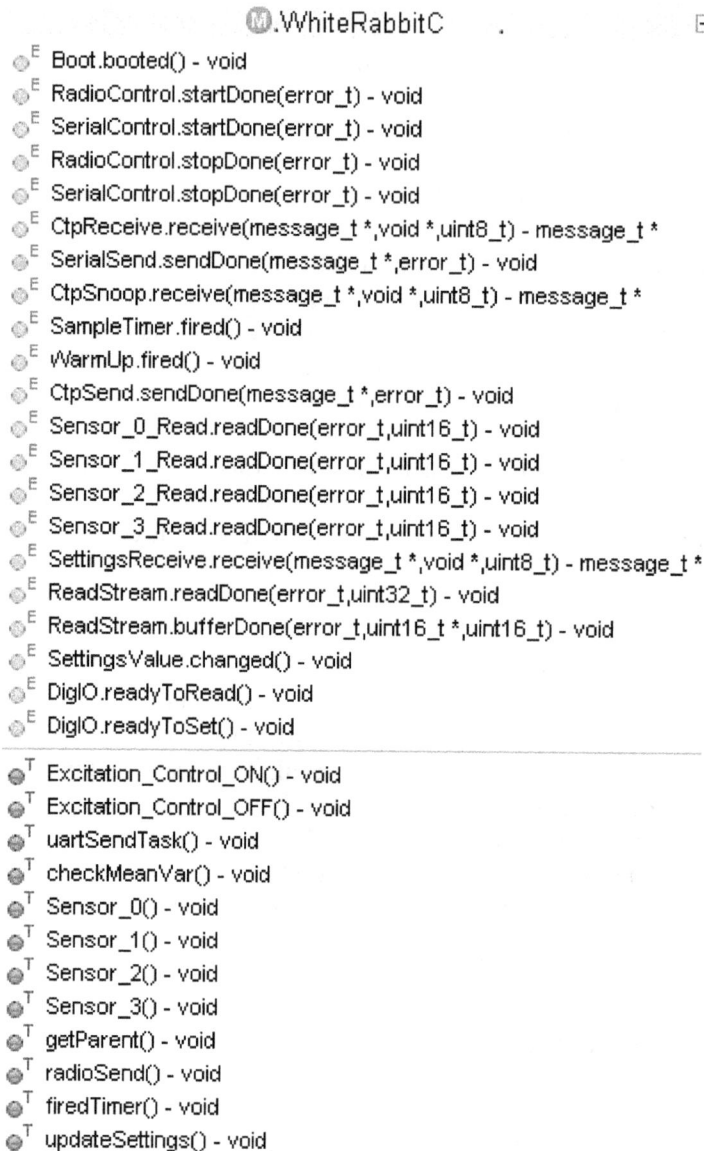

Ⓜ.WhiteRabbitC   .   E

- E Boot.booted() - void
- E RadioControl.startDone(error_t) - void
- E SerialControl.startDone(error_t) - void
- E RadioControl.stopDone(error_t) - void
- E SerialControl.stopDone(error_t) - void
- E CtpReceive.receive(message_t *,void *,uint8_t) - message_t *
- E SerialSend.sendDone(message_t *,error_t) - void
- E CtpSnoop.receive(message_t *,void *,uint8_t) - message_t *
- E SampleTimer.fired() - void
- E WarmUp.fired() - void
- E CtpSend.sendDone(message_t *,error_t) - void
- E Sensor_0_Read.readDone(error_t,uint16_t) - void
- E Sensor_1_Read.readDone(error_t,uint16_t) - void
- E Sensor_2_Read.readDone(error_t,uint16_t) - void
- E Sensor_3_Read.readDone(error_t,uint16_t) - void
- E SettingsReceive.receive(message_t *,void *,uint8_t) - message_t *
- E ReadStream.readDone(error_t,uint32_t) - void
- E ReadStream.bufferDone(error_t,uint16_t *,uint16_t) - void
- E SettingsValue.changed() - void
- E DigIO.readyToRead() - void
- E DigIO.readyToSet() - void

- T Excitation_Control_ON() - void
- T Excitation_Control_OFF() - void
- T uartSendTask() - void
- T checkMeanVar() - void
- T Sensor_0() - void
- T Sensor_1() - void
- T Sensor_2() - void
- T Sensor_3() - void
- T getParent() - void
- T radioSend() - void
- T firedTimer() - void
- T updateSettings() - void

Fig. 11.2. SensIV Software Component.

This implies that to be able to leverage the SDL modeling language we should be considering the use of several components with very simple behaviors that can be captured with the use of 1 task. This is not something that comes naturally in the design of the software. After reflecting on our design it is possible to consider components that contain only 1 task but this significantly breaks up the code into many small components each having a very specific role. For example, SensIV uses 4 tasks to read the sensory data. It does this because it takes some time to read the data and the read operations are asynchronous. This results in very small and detailed state machines since each component requires one as opposed to a single state machine in SensIV that captures the data acquisition cycle and depicted in Fig. 11.3.

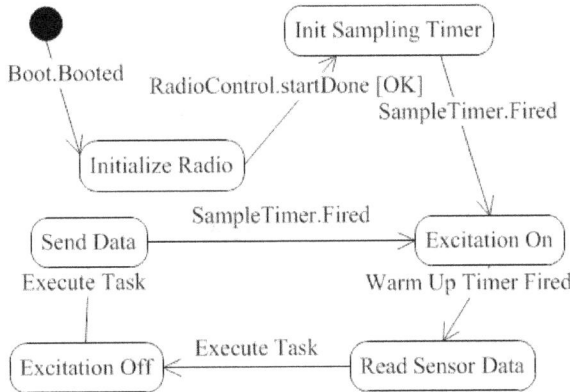

Fig. 11.3. SensIV State Diagram.

We can follow the process presented in HL-SDL to create an SDL process diagram for the SensIV software but one has to keep in mind that the SDL analysis of this result would not be helpful since several tasks are being used to achieve this. An SDL process is built from the combination of a state and activity diagram. Fig. 11.3 and Fig. 11.4 show the state and activity diagram of SensIV software related to the data acquisition logic.

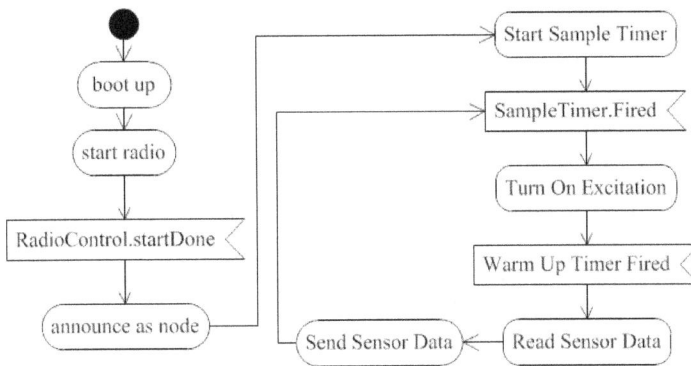

Fig. 11.4. SensIV Activity Diagram.

An SDL process diagram can be created from the combination of the state and activity diagram as depicted in Fig. 11.5.

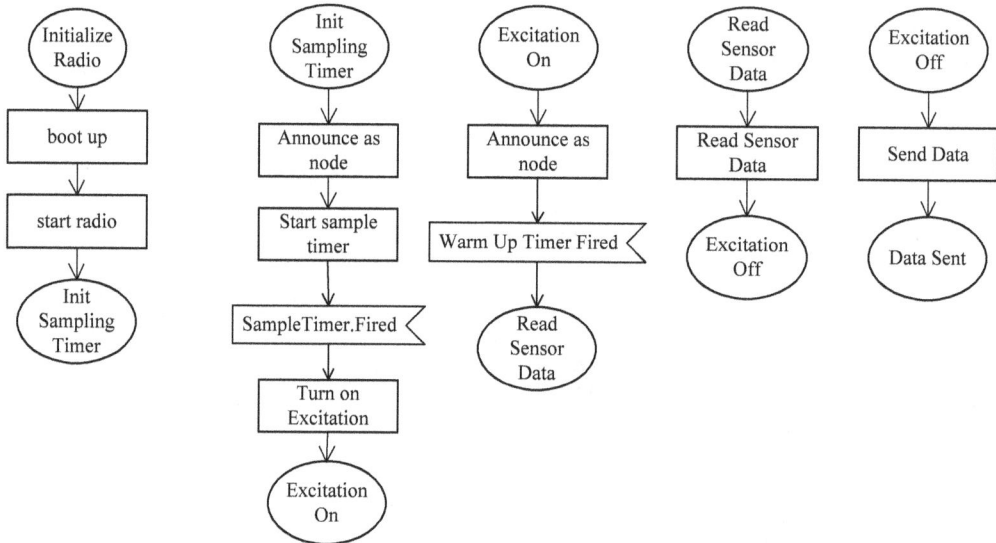

Fig. 11.5. SensIV SDL Process Diagram.

11.3.2. Insense

Dearle et al. [9] use the Insense modeling language to create a component-based model for a WSN. Components in Insense are concurrent and they communicate synchronously via directional channels that are used to abstract away from low-level synchronization and communication issues. Insense is built in the Contiki operating system, which similar to TinyOS is a popular operating system used for WSNs. The Insense model has a translator that produces C source code that can be used to calculate important details such as worst case execution time.

11.3.2.1. Insense Model Elements

Insense captures the key system elements of the WSN as components. The components can represent software or hardware entities of a WSN. The components capsulate a specific behavior of an element in the WSN system and has interfaces to interact with other components of the system. Each component has no dependences on the other components. However, a component can initiate the activity of another component. The component stops execution either by an external signal from the other component or by itself. Each component definition contains four main parts:

- The channel part, which contains the input and output channels of each component and other components which interact with the component;
- The component variables;
- The component constructor;
- The behavior part, which captures the component behavior.

The component starts to execute the behavior once the other component creates the instance of it. The flow of the data is declared by using the command send, receive, input, and output as shown in Fig. 11.8. In order to bind the components with each other, the channels are declared in the design as well. For example, *connect sensor. output to sensor reader. input* where reader and sensor are components and input and output are the ports. Based on that, the communication between the components is synchronized. The language supports deterministic and non-deterministic selection of the nodes by using logic statement, such as If and select statements. Insense provides components in order to model the hardware such as temperature sensor or humidity sensor.

11.3.2.2. Insense Model for SensIV

Insense models SensIV by using components which represents the hardware and the software components of the one node. In this section, we only explain some components of the SensIV system due to the size limitation of this chapter (see Fig. 11.6).

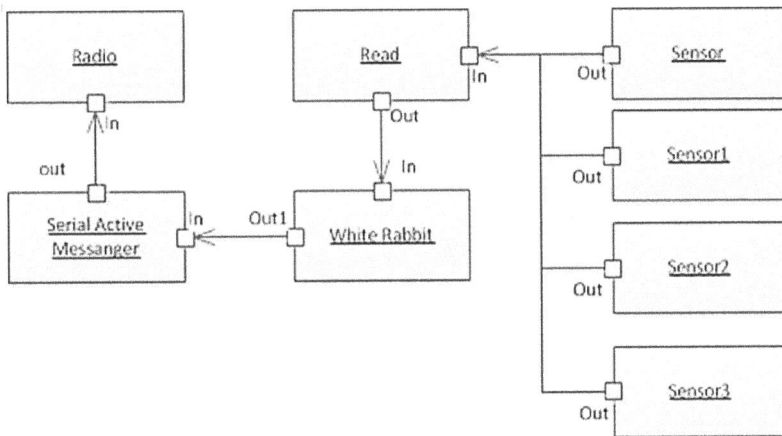

Fig. 11.6. Captured SensIV Components by Insense.

The model captures the interface between the captured nodes (see Fig. 11.7) and the behavior of the following components:

- **WhiteRabbit component:** requests the data from the sensors, receives the four values, and generates one data packet (see Fig. 11.8).

- **Read component**: triggers the sensors to read, receives the values, and send the values to WhiteRabbit (see Fig. 11.9).
- **Sensor component:** represents a hardware component and captures the behavior of the thermal sensor (see Fig. 11.10).
- **Serial Active Messenger component:** sends the data packet to the radio and invokes the radio to send them over (see Fig. 11.11).
- **Radio component**: sends the data through the wireless signal (see Fig. 11.12).

```
type SensorRD  is interface( out float out )
type SensorRD1  is interface( out float out )
type SensorRD2  is interface( out float out )
type SensorRD3  is interface( out float out )
type Read_Interface is interface( in float  input; out float output)
type White_Interface is interface( in integer input; out vector output )
type Serial_Interface is interface( in vector input; out vector output )
type Radio_Interface is interface( in vector input)
```

Fig. 11.7. Declaration of Component Interfaces.

```
component WhiteRabbit presents White_Interface {
    size=4
    index=0

    constructor() { ... }

    behaviour {
    Read= new read()
    receive next from input
    store[index] := next
        index := index + 1
        if( index >= size ) {
            SerialActiveMessanger = new SerialActiveMessanger()
                    connects White_Rabbit.ouput to SerialActiveMessanger.input
            send store on output
            index := 0
        }
    }
}
```

Fig. 11.8. WhiteRabbit Component Model.

```
component Read presents Read_Interface {

    constructor() {
        sensor = new sensor ()
        sensor1 = new sensor1 ()
        sensor2 = new sensor2 ()
        sensor3 = new sensor3 ()
    }

    behaviour {
        receive Temp from input
        connects Read.ouput to WhiteRabbit.input
        send Temp on output
        index := 0
    }
}
```

Fig. 11.9. Read Component Model.

```
component Sensor presents SensorRD {
    constructor() {...}

    behaviour {
        connects sensor0.ouput to Read_Interface.input
        send Temp on output
    }
}
```

Fig. 11.10. Sensor Component Model.

```
component SerialActiveMessanger presents Serial_Interface {

    constructor() {...}

    behaviour {
        receive store from input
        Radio= new Radio()
        connects SerialActiveMessanger.ouput to Radio.input
        send store on output
    }
}
```

Fig. 11.11. SerialActiveMessanger Component Model.

```
component Radio presents Radio_Interface {

    constructor() {...}

    behaviour {
        receive store from input
    }
}
```

Fig. 11.12. Radio Component Model.

We would like to point out that WhiteRabbit develops other activities. However, we include the activity of sensing and sending the data only due to the size limitation of this chapter.

11.3.3. Mathworks

The framework in [10] aims to design, simulate, and generate the code for WSNs. The node behavior is modeled as a parameterized Stateflow block. Nodes in the Mathworks approach also contain timing and random number generators that are used for simulation. Additionally, the communication medium, which is used to define the connectivity between the nodes, is represented at a lower abstraction level and is implemented in the C language. By leveraging Mathworks tools, such as animated state charts, chart displays, scopes, and plots, analysis of the WSNs can be performed. According to the results, the model can be refined. The final stage is to generate the WSN code using the Target Language Complier (TLC) which can generate C code for MANTIS and nesC code for TinyOS. The Mathworks approach has been used successfully to generate the code for Energy Efficient and Reliable In-Network Aggregation (EERINA) algorithm for clustered sensor networks [11].

11.3.3.1. Design Representation

The authors of Mathworks have designed a sensor node block and a communication medium block. The sensor node block contains a timer generator and a parameterized state flow which implements the algorithm deployed inside each node. The sensor node library has been implemented separately and each node contains an instance of the library. Therefore, all nodes run the same algorithm. However, changing the algorithm can be done by creating a new library and then creating an instance of the new library.

The communication medium block was implemented on C language based S-Function [12] which is used to specify the connectivity of each node and implement the communication medium logic. The data packets are inputs and outputs of the communication block. The data packets are processed first by the communication medium block then it is fed to the appropriate output.

11.3.3.2. System Analysis

Mathworks takes advantage of the analysis tools, such as the animated state charts, scopes, and displays, to perform functional analysis of the algorithm. The design is enhanced based on the analysis results. However, the authors did not present any results from the analysis or an example of the enhancements that could be done on the model based on the analysis results.

11.3.3.3. Code Generation

The embedded coder is used in order to generate ANSI C code which represents the state charts of the nodes. The last stage is the Target Language Complier (TLC) which generates the code for the deployment. TLC generates 2 types of code: nesC code for TinyOS and C code for MANTIS [13] (MANTIS is another operating system for WSNs).

TLC contains a group of mapping rules which controls the code transformation process from the ANSI C code to the target code. In addition, TLC adds the information and details to the code which are required by the target platform in order to execute the code. The authors comment that generating the C code for MANTIS was easier than generating nesC for TinyOS because of the similarity between C and ANSI C code. The generated code did not need any kind of modification. On the other hand, the generated code size is bigger than the manually implemented code.

11.3.3.4. Mathwork Model for SensIV

As mentioned above, Mathworks have used state charts and activity diagram to capture the system behavior. SensIV state chart and activity diagrams are shown in Fig. 11.3 and Fig. 11.4 respectively. From communication medium prospective, SensIV nodes are deployed within the range of 500 m of each other. Therefore, potentially all nodes can communicate with each other. Additionally, CTP chooses next hop based on the link estimation. Therefore we have no expectation of the connectivity of the nodes connectivity since all nodes can communicate with each.

11.3.4. Model Driven Engineering Approach (MDEA)

Losilla et al. [2] use UML and a Model Driven Engineering (MDE) approach that includes three the 3 modeling layers shown in Table 11.1. The research team was working in parallel on Model-to-Model (M2M) Transformations and Model-to-Text

(M2T) Transformations. In M2M transformations the model transformation takes place between the WSN-DSL layer to the PIM-UML layer and then from the PIM-UML layer to nesC. In the M2T transformation the nesC code is generated from the nesC meta-model.

Transformation rules control transforming from one modeling layer to another. Moreover, refinement can occur after every transformation to improve the generated model. The MDEA approach is supported by the Eclipse IDE as well as a number of Eclipse plug-ins (e.g., MOFScript) that are responsible for automating the transformation process.

Table 11.1. MDEA Layers.

Layer	Information Captured	Example	Annotations
Domain Specific Language	Functional and non-functional requirements	Nodes functionality, data stores locations, physical distributions, communication mechanism, and communication rate	Class diagrams
UML PIM	Data flow	System states and Flow of the algorithm	Activity diagrams and state machine diagrams
nesC Meta Model	Software components	Radio methods, LEDs, and timer methods	Component diagrams

11.3.4.1. WSN-DSL Model

WSN-DSL model captures the functional and non-functional requirements of the system. The information captured by this layer along with the corresponding SensIV entities is shown in Table 11.2.

The SensIV class diagram is shown if Fig. 11.14 as derived by leveraging the DSL meta-model defined in MDEA for sensor networks (reproduced in Fig. 11.13).

11.3.4.2. UML-PIM Model

The UML-PIM (Platform) model has been created by using UML 2.0. The aim of developing this stage is to cover the semantic gap between the WSN-DSL model and the nesC meta-model. The UML-PIM model explains the control and the data flow of the system through using the activity diagrams and using state machine diagrams. An example of the transformation rules between WSN-DSL and UML-PIM is each node group is mapped to a component. The state diagram and the activity diagram for SensIV can be used as the UML-PIM model and have been presented in Fig. 11.3 and Fig. 11.4 respectively.

Table 11.2. SensIV WSN-DSL Layer.

Item	Description	SensIV
Node groups	All nodes with the same behavior	Group 1: The sensors (10 nodes) Group 2: The base station (1 node)
Functional Units	Information about the function of each group	Group 1: Sensing Group 2: Network monitoring and discrimination commands
Functional Units Types	Methods and components of each group	mainC, LedsC ,TimerMiliC, WarmUP WatchDog, DemoSensor0, DemoSensor1 DemoSensor2, DemoSensor3, Radio(ActiveMessageC), DisseminatorC ActiveMessageC, SerialActiveMessage SerialAMSenderC, UARTMessagePoolP(PoolC) UARTQueuePC(QueueC), Batter
Resources	Sensors and ports	Thermal sensors, Ethernet ports, radio unit, and solar cell

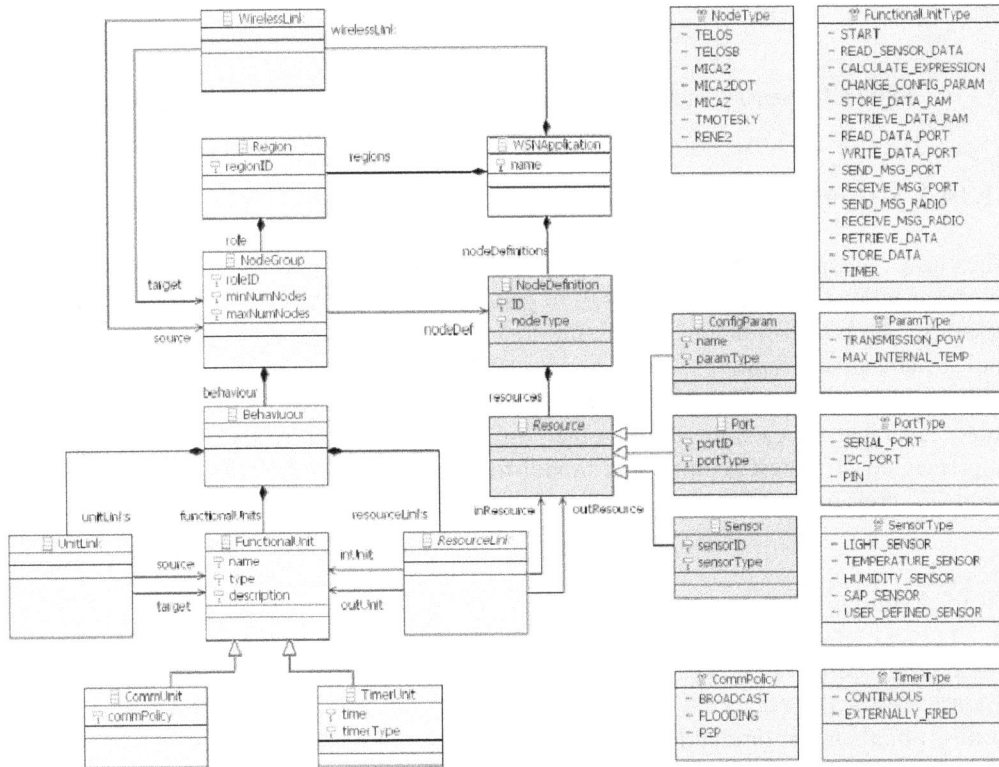

Fig. 11.13. WSN-DSL Meta-Model, reproduced from [2].

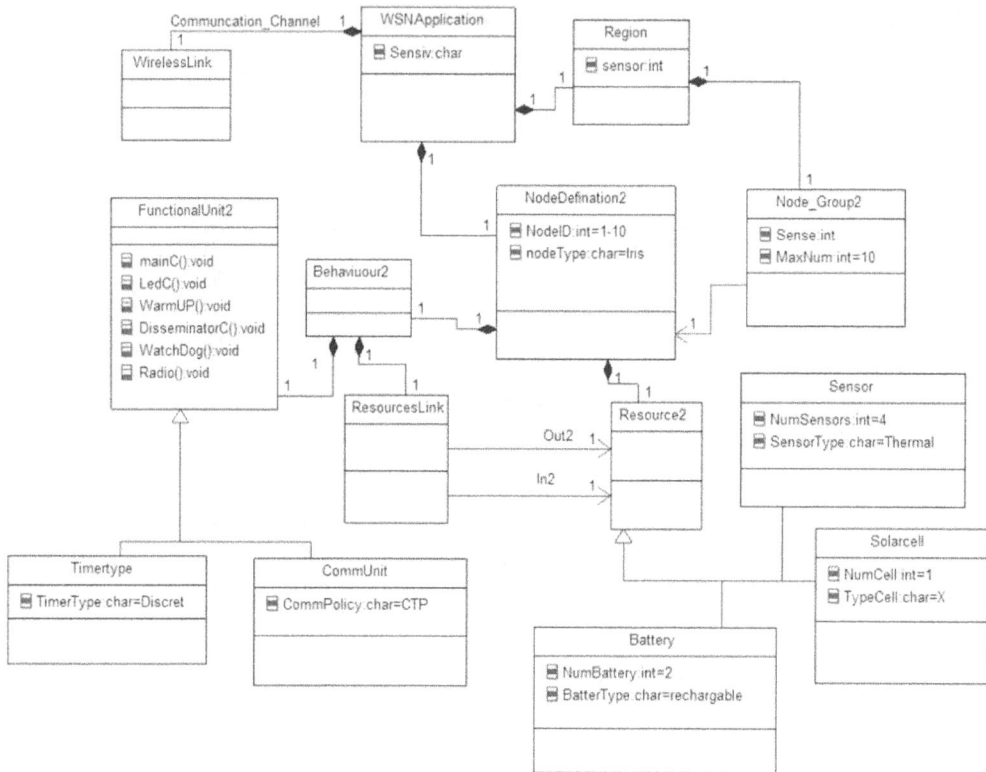

Fig. 11.14. SensIV Node Class Diagram based on the WSN-DSL metal Model.

11.3.4.3. nesC Meta-model

The nesC meta-model represents the components of the nesC code and their interconnections. It was used by MDE to generate the nesC code that is used by TinyOS systems. The transformation is done through the M2T Eclipse plug-in MOFScript. The tool offers other facilities such as model checking, parsing, and querying. The transformation process is controlled by a group of rules as well. For example, one of the rules states that for each interface in the nesC meta-model, all of its commands must be implemented. The nesC component model for SensIV is shown in Fig. 11.15 based on what was presented as the nesC meta-model in MDE.

11.3.5. Promela Model

The Promela Model [14] was created in order to check a WSN system which has 3 main characteristics: the nodes are in a dynamic movement state, the communication link is bi-directional and is unreliable, and the routing protocol ensures that each node is aware of the neighbor nodes. The model is created in the Promela modeling language, the input language for the model checker Spin. The Spin model checker has been used in the verification of both hardware and software systems. The WSN Promela Model verifies the following correctness properties in Spin:

- Network sensors are **always** connected.
- There is at **least one** communication way for each sensor.
- **Eventually** each sensor will be connected to the rest of the network.

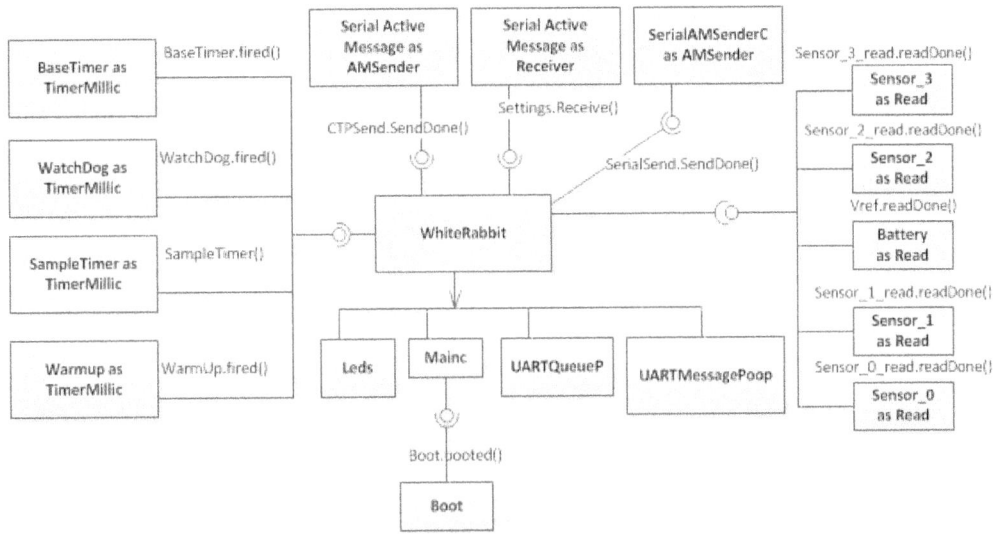

Fig. 11.15. nesC Component Model for SensIV.

11.3.5.1. Promela Model Elements

The Promela Model captures the following details of a distributed WSN system:

- The model captures each node behavior as a process. Therefore, there is availability to dynamically create/remove a process from the model based on the dynamic behavior of the nodes. For example, a new node joins the network.
- The nodes update their physical location to a central unit called the Location Manager (LM). The LM keeps data about each node such as the ID, the physical location, the movement location, and the distance between each node.
- The model contains a model for the communication channel, which captures the interaction between each node and the environment. An example for such interaction is when a node receives a notification in order to capture the current temperature. Also, when the node notifies the neighbor node when the temperature exceeds a specific level.
- The internal computation of each sensor is abstracted out since the main focus is the correctness of the nodes interaction with each other.
- The communication channels are dynamically terminated and created based on the nodes physical location and the distance between each node and the neighbor node.

11.3.5.2. Promela Model and SensIV

The Promela Model approach is intended for a different type of sensor network application than SensIV. There are several reasons why the Promela Model is not appropriate for SensIV:

- The SensIV nodes are in a static state and not dynamic. Therefore, none of the correctness properties are valid for SensIV.
- The nodes do not notify the outer environment when the temperature exceeds a specific value.

Although the Promela Model approach [14] does not suite our case study example, SensIV, we believe that in general model checking techniques, such as Spin, can be used to address the modeling and analysis of SensIV provided that a new model was created.

11.3.6. SensorML

SensorML [15] is an XML based language that is used to model the sensor specifications such as hardware, physical location, and internal process. The SensorML is designed for the following purposes:

- Support the geo-location of sensors in order to search online for WSNs.
- Capture sensor information, such as the manufacturer, sensor types, and physical locations
- Capturing the flow of data between components in sensor system.

In SensorML, the system components are captured as processes. The components modeled can be physical and non-physical components. Physical components include detectors, actuators, information about the node location, and interfaces while non-physical components include the mathematical equations used to transform raw transducer data to calibrated sensor values.

11.3.6.1. SensorML Elements

A SensorML specification of a sensor system is composed of the following SensorML elements.

- The **process model** is defined as an operation which takes one or more inputs and according to the data flow generates one or more outputs based on some adjustable parameters.
- The **process chain** block consists of interconnected linked processes. The process chain has its own inputs and outputs and it can participate with other process chains. The process chains are used to model composite models which do not exist in the

physical domain. The chain contains the description of the connection between the components it has within the chain.

- A **component** is defined as the basic unit of modeling. A component can be part of the system or can be part of the process chain.
- A **system** contains several physical and non-physical processes that act together to form the system output. A system provides a list of processes, the links between the processes, the components, and connections properties. In addition, the system can provide relative positions of the components through the positions property.

11.3.6.2. SensorML Model for SensIV

This section contains a part of the SensorML file for one of SensIV nodes (see Fig. 11.16 and Fig. 11.17). The parts shown here are as follows:

- The node manufacture: Crossbow;
- The mote platform: Iris;
- The sensor type: Temperature (thermal) sensors.

11.3.7. SystemC-AMS

SystemC-AMS [16] is an open source C++ based language. SystemC-AMS captures the system behavior through the use of modules. The system behavior is modeled by using a data flow diagram. The data flow takes place based on a set time steps. SystemC-AMS is suited for communication systems which contain sampling components and communication channels for WSNs. The modeling technique represents each node by using a module. The module contains a model for the sensor, the A/D converter, the microcontroller, the RF, and the transceiver. The analysis developed by SystemC-AMS focuses on the SNR and BER of the ADC process and the communication channel, respectively. Therefore, the model tends to capture the electronic design of the system.

We would like to point out that parameters captured in SensIV model, are taken from the datasheet of the electronic devices in our design. Designing SensIV did not involve the design of the electronic structure of the components. Our designing process involved the following:

- Developing the software deployed in the sensors;
- Assembling the hardware components together, such as the acquisition board, the thermal sensors, and the wireless motes;
- Deployment of the nodes in the field.

We explain the SystemC-AMS electronic model for SensIV in order to help the reader to have a good understanding of using SensIV as a modeling technique for WSNs.

```
<sml:SensorML>
   <sml:identification>
      <sml:IdentifierList>
         <sml:identifier name="Mote Type">
            <sml:Term definition="urn:ogc:def:identifier:moteType">
               <sml:value>Iris</sml:value>
            </sml:Term>
         </sml:identifier>
            ...
      </sml:IdentifierList>
   </sml:identification>
   ...
   <sml:inputs>
      <sml:InputList>
         <sml:input name="Temp1">
            <swe:Quantity>
               <swe:uom code="" xlink:href="degreeC" />
            </swe:Quantity>
         </sml:input>
      </sml:InputList>
   </sml:inputs>
   ...
   <sml:member>
      <sml:System>
         <sml:classification>
            <sml:ClassifierList>
               <sml:classifier name="sensorType">
                  <sml:Term definition="urn:ogc:def:identifier:sensorType">
                     <sml:value>Temperature</sml:value>
                  </sml:Term>
               </sml:classifier>
            </sml:ClassifierList>
         </sml:classification>

         <sml:components>
            <sml:ComponentList>
               <sml:component name="Temperature">
                  ...
               </sml:component>
               ...
            </sml:ComponentList>
         </sml:components>
         ...
```

Fig. 11.16. SensorML Model for SensIV (Part A: Components).

```
...
    <sml:connections>
        <sml:ConnectionList>
            <sml:connection>
                <sml:Link>
                    <sml:source ref="Temperature\Temp1" />
                    <sml:destination ref="Temperature\Temp1" />
                </sml:Link>
            </sml:connection>
            ...
        </sml:ConnectionList>
    </sml:connections>
</sml:System>
</sml:member>
</sml:SensorML>
```

Fig. 11.17. SensorML Model for SensIV (Part B: Connections).

11.3.7.1. SensIV Thermal Sensors Model

The sensor's basic task is to transfer the analog signal, which represents the temperature, to an analog voltage. The generated voltage of the temperature is loaded on the resistor. The value of resistor R = 10 kΩ and value of f = 2.4 GHz (see Fig. 11.18). The output voltage is proportional to the temperature.

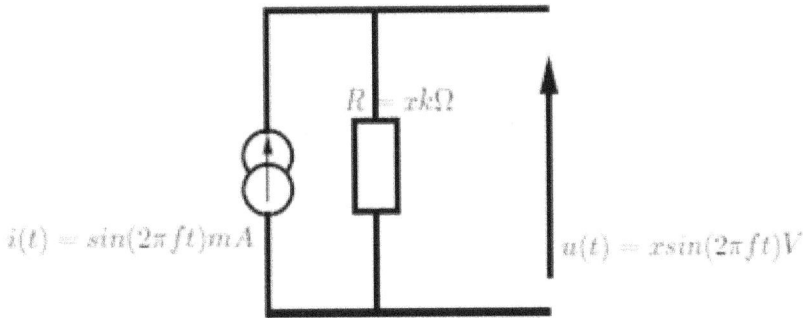

Fig. 11.18. Electronic Model of the Thermal Sensors [16].

11.3.7.2. A/D Converter

The analog to digital converter converts the signal received from the sensor to a digital form. SensIV A/D converter is captured by second order FIR filter. The sampling rate for T = 0.005 sec. (Micaz Datasheet) and number of bits is 16 bit.

11.3.7.3. Microprocessor

The microprocessor used in SensIV nodes is ATMEGA128 [17]. The microprocessor is a 16 bit wide instruction processor and has 128 Kbyte flash memory. SystemC-AMS uses a C language complied binary file to capture the behavior of the processor.

11.3.7.4. Transceiver

The RF transceiver uses Quadrature Phase Shift Keying transmission with carrier frequency 2.4 GHz and the data frequency 2.4 MHz frequency model. Noises of the communication channel are introduced to the model in order to calculate the BER communication channel.

11.3.8. UM-RTCOM Model

UM-RTCOM [18] is a real-time component based modeling framework written in CORBA that is better suited for the modeling of Actuator WSNs. It is composed of sensors, actors, and a coordinator (the coordinator is located in the base station). Actors gather data from groups of sensors based on their physical locality and respond to a phenomenon identified by the coordinator. Sensors communicate with actors and actors communicate with the coordinator using channels. Communication via a channel is modeled as a tuple. "A tuple is a sequence of fields with the form: $(t_1, t_2, ..., t_n)$ where each field t_i can be: a TC identifier (or) a value of any established data type of the host language where the model is integrated" [9] A UM-RTCOM model can also be used for several kinds of analysis, such as Worst Case Execution time (WCET), deadlock freedom, and verification of liveness properties. The communication channel protocol modeled in UM-RTCOM has been tested in an actual sensor network deployment by Barbaran et al. [19]. The deployment shows the improvement of the middleware overhead compared to another deployment where the motes send the sensed data periodically to the actors.

11.3.8.1. System Components

UM-RTCOM uses components to capture the system behavior. The model declares for each component interfaces, the services which are offered by each component, and the connectivity of each component. The components can invoke each other based on event triggers, time triggers, or service requests. The 3 types of components in the UM-RTCOM model are generic components, active components, and passive components. These are defined as follows:

- **Generic components:** A generic component is considered as a container for the other types of components.

- **Active components:** An active component represents the data flow inside the components. Also, it represents components which are responsible for handling the data flow inside the node. Furthermore, active components have the ability to invoke other components.
- **Passive components:** A passive component represents the shared resources. They cannot invoke any action or any other components. However, they offer services to other components.

11.3.8.2. Virtual Machines

UM-RTCOM models the network element nodes and the base Station by using the virtual machine. The virtual machine communicates with each other by using the tuple channel (TC). All the data exchanged between the nodes and the base station in done through tuples. TC gives the capability of one-to-many and many-to-one communication. All the network elements have an access to the communication channel and the elements are capable to see which tuple has been consumed by which node or base station. The network elements manipulate the channel tuple based on the tuple attributes priority, deadline and remove.

11.3.8.3. Nodes Locations

UM-RTCOM takes into consideration the physical location of the nodes in order to facilitate the coordination between nodes, avoiding redundant data, and to determine suitable reaction in case the reaction ability is available.

UM-RTCOM is capable of capturing the physical location of the nodes. The location of the node is declared as one of the communication channel parameters. For instance, the communication channel code between the base station and the nodes 2D location is [com_type = one_to_many, location = S_E] where S_E referees to south east.

11.3.8.4. UM-RTCOM Model for SensIV

UM-RTCOM models SensIV nodes by using one virtual machine and the base station by using another virtual machine. The nodes send the sense information to the base station in the forms of tuples and the communication type will be many-to-one since the data is collected from many nodes to the base station. The base station sends dissemination commands to change the sampling rate. In this case, the communication type will be many-to-one. The tuple which carries the dissemination commands has a higher priority than the tuple which carries the sensed data, such that the TC transfers the tuple to the other nodes. The sensor nodes are represented by using the generic components which contains 2 active components (the CPU and the timers). The radio units and the temperature sensors are represented by using the passive components.

11.3.9. eXtended Reactive Modules (XRM)

XRM [20] is an extension language of Reactive Modules (RMs). Demaille et al. used WSNs as a case study to evaluate the XRM modeling language. The case study successfully used XRM to model multiple nodes as modules and was able to support several network issues such as communication capability, memory, and energy consumption. Model checking of XRM is possible via a transformation to the original RM language. RM modules can be used with the model checking tools PRISM. PRISM overcomes the problem of state space explosion phenomenon of the traditional model checking techniques. The new language deals with different WSN designing issues such as WSN scalability, nodes locations, power consumption, and package delivery probability. XRM captures each node in the network by using a module. The modules contain variables and methods which represents the node behavior.

11.3.9.1. WSN Scalability

XRM deals with scalability by generating the modules and using the *gen* function. XRM generates modules according to the number of nodes in the network.

11.3.9.2. Node Locations

The node physical location is specified using the variables X and Y at each module. Physical locations of the nodes are represented by using a rectangular grid. Each node module contains a declaration of each of the neighbor nodes to which it transmits and receives.

11.3.9.3. Package Delivery Probability

Each node can be on one of the following four states:

- **Sensing:** The node switches to the state sensing every 5 minutes;
- **Broadcasting:** The node sending a data package to the neighbor's nodes which are declared in the module;
- **Listening:** The node goes into the listening state after it broadcasts the information.
- **Sleep:** The nodes switches to sleep state for the majority of time in order to save energy.

The data packages are lost when a node transfers the data while the neighbor is at sleeping state. Based on this scenario, XRM calculates the package delivery probability for the network.

11.3.9.4. Power Consumptions

The battery usage of each node is modeled as a local variable of the module. The node activities, such as transmitting, receiving, or sensing, consume some therefore, the value stored in the battery variable is reduced according to the activity. The value of zero represents the death of the node. The modules contain a declaration of the consumed power of each activity. Once the node is down because of a dead battery, the module is not considered in the simulation anymore. Every time the node switches from one state to another, the equivalent part of the energy is reduced.

11.3.9.5. XRM Model for SensIV

The XRM model contains 10 modules to represent the 10 nodes. Fig. 11.19 shows an example for node 6. The power consumption values were measured from a real-time test for the nodes and were published in [21]. We have documented the physical location for each of the nodes in the SensIV system. The nodes switch to the sensing mode every 5 minutes.

```
// Grid (even) dimensions.
const int X = 18, Y = 35;
// Initial energy, percentage of lost cells.
const int POWER = 2000
// States.
const int OFF = 0, SLEEP = 1, SENSE = 2,
LISTEN = 3, BROADCAST = 4;
// Energy consumptionfor each state.
const int COST_SLEEP = 0.3, COST_SENSE = 13,
COST_LISTEN = 8, COST_BROADCAST = 12;
```

Fig. 11.19. XRM Model for Node 6.

11.4. Modeling At the Node and Sensor Level

In this section, we consider how the different modeling techniques represent WSN elements, including nodes, sensors, and hardware. In particular, we consider the modeling of node/sensor behavior, sensor data, and hardware components (see Table 11.3). The node behavior column tries to capture which particular characteristics that an approach focused on modeling. The sensor and hardware modeling column considers if the actual WSN hardware, such as the ADC, microprocessor, and the wireless channel are included in a model. Hardware modeling also considers the types of sensors that a modeling technique can represent.

Table 11.3. Modeling of WSN Elements.

Approach	Node Behaviour	Sensor & Hardware Modeling
HL-SDL [7]	Concurrency, event-driven	-
Insense [9]	Concurrency, real-time	Sensor type
Mathworks [10]	Procedural, state space	-
MDEA [2]	Procedural, state space	Times, ports, wireless channel
PM [14]	-	-
SensorML	Event-driven	Sensor types
SystemC-AMS [16]	procedural	ADC, microprocessor, wireless channel
UM-RTCOM [18]	Concurrency, real-time, event-driven	-
XRM [20]	procedural, state space	-

11.4.1. Node Behavior

Most of the modeling techniques use a form of component-based modeling to represent a sensor node. The WSN behavior is modeled by specifying the component's internal behavior, component to component interactions, and the communication channel's characteristics. It should be noted that the approaches are divided into two distinct types. Those that focused on the augmentation of the models to capture particular features such as concurrency, event-driven behavior, and real-time behavior and those that leveraged standard models like state space and procedural coding that can be used for code generation or performance analysis. This separation also lets us clearly see those approaches that have included concurrency, event-driven behavior, and real-time behavior, since these three features are crucial for WSN design.

The only technique that we felt did not model node behavior was the work using the Promela Model Checker [14]. The authors of this work decided to simply focus on the modeling of network connectivity as opposed to including any significant modeling of the node behaviors. Promela though can be used to model and analyze node behaviors.

11.4.2. Modeling Sensors and Hardware

Most of the modeling techniques surveyed can be used to create a platform independent model. However, even in a platform independent model there is a necessity to include some of the hardware details. One of the reasons for including the hardware details is that the software in the nodes of a WSN is tightly coupled to the hardware elements of the node. Therefore the binding of software and hardware components should be represented in the model. An example of a hardware-software binding is the interaction between the sensor (e.g., humidity, temperature or moisture) and the software component that handles the readings. The sensor type is modeled as a component that uses a communication channel to transfer data to the software components.

Another reason that hardware information may need to be represented is to be able to generate source code from the model. Generated source code is interacting with the node hardware (timers, ports, sensor types) and therefore the model has to be aware of the hardware components in order to generate the correct code [2, 7]. Finally, hardware representation also helps in the analysis stage. For instance the ADC circuit has to be modeled to calculate the SNR, the communication channel has to be modeled to calculate the BER value.

11.5. Modeling at the System Level

This part of the paper focuses on modeling contributions to the distributed nature of sensor networks. The modeling techniques deal with various distribution issues, such as network behavior and topology modeling. The modeling techniques that deal with the network system are shown in Table 11.4.

Table 11.4. Modeling Technique at the System Level.

Approach	Network Behaviour	Topology Modeling
HL-SDL [7]	-	-
Insense [9]	-	-
Mathworks [10]	Node/base station interaction	Single hop, static topology
MDEA [2]	Node/base station interaction	-
PM [14]	Nodes connectivity	Multi hop-dynamic topology
SensorML [15]	-	-
SystemC-AMS [16]	-	Single hop, static topology
UM-RTCOM [18]	Nodes/actors/base station interaction	Single hop, static topology
XRM [20]	Power management-wake up states	Single hop, static topology

11.5.1. Network Behavior

Modeling network behavior in a WSN is crucial because many important performance values are based on the network. For example, the trade-off between packet loss and power. Due to the fact that the node has limited power resources, it is common to use a power management algorithm that controls the wake up state of a node from active to sleeping and vice versa. Packages can be lost if this is not done properly. XRM for example, calculates the package delivery probability. This can be helpful for applications in which package delivery is an important factor.

Also related to power management, modeling the power consumed in the wireless communication process between the nodes is an important factor in increasing the life-time of the WSN. XRM models the power consumed by each wireless communication channel. Every time the node model is provoked to send or receive a signal, a specific amount is subtracted from the energy level. Another example where modeling at the network behavior is important is in capturing the deployment and interaction of the software components across the network. For example, MDEA divides the software elements into two groups, those residing on the nodes and those residing on the gateway. The generated code should guarantee the interaction between the node and the gateway.

In a similar fashion UM-RTCOM models the network elements (sensors, actors, and the gateway) as three virtual machines (VMs), where each VM models a single element. The system behavior is modeled by the interaction between the three VMs.

11.5.2. Topology Modeling

This section focuses on how the topology is modeled, in other words how the physical locations of the nodes have been modeled. The topology of WSN systems can be dynamic or static. The static topology represents the nodes in a fixed location while the dynamic topology represents the nodes while they are in a moving state. PM captures the dynamic topology by recording the physical location of the nodes in a Location Manager (LM). While the nodes change their physical location, they send the updated location to the LM. Through the use of model checking, the nodes connectivity can be checked. Additionally, based on the aim of modeling, the technique models the number of hops in the network design. For instance, SystemC-AMS analyzes the communication channel between two nodes, such that the model deals with single hop communication issues between two nodes.

XRM is an example of modeling for static topologies. The topology is modeled as a grid location. Each node location is captured as a 2D variable. In MathWorks, the framework is able to model the static topology declaration of the nodes connectivity. In the UM-RTCOM model, single hop communication is used between the network nodes because of the application requirements and nature of the problem. The behavior is modeled by the interaction between the three VMs.

11.6. Supporting Tools

Almost all of the modeling techniques surveyed offers some tools to support the design of WSNs. In this section we discuss support tools that include code generation, execution and analysis, and model checkers (see Table 11.5).

11.6.1. Code Generation

Code generation is the process of generating source code from a model or other source code representation. Tools for generating source code from WSN models are beneficial with respect to design for two main reasons:
- Implementing the source code for the nodes is tedious, time consuming and requires a lot of time and effort from the developers.

- Debugging the design at the source code level is also a very challenging and time consuming process.

Table 11.5. Modeling Techniques Supporting Tools.

Approach	Code Generation	Model Checking	Execution and Analysis
HL-SDL [7]	NesC	-	WECT
Insense [9]	C	Spin (Channel Protocol)	WCS
Mathworks [10]	NesC, C	-	Functional analysis
MDEA [2]	NesC	-	-
PM [14]		Spin (Connectivity)	
SensorML [15]	JavaBeans	-	-
SystemC-AMS [16]	-	-	BER, SNR
UM-RTCOM [18]	-	-	Deadlock, WCET
XRM [20]	-	Prism, APMC	Execution, debugging

Modeling can help to solve code implementation problems by designing the system at higher abstraction layers and generating the target code from that layer. The simplicity of the code generation process depends on the degree of similarity between the modeling notation and the generated code notation. The Mathwork technique generates nesC code and C code from ANSI C modeling notation.

Code is developed through minor changes in the modeling notation versus the generation for nesC needs a lot of the changes for ANSI C to generate the proper code.

One criticism of code generation tools is that the code produced is not as efficient as hand-written source code. Manual optimization by the user is one solution to achieving better performance from generated source code. An example of manual optimization of WSN source code is modifying the communication between the components from asynchronous in the model to synchronous in the target
platform. In addition to manual optimization of the generated source code, simulation can be used at the model level to refine the model (with respect to performance) prior to code generation.

Our survey reviewed four modeling techniques which are capable of generating source code: MDEA, HL-SDL, Insense and MathWorks. MDEA and HL-SDL can generate nesC code for WSN. MathWorks generates nesC as well as C code that executes under the MANTIS operating system while Insense generates C code.

11.6.2. Model Checking

Model checking is a formal methods technique for software engineering [22]. A model checker takes as input a model of a system and a property specification. The model is converted into a finite state model and the model checker uses an exhaustive state space search to verify the specification. The model checker will determine if the model satisfies the specification. If it does not, than a counter example (error trace) may be provided.

Applying model checking to WSN models allows the designer to verify that the design is correctness as well as detect potential errors. In response to errors, the model can be modified and the design improved prior to implementation. Model checking for WSNs can be classified as direct and indirect model checking. Direct model checker occurs when a model checker exists that can take the WSN model as input. An example of direct model checking is in PM where the modeling language, Promela, is also the input language for the model checker Spin [23].

Indirect model checking occurs when no model checker exists for the WSN modeling language and model transformation is required in order to transform the WSN model to a language that can be input to a model checker. For example, XRM models need to be transformed into RM in order to be used with the model checkers PRISM and APMC [22].

A drawback of indirect model checking approaches such as the one used in XRM is that the model checking results are given with respect to the RM model and need to be transformed back into an XRM form. The challenge of transforming between the WSN modeling language and the model checker input language is known as the semantic gap problem. Another example of indirect model checking is in IM where the authors manually created a Promela model of the component communication channel in order to verify the correctness of the communication protocol. Their verification identified an error that lead to a modification to the original Insense model.

11.6.3. Model Execution and Analysis

In addition to code generation and model checking, we also consider other tool support including tools that execute and analyze the WSN models. Model execution refers to the execution or interpretation of the WSN design at the model level. XRM is the only techniques in our survey supports model execution. The XRM compiler, a domain specific compiler, allows for model execution and debugging as well as model optimizations (e.g., dead code removal).

Model analysis includes a variety of static and dynamic techniques and a number of the approaches in our survey include some kind of analysis tool. The analysis predicts the behavior of the system through calculating some the parameters. The modeling techniques can analysis the system for the system performance such as (Network performance, ADC process, communication process performance). This section contains an explanation of parameters can be calculated by the modeling techniques included in the survey:

- The **deadlock** happens in large scale networks which contains many components. Because the network component are interacting with each other and because every component has a buffer, there is a possibility when all buffers are full and all nodes are waiting for each other to transmit but no node will because of the full buffer. The deadlock analysis shows the probability of that scenario happening so the designer should avoid that in the designing process [24].

- **WCET** is the maximum time length taken to execute the process. WCET is important on the real-time schedulability analysis. While the WSN components are interacting with each other, the commands call and wait are used. For the schedule analysis, it is important to calculate the WCET value. The value basically depends on two factors, the software implementation such as the maximum number of iterations, if conditions, etc and the target hardware platform. Both factors should be annotated in the model in order to calculate the WCET. The hardware factor is represented by timing information of the hardware, such as the time response provided by the manufacture [25].

- **WCS** is the maximum space taken by the components. WCS can be calculated by adding the space requirements of the parameter used by each component. The parameters can be local, which are used by the component only or global parameters, which are used by the components to interact with other components. Signal to Noise Ratio (SNR) is the ratio of the level of the desired signal to the background noise level. SystemC-AMS calculates the SNR ratio of the ADC process [26].

11.7. Related Work

Akyildiz et al. [1] survey talks about the WSN system design elements such as the routing protocols, the sensors, and application layer. The authors also provide the factors which affect the implementation of each component. In addition, the survey contains information about open research topics for each design element. Our survey talks about the software modeling techniques which are used to capture those system elements.

The survey by Akkaya et al. [27] talks about the routing protocols for wireless sensor networks. According to their classification there are 3 routing categories; data-centric, hierarchical, and location based. The survey discusses each protocol and classifies the protocol based on the 3 mentioned routing categories. The modeling techniques which have been included in our survey have captured the sensors connectivity only. However, none of them have a representation for the routing behavior.

11.8. Conclusion and Future Direction

Modeling helps to resolve some of the WSN software implementation challenges before deployment.

As depicted in Table 11.1, several approaches have focused only on the modeling of software elements in a sensor node while others also model the sensor network. Both of these are important to be modeled from a sensor network perspective. Also many of the modeling languages support components as one of its basic modeling elements.

Component based modeling is a fundamental way to partition software entities because it can be used to support multi-threading design and analysis. Most of the approaches reviewed can also model the communication channel. A few like PM and SensorML can model a process. Process modeling is important in capturing a systems behavior. As depicted in Table 11.2, the modeling techniques are targeted to analyze specific software challenges like concurrency, real-time, and event modeling. We also see that they may be focused on simply modeling the sensor information or hardware to facilitate code design as is the case with MDE and SystemC-AMS.

Modeling at the system level is also another feature that some modelers support. As seen in Table 11.4, several modeling techniques like UM-RTCOM, XRM, PM, MDEA, and MathWorks can all model the sensor network but there is a focus for each on what behavior they model. They all model node activity and take into account node to node communication but not all can explicitly model the network topology as is the case with MDEA.

Support for analysis and code generation tools is vital for a modeling technique. Table 11.3 reflects, the fact that certain modeling technique can do code generation while others cannot. This depends primarily on the focus of the developers of the modeling technique and the maturity of the approach. It should be noted that very few of the techniques support model checking. We speculate that this is largely due to the gap between the designing process and the model checking. This gap exists because the design takes place in domains such as CORBA and UML. In order to check the model with model checking methods, the design has to be re-modeled again in the model checking domain.

As a future direction for WSN modeling, enhancements for code generation tool are required. Such enhancements can improve the quality of the generated code in terms of the code size and avoidance of manual optimization for the generated code. Additionally, the modeling domain should be selected such that model checking can be done without redoing the model in the model checking domain.

Moreover, the modeling domain should support analysis at the design stage, which helps the software system developer detect and correct software system problems at an earlier stage of the sensor system design. Some of the reviewed papers have performed analysis for WCET, WCS, deadlock, SNR for sensor interfaces, and BER for the communication channel. To the best of our knowledge, package delay and data losses have not been considered in the analysis but these factors are important for sensor networks.

The modeling techniques also could not capture the behaviour of the routing protocol used in the SensIV system. MDEA approach has the facility to mention the routing protocol in the WSN-DSL protocol. However, none of the modeling techniques can capture the behaviour of the routing algorithm. The modeling techniques which are capable of capturing the distributed nature of the network have focused on the interaction between 2 neighbour nodes. Therefore, there is a necessity to state the nodes connectivity since there is no possibility to capture the CTP protocol.

Reference

[1]. I. F. Akyildiz, W. Su, Y. Sankarasubramaniam, and E. Cayirci, Wireless sensor networks a survey, *Computer Networks*, Vol. 38, March 2002, pp. 393-422.

[2]. F. Losilla, C. Vicente-Chicote, B B. lvarez, and A. Iborra and P. Snchez, Wireless sensor network application development: An architecture-centric mde approach, in *Software Architecture Volume 4758 of Lecture Notes in Computer Science,* 2007, pp. 179-194.

[3]. R. Liscano et al., Network Performance of a Wireless Sensor Network for Temperature Monitoring in Vineyards, in *Welcome to the 8th ACM PE-WASUN 2011*, Miami, Fl, USA, November, 2011.

[4]. LM135 Datasheet (http://www.national.com/ds/LM/LM135.pdf).

[5]. Tinyos 2 tep 123. The collection tree protocol, http://www.tinyos.net/tinyos-2.x/doc/txt/tep123.txt

[6]. P. Levis, TinyOS Programming, *Cambridge University Press*, 2009.

[7]. D. Dietterle, J. Ryman, and and R. Kraemer. K. Dombrowski, Mapping of high-level SDL models to efficient implementations for TinyOS, in *Proc. of the Euromicro Symp. on Digital System Design (DSD' 2004)*, Sept. 2004, pp. 402-406.

[8]. ITU-T, Specification and Description Language (SDL) z. 100, 11/99, 1999.

[9]. A. Dearle, D. Balasubramaniam, J. Lewis, and and R. Morrison, A component-based model and language for wireless sensor network applications, in *Proc. of the 32nd Annual IEEE Int. Conf. on Computer Software and Applications (COMPSAC)*, Aug. 2008, pp. 1303-1308.

[10]. M. Mozumdar, F. Gregorett, L. Lavagno, L. Vanzago, and S. Olivieri, A framework for modeling, simulation and automatic code generation of sensor network application, in *Proc. of the 5th IEEE Comm. Soc. Conf. on Sensor, Mesh and Ad Hoc Communications and Networks (SECON' 08),* June 2008, pp. 515-522.

[11]. A. Bonivento, L. Lavagno, A. Sangiovanni-Vincentelli, and L. Vanzago. L. Necchi, EERINA: an energy efficient and reliable in-network aggregation for clustered wireless sensor networks Conference, in *Wireless Communications and Networking*, 2007, pp. 3364-3369.

[12]. Simulak, Simulation and Model-Based Design, at http://www.mathworks.com/products/simulink/

[13]. S. Bhatti et al., MANTIS OS: An embedded multithreaded operating system for wireless micro sensor platforms, in *Mobile Networks and Applications,* 2005, pp. 563–579.

[14]. V. Oleshchuk, Ad-hoc sensor networks: modeling, specification and verification, in *Proc. of the 2nd IEEE Int. Work on Intelligent Data Acquisition and Advanced Computing Systems Technology and Applications*, Sept. 2003, pp. 76-79.

[15]. M. Botts, OpenGIS sensor model language (SensorML) implementation specification, July 2007.

[16]. M. Vasilevski, N. Beilleau, H. Aboushady, and and F. Pecheux, Efficient and refined modeling of wireless sensor network nodes using System-AMS, in *Conf. on Ph. D. Research in Microelectronics and Electronics (PRIME' 08)*, 2008, pp. 81-84.

[17]. Crossbow (http://www.openautomation.net/uploadsproductos/micaz_datasheet.pdf).

[18]. M. Diaz, D. Garrido, L. Llopis, B. Rubio, and J. Troya, A component framework for wireless sensor and actor networks., in *Proc. of the IEEE Conf. on Emerging Technologies and Factory Automation (ETFA '06)*, Sept. 2006, pp. 300-307.

[19]. J. Barbaran, M. Diaz, I. Esteve, D. Garrido, and L. Llopis B. Rubio and J. Troya, Tc-wsans: A tuple channel based coordination model for wireless sensor and actor networks, in *Proceedings of the 12th IEEE Symposium on Computers and Communications (ISCC' 07),* 2007, pp. 173-178.

[20]. A. Demaille, S. Peyronnet, and and B. Sigoure, Modeling of sensor networks using XRM, in *Proc. of the 2nd Int. Symp. on Leveraging Applications of Formal Methods, Verification and Validation (ISoLA'06)*, November 2006, pp. 271-276.

[21]. J. Zheng, C. Elliott, A. Dersingh, R. Liscano, and M. Eklund, Design of a Wireless Sensor Network from an Energy Management Perspective, in *Proceedings of the 8th Annual Research Conference on Communication Networks and Services (CNSR' 10)*, 2010, pp. 80-86.

[22]. E. M. Clarke Jr., D O. Grumberg, and A. Peled, Model Checking, *The MIT Press*, 1999.

[23]. O. Sharma et al., Towards verifying correctness of wireless sensor network applications using Insense and Spin., in *Model Checking Software vol. 5578 of Lecture Notes in Computer Science*, 2009, pp. 223-240.

[24]. F. L. Lewis, Wireless Sensor Networks. Smart Environments: Technologies, Protocols, and Applications (ed. D. J. Cook and S. K. Das), *John Wiley*, New York, 2004.

[25]. M. Diaz et al., Integrating real-time analysis in a component model for embedded systems., in *Euromicro Conference*, September 2004, pp. 14-21.

[26]. J. Blieberger and R. Lieger, Worst-case space and time complexity of recursive procedures, Real-time System, *Springer Netherland*, No. 0922-6443, pp. 115-144, 1996.

[27]. K. Akkayal and M. Younis, A survey on routing protocols for wireless sensor networks, *Ad Hoc Networks*, Vol. 3, No. 3, pp. 325-249, May 2005.

[28]. Manuel Díaz, Daniel Garrido, Luis Llopis, Bartolomé Rubio, and José M. Troya, A Component Framework for Wireless Sensor and Actor Networks, in *Proceedings of the IEEE Conference on Emerging Technologies and Factory Automation (ETFA '06)*, 2006, pp. 300-307.

[29]. B. Lu. and J. Nickerson, A language for wireless sensor webs, in *Proceedings of the 2nd Annual Conference on Communication Networks and Services Research*, 2004, pp. 293-300.

Chapter 12
Multi-Sensor Wireless Network System for Hurricane Monitoring

Chelakara Subramanian, Gabriel Lapilli, Frederic Kreit, Jean-Paul Pinelli, Ivica Kostanic

12.1. Introduction

Understanding and predicting the damage caused by hurricanes on houses is an important issue in coastal areas. Mathematical models of pressure distribution due to wind effects on structures are essentially based on wind tunnel experiments (e.g. ASCE 7-10). Even though these models give good approximations for non-extreme winds, they might be inadequate for hurricane winds.

Several studies [1-7] extended the wind tunnel scaled model experiments to full-scale building models and attempted to validate the wind tunnel model results. The first attempts in this direction required buildings being dedicated to experiments. These tests were very useful, but were unable to achieve the goal of providing accurate measurements that were applicable to real life coastal structures or structures in urban settings since the buildings in the tests were only models under ideal surroundings. Recently, researchers [7-13] started investigating the effect of wind on low-rise structures by installing sensor systems directly on private houses in residential areas.

The data generated by the sensor system can be used to create a wind-induced pressure model for this type of building. Such a model will allow for future corrections to building code provisions, and predicting pressures with more accuracy than what is possible today. Another use for the data from full-scale tests is the evaluation of structural damage, and its correlation with the recorded pressures.

In the following sections a wireless sensor system is described. Comparisons of pressure sensors measurements are made against National Weather Service data and wind tunnel test data. A minivan highway test is then described which was used to assess the sensor case shape effect on pressure measurements. Computational fluid dynamics (CFD)

Chelakara Subramanian

Mechanical & Aerospace Engineering, Florida Institute of Technology, Melbourne, FL, USA analyses were also performed to complement the wind tunnel test and highway van tests and to determine the optimum mesh shapes and sizes, boundary conditions, and the best turbulence model that will reproduce the measured pressures. University of Florida (UF) Hurricane Simulator tests are described next, which were used to assess the effects of wind gusts on pressure and pressure-velocity auto/cross correlations between different sensors. A CFD simulation of the UF Hurricane Simulator test was also performed to assess the sensitivity of the turbulence models in capturing the true pressure and velocity variations on the test object due to the gusts. Finally, vibration tests using a shaker table are described to analyze the effects of structural vibration on the sensor pressure measurements. Specific conclusions are given for each tests mentioned above. Overall conclusions are also provided at the end.

12.1.1. The Wireless Sensors System

Since the sensors are deployed on selected private residential houses the measuring devices must be non-intrusive to the private houses. For example, it is not possible to drill holes on the roofs of the houses for experiments. So, the sensors fit into pre-installed brackets on the roof. The architecture of the sensors system, including a detailed description of the associated software, and its networking capabilities are described in [14]. To summarize, the overall networked system consists of three major sub-components. The first sub-component comprises a set of remote sensor units and associated base unit that are installed on the structure under monitoring. This sub-component is referred to as the *house installation* and it is responsible for performing measurements and data acquisition (up to 30 pressure sensors, one is located inside the house to measure pressure differences, plus one anemometer). The second sub-component is the *communication network* connecting individual house installations to a central server. Finally, the third component is the *central server* itself, which processes measured data and enables a network operator to control the operation of the base and sensor units.

12.1.1.1. Remote Sensor Unit

A block diagram of a sensor unit is presented in Fig. 12.1. The system supports different types of sensor units. One type is used for collecting air pressure and temperature data, while another type is used for sensing wind speed and direction. Both sensor unit types have similar architecture and the only differences are in the type of sensor that is connected to the unit and in the design of the sensor case.

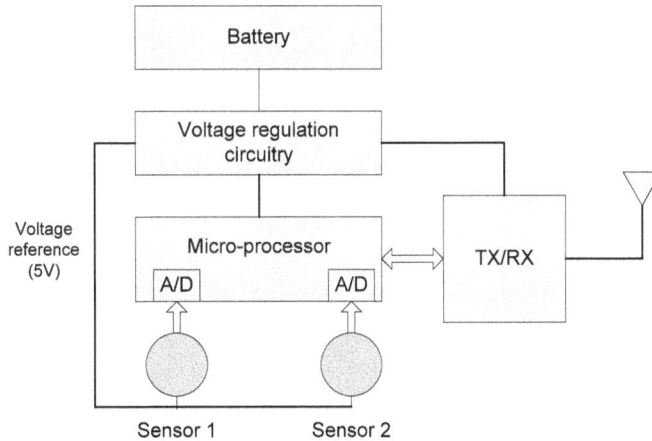

Fig. 12.1. Outline of the sensor unit.

Four common components to all sensor units include:
- 6 V battery (Power Sonic-630), 3.5 Ah;
- Micro-processor (PIC16F867);
- Transceiver (Radiometrix BIM3-64);
- Voltage regulation unit (MAX8881).

The pressure sensor has an absolute pressure transducer MPX-4115 from Motorola whose range has been rescaled electronically to read from 580 to 1100 mbar. The achieved measurement accuracy is 1 mbar and the precision is on the order of 0.24 mbar [10], which is adequate for the atmospheric measurement.

The temperature sensor is a LM34 by National Semiconductor, which has a 1.0 °F accuracy over a range from -50 °F to +300 °F. The repeatability of the sensor was ±0.5 ^{0}F.

A Young® anemometer is used to measure the wind speed from 0 to 60 m/s with an accuracy of 3 m/s, and the wind direction from 0° to 360° with an accuracy of 3°. The repeatability of the anemometer is ±0.5 m/s.

The architecture of the base unit is similar to that of the sensor unit. The micro-processor is programmed to poll the sensor units, collect their measured data, and relay the data to the collection laptop through a serial communication port. The transceiver is connected to a 6 dB omni-directional antenna.

A data acquisition software [11] has been designed to collect and archive the data measured by the different remotes. The data are then transmitted to the server via wireless internet.

12.2. Reliability of the Pressure Sensors

The reliability of pressure sensors is established by comparing the measured values with other sensors in the field as well as in a wind tunnel.

12.2.1. Comparison with Established References

The best way to study the reliability and precision of the sensors measurements is to compare them to a standard reference.

12.2.1.1. Comparison with the National Weather Service (NWS) Pressure Measurements

On February 2009 a test was conducted in the Wireless Center for Excellence (WiCE) lab at Florida Tech. Three sensors were randomly placed in the lab, and data were recorded continuously for approximately 3 days. No wind velocity or direction data were recorded, because the sensors were inside the lab.

Simultaneously, pressure data were collected from the nearby weather stations at Melbourne International Airport, FL, which are operated by NWS.

Data recording started on 2-13-2009 at 10:53:39 and lasted until 2-16-2009 at 11:12:54.

The results showed (Fig. 12.2) a good agreement. The statistical distribution of the difference of the two measurements showed a Gaussian characteristic with a standard deviation of 0.063 mbar, attributable to the distance between the measuring locations. This first test proved that the wireless sensors are able to follow the variation of atmospheric pressure as reported by the local weather service reliably.

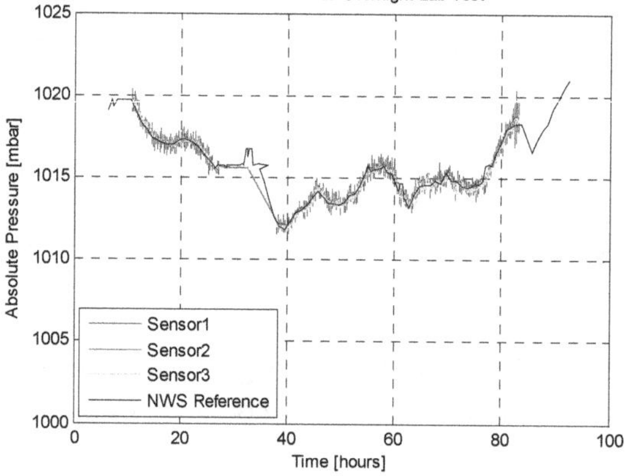

Fig. 12.2. Laboratory pressure measurements compared to NWS data.

12.2.1.2. Comparison with the MET3A

The Florida Tech Department of Oceanography has a Paroscientific MET3A weather station located on the roof of the Links building (on the Florida Tech campus). The Paroscientific MET3A weather station has the following properties:

- atmospheric pressure measurement with accuracy of 0.08 mbar;
- temperature measurement with accuracy of 0.1 °C;
- humidity measurement with accuracy better than ±2 % RH at 25 ° C.

The data are recorded and averaged to obtain a value every half hour. The sensors were deployed as close as possible to the MET3A weather station. This way an accurate comparison between the two measurements could be done without any concern regarding the distance between the measuring locations.

Data recording started on 06-05-2009 at 16:03:44 and lasted until 06-07-2009 at 14:42:17. Twenty-seven sensors and one anemometer were used for this test. Fig. 12.3 shows pressure readings for the sensors, along with the MET3A data, referenced as sensor 99, collected over two consecutive days. The wind velocity was very low during the test (always below 5 m/s). Such a small wind velocity has no significant impact on the pressure measurement. Therefore the pressure variations recorded by the sensors were only due to the change in daily atmospheric pressure variations.

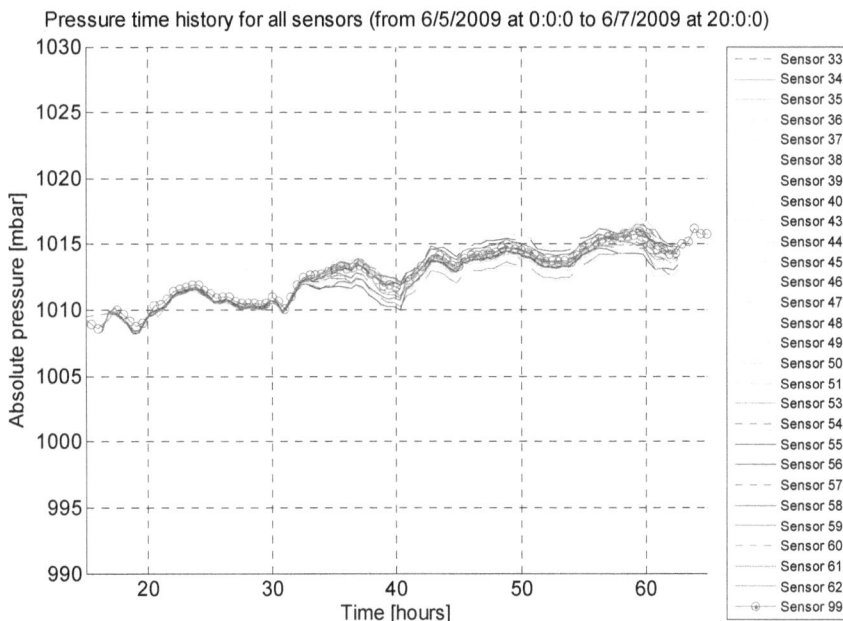

Fig. 12.3. Pressure time history for all sensors and the MET3A.

This test showed a reasonably (within the limits of low- speed signal to noise ratio) agreement between the pressure recorded by the wireless system and the pressure reading of the MET3A. This test was then the second proof of the sensors' ability to reliably record the variation of atmospheric pressure in comparison to other secondary references.

12.2.2. Wind Tunnel Tests

The reliability of the sensor measurements was also evaluated through a controlled experiment in a wind tunnel, and by comparison with an analytical solution.

12.2.2.1. Wind Tunnel Test Description

A total of 11 sensors and 1 anemometer were used for this test. Sensors 41-48 were connected to static tubes set flushed on a board, at regular intervals from each other as shown in Fig. 12.4. The purpose of these tests was to study the reliability and accuracy of the sensors independently of the cases (i.e. containers) used in the field deployments.

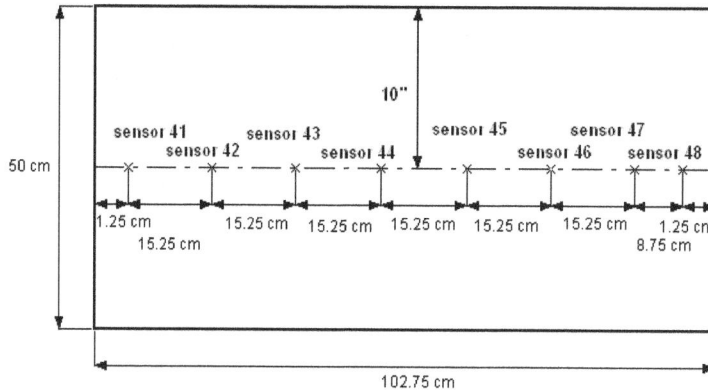

Fig. 12.4. Location of the sensors on the board.

The board was then placed in the test section of the Florida Institute of Technology low-speed open circuit wind tunnel, with dimensions of $54 \times 54 \times 165$ cm. The tunnel has a maximum speed limit of 22 m/s and a nominal free stream turbulence intensity of 0.2 percent. The leading edge of the board was 66 cm from the beginning of the test section and was inclined at 15° from the horizontal (see Fig. 12.5). Sensors 50 and 51, outside the wind tunnel, were used as static references.

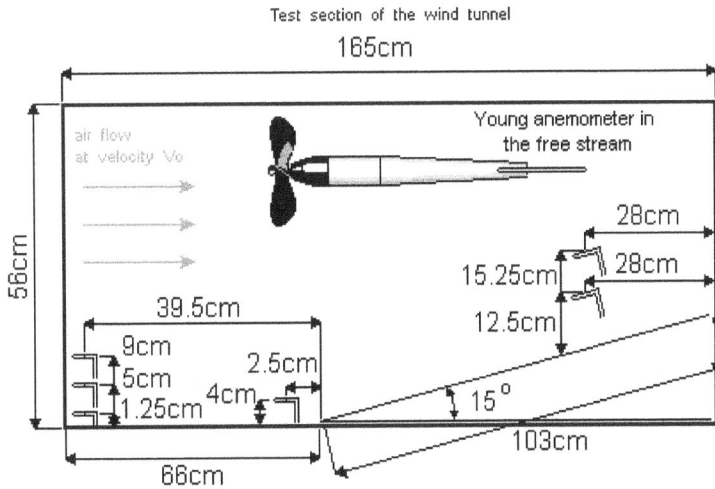

Fig. 12.5. Wind tunnel sensor set up.

Sensor 49 was connected to the static pressure channel of a pitot-static tube to record the changes in static pressure at different locations in the free air-stream. Finally, a Young® anemometer (with its circuit board mounted on top of the wind tunnel test section) recorded the wind velocity. It was placed as high as possible in the test section to avoid disturbing the pressure measurements.

Data recording started on 10-03-2009 at 13:53:59 and lasted until 10-10-2009 at 16:34:58.

Seven series of tests were conducted in the wind tunnel. For each of 6 test series, the static tube (sensor 49) was placed at different locations in the free air-stream as indicated in Fig. 12.5. For the seventh series of tests, the static tube was replaced in position 1 (39.5 cm ahead of the board and 1.25 cm high) and the anemometer was removed from the test section.

12.2.2.2. Wind Tunnel Test Experimental Results

The pressure recorded by sensors 50 and 51 were used as references. Therefore, to obtain the pressure variation the pressure records of any sensor can be subtracted from the pressure record of sensor 50 or 51. Sensor 50 was chosen for the reference and is not shown in the plots.

The pressure variation as a function of the wind velocity is plotted in Fig. 12.6.

Pressure Variation vs. Wind velocity for Test 1

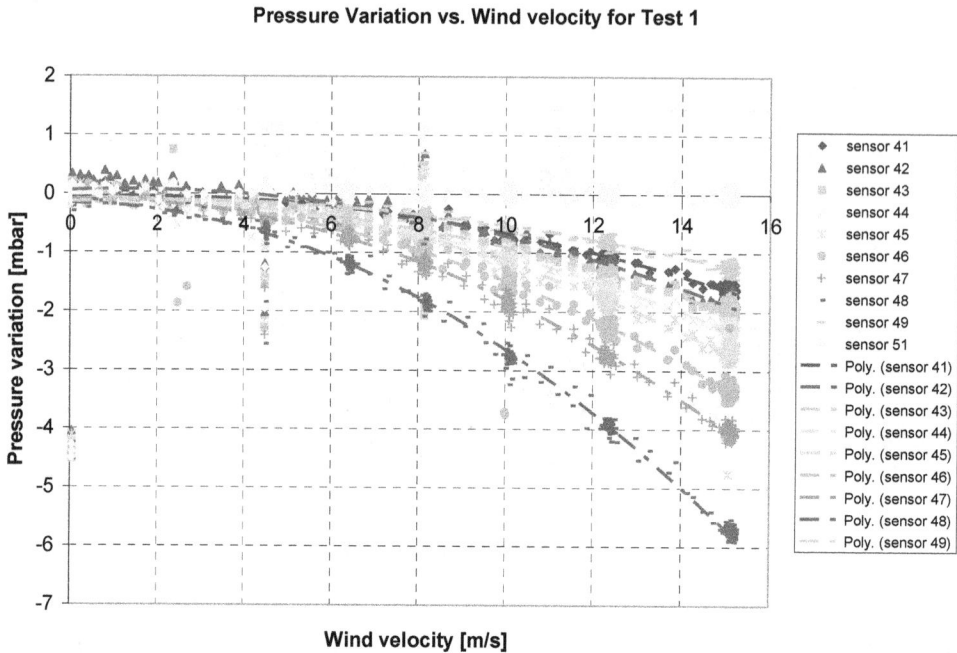

Fig. 12.6. Pressure variation as function of the wind velocity for test 1.

On the ramp the velocity increases continuously. As the velocity increases, the surface pressure variation is increased. This is evident from the sensor taps farther away from the start of the slope where the pressure variation is the largest. Sensor 51 showed no pressure variation because it measures the same values as the reference.

For all the tests, the trend line of the free stream static pressure sensor 49 was higher than any other sensor. This behavior suggests that the free stream pressure variation is smaller than that of the surface pressure sensors.

The pressure coefficient changed along the board. For calculating the pressure coefficient of each sensor, the following formula was used:

$$C_p = \frac{P_s - P_\infty}{\frac{1}{2} * \rho * V^2},$$

$$(12.1)$$

where C_p is the pressure coefficient, P_s is the surface pressure recorded by the sensor, $P\infty$ is the pressure in the free stream, ρ is the density of air 1.225 kg/m^3 and V is the free stream wind velocity recorded by the anemometer.

It was then possible to plot the pressure coefficient in relation to the tap location on the ramp (Fig. 12.7).

Pressure coefficient vs. location of the tap on the ramp

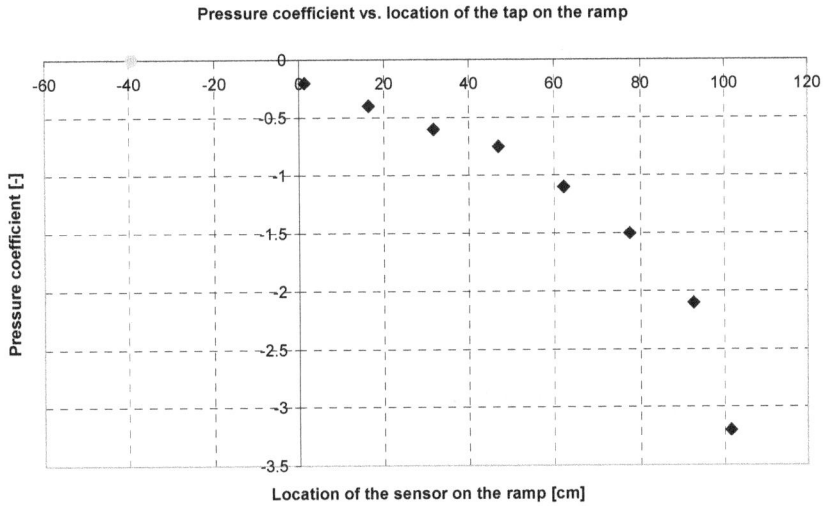

Fig. 12.7. Pressure coefficient in relation to the sensor location on the ramp for test 1.

The pressure coefficient of sensor 49 is equal to 0 because this sensor is in the free stream. The pressure coefficient was found to decrease progressively along the board.

12.2.2.3. Wind Tunnel Analytical Study

The experimental wind tunnel results were compared to the potential flow analytical solution. The wind tunnel test section is small enough that as the ramp rises the section is reduced. The dimensions of the wind tunnel and the ramp are shown in Fig. 12.8.

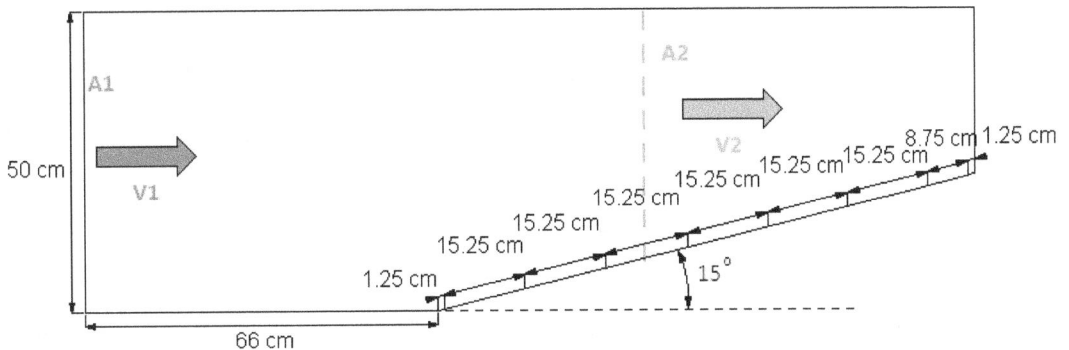

Fig. 12.8. Wind tunnel test dimensions.

Then, using the conservation of mass principle and ignoring the viscous effects, based on the geometry of the model, it was possible to determine the wind velocity at the location of every sensor along the slope. From there, using the Bernoulli's equation it was possible to calculate the static pressure corresponding to the wind velocity.

This permitted to establish the pressure difference between the free stream and the location of the static tubes. Finally using (1), the pressure coefficient at every point along the slope was calculated.

12.2.2.4. Wind Tunnel Test CFD Study

In addition to the theoretical calculation and experimental results, a CFD simulation was run using the Fluent software. The model was created in Gambit using the dimensions of the Florida Tech wind tunnel and of the ramp model used in the October 5th, 2009 test.

In the CFD model the ramp was raised 1.25 cm from the floor of the wind tunnel's test section to represent the thickness of the board used in the tests. The distance behind the slope was also increased to 127 cm (instead of 7.5 cm in the tests) to avoid any negative influences the end of the model could have on the pressure and velocity calculations.

The geometry in Fig. 12.6 was meshed and the boundary conditions were defined as follows:
- The floor of the wind tunnel, the ramp, and the roof of the test section were defined as impervious walls;
- The beginning of the test section was defined as a constant velocity inlet;
- The exit surface (on the right side) was defined as an outflow to not restrain the pressure or the velocity values.

The model was exported to Fluent and tested with the following initial velocities V1 of 5, 10, 15, 20, 30, and 40 m/s. The fluid viscosity was not neglected because it is the essential element responsible for the boundary layer effects, so the viscous calculations and turbulence model "k-epsilon" (for turbulent flow Reynolds number) was used. For each velocity considered, distributions of the velocity contours and static pressure were obtained. All contours showed the same trend in each test, so only the contours at 20 m/s are shown in the following figures. The wind tunnel test free stream speed was about 20 m/s.

At the beginning of the ramp the board's thickness creates a stagnation effect (shown in red in Fig. 12.10).

In the velocity contours, the velocity at the end of the ramp suddenly increases due to the abrupt end of the geometry. The vena-contracta like region appears downstream of the separation point where the velocity reaches a maximum and the suction peak also occurs (Fig. 12.9). The flow expands further downstream.

Thus, when studying static pressure on an inclined plane, at the top of the ramp a large decrease in static pressure should be expected. This assumption is confirmed by the results observed in Fig. 12.9.

In the middle of the board, the pressure decreases gradually, as a result of the decrease in the cross section of the wind tunnel caused by the inclined plane.

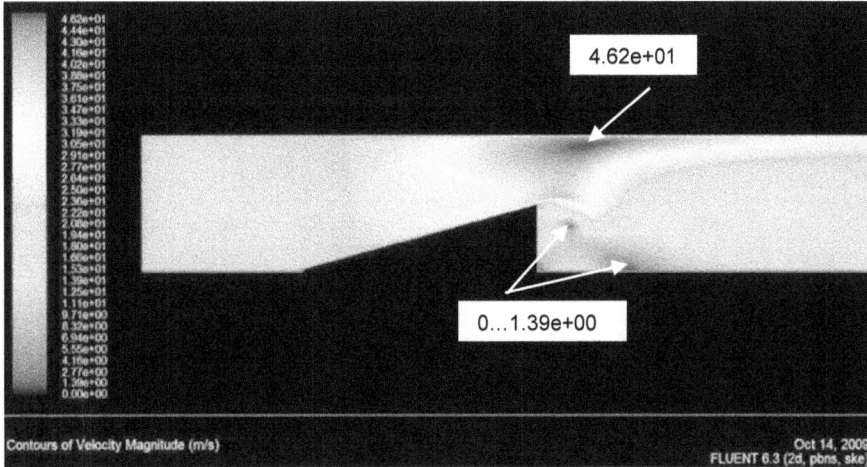

Fig. 12.9. Velocity contours for the 20 m/s computation.

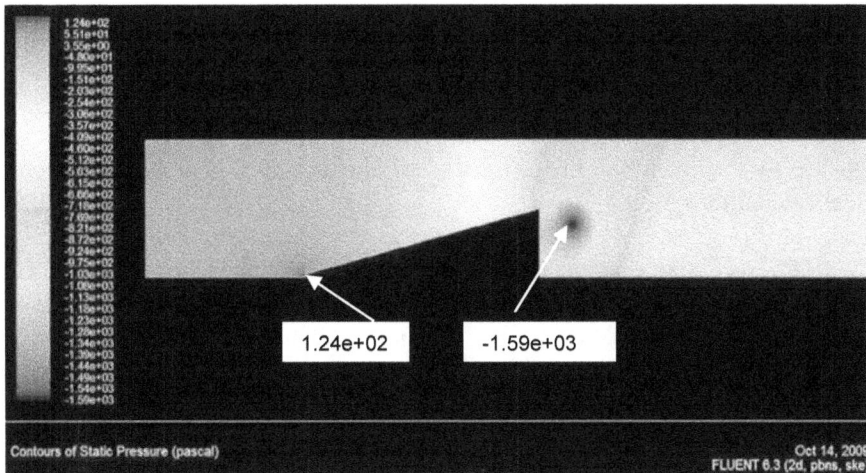

Fig. 12.10. Static pressure contours for the 20 m/s computation.

The pressures at the tap locations were recorded for each velocity tested. Likewise, the pressure coefficient was calculated at each tap location and plotted in relation to the airflow velocity.

The pressure coefficient is calculated for the static tube position on the ramp with reference to the static tube in the free stream.

12.2.2.5. Comparison between the Experimental, Numerical, and Analytical Results

The experimental (for the test without the anemometer, Test 7), numerical (from CFD), and theoretical pressure coefficients at the tap locations on the ramp are plotted on the same graph for comparison in Fig. 12.11.

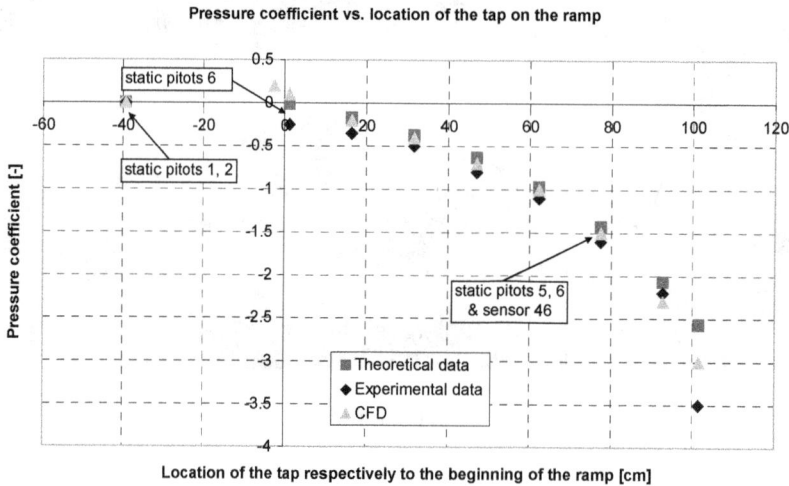

Fig. 12.11. Comparison of pressure coefficients.

The difference at the beginning of the ramp between the experiment and the analytical study is due to the fact that the thickness of the board was not considered in the theoretical calculations.

There is a large difference in pressure coefficients from experimental, numerical, and theoretical pressure coefficient at the end of the ramp. The difference between the theoretical and experimental values is due to the abrupt end of the ramp in the experimental tests that was not considered in the theoretical calculations. The difference between the experimental and numerical case may be attributed to the vena-contracta effect mentioned earlier.

Despite the differences at the beginning and end of the ramp, the middle portion of the ramp showed differences in the C_p values smaller than 0.2 between the experimental data, the numerical data and the theoretical data. This proves the reliability of the sensors' electronics to measure pressure decreases corresponding to velocity increases. It also validates the assumptions and boundary conditions of the CFD simulation to capture reasonably well the surface pressure change.

12.2.3. Repeatability of the Measurements

The purpose of the wind tunnel test was also to study the repeatability and reliability of measurements made with the sensors' electronic circuit. To this end, sensors were identified in a wind tunnel with the pressure tap at a known cross section, on a known angle slope, at known intervals. For every sensor thus the test parameters are known, the pressure was completely predictable by theoretical analysis (mass conservation and Bernoulli's equation). Then sensors' measured pressure can be compared to theory to assess the reliability.

But in the wind tunnel, due to wall confinement, the local free stream pressure changes on the ramp model. However, due to non availability of many static tubes, it was not possible to simultaneously measure the pressure of the free stream at all previously defined locations. So, one test was made for each location of the static tube. Thus, six tests, each with a different static tube location were conducted. One more test was made without the anemometer to study its influence on the measurement of the sensors. In summary, seven tests were made with sensors 41-48 at the same location on the board. This ensured the sensors' ability to give the same measurement when tested under the same conditions and wind speeds.

For studying the accuracy of the sensors, the equations of the trend lines were traced for each test and for all of the sensors used. Pressures corresponding to wind velocities were calculated from 0 to 20 m/s (in increments of 2 m/s) by using the trend line equations from the seven tests. Then the pressure average for a given wind velocity and a given sensor were calculated. More importantly, the standard deviation of the pressure in relation to the velocities was calculated for each test and for each sensor.

Next, the average of the standard deviation was calculated for each sensor and for all velocities. Finally, the average standard deviation at all velocities was calculated (See Table 12.1).

Table 12.1. Standard deviation over the wind tunnel tests for all sensors on the slope.

Sensor #	Standard deviation of the pressure [mbar]
41	0.050291874
42	0.044104608
43	0.085409689
44	0.075641291
45	0.073300046
46	0.113405777
47	0.112945412
48	0.115262638
Average	0.083795167

The averaged standard deviation was 0.08 mbar. This value is only one-third of the electronic components' resolution of the sensors (calculated as 0.24 mbar), and is encouraging. Nevertheless, accuracy implies that the sensors must be tested under identical experimental conditions. This means that test 7, when the anemometer was removed from the wind tunnel test section, may negatively influence the standard deviation and may increase its value. For this reason, the standard deviation was calculated for each sensor only for the first six tests. The average was computed for all the velocities, and finally the standard deviation value was averaged for all sensors (See Table 12.2).

Table 12.2. Standard deviation of all sensors for the first 6 tests.

Sensor #	Standard deviation of the pressure [mbar]
41	0.042224024
42	0.033629211
43	0.083377513
44	0.072421338
45	0.033980008
46	0.062223171
47	0.026752687
48	0.066874907
Average	0.052685357

The standard deviation decreased from 0.08 mbar to 0.053 mbar. This shows that the presence of the anemometer inside the test section influences the measurements of the taps. By observing and comparing the two Tables (Table 12.1 and Table 12.2), the values of the standard deviation do not differ much for sensors 41-44. However, for sensors 45-48 the values of the standard deviation were completely different, proving that the anemometer influences the tap's measurements at the top portion of the ramp.

The pressure measurements as a function of the wind velocity were analyzed for the first 6 tests. The standard deviation was 0.053 mbar (less than ¼ of the sensors' electronic precision).

This analysis showed that the sensors reading are repeatable within the limits of their precision.

In the next section the authors describe some tests performed to determine the effect of sensor case shape factor on the measured pressure.

12.3. Study of Sensor Shape Factor on Measured Pressure

12.3.1. Highway Test

A battery powers each of the wireless sensors. Its size requires the aerodynamic sensor case to have a certain height and diameter (48 mm and 260 mm, respectively).

Sensor case height is an issue because it creates flow disturbances that may affect the pressure measurements, and it may also affect the measurements of other sensors located in the sensor's zone of influence. The following test was performed to identify, characterize and quantify this effect.

12.3.1.1. Test Description

For this test the dimensions of a ramp model (to simulate the wind flow on a roof slope) are shown in Fig. 12.12.

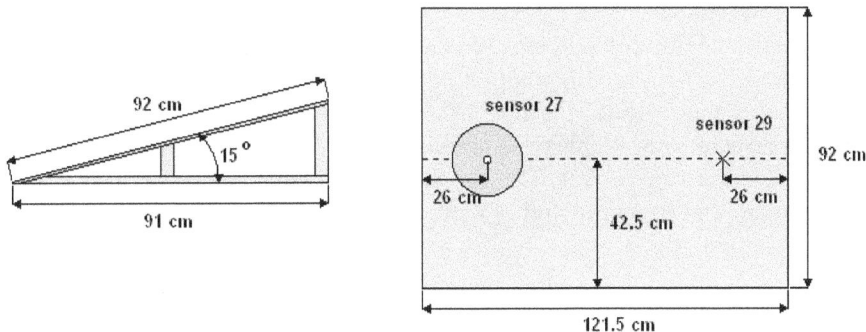

Fig. 12.12. Roof model dimensions.

Data recording started on 04-18-2009 at 10:50:31 and lasted until 04-18-2009 at 13:07:06.

Six sensors were used for this test (sensors 5, 15, 27, 28, 29, 30) in addition to the Young® Anemometer 31. One complete sensor with its case, sensor 27, was placed on the left side of the ramp model. Except for this sensor, the other sensors were connected via small Tygon tubes to surface pressure taps. These were then placed at different locations around the sensor as shown in Fig. 12.13.
- Sensor 30 was upstream of the sensor 27.
- Sensor 5 was downstream of the sensor 27.
- Sensors 15 and 28 were on the case of sensor 27, on either sides of its pressure cap.
- Sensor 29 was on the right side of the ramp model, at the same height and distance from the leading edge as sensor 27.

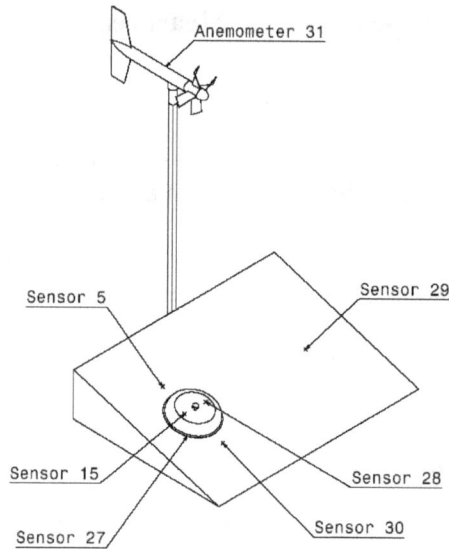

Fig. 12.13. Diagram of the sensors locations during the highway test.

The pressure taps of sensors 5, 29, and 30 were flushed with the ramp board.

The assembly was mounted on the roof of a minivan as shown in Fig. 12.14. The van was driven on an interstate highway to simulate wind velocity. The ramp faced forward to simulate the windward effect. During the second test, the ramp was rotated 180 degrees to simulate the leeward wind effect.

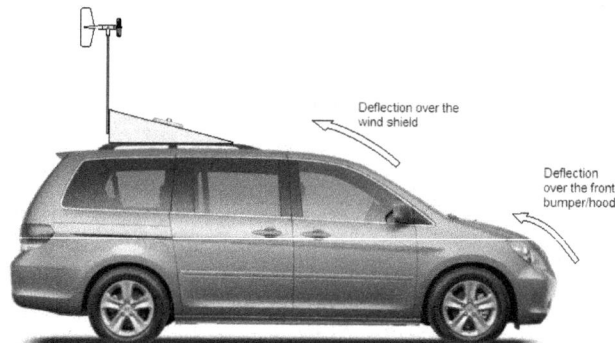

Fig. 12.14. Minivan with the ramp.

12.3.1.2. Data Analysis

After observing that the pressure variation conforms to the change in velocity (except for velocities < 5 m/s), the pressure coefficient was then calculated from (1). Note that in

Fig. 12.15 the pressure coefficients for velocity below 5 m/s are not resolved accurately because of poor signal to noise ratio at very low wind speeds.

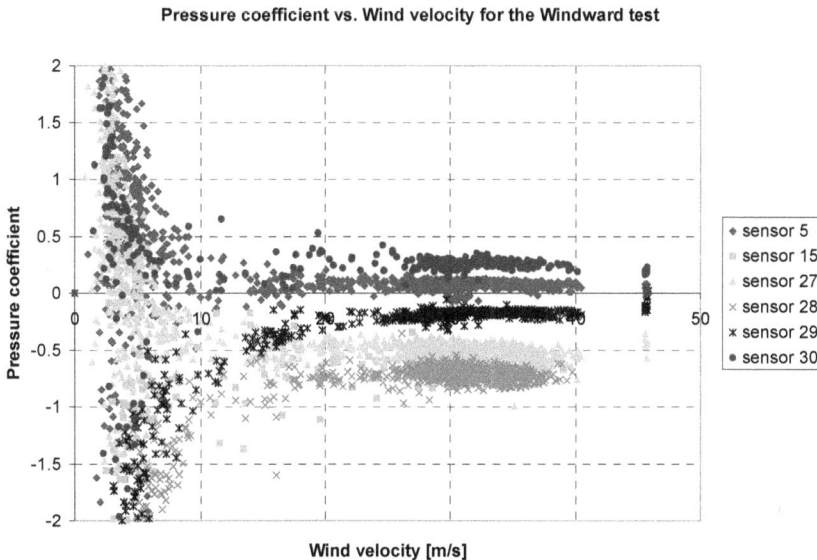

Pressure coefficient vs. Wind velocity for the Windward test

Fig. 12.15. Pressure coefficient for each sensor in relation to wind velocity for the windward test.

In Fig. 12.15, the pressure coefficients from the windward test show that the pressure coefficients of sensors 5 and 30 are positive, being close to the forward and rear stagnation points of Sensor 27 case. On the other hand, the pressure coefficients of sensors 15, 27, 28, and 29 are negative because they experienced flow acceleration. The pressure coefficients of sensors 15, 27, and 28 are very similar, as expected as they are close to each other. The pressure coefficient of sensor 29 is higher than the other three gages because it is not influenced by the local acceleration caused by the sensor 27 form factor.

For the leeward test (Fig. 12.16), the pressure coefficients of all sensors are identical. All of the sensors experienced the same pressure variation being in the separated zone.

12.3.1.3. CFD Model Comparison

Two CFD models were designed to model airflow over the ramp model. The two geometries were respecting the dimensions of the ramp model explained earlier but, without considering the front shape of the minivan. The first model was a simple ramp to study the pressure on the surface without the sensor. In the second model, the sensor shape was added to study the pressure on the sensor case.

Pressure coefficient vs. Wind velocity for the Leeward test

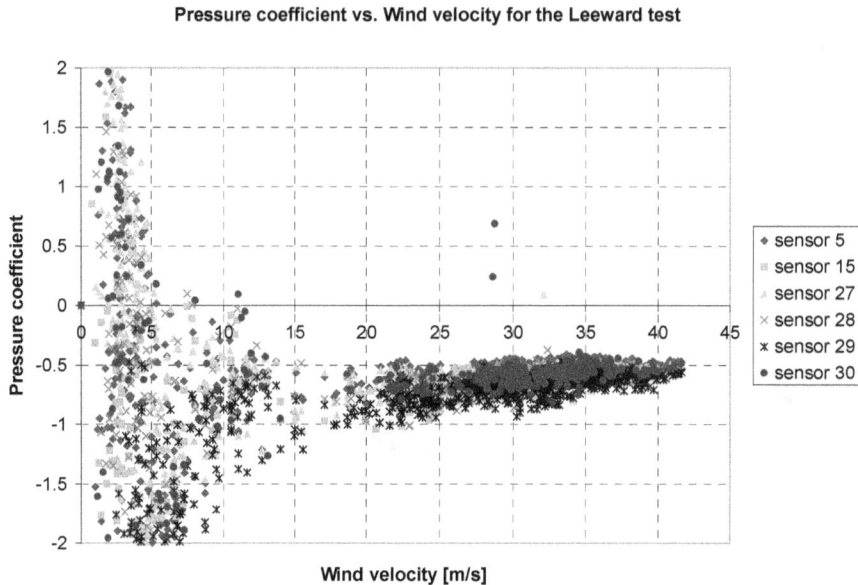

Fig. 12.16. Pressure coefficient for each sensor in relation to wind velocity for the leeward test.

The boundary conditions were defined similar to the wind tunnel CFD model and a "k-epsilon" turbulence model was used. In this only the 2-D simulation at the sensor port was performed.

These models were exported to Fluent and ran for the following freestream velocities V1 of 5, 10, 15, 20, 30, and 40 m/s.

The pressure readings were taken at the same locations as the static tubes and the sensor 27. The pressure coefficient was then calculated using Equation (12.1).

The pressure coefficients results found from the experimental highway test and those found with the CFD simulation are compared in Fig. 12.17.

Unlike the wind tunnel test where the CFD and experimental results were matching well, in the highway test the pressure coefficients were very different. In Fig. 12.17 the CFD pressure coefficients for the location in front of the sensor (with case)varied the most, by a factor of 10. At the other locations the agreement is not as bad.

The difference between the CFD calculations and the tests results may be due to:
*   The fact that the velocity from the anemometer was measured at the end of the ramp. Therefore this is not the same as the free stream velocity used in the simulation;
*   The front shape of the van that was deflecting and accelerating the air flow arriving on the ramp was not modeled in the CFD;
*   The fact that the CFD calculations is for a 2-D flow while the experiment was a 3-D flow.

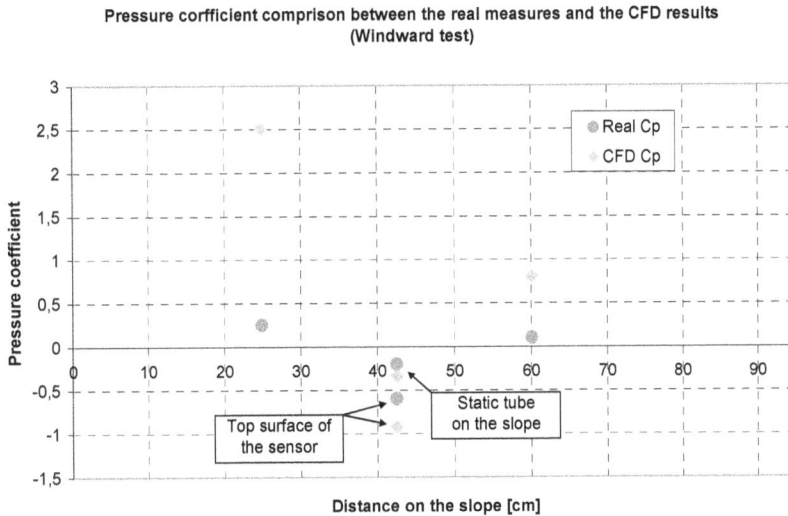

Fig. 12.17. Windward test - Pressure coefficient comparison between the experimental measurements and the CFD results.

12.3.1.3.1. 3-D CFD Simulation of Windward Test

A 3-D CFD model was created in three dimensions for the windward case. The sensor shape was simplified; the base plate underneath the sensor case as well as the pressure cap was not modeled, and just the sensor upper case on an inclined plate was modeled, without van front shape (Fig. 12.18).

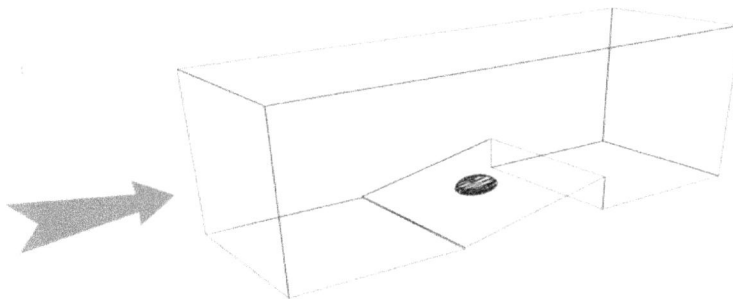

Fig. 12.18. Diagram of 3D CFD model.

The fine mesh was created in Gambit, with a Tri/Hybrid type and roughly 1.6 million cells, featuring a high resolution in the area of importance, while maintaining a coarser size for the rest of the volume. The box has a volume of 120" long, 50" high and 40.5" wide (3.048 m × 1.27 m × 1.028 m, respectively).

All the faces were set as walls, except for the inlet (set as "Velocity inlet"), and the outlet (set as "Outflow"). The direction of the flow is marked with an arrow in Fig. 12.18.

The Fluent simulation was run using the k-ε model for turbulent Reynolds number, for five different inlet velocities: 5, 10, 20, 30 and 40 m/s. The pressure was measured at the same points observed in Fig. 12.13. Sensor 30 is named as point 1, Sensor 15 as point 2, the measurement under the pressure cap (sensor 27) as point 3, Sensor 28 as point 4 and Sensor 5 as point 5.

Fig. 12.19 to Fig. 12.23 are the plots for the five velocities simulated. The experimental Cp values were taken as the averages of Fig. 12.15. Pressure coefficient for each sensor in relation to wind velocity for the windward test Fig. 12.15. The error for every point i was calculated as:

$$E_i = \frac{\left|Cp_{CFDcalc_i}\right|}{max\left(\left|Cp_{Exp}\right|\right)},$$

(12.2)

where $C_p\_CFDcalc$ is the simulated Pressure Coefficient at every point. Cp_Exp is the experimental C_p measured during the van test.

While observing the 5 m/s test (Fig. 12.19), a very large difference in C_p is noticeable between the experimental values and the simulated ones. The error curve (dotted line) shows this behavior. This can be associated with two facts:
- One, that the flow might not be fully developed as turbulent;
- Two, the atmospheric wind or passing vehicles effect is in the same order of magnitude than the velocity that is intended to be measured. Therefore, its uncertainty is high.

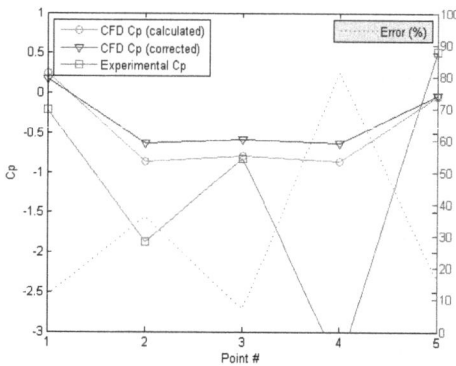

Fig. 12.19. CFD/Experimental comparison – 5 m/s.

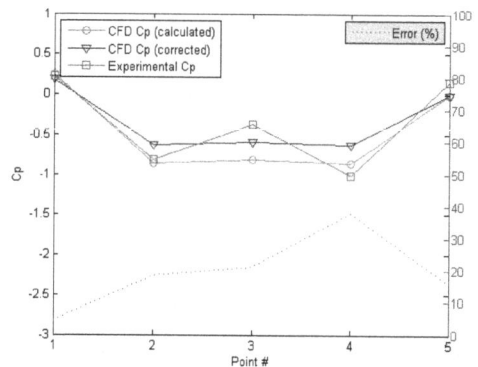

Fig. 12.20. CFD/Experimental comparison – 10 m/s.

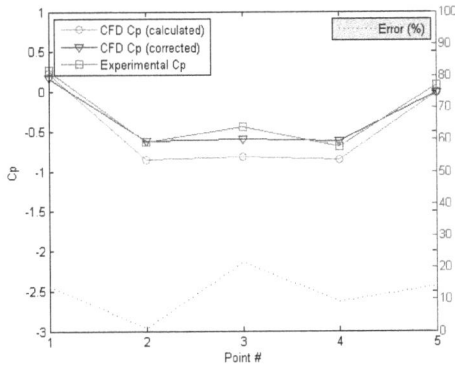

Fig. 12.21. CFD/Experimental comparison
– 20 m/s.

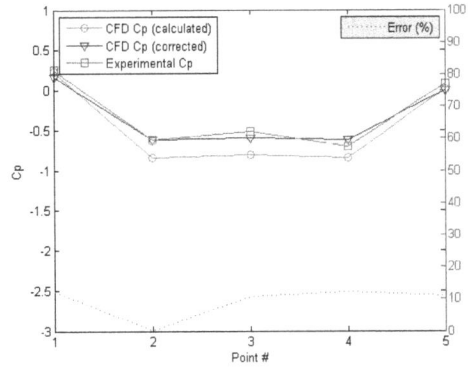

Fig. 12.22. CFD/Experimental comparison
– 30 m/s.

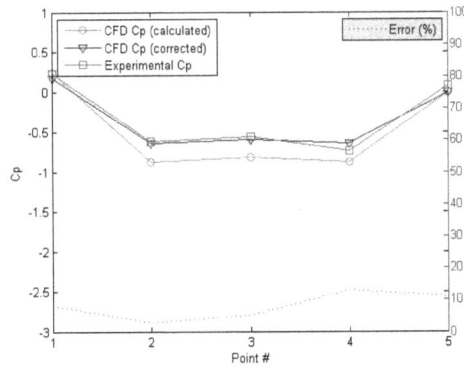

Fig. 12.23. CFD/Experimental comparison - 40m/s.

When observing the plots for 10, 20, 30, and 40 m/s, the results show a better agreement. However, there is a scaling difference between the values calculated with CFD and the experimental values. It was found that the freestream velocity is overestimated by 17 % in the CFD calculation of pressure coefficient. This difference has an explanation: the simulation cross-section area at the upper tip of the board is 18.7 % smaller than that in the inlet section, increasing the velocity in this point by roughly the same amount.

12.3.1.4. Conclusions on Shape Factor Studies

Through this experiment, the authors conclude that:
- Sensor case shape does influence the pressure measured by the transducer if it is placed on the windward side;

- There is no sensor shape effect when it is on the leeward case because of the leading edge flow separation. In the field test, sensors lying in the separated region will not be affected by the sensor shape;
- Speeds below 5 m/s produce large differences in the CFD estimates of C_p.

Some field tests were also performed to verify the above sensor shape factor on the Cp distribution.

12.3.2. University of Florida (UF) Hurricane Simulator Test

12.3.2.1. Test Description

A series of tests at the University of Florida hurricane simulator were conducted to assess the effects of wind gusts on pressure and pressure-velocity auto/cross correlations between different sensors. In particular, the tests study the influence of sensor shape on both the measurements of the sensor itself and on the measurements of a sensor located behind it. A CFD simulation of the UF Hurricane Simulator test was also performed to assess the sensitivity of the turbulence models in capturing the true pressure and velocity variations on the test object due to the gusts.

The roof model was made of two slopes at 30° as shown in Figs. 12.24 and 12.25. One slope faced the airflow generated by the hurricane simulator, and was designated as the windward side. The opposite side of the roof was then protected from the airflow and called the leeward side.

It should be noted that in the actual tests straight walls were mounted on the side the roof model (these are not represented on the pictures for clarity). The walls were there to channel the wind flow and to force it towards the roof model. Without the walls the airflow would naturally flow around the roof model, inducing side-roof effects and producing smaller pressure changes along the roof model. Constraining the fluid flow permitted study of what was occurring on the middle section of a roof.

A total of 29 sensors and 1 anemometer were used for this. Sensors 41, 42, 43, 44, 45, 50, 51, 52 and 53 were mounted on the windward side of the roof model. The sensors 54, 55, 56, 57, 58, 59, 60, 61 and 62 were installed on the leeward side of the roof model (see Fig. 12.26), also following the placement of a typical hurricane house.

In addition to the wireless sensors mentioned above which were in their metallic cases, additional sensors were connected to static tubes flushed with the roof. These were sensors 33, 34, 35, 37, and 38. Their pressure transducers were connected to pressure taps mounted on the windward side. Similarly sensors 39, 40, 46, 47, and 48 were connected to taps mounted on the leeward side (See Fig. 12.26.).

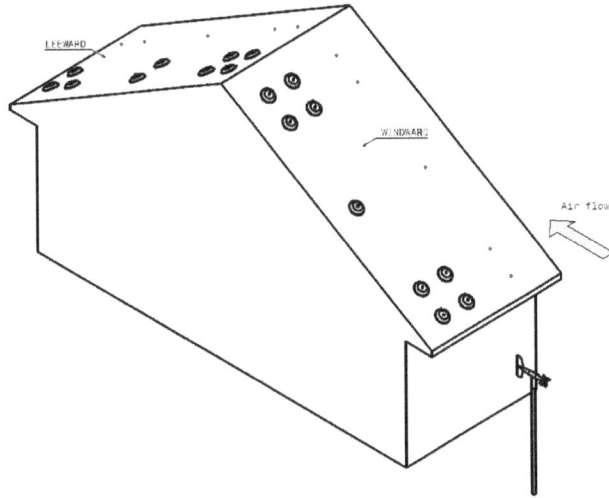

Fig. 12.24. Complete view of how the sensors were mounted on the roof model.

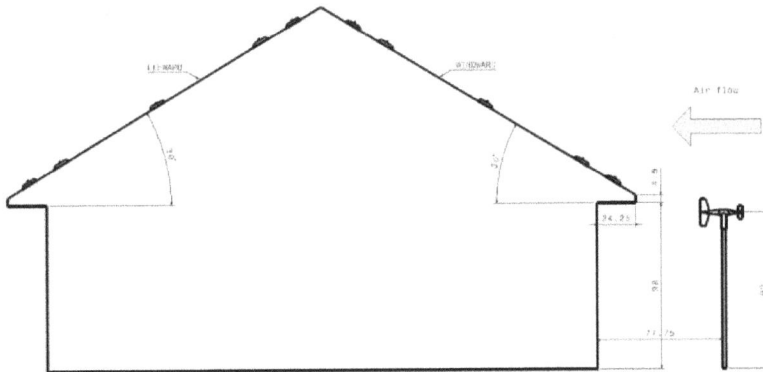

Fig. 12.25. Side view of the roof model.

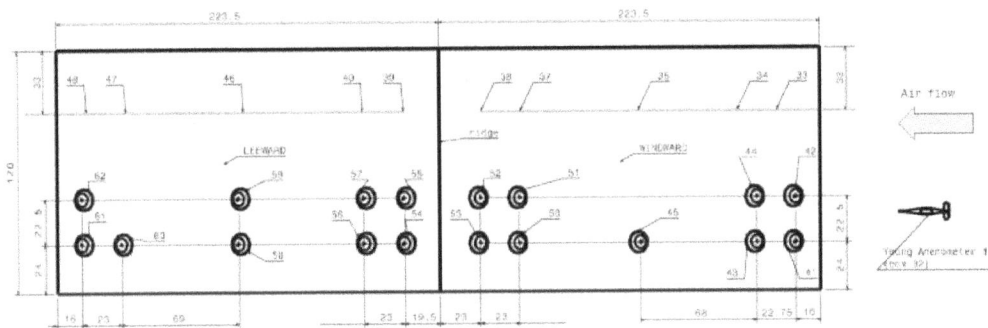

Fig. 12.26. Setup of the sensor/taps on the roof model.

Sensor 49 was placed in the control room, and was protected from the hurricane simulator's wind flow.

The anemometer was mounted on a poll 1.96 m in front of the roof model at a height of 2.34 m (Fig. 12.25), to be able to record the wind flow speed directly. This measurement was more accurate than a measurement taken on top of the roof because near the ridge separation occurs and the wind velocity is accelerated compared to the free stream velocity. Data recording started on 11-10-2009 at 10:07:39 and lasted until 11-10-2009 at 15:01:28 for the hurricane simulator tests.

To characterize and quantify the pressure variation along the roof as measured by the sensors and the static pressure taps, static tubes were located at the same height as the rows of sensors. These tests were meant to determine:
- The pressure change along the roof models (with the taps on one side).
- The difference in pressure measurements between the actual sensors and the taps located at the same height.

Five tests were made as follows:
- **Test 1**: a 1 minute progressive acceleration with 5-second steps of velocity, meaning that for one minute the speed was increased every 5 seconds.
- **Test 2**: a "3 by 20-second step test", meaning one minute steps with different wind velocities (30, 50, and 80 mph).
- **Test 3**: a "1 minute sinusoidal waves test", meaning that after reaching a velocity of 65 mph the fans blow wind making varying velocities between 45-80 mph with a 2 Hz frequency.
- **Test 4**: a "1 minute full speed test", meaning that the fan produced an 80 mph wind velocity for 1 minute.
- **Test 5**: a "3 by 20 second step test with water injection", meaning that Test 2 was run with the addition of water while the fans were blowing air to simulate the rain that could occur during a hurricane.

After these five tests, the first row of sensors (sensor 41 and 42) were removed from the roof and replaced by a static pressure tap connected to 42. Sensor 41 was placed beside sensor 49 to continue recording the atmospheric pressure. Having a tap present was important to monitor the pressure at the height of these sensors and to compare it with the pressure recorded previously. This way it was possible to run more tests and to evaluate the influence of the presence of the first row of sensors upon the other sensors.

Four tests were made with this new configuration:
- **Test 6**: a "3 by 20 second step test".
- **Test 7**: a "1 minute sinusoidal waves test".
- **Test 8**: a "1 minute full speed test".
- **Test 9**: a "3 by 20 second step test with water injection".

After these tests were complete, the second row of sensors (sensors 43 and 44) were removed and replaced by a tap connected to the sensor 44. Sensor 43 was placed beside sensors 49 and 41 to continue recording the atmospheric pressure. More tests were run to evaluate the influence of the presence of the second row of sensors upon the other sensors.

Four tests were made with this new configuration:
- **Test 10**: a "3 by 20 second step test".
- **Test 11**: a "1 minute sinusoidal waves test".
- **Test 12**: a "1 minute full speed test".
- **Test 13**: a "3 by 20 second step test with water injection".

12.3.2.2. Data Analysis

The comparison between the pressure data collected by the static tubes and the UF sensors data matched in time and amplitude. Using the wind velocity measurements recorded simultaneously by the wind anemometer 1, the pressure variation in relation to freestream dynamic pressure was calculated and plotted (see Fig. 12.27 and Fig. 12.28).

First, as expected the pressure variations recorded by the static tubes on the leeward side were much smaller than the variations measured on the windward side. Pressure variations are greater on the windward side because that side directly experiences the accelerating wind. The separation from the ridge causes the leeward pressures to be constant.

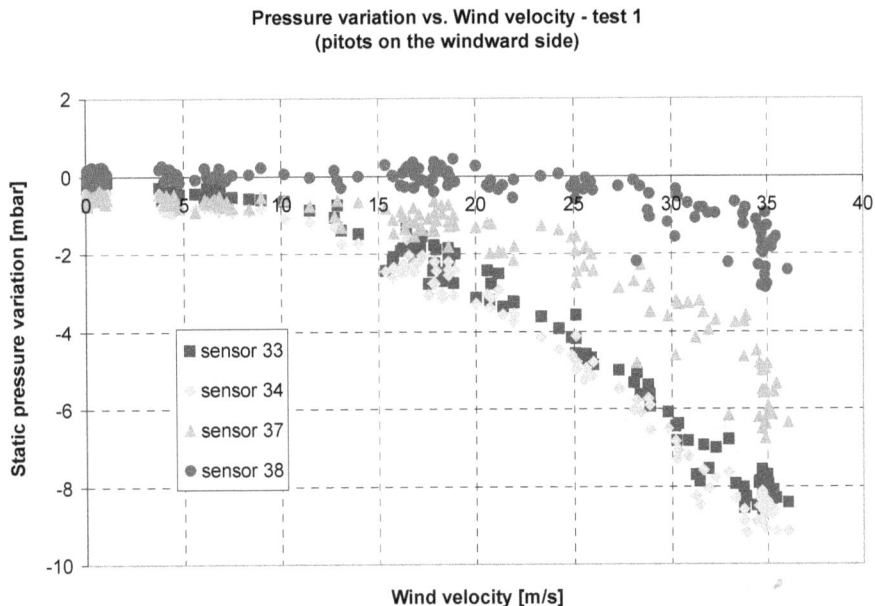

Fig. 12.27. Pressure variation recorded by the static tubes on the windward side of the roof.

Pressure variation vs. Wind velocity - test 1
(pitots on the leeward side)

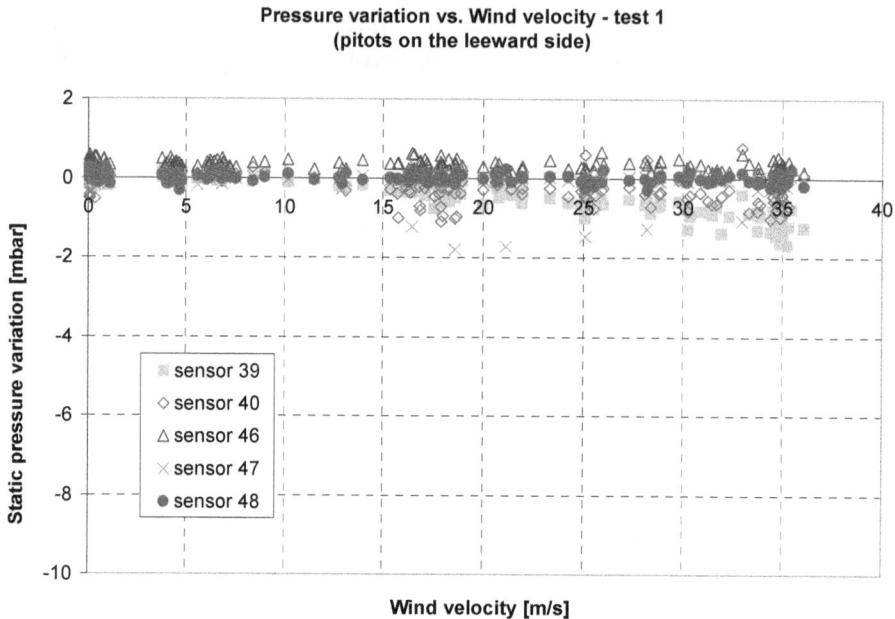

Fig. 12.28. Pressure variation recorded by the static tubes on the leeward side of the roof.

In Fig. 12.27, the pressure variation near the leading edge of the roof (sensors 33 and 34) was larger in magnitude than the value recorded at the top (sensors 37 and 38). Normally, it is expected that pressure difference increases further up the slope, and here the opposite occurred. This is probably the results of the walls surrounding the roof model. The test geometry forces the airflow to re-circulate in a zone in front of the roof. Further flow analysis was performed to study the scales of motions occurring on the roof flow.

12.3.3. Flow Analysis

12.3.3.1. Pressure Autocorrelation

The autocorrelation of each sensor determines the level of randomness of the pressure measured by the sensor, i.e. the time it takes to the signal to lose relation to itself in time. This gives an estimate of the time scales of motion needed for capturing the variations of pressure. A free stream wind speed of 35 m/s test was used for this analysis.

Fig. 12.29, Fig. 12.30 and Fig. 12.31 show the autocorrelation for the windward facing sensors. It can be seen that he autocorrelation decays faster near the eave than near the ridge. This can be due to eddies that caused by separation from the edge, creating a more unsteady behavior. In all cases, however, it can be seen that the decay time is more than 2 seconds. This suggests that using 3-second averages for pressure analysis is adequate for analyzing the mean pressure variations.

The difference between adjacent located sensors (lines in each graph) can be explained by side-wall effects.

On the leeward side, a different behavior is observed. For a full speed test, there is a poor autocorrelation of the signals (Fig. 12.32 - Fig. 12.34), despite the position of the sensor. This can be explained by the fact that the leeward side is affected by the vortices separated from the ridge, causing a random, less correlated pressures distribution.

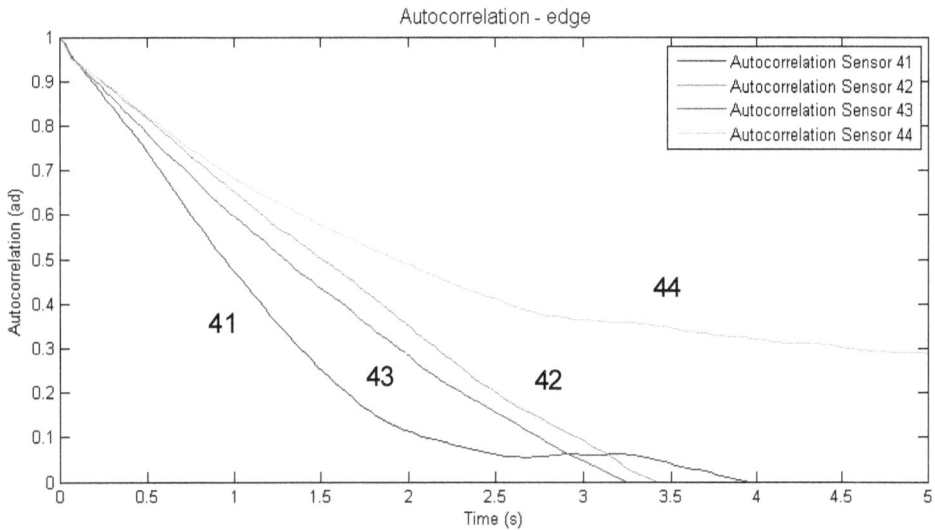

Fig. 12.29. Autocorrelation near edge (35 m/s windward).

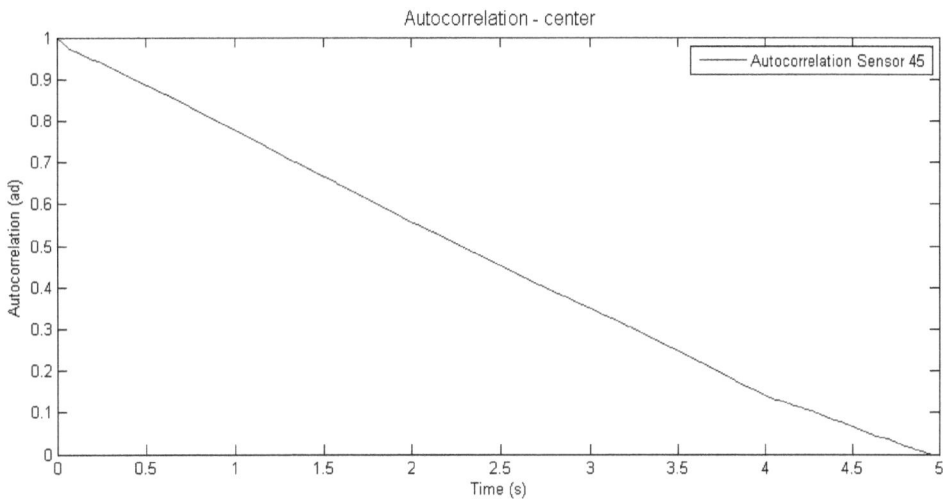

Fig. 12.30. Autocorrelation - center (35 m/s windward).

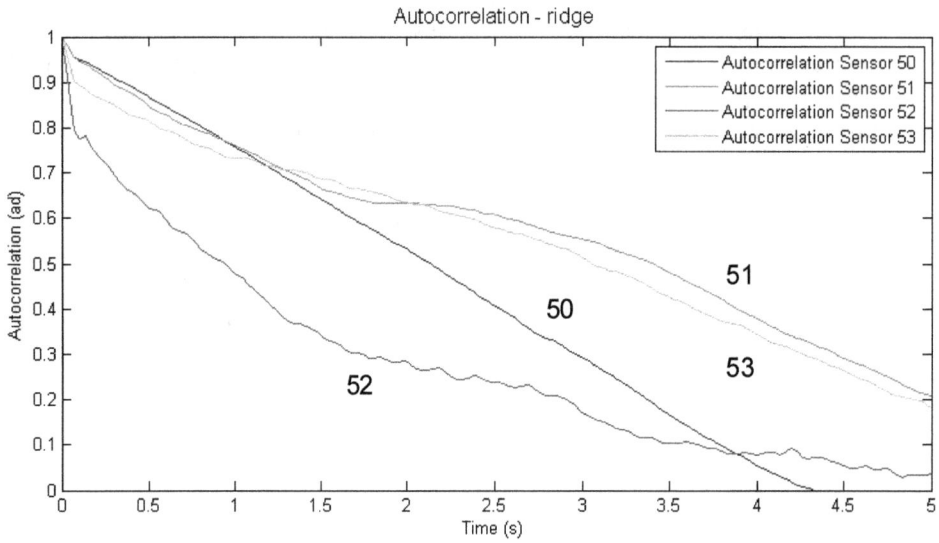

Fig. 12.31. Autocorrelation near ridge (35 m/s windward).

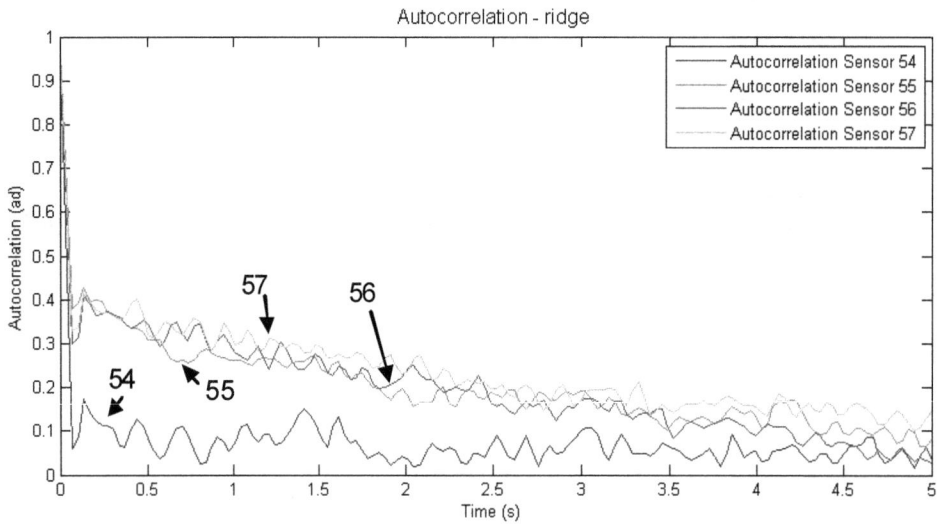

Fig. 12.32. Autocorrelation near ridge (35 m/s leeward).

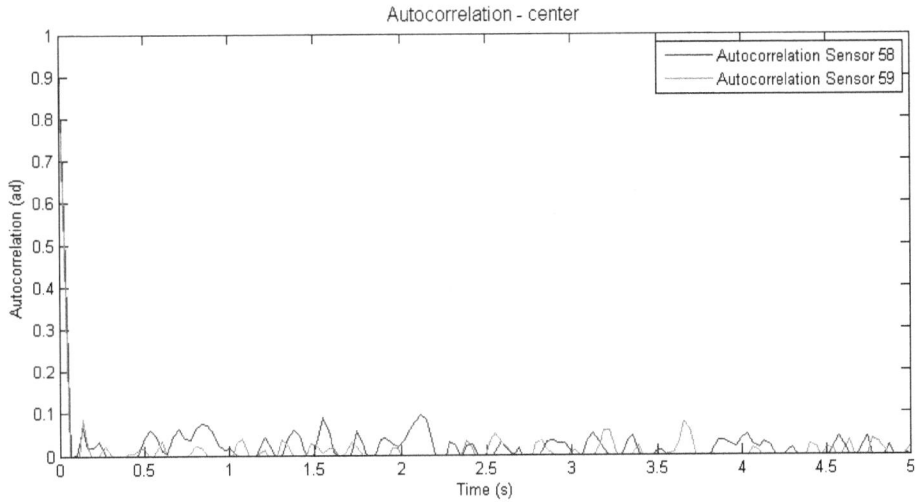

Fig. 12.33. Autocorrelation near center (35 m/s leeward).

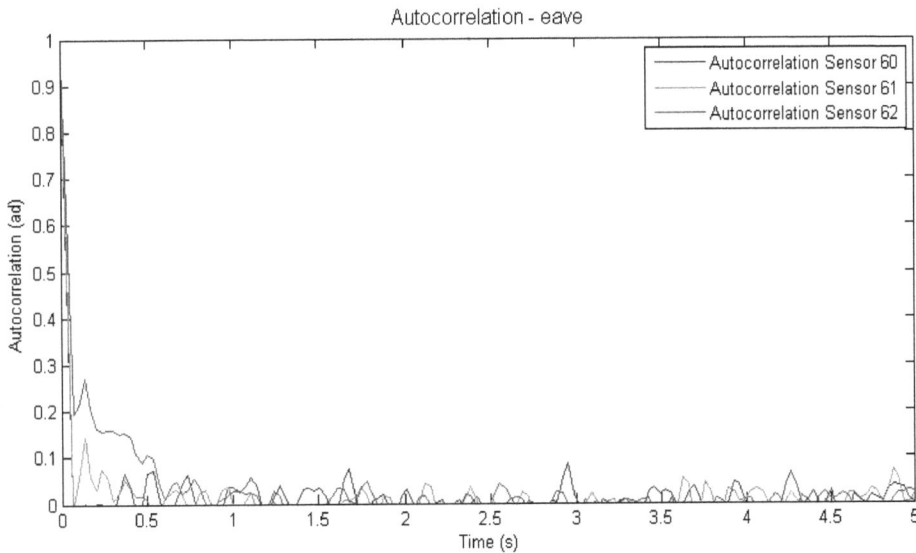

Fig. 12.34. Autocorrelation near edge (35 m/s leeward).

12.3.3.2. Pressure/pressure Cross-Correlation

For analyzing how the gusts pass, and how far the edge effects occur, the spatial cross correlation is performed between different sensors. Only the results of full speed test (35 m/s) is displayed in this case.

In Fig. 12.35, the correlation for 3 groups of sensors is plotted at three different distances. Sensor 41 is near the edge (on the eave), while 43 is just 22 inches

397

downstream. Sensors 45 and 50 are farther apart, at 80 and 160 inches respectively. From this graph the relationship between signals is evident. There is a time lag of around 0.8 seconds between the nearest sensors, and it increases as one moves downstream.

When the sensors are separated some correlation exists, but is uncertain and random. In the presence of reversed flows, or separated vortices creating distinct signals that is not discernable from this data. Therefore, for predicting how the vortex shedding phenomena occurs, and how it could relate with the damages, a higher spatial resolution for the pressure map must be used.

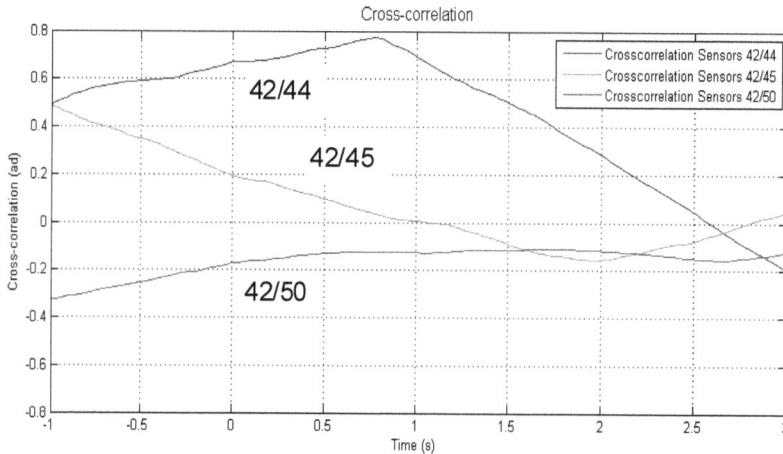

Fig. 12.35. Cross-correlation (windward 35 m/s).

For the leeward side (Fig. 12.36), the assumption of randomness stated in the autocorrelation analysis (Fig. 12.32 in Section 12.3.3.1) is reaffirmed. Only a small correlation value (~20 %) is observed between the adjacent sensors, but it is constant in time, meaning that there is no clear time dependence of the flow on this side. The correlation becomes zero when the sensors are further separated.

With further research, this type of analysis will help to determine the required spacing between sensors to obtain a pressure map that is close to reality.

12.3.3.3. Pressure Spectrum

Irrespective of the position of the sensor on the roof, the frequency spectrum shows a shape similar to that shown in Fig. 12.37. This is obtained from 20 seconds of data at a rate of ~20 samples per second. Therefore, according to Nyquist criteria, the authors will consider the frequencies under 10 Hz.

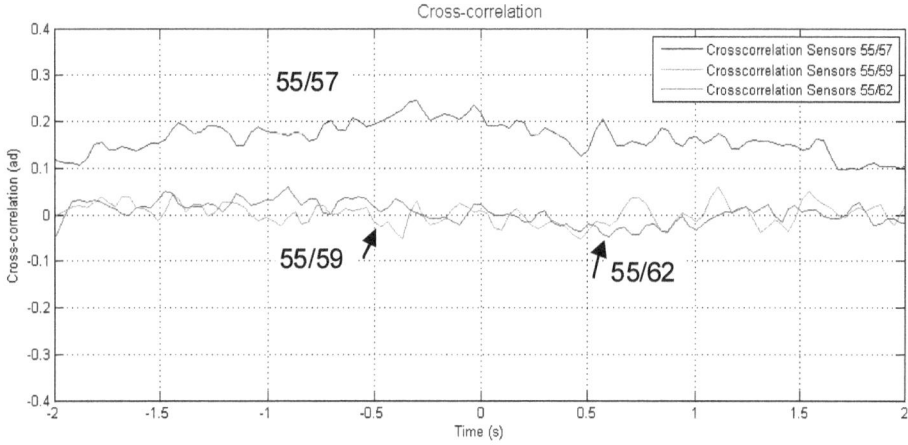

Fig. 12.36. Cross-correlation (leeward 35 m/s).

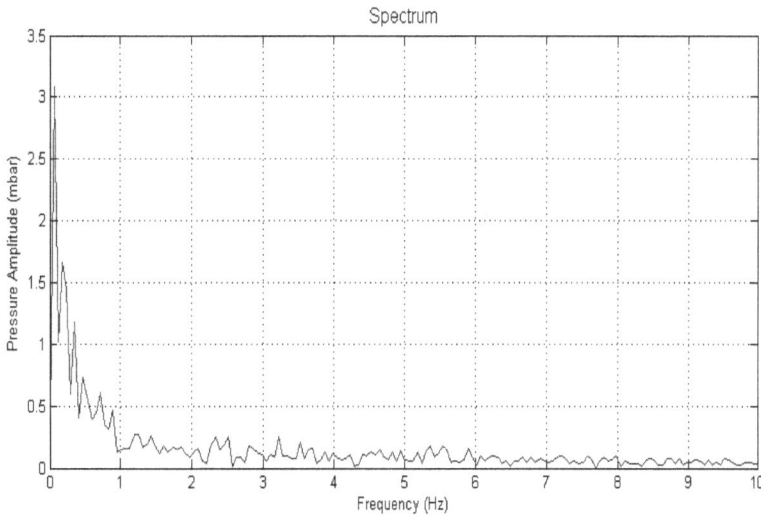

Fig. 12.37. Characteristic spectrum during full-speed test (35 m/s) - Sensor 44.

The motions with lowest frequencies (up to 0.5 Hz) are the ones that contain the gusts and contribute to the major variations of pressure. The first peaks observed correspond to the mean motion (even though the mean has been subtracted, but as this is not a steady signal around that value, some variations appear). After 2 Hz, the amplitudes remain constant, probably attributable to noise in the measurement.

No significant change is observed in the above spectrum between several tests and speeds. The spectrum is the same even for the flow over the eave. No special frequencies are excited in this position, only random variations of pressures occur, depending on the experimental conditions of the test (type of wind load applied and velocity), but not depending on its proximity to the edge. Fig. 12.38 shows an example of this, no significant amplitude occurs for frequencies between 0.5 and 2 Hz. This appears to be a general trend for all the experiments conducted at this speed of 35 m/s.

The frequency spectrum for the leeward side sensors (Fig. 12.39) suggests that the pressure variations are not very significant. Sensors near the ridge show some dependence of the mean flow, but the spectrum amplitude is only 25-30 % of that seen for the windward side.

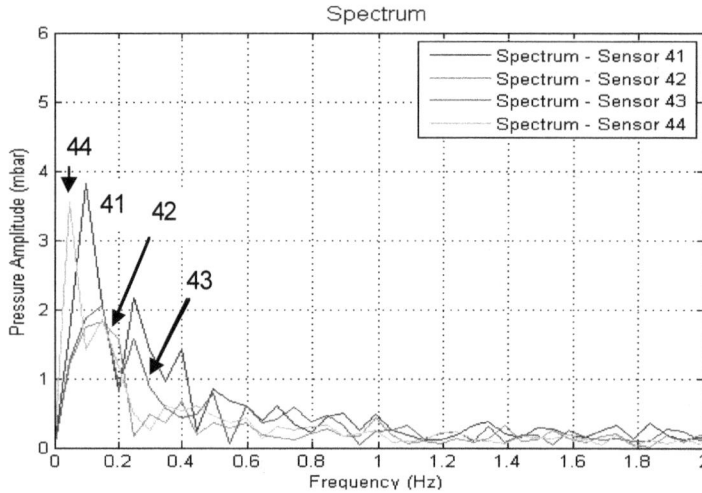

Fig. 12.38. Pressure spectrum (windward at 35 m/s).

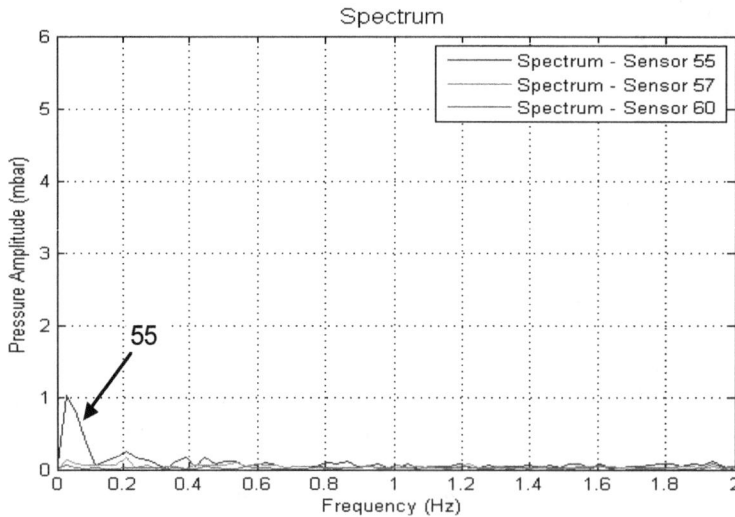

Fig. 12.39. Pressure spectrum (leeward at 35 m/s).

12.3.3.4. Gust effects – Velocity/Pressure Cross-correlations

For considering the effect of gusts, the velocity/pressure cross-correlation is performed on a 1-second averaged signal. As mentioned before, the mean flow contains the greatest amount of energy, and it is observed in the frequencies below 1 Hz. This analysis, shows how the gust moves forward and how the pressure changes along a line on the roof, for three types of gusts created in the simulator.

Fig. 12.40 shows that for a gust that starts from standstill (initial velocity = 0), similar variations in real time are seen between all the sensors, even for the leeward side sensors. This may indicate the presence of a uniform flow, with large structures moving (if there are eddies created, their order of magnitude is of the same size as that of the roof). The last sensor in the leeward side (shown red in Fig. 12.41), does not display much relationship because the velocity is too small, and the roof has a big slope (30 degrees), which could be separating the flow from the roof before this sensor, creating a recirculation region.

In Fig. 12.42 and Fig. 12.43, the correlation created suggests that near the ridge, the pressure has high correlation with velocity on both sides (negative correlation means the pressure decreases as velocity increases). However, on the eave and up to the middle, a different behavior is observed, creating first a low-pressure front and then switching to a higher pressure after 3-5 seconds.

For Fig. 12.44 and Fig. 12.45, the highest speed gust, the relationship is not so clear, because different turbulence structures are excited. This set of tests is unable to explain high-speed correlation between velocity and pressure. However, the basis is set for future studies, to collect data that explain this behavior more clearly.

It is important to note that the spectrum analysis does not show any unexpected amplitudes or frequencies under these conditions.

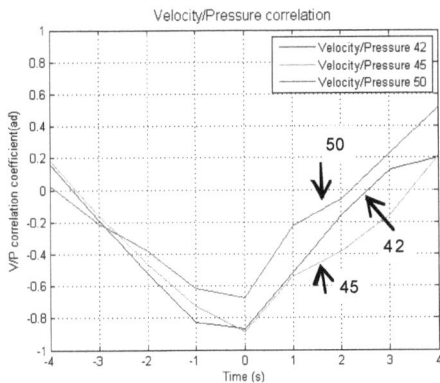

Fig. 12.40. V/P correlation (windward 0-12 m/s).

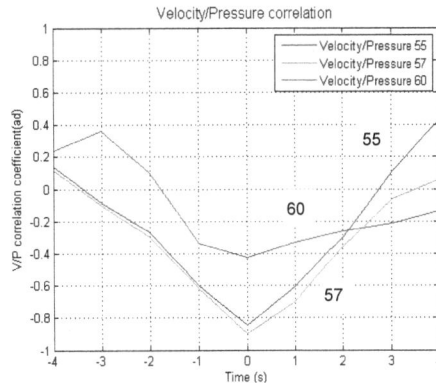

Fig. 12.41. V/P correlation (leeward 0-12 m/s).

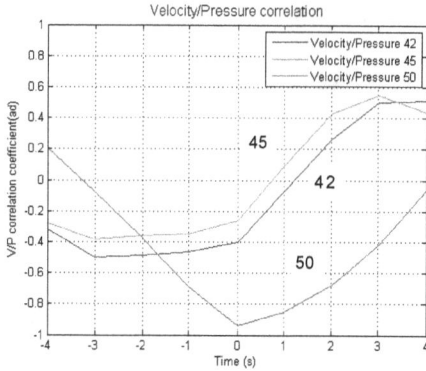

Fig. 12.42. V/P correlation
(windward 12-28 m/s).

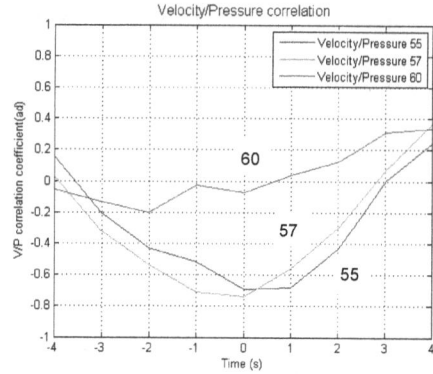

Fig. 12.43. V/P correlation
(leeward 12-28 m/s).

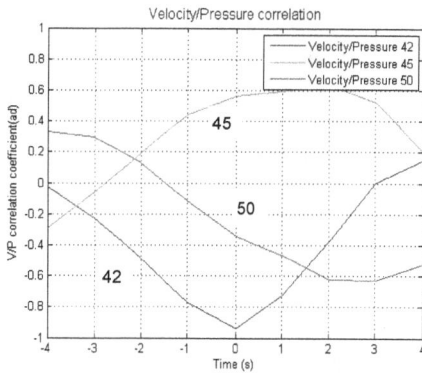

Fig. 12.44. V/P correlation
(windward 28-35 m/s).

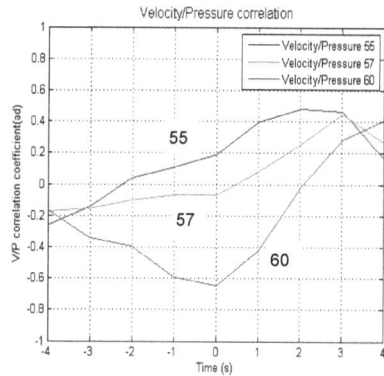

Fig. 12.45. V/P correlation
(leeward 28-35 m/s).

For multiple gusts, (Fig. 12.46 and Fig. 12.47) the influence of the anemometer's inertia is evident, especially on the windward side sensors, where the sensor closest to the anemometer shows the peak of correlation at -1 second, while the rest is drifted forward in time. In this case, the relationship between pressure and velocity is evidenced. However, sensor 57 shows a higher correlation than the others.

12.3.4. CFD Simulation of UF Wind Tunnel Test

For assessing the ability of different turbulence models to predict best the measured pressure distribution, full speed test simulations (35 m/s) were performed in Fluent 6.3, using different eddy viscosity models. The fluid used was air with Fluent's default properties.

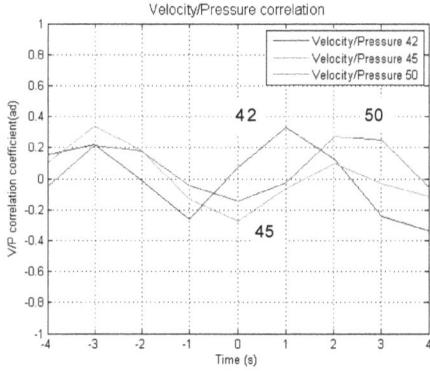

Fig. 12.46. V/P correlation (windward multiple gusts 20-30 m/s).

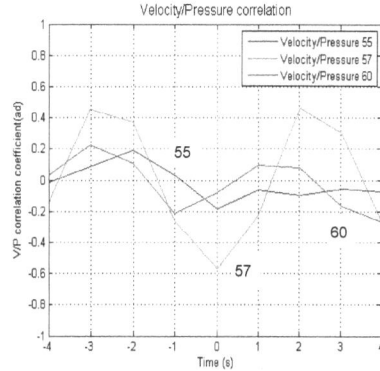

Fig. 12.47. V/P correlation (leeward multiple gusts 20-30 m/s).

12.3.4.1. Geometrical Setup

A recreation of the test setup was made in Gambit. Because of this flow nature, the model was split in to two symmetric halves, coinciding with the center line of the roof. The basic shape is a pseudo eighth of sphere around the test model. It includes the parapets and all the basic shapes used (Fig. 12.48). The pseudo sphere has a radius of approximately 70 meters.

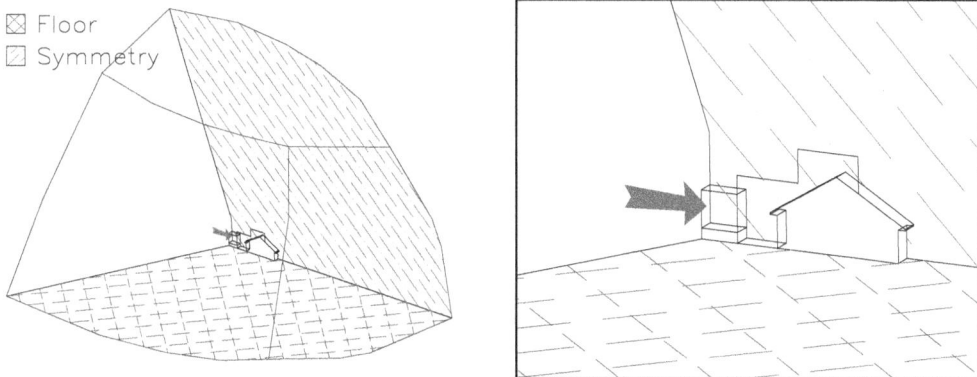

Fig. 12.48. Diagram of 3D CFD geometry.

The boundary conditions were set as:
- One "velocity inlet" (indicated with the arrow in Fig. 12.48);
- The floor and structure were set as "wall" type;
- The outer surface of the dome was set as "pressure outlet";
- The middle plane was set as "symmetry".

The final mesh created in Gambit (Fig. 12.49) is a Tri type with roughly 1.7 million elements. Higher resolution was used in the zone of relevance (near the house model), and a coarser mesh was employed near the pressure outlet.

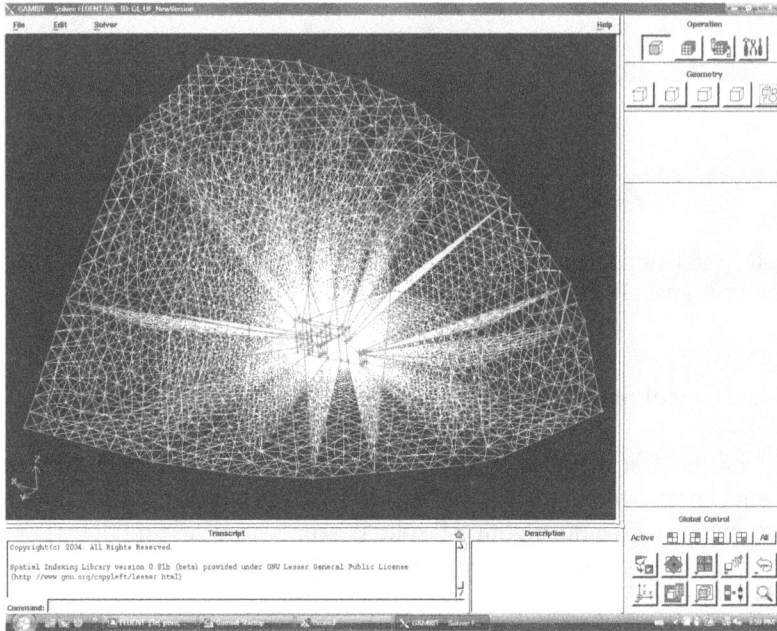

Fig. 12.49. Final mesh in Gambit.

12.3.4.2. Simulation Results

As a starting point for comparisons, the experimental pressure coefficients were calculated and averaged for a full speed test. An interpolated plot of them is shown in Fig. 12.50 (note that this is an interpolation between very few data points, special care must be taken when comparing values).

The results of the calculations using k-ε turbulence model are presented in Fig. 12.51 and Fig. 12.52.

The results obtained using this model lack consistency due to the fact that the flow does not feature fully developed turbulence in the entire domain. In reality, an unsteady behavior is observed with large eddies detaching from the roof. The assumption of the k-ε model entails a steady state mean motion.

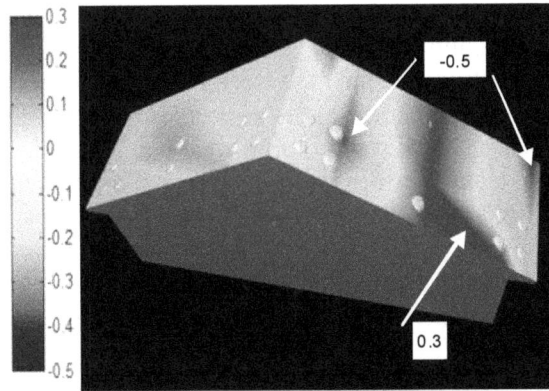

Fig. 12.50. Experimental pressure coefficient (35 m/s test average). Flow is from right.

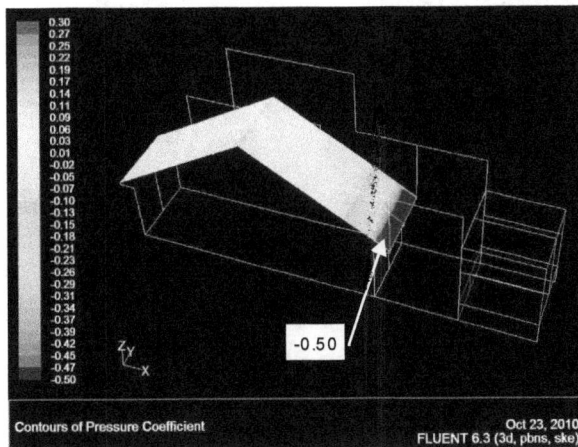

Fig. 12.51. CFD pressure coefficient contours (k-ε model at 35 m/s). Flow is from right.

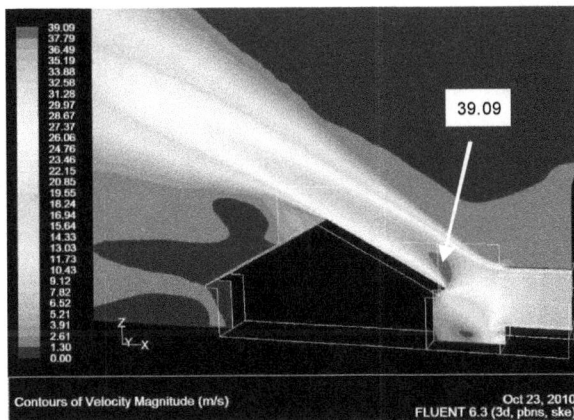

Fig. 12.52. Velocities in symmetry plane (k-ε model at 35 m/s). Flow is from right.

Another simulation was run using Large Eddy Simulation (LES) method, creating better results in terms of the magnitude of the values obtained. Comparing Fig. 12.50 with Fig. 12.53, a relatively better agreement is seen when paying attention to the points where the sensors were placed, except in the area near the ridge. These simulations can be extensively improved in terms of the turbulence characteristics applied. Moreover, for detailed matching of results the sensor shape factor correction must be applied to the real measurements in Fig. 12.50.

Fig. 12.54 shows the velocity distribution obtained by this method, displaying clearly the large-size eddies generated.

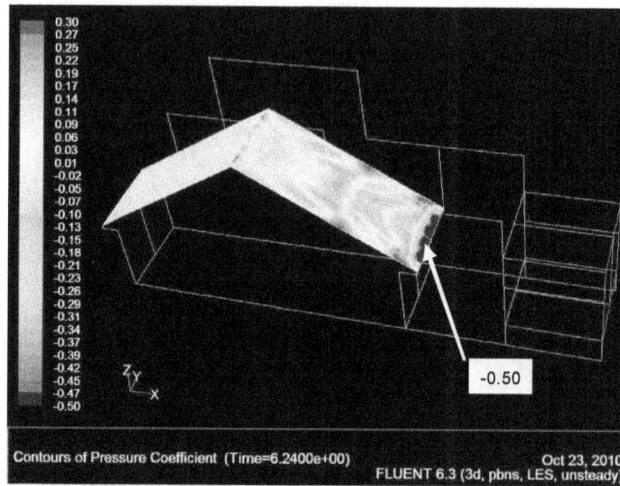

Fig. 12.53. CFD pressure coefficient (LES model at 35m/s). Flow is from right.

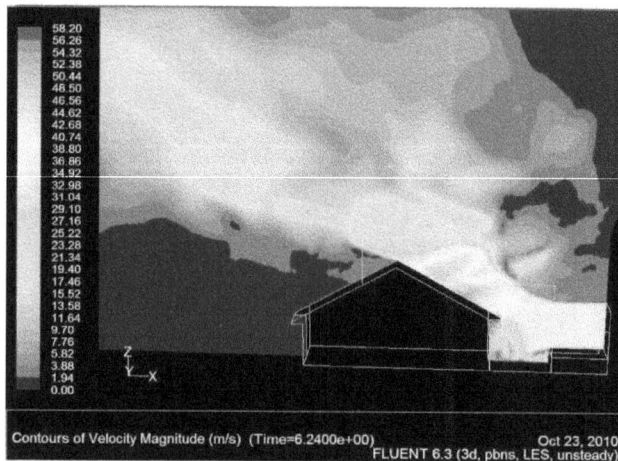

Fig. 12.54. Velocities in symmetry plane (LES model at 35m/s). Flow is from right.

12.4. Vibration Study

During hurricane deployments the roof structures may be excited to its natural frequencies at certain wind speeds. In the UF tests, noise was observed in the pressure records. Analysis showed that it was due to mechanical vibrations, and not due to electromagnetic disturbances. So, the authors performed a vibration study. The next section describes the experiment to determine the sensitivity of the pressure sensor to mechanically induced vibrations.

12.4.1. Noise Source Analysis

During the test, the wooden structure of the roof model, on which the sensors were mounted, was shaking. It was then probable that the noise observed on the pressure records was due to mechanical vibrations. In fact inside the sensor, the pressure transducer is composed of a flexible membrane. The movement of this membrane is then converted into an electrical current whose intensity corresponds to the amplitude of the deformation. Therefore, any deformation of the membrane is considered by the sensor as a pressure variation. The question arose regarding whether mechanical vibration could cause this membrane to move. A vibration test was performed and showed that mechanical vibration can cause the same type of incoherent pressure records, observed in the UF tests.

12.4.2 Test Description

A total of 3 sensors were used for this test: sensors 41, 42, 44. The remote 41 was mounted on a vibrating structure that was placed on a vibration table (see Fig. 12.55). The vibration table produces horizontal vibrations. However, during the UF test, the vibrations were perpendicular to the surface of the roof model. In order to reproduce the type of vibrations experienced by the remotes during the UF test, the sensor was mounted vertically.

Only the remote 41 was submitted to vibrations, the other sensors were used as references. Seven tests were made with this set of sensors changing only the frequency of the vibrations and the amplitude when it was possible:
- Test 1: 5 Hz with an amplitude of 0.3 V;
- Test 2: 5 Hz with an amplitude of 0.5 V;
- Test 3: 10 Hz with an amplitude of 0.3 V;
- Test 4: 10 Hz with an amplitude of 0.5 V;
- Test 4: 20 Hz with an amplitude of 0.3 V;
- Test 6: 30 Hz with an amplitude of 0.3 V;
- Test 7: 50 Hz with an amplitude of 0.3 V.

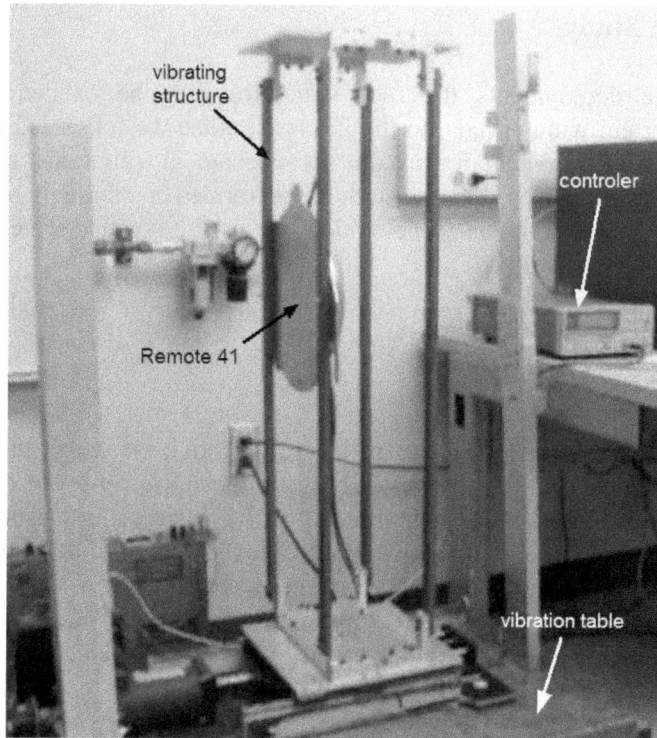

Fig. 12.55. Vibration test set up - sensor 41 mounting.

12.4.3. Results and Analysis

One may observe in Fig. 12.56 that for the test made with 5 Hz vibrations the sensor did not record any pressure difference. The records of sensor 41 match those of the two other remotes. This test shows that a frequency of 5 Hz does not create vibrations that produce noise in the pressure records.

Similarly, for the tests 3 and 4 the records of the sensor 41 are matching those of sensor 42 and 44. Therefore, one may also conclude that a frequency of 10 Hz does not produce vibrations that disturb the pressure measurements.

It is important to note that for frequencies higher than 10 Hz, it was not possible to test the structure with more than 0.3 V of amplitude. For this reason only one amplitude (0.3 V) was tested for the following frequencies.

One may clearly observe in the pressure records of the sensor 41 in Fig. 12.57 that the sensor experienced noise. The remotes 42 and 44, on the contrary, did not show any pressure change. Considering that all the sensors were in the same closed room, they should all show the same pressure variation. This test proves that mechanical vibrations can induce noise in the pressure measurement made by the sensor.

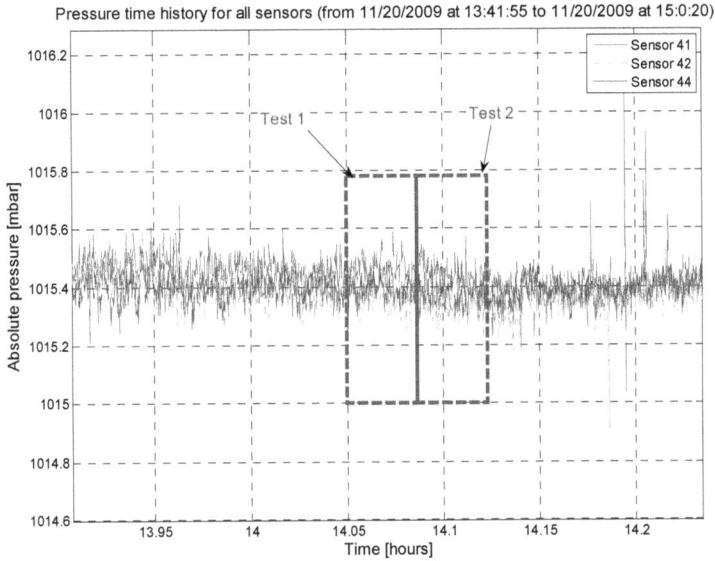

Fig. 12.56. Pressure variation records for the 5 Hz tests.

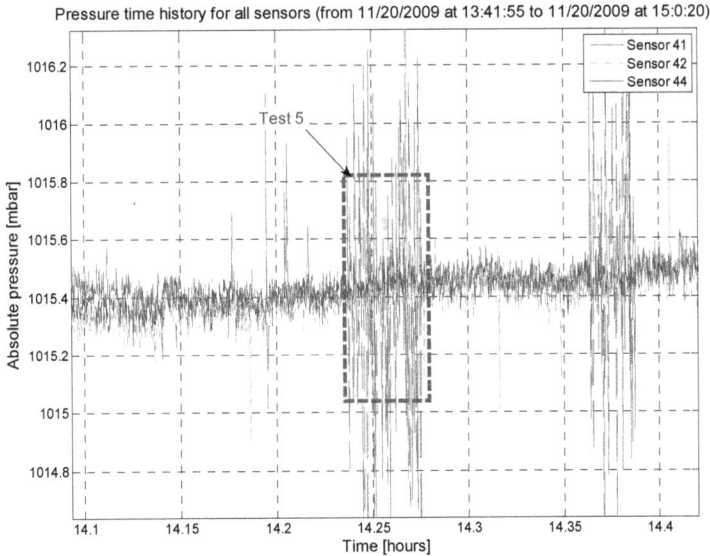

Fig. 12.57. Pressure variation records for the 20 Hz tests.

Similarly, for 30 Hz vibrations, the sensor 41 experienced the same kind of noise. The test 6 proves that the pressure sensor is also sensitive to 30 Hz vibrations.

Finally, for the seventh test, no disturbance was recorded by the pressure sensor. The pressure variations recorded by the sensor 41 match those recorded by 42 and 44. This test shows that above 30 Hz, high frequency vibrations do not influence the pressure sensor and the pressure records.

The vibration tests proved that mechanical vibrations can cause the pressure sensor to record erroneous pressure variations. For 20 Hz and 30 Hz the tests showed that the pressure sensor mounted on the vibrating structure recorded pressure changes while the reference sensors were showing a constant pressure. These are the same type of pressure variation recorded during the UF test by the sensors fixed to the top of the roof model.

12.5. Conclusions

A series of tests showed the ability of the multisensory wireless system to measure reliably the ambient pressures accurately with respect to local NWS reported pressures.

Wind tunnel tests showed that the pressures measured by the sensors, when not encased, were comparable to those estimated from the Bernoulli's equation. The reliability of the sensors was further confirmed through CDF simulations.

However, high wind speed tests performed on a highway demonstrated a significant influence of the form factor of the sensor case on the pressure measurement for windward winds. Leeward wind tests, on the other hand, showed no effect of the sensor shape factor. The 3-D CFD simulations also showed similar results. The 3-D simulations provided more flow details than the measurements.

Additional full scale tests in the UF hurricane simulator showed that on the windward side, the pressure variations can be adequately captured using a low-sampling rate (4 Hz) signal. The magnitude of the pressure variations is greater for the windward sensors than for the leeward sensors by a factor between 3 and 10. Regarding the spatial distance between sensors, the 22 inches distance is adequate for capturing the pressure variation at velocities of up to 35 m/s on the eave; the 80 inches separation is too large and there are gust structures that could be missed. Because of this, various rows of sensors at increasing distances between them must be analyzed to determine accurate correlation, to define a proper distance between sensors in order to resolve the range of structures that needs to be captured.

The high-speed (above 25 m/s) and gust analysis show that in general terms, a velocity/pressure correlation exists and can be measured, but further tests need to be made to explain certain local effects observed in the graphs.

The measurements further showed that on the windward side of a roof the taps recorded pressure profiles that did not follow the expected trend because of high free stream intensity. In order to simulate this flow accurately in CFD, a 3-D large eddy simulation model is necessary, and the turbulence characteristics extensively adjusted. In addition, the pressure sensors installed on the roof model were extremely sensitive to mechanical vibrations.

The influence of such vibrations was studied with a shaker table. The vibration tests showed that the remote sensors experience noise when submitted to vibrations from 20 Hz to 30 Hz. Unfortunately, even if the cause is now known, the spectrum analysis demonstrated that the vibration frequencies were in the same range as the pressure data sampling frequencies. This made filtering and retrieving of the clean pressure data difficult. To avoid this problem during deployments, the sensor natural frequency must be kept much higher than the roof system natural frequency.

In conclusion, even if the hurricane simulator test did not obtain very clean data, several tests proved the system to be reliable and able to measure pressure variations greater than 0.24 mbar accurately. Future studies will address reshaping of the sensor case for low form factor, increasing the sensor data sampling rate to 100 Hz, and expanding the sensing platform for strain gage and accelerometer measurements.

Acknowledgements

The authors acknowledge the help of Dr. Forrest Masters and his team who facilitated the use of the hurricane simulator at the University of Florida. The authors are also grateful to Mr. Sylvin Dozolme and Mr. Nicolas Fillaud for their help with CFD calculations. This material is based upon work supported by the National Science Foundation under Grant No. 0625124, and by the Division of Emergency Management of the Florida Department of Community Affairs under Grant No. 10-RC-26-12-00-22-254. Any opinions, findings, and conclusions or recommendations expressed in this material are those of the authors and do not necessarily reflect the views of the National Science Foundation or the Florida Department of Community Affairs.

References

[1]. F. Kreit, G. Barberio, C. Subramanian, I. Kostanic, J.-P Pinelli, Performance Testing of the Wireless Sensor Network System for Hurricane Monitoring, in of the in *Proceedings 2010 The 1st International Conference on Sensor Device Technologies and Applications*, 2010 IEEE Computer Society, Mestre, Venice, Italy, July 18 - 25, 2010, pp. 63-72.

[2]. K. J. Eaton and J. R. Mayne, The measurement of wind pressures on two-story houses at Aylesbury, *J. Ind. Aerodyn.*, 1, 1975, pp. 67-109.

[3]. G. M. Richardson, A. P. Robertson, R. P. Hoxey, and D. Surry, Full-scale and model investigations of pressures on an industrial/agricultural building, *J. Wind Eng. Ind. Aerodyn.*, 36, 1990, pp. 1053-1062.

[4]. M. L. Levitan, and K. C. Mehta, Texas Tech field experiments for wind loads part I. Building and pressure measuring system, *J. Wind Eng. Ind. Aerodyn.*, 43, 3, Oct, 1992, pp. 1565-1576.

[5]. M. L. Levitan, and K. C. Mehta, Texas Tech field experiments for wind loads part II. Meteorological instrumentation and terrain parameters, *J. Wind Eng. Ind. Aerodyn.*, 43, 3, Oct, 1992, pp. 1577-1588.

[6]. M. L. Porterfield, and N. P Jones, The development of a field measurement instrumentation system for low-rise construction, *Wind and Structures, An International Journal*, 4, 3, June, 2001, pp. 247-260.

[7]. B. J. Michot, Full-Scale Wind Pressure Measurement Utilizing Unobtrusive Absolute Pressure Transducer Technology, M. S. Thesis, *Civil Engineering Department, Clemson University,* S. C., Dec 1999.

[8]. C. S. Subramanian, J.-P. Pinelli, C. Lapilli, and L. Buist, A Remote Multi-Point Pressure Sensing System, *43rd AIAA Aerospace Sciences Meeting and Exhibit* 10 - 13 January 2005, Reno, Nevada.

[9]. C. S. Subramanian, J-P Pinelli, C. D. Lapilli, and L. Buist, A Wireless Multipoint Pressure Sensing System: Design and Operation, *IEEE Sensors Journal*, Vol. 5, No. 5, October 2005.

[10]. C. S. Subramanian, J.-P. Pinelli, I. Kostanic, L. Buist , B. Van der Veek, A. Velazquez, F. Kreit, and C. Otero, Development and testing of a second generation wireless hurricane wind and pressure monitoring system, HURSENSOR Report, *Florida Institute of Technology*, 2009.

[11]. C. Subramanian, J.-P. Pinelli., and I. Kostanic, Measurement and Characterization of Hurricane Wind Loads on Structures Using a Wireless Sensing Networking System, in *Proc. of the 11th American Conference on Wind Engineering*, Puorto Rico, June 22-26, 2009.

[12]. A. Velasquez, WINDS-HM Wireless networking sensors for hurricane monitoring, MS Thesis, *Florida Institute of Technology, Civil Engineering Department,* December 2009.

[13]. F. Kreit, Performance Testing of the Wireless Sensor Network System for Hurricane Monitoring, MS Thesis, *Florida institute of Technology, Mechanical & Aerospace Engineering Department,* December 2009.

[14]. C. Otero, A. Velazquez, I Kostanic, C. Subramanian, and J.-P. Pinelli, Real-Time Monitoring of Hurricane Winds using Wireless Sensor Technology, *Journal of Computers*, Vol. 4, No. 12, Dec. 2009, pp. 1275-1285.

Index

A

B

C

D

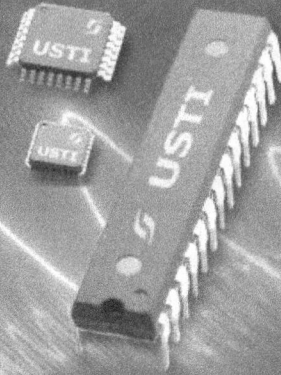

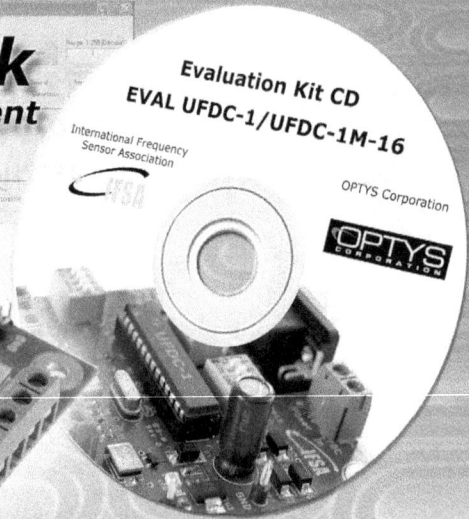

www.ingramcontent.com/pod-product-compliance
Lightning Source LLC
Chambersburg PA
CBHW080650220326
41598CB00033B/5162